COMBINATORIAL LIBRARY DESIGN AND EVALUATION

COMBINATORIAL LIBRARY DESIGN AND EVALUATION

Principles, Software Tools, and Applications in Drug Discovery

edited by
Arup K. Ghose
Vellarkad N. Viswanadhan

Amgen Inc.
Thousand Oaks, California

MARCEL DEKKER, INC.

NEW YORK • BASEL

ISBN: 0-8247-0487-8

This book is printed on acid-free paper.

Headquarters
Marcel Dekker, Inc.
270 Madison Avenue, New York, NY 10016
tel: 212-696-9000; fax: 212-685-4540

Eastern Hemisphere Distribution
Marcel Dekker AG
Hutgasse 4, Postfach 812, CH-4001 Basel, Switzerland
tel: 41-61-261-8482; fax: 41-61-261-8896

World Wide Web
http://www.dekker.com

The publisher offers discounts on this book when ordered in bulk quantities. For more information, write to Special Sales/Professional Marketing at the headquarters address above.

Current printing (last digit):
10 9 8 7 6 5 4 3 2 1

PRINTED IN THE UNITED STATES OF AMERICA

Preface

It is widely believed that the pharmaceutical and biotechnology industry will be one of the most active industrial fields in this new century because of the information explosion in the field of genomics. As a result, the number of target proteins that can yield important therapeutic agents is expected to increase dramatically in the near future. Pharmaceutical drug discovery research is currently undergoing another tremendous change due both to automated combinatorial/parallel organic synthesis and to high throughput biochemical screening. Although it is one of the most important technologies in the history of modern drug discovery, the pitfalls of combinatorial explosion must be avoided. The costs of high throughput screening or automated synthesis per compound may be low, but they will become fairly expensive when multiplied by millions. This requires the development and adoption of technologies for making combinatorial synthesis and library design cost-effective. We have collected here various technologies that should be used during the combinatorial library design.

The introductory chapter, presented by Hobbs and Guo, contains the chemist's view of combinatorial library design, which is by far the most important aspect of a library design. As this aspect is discussed in considerable depth, this chapter will also be helpful to computational chemists in learning the chemical aspects of a combinatorial library.

Part II contains the various design principles in drug discovery. For non-computational chemists these chapters will be very useful for understanding the rest of the book. Computational chemists may also find this section useful. It begins with Ghose et al.'s fundamental aspects of pharmacophore modeling, followed by Debnath's classical QSAR methods, Crippen and Wildman's 3D QSAR review, Verkhivker et al.'s critical analysis of structure-based drug design, and Tenette-Souaille et al.'s fast continuum electrostatics methods for structure-based drug design. The last chapter in this section, by Oprea et al., broadly covers protein structure–based 3D QSAR as well as the concept of optimizing protein binding and pharmacokinetics simultaneously.

In Part III we include chapters on all software-tools-related materials currently available for library design. Chapter 8, by Viswanadhan et al., deals with the definition and application of "druglikeness" to the design of combinatorial libraries. The next chapter summarizes recent neural net–based approaches for discriminating druglike and non-druglike molecules. Chapter 10 by Brown et al. discusses the tools available in the Cerius2 package for designing diverse, druglike, cost-effective combinatorial libraries. Clark discusses in Chapter 11 the tools generally used in diversity analysis, in particular those available in Sybyl software. Chapters 12–14, by Tropsha, Gillet and Willet, and Good et al., describe three different approaches for library design. The last chapter in this section, by Horvath, presents a novel approach to library design based on fuzzy logic.

Part IV deals with the applications of library design in the industrial setting. Here the authors gave specific examples of libraries and the various tools that they used during the design process. Chapter 16, by Schnur and Venkatarangan, details a validation strategy and applications of cell-based diversity metrics. Chapters 17 and 19, by Joseph-McCarthy and Tondi and Costi, describe applications based on protein structure–based combinatorial library design. In Chapter 18, Singh and Treasurywalla describe a genetic algorithm and its application for lead generation. Finally, in Chapter 20 Senderowitz and Rosenfeld describe the design and application of libraries mimicking biological motifs.

The editors would like to express their heartfelt gratitude to their families and mentors. Both want to thank their mentors, in particular Prof. Gordon M. Crippen, who also took the time to contribute a fine chapter for this book. Arup K. Ghose would like to thank Prof. A. U. De of Jadavpur University, who gave him his first lesson in drug discovery research, and his wife, Chandralekha, and two daughters, Monalisa and Kristy, for letting him take time from their lives to do this job. Vellarkad N. Viswanadhan thanks Professors Wayne L. Mattice of the University of Akron, K. Sundaram of the University of Madras, and Dr. John N. Weinstein of the National Institutes of Health, who inspired and mentored his work in molecular structure and design. He further thanks his wife, Raji, and daughters,

Katya and Kalpa, for being patient and understanding about his weekend trips to the office. Last, the editors are highly grateful to their employer, Amgen Inc., for giving them the opportunity to work in this exciting field, and their colleagues from the Molecular Structure and Design and Small Molecule Drug Design departments, in particular Dr. John J. Wendoloski, who have been outstanding as collaborators in their efforts to discover human therapeutics.

Arup K. Ghose
Vellarkad N. Viswanadhan

Contents

Preface   *iii*
Contributors   *xi*

PART I   INTRODUCTION

1. Library Design Concepts and Implementation Strategies   1
    Doug W. Hobbs and Tao Guo

PART II   DESIGN PRINCIPLES

2. Fundamentals of Pharmacophore Modeling in Combinatorial
    Chemistry   51
    Arup K. Ghose, Vellarkad N. Viswanadhan, and
    John J. Wendoloski

3. Quantitative Structure–Activity Relationship (QSAR): A Versatile
    Tool in Drug Design   73
    Asim Kumar Debnath

4. Quantitative Structure–Activity Relationships (QSAR):
    A Review of 3D QSAR   131
    Gordon M. Crippen and Scott A. Wildman

5. Binding Energy Landscapes of Ligand–Protein Complexes
    and Molecular Docking: Principles, Methods, and Validation
    Experiments   157
    *Gennady M. Verkhivker, Djamal Bouzida, Daniel K. Gehlhaar,
    Paul A. Rejto, Sandra Arthurs, Anthony B. Colson,
    Stephan T. Freer, Veda Larson, Brock A. Luty, Tami Marrone,
    and Peter W. Rose*

6. Fast Continuum Electrostatics Methods for Structure-Based
    Ligand Design   197
    *Catherine Tenette-Souaille, Nicolas Budin, Nicolas Majeux,
    and Amedeo Caflisch*

7. Qvo Vadis, Scoring Functions? Toward an Integrated
    Pharmacokinetic and Binding Affinity Prediction
    Framework   233
    Tudor I. Oprea, Ismael Zamora, and Peder Svensson

PART III   CURRENT METHODS AND SOFTWARE TOOLS

8. Knowledge-Based Approaches for the Design of Small-
    Molecule Libraries for Drug Discovery   267
    *Vellarkad N. Viswanadhan, Arup K. Ghose, Alex Kiselyov,
    John J. Wendoloski and John N. Weinstein*

9. Druglikeness Profiles of Chemical Libraries   291
    Jens Sadowski

10. Tools for Designing Diverse, Druglike, Cost-Effective
    Combinatorial Libraries   301
    Robert D. Brown, Moises Hassan, and Marvin Waldman

11. Relative and Absolute Diversity Analysis of Combinatorial
    Libraries   337
    Robert D. Clark

12. Rational Combinatorial Library Design and Database Mining
    Using Inverse QSAR Approach   363
    Alexander Tropsha

13. Dissimilarity-Based Compound Selection for Library Design   379
    Valerie J. Gillet and Peter Willett

14. Pharmacophore-Based Approaches to Combinatorial Library
    Design   399
    *Andrew C. Good, Jonathan S. Mason, Darren V. S. Green,
    and Andrew R. Leach*

15. High Throughput Conformational Sampling and Fuzzy
    Similarity Metrics: A Novel Approach to Similarity Searching
    and Focused Combinatorial Library Design and Its Role
    in the Drug Discovery Laboratory   429
    Dragos Horvath

PART IV   APPLICATIONS

16. Applications of Cell-Based Diversity Methods to Combinatorial
    Library Design   473
    Dora Schnur and Prabha Venkatarangan

17. Structure-Based Combinatorial Library Design and Screening:
    Application of the Multiple Copy Simultaneous Search
    Method   503
    Diane Joseph-McCarthy

18. Genetic Algorithm-Directed Lead Generation   531
    Jasbir Singh and Adi M. Treasurywala

19. Enhancing the Drug Discovery Process by Integration of
    Structure-Based Drug Design and Combinatorial Synthesis   563
    Donatella Tondi and Maria Paola Costi

20.   Design of Structural Combinatorial Libraries That Mimic
      Biological Motifs   605
      Hanoch Senderowitz and Rakefet Rosenfeld

Index   *625*

Contributors

Sandra Arthurs Department of Computational Chemistry, Agouron Pharmaceuticals, Inc., A Pfizer Company, San Diego, California

Djamal Bouzida, Ph.D. Department of Computational Chemistry, Agouron Pharmaceuticals, Inc., A Pfizer Company, San Diego, California

Robert D. Brown, Ph.D. Director, Molecular Simulations, Inc., San Diego, California

Nicolas Budin, Ph.D. Department of Biochemistry, University of Zurich, Zurich, Switzerland

Amedeo Caflisch, Ph.D. Department of Biochemistry, University of Zurich, Zurich, Switzerland

Robert D. Clark, Ph.D. Department of Software Research, Tripos, Inc., St. Louis, Missouri

Anthony B. Colson   Department of Computational Chemistry, Agouron Pharmaceuticals, Inc., A Pfizer Company, San Diego, California

Maria Paola Costi, Ph.D.   Dipartimento di Scienze Farmaceutiche, Università degli Studi di Modena e Reggio Emilia, Modena, Italy

Gordon M. Crippen, Ph.D.   College of Pharmacy, University of Michigan, Ann Arbor, Michigan

Asim Kumar Debnath, Ph.D.   Biochemical Virology Laboratory, Lindsley F. Kimball Research Institute, The New York Blood Center, New York, New York

Stephan T. Freer, Ph.D.   Department of Computational Chemistry, Agouron Pharmaceuticals, Inc., A Pfizer Company, San Diego, California

Daniel K. Gehlhaar   Department of Computational Chemistry, Agouron Pharmaceuticals, Inc., A Pfizer Company, San Diego, California

Arup K. Ghose, Ph.D.   Department of Molecular Structure and Design, Amgen Inc., Thousand Oaks, California

Valerie J. Gillet, M.Sc., Ph.D.   Department of Information Studies, University of Sheffield, Sheffield, England

Andrew C. Good, D.Phil., B.Sc.   Department of Structural Biology and Modeling, Bristol-Myers Squibb, Wallingford, Connecticut

Darren V. S. Green, B.Sc., Ph.D.   Department of Computational Chemistry and Chemoinformatics, Glaxo Wellcome, Stevenage, England

Tao Guo, Ph.D.   Department of Chemistry, Pharmacopeia, Inc., Princeton, New Jersey

Moises Hassan, Ph.D.   Department of Rational and Combinatorial Drug Design, Molecular Simulations, Inc., San Diego, California

Doug W. Hobbs, Ph.D.   Department of Chemistry, Pharmacopeia, Inc., Princeton, New Jersey

Dragos Horvath, Ph.D.   Director of Molecular Modeling, CEREP, Rueil-Malmaison, France

Diane Joseph-McCarthy, Ph.D. Biological Chemistry Department, Genetics Institute, Wyeth Research, Cambridge, Massachusetts

Alex Kiselyov Department of Molecular Structure and Design, Amgen Inc., Thousand Oaks, California

Veda Larson Department of Computational Chemistry, Agouron Pharmaceuticals, Inc., A Pfizer Company, San Diego, California

Andrew R. Leach, B.Sc., D.Phil. Department of Computational Chemistry and Chemoinformatics, Glaxo Wellcome, Stevenage, England

Brock A. Luty, Ph.D. Department of Computational Chemistry, Agouron Pharmaceuticals, Inc., A Pfizer Company, San Diego, California

Nicolas Majeux, Ph.D. Department of Biochemistry, University of Zurich, Zurich, Switzerland

Tami Marrone, Ph.D. Department of Computational Chemistry, Agouron Pharmaceuticals, Inc., A Pfizer Company, San Diego, California

Jonathan S. Mason, B.Sc., Ph.D. Department of Structural Biology and Modeling, Bristol-Myers Squibb, Princeton, New Jersey

Tudor I. Oprea, M.D., Ph.D. Department of EST Informatics, AstraZeneca R&D Molndal, Molndal, Sweden

Paul A. Rejto, Ph.D. Department of Computational Chemistry, Agouron Pharmaceuticals, Inc., A Pfizer Company, San Diego, California

Peter W. Rose, Ph.D. Department of Computational Chemistry, Agouron Pharmaceuticals, Inc., A Pfizer Company, San Diego, California

Rakefet Rosenfeld,* Ph.D. Department of Molecular Modeling, Peptor, Ltd., Rehovot, Israel

Jens Sadowski, Ph.D. Department of Structural Chemistry, AstraZeneca R&D Molndal, Molndal, Sweden

Dora M. Schnur,† Ph.D. Pharmacopeia, Inc., Princeton, New Jersey

---

* *Current affiliation:* QBI Enterprises, Ltd., Nes Ziona, Israel
† *Current affiliation:* Department of Computer-Aided Design (CADD), Bristol-Myers Squibb, Princeton, New Jersey

Hanoch Senderowitz, Ph.D.   Department of Molecular Modeling, Peptor, Ltd., Rehovot, Israel

Jasbir Singh, Ph.D.   Department of Medicinal Chemistry, Cephalon, Inc., West Chester, Pennsylvania

Peder Svensson, Ph.D.*   AstraZeneca R&D Molndal, Molndal, Switzerland

Catherine Tenette-Souaille, Ph.D.   Department of Biochemistry, University of Zurich, Zurich, Switzerland

Donatella Tondi, Ph.D.   Dipartimento di Scienze Farmaceutiche, Università degli Studi di Modena e Reggio Emilia, Modena, Italy

Adi M. Treasurywala, Ph.D.   Consultant to the pharmaceutical industry, Mississauga, Ontario, Canada

Alexander Tropsha, Ph.D.   School of Pharmacy, University of North Carolina, Chapel Hill, North Carolina

Prabha Venkatarangan, †Ph.D.   Pharmacopeia, Inc., Princeton, New Jersey

Gennady M. Verkhivker, Ph.D.   Department of Computational Chemistry, Agouron Pharmaceuticals, Inc., A Pfizer Company, San Diego, California

Vellarkad N. Viswanadhan, Ph.D.   Department of Molecular Structure and Design, Amgen Inc., Thousand Oaks, California

Marvin Waldman, Ph.D.   Department of Rational and Combinatorial Drug Design, Molecular Simulations, Inc., San Diego, California

John N. Weinstein   National Cancer Institute, National Institutes of Health, Bethesda, Maryland

John J. Wendoloski, Ph.D.   Department of EST Chemical Computing, AstraZeneca, Boston, Massachusetts

Scott A. Wildman, M.S.   Department of Medicinal Chemistry, University of Michigan, Ann Arbor, Michigan

---

* *Current affiliation:*   Carlsson Research, Göteborg, Sweden
† *Current affiliation:*   R. W. Johnson PRI, Raritan, New Jersey

Peter Willet, M.A., M.Sc., Ph.D., D.Sc., F.I.Inf.Sc. Department of Information Studies, University of Sheffield, Sheffield, England

Ismael Zamora, Ph.D. Department of DMPK and Bioanalitical Chemistry, AstraZeneca R&D Molndal, Molndal, Sweden

COMBINATORIAL LIBRARY DESIGN AND EVALUATION

1

Library Design Concepts and Implementation Strategies

Doug W. Hobbs and Tao Guo

Pharmacopeia, Inc.
Princeton, New Jersey

I. INTRODUCTION

Bringing new drugs to market is an increasingly difficult task. The costs associated with research are rising, and the list of criteria necessary for acceptance continues to grow. According to statistics compiled by the Pharmaceutical Research and Manufacturers of America (PhRMA), 90% of the compounds that enter the development cycle fail to reach market (1,2). Many of the failures have been attributed to poor pharmacokinetic performance that was not discovered until late in development, after substantial research investment had already been made (3). At the same time, accelerating improvements in high throughput screening (HTS) and advances in molecular biology and genomics have led to a proliferation of new potential targets that compete for resources. These targets offer the opportunity for new and innovative products, as well as for leadership in the marketplace by firms able to exploit such products effectively and expeditiously. Thus, pharmaceutical companies are under the combined pressures to rapidly find leads for the many new targets available and to focus on better quality leads with improved chances of survival.

Combinatorial chemistry, which collectively refers to a variety of high throughput synthesis methodologies, was developed, in part, to assist in addressing these needs. One of the key benefits of combinatorial chemistry is the extreme efficiency with which large collections of compounds can be prepared. By optimizing the process by which libraries are constructed, and with the aid of automated equipment, a single chemist can now produce 10,000–100,000 compounds per year. While this is effective in supplying assays with an abundance of compounds so that new leads can be identified, it does not directly address the poor survival statistics of compounds as they progress through development. Even with the increased supply of leads, there remains a critical need to choose the best lead with the greatest potential for development. For this reason, it has become increasingly apparent that effective library design must include both high throughput synthesis techniques and thoughtful compound design to accelerate the development of new drugs.

This chapter provides a medicinal chemist's perspective of library design for drug discovery. It is useful for medicinal chemists practicing combinatorial chemistry, to consider the distinction between library design and compound design. Compound design refers to design decisions that are made with respect to a single compound. A medicinal chemist will typically develop a hypothesis or rationale supporting a particular structure prior to synthesis. Based on detailed knowledge of the biological target, individual compounds can be made to fit in an enzyme active site, or to reproduce a desired conformation for binding to a receptor. At the same time, choices are guided by an understanding of in vivo parameters such as ADME (absorption, distribution, metabolism, and excretion) considerations. Library design, on the other hand, must also consider explicitly the process by which the compounds will be generated. In many cases, the process itself places limitations on the types of structure that can be obtained. The goal of library design, therefore, is to balance the desire for efficient and rapid compound synthesis with the requirement for well-designed compounds.

This chapter is structured to introduce and discuss the main issues relating to library design. Since there are several excellent reviews covering nearly every aspect of combinatorial chemistry and related technologies, some of the discussions are brief and merely summarize the key issues. An attempt is made to draw attention to subtle issues that have not been highlighted in other reviews, but this is at the expense of comprehensive coverage of each topic. Sections II and III describe the available techniques for library construction. Section IV emphasizes compound design issues and how they influence—and are influenced by—the implementation strategy. The remaining sections present a limited number of examples taken from the literature to illustrate some of the library design principles in action.

A. Defining the Goal

The first step in designing a library is to define the purpose, or the issue that the library is intended to address. Libraries are designed for three common purposes, which are discussed in the subsections that follow.

1. To Expand the Diversity of a Compound Collection for General Screening

Libraries that fall into the first category are frequently referred to as "discovery" or "prospecting" libraries (4). While all combinatorial libraries are likely to be screened against many targets, discovery libraries are specifically built with broad screening as their main objective. Since these facilities are intended to be screened against a large number of potentially unrelated targets, a discovery library benefits from being as large and diverse as possible. A general discussion of discovery libraries and principles used in their design is presented in Section V.

2. To Find a Lead for a Specific Target

Commonly, the goal of library screening is to find a lead against a specific target or family of targets. Libraries designed for this purpose are referred to as "focused," or "targeted," libraries. Typical design strategies involve the use of privileged structure motifs or specific recognition elements that are suspected to be required for binding. In addition, the integration of combinatorial chemistry and molecular modeling has proven very powerful, leading to structure-based, rationally designed libraries (5). A couple of examples of targeted libraries are presented in Section VI.

3. To Improve on an Existing Lead

In addition to finding new or novel leads, combinatorial chemistry is finding increased application in lead optimization. Here, optimization refers not just to improvements in compound affinity, but also with respect to ADME properties. Early in the history of combinatorial chemistry, it was believed that ADME optimization was the exclusive domain of the medicinal chemist. More recently, it has become clear that the distinction between medicinal chemistry and combinatorial chemistry is arbitrary, and the tools and principles of combinatorial chemistry can be applied to drug discovery all the way from lead generation to preclinical assessment. The use of combinatorial chemistry to address lead optimization is discussed in Section VII.

There are many tools available for library construction. In many cases, the desire to use a particular technique or piece of equipment will influence the structure, quantity, and number of compounds produced. Therefore, we next describe the common methods of combinatorial chemistry so that their influence on library designs can be assessed.

II. HIGH THROUGHPUT SYNTHESIS TECHNIQUES

Combinatorial chemistry collectively refers to a variety of tools and techniques. Technically, the term "combinatorial" implies a degree of experimental design whereby relationships between variables can be elucidated through the evaluation of their combinations. In common usage, combinatorial chemistry also refers to almost any method of high throughput synthesis. High throughput synthesis has been enabled largely by the development of techniques that simplify the most time-consuming aspects of organic synthesis.

A. Solid Phase Organic Synthesis (SPOS)

The syntheses of peptides (6) and nucleotides (7) on solid support are well-developed fields and have proven to be reliable and time-saving approaches to the preparation of these biopolymers. Application of the same principles to the construction of small diverse organic compounds has been the subject of intense investigation and several recent reviews cover research in the area (8–14).

The subsections that follow include some of the strengths and limitations of the solid phase approach as applied to organic synthesis.

1.   Solid Supported Synthesis Simplifies Workup

Since the desired products remain attached to an insoluble particle throughout the synthesis, purification can be performed by simply filtering soluble reagents and reactants away from the insoluble particle.

2.   Excess Reagent Can Be Used to Drive Reactions to Completion

Because separation of the resin from the reaction milieu is so easy, reactions can be driven to completion by the use of excess reagents. Two different methods are commonly practiced. For some reactions, especially those that involve an equilibrating step, excess reagent/reactant is added all at once. An example is the formation of an imine from a resin-bound amine and an aldehyde. According to Le Châtelier's principle, the equilibrium can be driven to the imine by increasing the concentration of the aldehyde reactant. The other method for driving reactions to completion is to perform the reaction in several cycles. This is practical because the workup between reaction cycles is accomplished by simply washing the resin. The value of this technique has been amply demonstrated during the synthesis of long peptides, where it is critical to achieve efficiency exceeding 99% to maximize the yield of the product and minimize the formation of by-products, at each coupling step.

3.   Solid Supports Are Convenient to Manipulate

Since reaction quenching and workup involves filtration, and purification is deferred until the compound is cleaved from the support, it is common for an entire multistep synthetic sequence to be performed in a single filter-bottom reaction vessel. In addition, the demonstrated success of synthesizers for peptides and nucleotides has amply proved that reliable instrumentation can be developed to handle reagent addition, washing, and cleavage in an automated way.

4. The Use of Insoluble Supports Makes Pooled Synthesis Practical

Pooled synthesis refers to the simultaneous synthesis of many compounds in a single reaction vessel. This is a very important advantage, since it is through pooled techniques that the greatest efficiency in compound synthesis can be achieved. This technique is discussed further in Section III.B.

5. The Solid Support May Limit the Range of Chemistry

The solid support places various constraints on the chemistry. The most common supports are gelatinous beads made from low cross-linked polymers. The structure of the polymer, degree of cross-linking, physical size, and swelling characteristics all influence the chemistry in not entirely predictable ways. One of the main limitations of SPOS therefore, is the effort that must go into reaction development before library synthesis can begin. This limitation is counterbalanced by research into new polymeric supports (15,16) and the active reporting in the chemical literature of new chemical transformations that have been demonstrated on solid phase (17).

6. Reaction Rates are Slower

Diffusion into and out of the polymer support can significantly decrease reaction rates. However, other than needing to know when the reaction is complete, this is rarely an issue in practice. Even with traditional organic synthesis methods, it is commonplace to allow chemical reactions to proceed longer than absolutely necessary. As long as the conditions are not particularly vigorous, there is usually no harm done by allowing a reaction that is complete in 4 hours to run overnight.

7. There Must Be a Linker, and this May Limit Reaction Conditions

The linker is a moiety that allows the product to be cleaved from the support. The combination of support and linker can be viewed as a kind of protecting group. Because of this conceptual analogy, the linkers are often adapted from known protecting groups. As with any other protecting group, the chemistry proposed for compound synthesis must be compatible with the linker.

8. Cleavage from Resin Generates a Functional Group that May Not Be Desirable in Products

While the linker serves to anchor the compound to the support, the linkage represents the functional group that is being protected on the molecule of interest. The

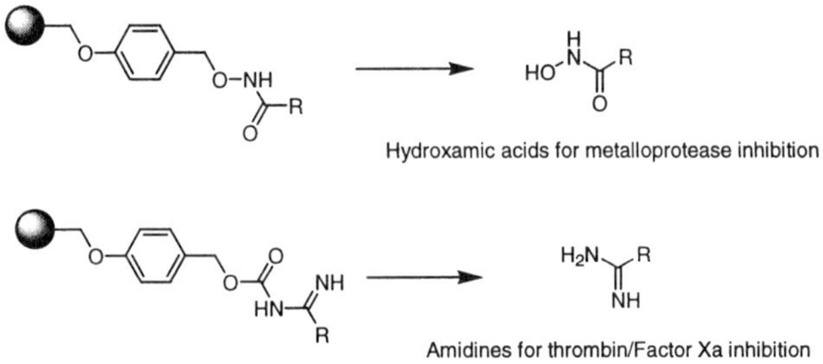

Hydroxamic acids for metalloprotease inhibition

Amidines for thrombin/Factor Xa inhibition

Figure 1   The linker cleaves to liberate a functional group important for binding to the target.

linker and the linkage have a large impact on library design. To state the problem simply: to use SPOS, there must be a functional group for attachment to the solid support. If the molecule of interest does not provide an attachment site, then one must be introduced. The extra functional group is incorporated into all the final library products having unknown or undesirable consequences on biological activity. One of the most active areas of research in SPOS is the development of strategies for overcoming this limitation (18).

In designing a library, the requirement for an attachment site within the molecule can be handled in several different ways. Figure 1 shows two examples of cleavage liberating a functional group that is necessary for inhibition. In the case of metalloproteases, the hydroxamic acid is necessary for binding to the active site metal ion (19), and for thrombin and factor Xa the basic amidine is important for binding to the specificity pocket (20,21). In these cases, since the liberated functional group is necessary for binding, it is not a liability in the products.

Another approach is to use a linker that can liberate different functional groups upon cleavage. This is illustrated in Fig. 2 with acid-cleavable and photo-

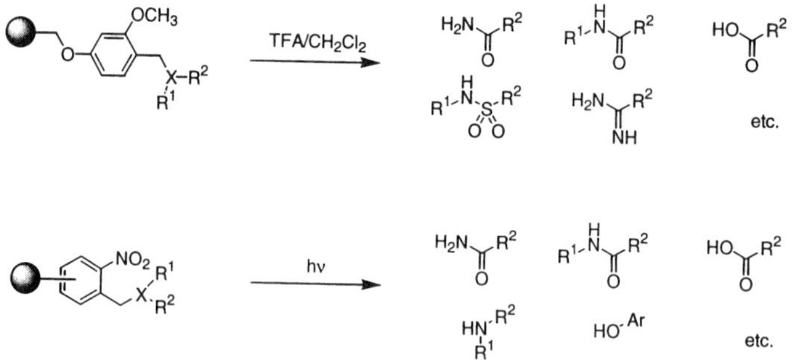

Figure 2   A single linker can liberate different functional groups upon cleavage.

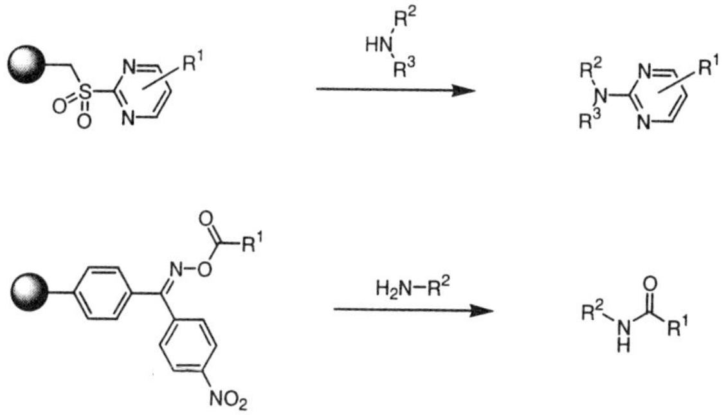

Figure 3 Amine nucleophiles can be used as both cleavage reagent and diversity element.

cleavable linkers. In these cases, the linkage type itself can be exploited as a diversity element in the library design (22).

In certain cases, the cleavage reagent can be used to generate diversity. In Fig. 3 amine nucleophiles are used to cleave the library member from resin. This approach is limited, however, since to avoid contamination of the product, the amine used for cleavage must be volatile or otherwise easily removed (23–25).

Finally, many researchers have chosen to focus on "traceless" cleavage methods, where the point of attachment to resin is obscured during the cleavage process (Fig. 4) (26–29). It should be noted, however, that even the traceless meth-

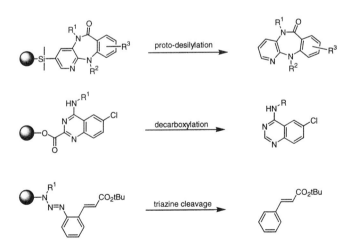

Figure 4 "Traceless" cleavage leaves no residual functional group in the product.

ods require a specific functional arrangement that will be present in all library members.

9. Monitoring Reactions Is Less Convenient

Since the desired compound is bound to the support, such chromatographic techniques as thin-layer chromatography (TLC) and high-performance liquid chromatography (HPLC) cannot be used effectively to monitor the reaction progress. Instead, colorometric tests are used to test for the presence of particular functional groups on resin (30–32), and both NMR and IR techniques have been adapted so that resin-bound materials can be examined (33–37).

10. Relatively Small Amounts of Compound Are Produced

While in theory, one can use as much resin as necessary to produce any amount of compound desired, large-scale preparation of compounds through SPOS is not very cost- or space-efficient. However, with most screening operations moving toward increased miniaturization, this is not usually a critical limitation. Many biological assays can be performed with picomolar amounts of material, and the formats for high throughput screening are steadily moving toward the use of smaller volumes. Conventional analytical methods such as HPLC and mass spectrometry are already sensitive enough to provide purity and identity determination with the amount of compound obtained from a single bead, and specialized detectors for HPLC, such as CLND (chemiluminescent nitrogen detector) and ELSD (evaporative light scattering detector), may have utility in providing quantitative data as well (38–40).

B. Solution Phase Synthesis (41,42)

The number of synthetic methods available for construction of any molecule vastly exceeds the number of methods known to proceed successfully on solid support. Thus when medicinal chemists want to construct a compound in the absence of proven methods for doing so, they may feel torn between spending the time to develop a synthetic route on solid support and resorting to methods more familiar. Solution phase synthesis has a familiarity that allows rapid implementation by most medicinal chemists, with little or no additional time spent learning specialized techniques. For example, while reaction monitoring remains cumbersome for solid phase reactions, for solution-based reactions TLC and HPLC techniques work as expected. This makes it convenient to run reactions with less up-front development and to follow reaction progress by means of standard methods.

Approaches to converting solution phase chemistry into a high throughput method focus on making product isolation more convenient. Most of the techniques rely on extracting components from the reaction milieu by taking advantage of a "handle," which is present in all compounds of the type one wants to remove. In a few cases, the handle is a moiety that has characteristic properties that override the properties of the products. This allows for a common purification strategy regardless of the compounds being synthesized. The synthesis of peptides on linear soluble polymers was one of the first demonstrated examples of this technique. Bayer and coworkers recognized the method as an alternative to solid phase synthesis as early as 1971 (43,44). Since then, soluble polymers have been used for combinatorial chemistry and organic synthesis primarily by Janda (45–48). A related method, called fluorous synthesis (see Fig. 5), shares many of the same advantages (49–51).

These strategies can be considered to be hybrid strategies that incorporate some of the advantages of both solution phase and solid phase chemistry. They also retain some of the limitations of the solid phase approach. Perhaps the biggest advantage of any solution method is the ability to perform heterogeneous reactions where the reagent is insoluble. Such reactions are not possible when the reactant is bound to a gelatinous polymer bead. On the other hand, one retains the need for a linker and linkage, and in addition, the handle itself can impose restrictions on the chemistry. For example, reactions using fluorous components must take place in solvents that are able to dissolve the fluorous compounds as well as

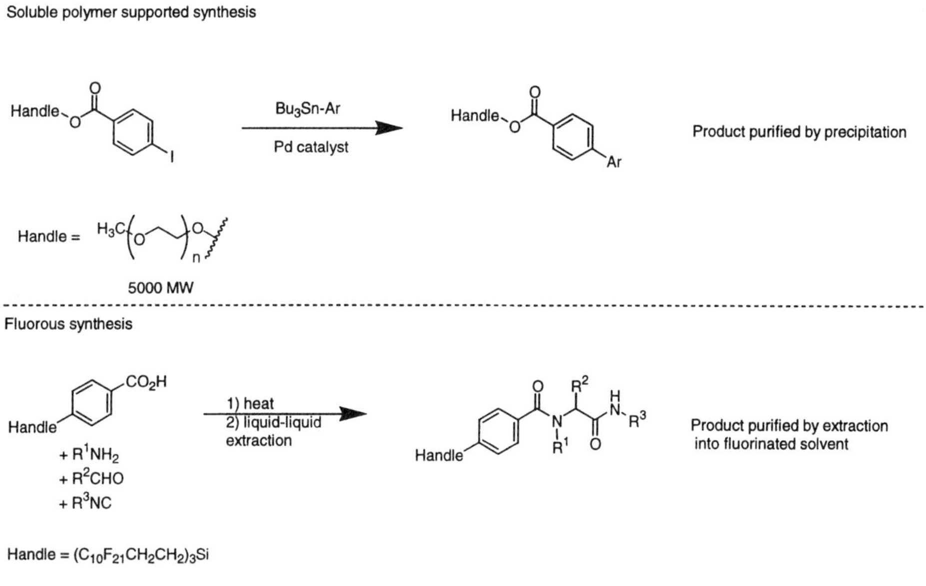

Figure 5   Soluble polymer supported solution synthesis and fluorous synthesis.

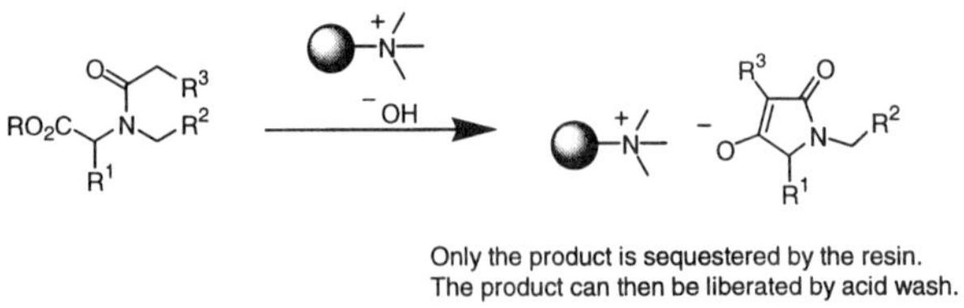

Only the product is sequestered by the resin.
The product can then be liberated by acid wash.

Figure 6   A functionalized resin acts as both reagent and purification medium.

to facilitate the desired reaction. Similar restrictions apply to soluble polymer supported synthesis.

The concept of a handle can be generalized further to include any common physical property that would allow all members of the library to be separated from by-products. This has been referred to as the principle of complementary reactivity (52). In Fig. 6, the product is characterized by a particularly acidic hydroxyl function that is not present in the starting material. This allows the selective capture of only the product by the basic resin. In this case, the resin acts as both reagent and purification medium (53).

Handles can also be attached to reagents or reactants. Resins that contain a variety of functionality for catalyzing reactions or capturing excess reactants can be made or purchased. A simple example is shown in Fig. 7, where an excess of either reactant can be selectively removed leaving the product in solution (52,54,55).

To summarize, the advantages of solution phase synthesis are the following:

1.   Solution Phase Reaction Kinetics

Since reactions take place in solution, reaction rates are faster and more predictable.

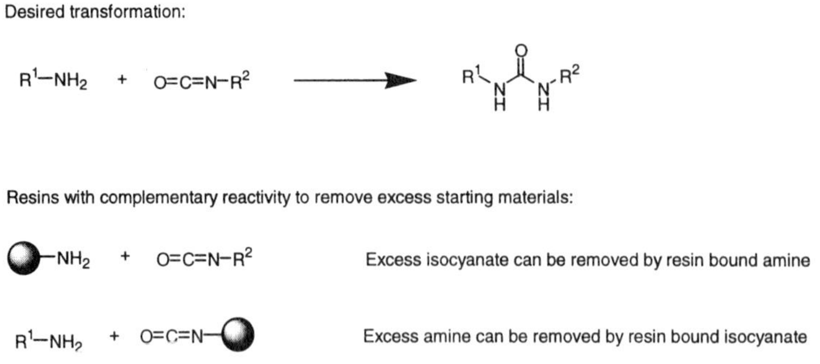

Desired transformation:

Resins with complementary reactivity to remove excess starting materials:

Excess isocyanate can be removed by resin bound amine

Excess amine can be removed by resin bound isocyanate

Figure 7   Excess reagents and reactants can be removed with reactive resins.

2.   A Wide Range of Solvents Can Be Used

There are a few exceptions, depending on which solution phase methodology one chooses to use, but generally, any solvent necessary can be used for the desired transformations.

3.   Larger Quantities of Compound Can Be Made

Solid phase synthesis becomes impractical if the quantities desired exceed a few hundred milligrams. Solution phase synthesis has no such restriction and is therefore more flexible in this regard.

4.   No Linker/Linkage is Required

While this is a commonly stated advantage of solution methods, its real value depends on the specific case. The development of traceless linkers and cleavage methods has given SPOS much of the same freedom (26). In addition, several methods for solution synthesis rely on the attachment of the desired product to a "handle." While the handle is designed to allow convenient purification, it also imposes the same linker/linkage disadvantage of resin-based chemistry.
If one uses resin-based scavengers or reagents, there are another advantages:

5.   Heterogeneous Reaction Conditions Can Be Used

This would appear to be an unsolvable limitation with SPOS techniques, and therefore it is a chief advantage of solution methods.

6.   Only One Reaction is Performed On Resin

Typical SPOS syntheses have more than one transformation, and the effect of the support must be determined for each step. For a polymer-assisted solution synthesis, the use of resin is less sophisticated. Either it is added as a reagent or it is used to capture excess reagent/reactant. In the latter case, the reaction that takes place on the solid support does not need to be optimized, and in fact, does not even have to be terribly efficient, since more resin can always be added.

III.   LIBRARY CONSTRUCTION STRATEGIES

The goal of high throughput synthesis is to speed up the process of synthesizing compounds for evaluation. This goal is largely achieved in two ways: by simplifying workup and purification via one of the synthetic approaches described in Section II, and by formatting synthetic operations to make them more amenable to simultaneous manipulation and automation. It is important to emphasize that the approach to high throughput synthesis itself is a parameter that needs to be optimized.

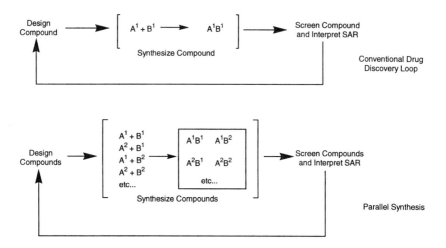

Figure 8   Parallel synthesis versus traditional compound optimization.

A.   Parallel Synthesis

If traditional compound optimization can be approximated by the information loop shown in Fig. 8 where information is gathered one compound at a time, then the goal of parallel synthesis is to format the system so that operations are conveniently performed, one batch at a time (56).

Parallel synthesis can be performed either in solution or on solid support. Current usage of the term is taken to mean that there is only one compound per reaction vessel, but several are made concurrently using the same methodology. The efficiency of the technique is gained by formatting the synthesis to make simultaneous manipulation more convenient. An example of a typical format for parallel synthesis is a block of polypropylene or Teflon that has multiple filter-bottom wells. Each well is essentially a different reaction compartment; however, the entire array of reactions can be heated, cooled, or agitated as a single unit. These reaction blocks can be manipulated manually or used with automated equipment. For manual operation, since the block is arranged in two dimensions, it is most convenient to explore two variables at a time and arrange the variables along the axes of the reaction block. This results in an array of compounds that explore every combination of the two variables. To explore a third variable, one can use another reaction block. In the example shown in Fig. 9, 24 compounds are prepared in a 6 × 4 matrix.

For manual operation, the two-variable experiment is most efficient, since it makes maximum use of common reagents. The use of 8- or 12-channel pipets, for example, allows convenient manual delivery of reagents in a 96-well synthesis block. Supporting equipment, such as rotisserie ovens, benchtop vortexers, and washing stations, has been developed to accommodate parallel synthesis in blocks and to increase efficiency by automating repetitive processes. This semiautomated style of parallel synthesis is inexpensive, modular, and allows for a very flexible workflow.

Fully automated synthesizers are also available from a variety of companies. These are complete units that provide liquid handling, temperature variation, atmosphere control, and agitation, all controlled by a single computer-driven interface. The controlling software offered by most of these workstations allows the user to program complex or tedious operations that can be saved and used by others. Almost every step in chemical synthesis can benefit from automated equipment. Typical examples are weighing of sample containers, dissolution of solid reagents/reactants, delivery of reagents/reactants to reaction vessels, workup, purification, and analysis of compound quality (57–59).

Parallel synthesis is particularly appropriate under the following conditions.

1. There is No Convenient Linkage Site in The Desired Structures

In this case parallel synthesis in solution may be most appropriate.

2. The Number of Synthons is Limited

For some compounds there are simply not very many reactants available for incorporation into the library.

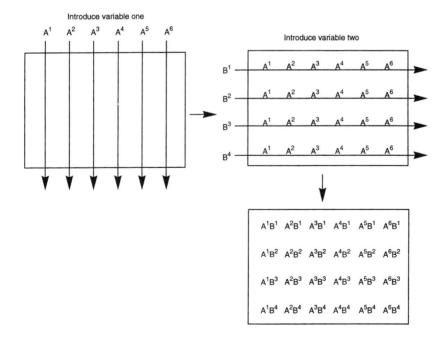

Figure 9 A 24-member library constructed in a two-dimensional array.

3. The Synthesis of Each Scaffold is Lengthy

When it is necessary to prepare several scaffolds for incorporation into the library, it may be more efficient to prepare a small library around each scaffold as it is made rather than to wait until they are all completed. In this way, screening can be performed on the first collection while preparation of the other scaffolds is in progress.

4. Purification is Necessary

In contrast to pooled synthesis (Sec. III.B), purification of the library members is a feasible operation. This allows the use of chemical transformations that do not yield highly pure products. In addition, larger amounts of pure compound are required for late-stage optimization and in vivo studies than can typically be obtained by pooled synthesis methods.

B. Pooled Synthesis

Combinatorial chemistry originated, in part, from the solid phase peptide synthesis methods developed by Merrifield (6,60). Many of the advantages of using insoluble supports were discussed in Section III.A. An additional benefit of solid phase chemistry came with the realization that each solid support represents an individual reaction compartment. This concept was a major breakthrough, since it enabled multiple reactions to be performed simultaneously by treating a mixture of resin-bound substrates with a reagent in a single reaction vessel. Several variation of pooled synthesis have been described (61–63). The split-pool synthesis approach outlined in Fig. 10 is among the most popular and efficient (64). In this approach, individual reactions are performed to attach the first synthon to the resin (e.g., A^1, A^2, A^3). The subsequent mix and split steps are necessary to distribute the "A" resins evenly, so that treatment in the next step with B^1 and B^2 generates all combinations of A and B. This continues as described in Fig. 10 until all combinations of A, B, and C have been obtained.

A modification of this approach, called "direct divide," was introduced to better control the distribution of resins from one set of reactions to the next set (Fig. 11) (65,66). This approach offers improved statistics with respect to achieving equal numbers of all combinations of products. The advantage shared by both approaches is that relatively few steps are required to produce many products. Pooled synthesis techniques have been used to prepare libraries of up to a million compounds (67,68).

Unlike parallel synthesis, pooled synthesis is not just an extension of classical chemical synthesis. A great many more compounds can be prepared simultaneously, and because of this, special accommodations need to be made to ensure efficient use of the technique.

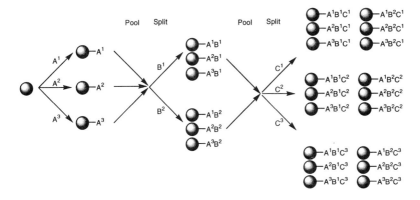

Figure 10 Split-pool synthesis method.

1. A Lot of Effort Is Required to Ensure Library Quality

Analysis of every compound is often impractical because of the number of compounds that can be produced. Since compounds are screened without purification, variation in yield and purity can confuse interpretation of the results. This means that there must be very high standards for product quality, and the library chemistry must be optimized extensively before the library is actually constructed. This is a key difference between libraries prepared by pooled synthesis and by parallel synthesis. With parallel synthesis, the quality of compounds can be determined after synthesis. Any compounds that do not meet criteria for quality can be dis-

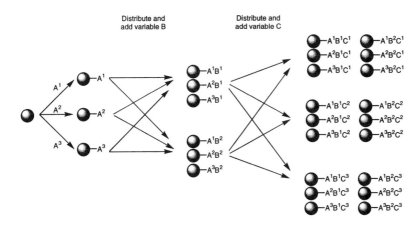

Figure 11 Direct divide is another pooled synthesis method.

carded. For libraries prepared by pooled synthesis, the situation is more complex. Statistical sampling of the completed library allows the assessment of the quality of some products. It is an artifact of pooled synthesis methodology, however, that any products that fail quality criteria are interspersed throughout the library and cannot be removed. Therefore, the best opportunity for ensuring library quality occurs before library construction, during early reaction development and optimization.

To most chemists, reaction optimization refers to optimizing the yield and purity of a single product. Reaction optimization for combinatorial chemistry, however, requires developing an understanding of the relationship between reaction parameters and reactant properties across a range of reactants. Consider the Diels–Alder reaction as a simple example. Successful condensation is not accomplished by simply mixing any diene and dienophile under conditions that were optimized for cyclopentadiene. In this case, there is an additional dependence on the electronic characteristics of the substrates, which need to be matched if cyclocondensation is to be successful. In practice, once the restrictions on the chemical transformations have been identified, pooled libraries can be constructed with a high degree of confidence in the chemistry. Sampling of the completed library by liquid chromatography/mass spectrometry can then be used as a final check to confirm overall library quality.

2. The Structure of Active Compounds Is Determined Indirectly

The power of the pooled synthesis approach is in its ability to produce large numbers of compounds very efficiently. However, one trade-off in this approach is that the identity of any individual compound on a bead becomes lost in the pooling process. When activity is observed in a biological assay, it can be difficult to determine the structure of the active compound. The small amount of compound produced may be insufficient for structural characterization. Several methods have been developed to overcome this limitation, including deconvolution techniques (69), indexing (70), and positional scanning (71,72). Excellent descriptions of these techniques can be found in recent reviews (73,74). Indexing and postional scanning techniques rely on preparing the library in subsets, each of which containing one synthon which is defined. By comparing the subsets for biological activity, one can estimate the impact of each synthon on overall activity. The structure of the active compound can then be inferred by extracting the most active synthon from each position and combining these synthons. In deconvolution approaches, portions of the library are saved at each step prior to mixing. Since the library is not mixed after the last step, the well in which biological activity is observed defines the last synthon. This preferred synthon is then placed on precursor resins, where the second-to-last synthon is known. In this way, preferred synthons for each position can be identified.

An alternative approach for the identification of active compounds from single beads is to prepare encoded libraries (74). In encoding strategies the synthetic history of each compound is described by a unique code. When biological activity is observed, the code can be read to determine the individual synthons from which the active compound was assembled. Early encoding methods used nucleotide sequences or peptide strands as the code elements. Similar to the way nature uses DNA to encode for complex proteins, specific sequences of nucleotides or amino acids were used to record each synthetic operation during library synthesis. These codes could be read by means of well-established sequencing and detection methods. More recent incarnations of encoding use nonsequential codes made up from haloaromatic alcohols (75,76), dialkylamines (77), reactants enriched in specific element isotopes (78), radiofrequency transponders (79,80), and laser optical synthesis chips (LOSCs) (81). One of the main advantages of encoding methods is the ease with which many structures can be identified. Instead of obtaining relatively few active compounds, dozens of compounds can be identified covering a range of activities. The greater number of compounds identified allows more complete interpretation of structure–activity relationships (SAR), which can guide subsequent efforts.

3. Multiple Variables Can Be Explored Simultaneously

Owing to the low throughput of serial compound synthesis, traditional drug discovery often follows a path similar to the logical tree in Fig. 12. The complex re-

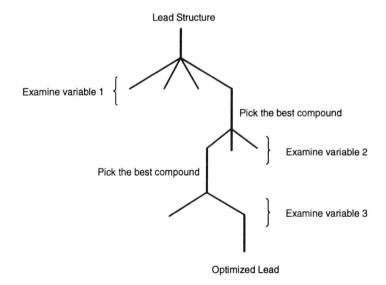

Figure 12   Schematic representation of "one variable at a time" compound optimization.

Library generic structure:

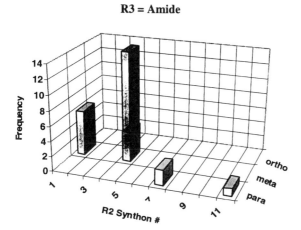

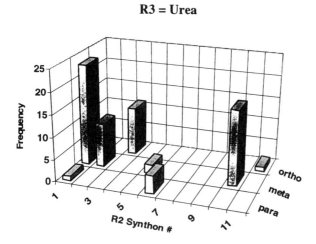

Figure 13 Large combinatorial libraries reveal the relationship between variables.

lationship between structure and biological activity is elucidated by evaluating the impact of structural variables one at a time. This approach, termed OVAT (one variable at a time), can be successful, but only when the variables are completely independent.

In the alternative situation commonly found, changes in one variable modify the effect of other variables on activity. Large libraries constructed through pooled approaches allow the evaluation of many structural changes and the relationship between them at the same time. Figure 13 shows a portion of the results obtained from screening an encoded library at Pharmacopeia (T. Guo, personal communication, 1999). The structure of active compounds was assigned based on the code sequence. The frequency refers to the number of times the code for a particular variable was found among decoded structures. Examining the chart at R^3 = amide, one can already see how the positional relationship on the aromatic ring (ortho, meta, para) affects which substituent is favored at R^2. For the ortho-substituted system, there is a strong preference for R^2 = synthon 3. In the meta substituted series synthon 3 is not found. Synthon 1 and perhaps 4 are preferred. The para-substituted series shows little activity. Examination of the chart at R^3 = urea shows a similar SAR pattern, with one interesting addition. When R^3 = urea and the substituent relationship is meta, synthon 10 now appears as one of the preferred components at R^2. This synthon is notably absent in the series for R^3 = amide. These findings suggest that there is a relationship between the positional isomer, R^2, and R^3. This synergistic relationship would have been difficult to uncover by means of OVAT optimization methods. These findings were confirmed through resynthesis of the decoded compounds and thorough biological analysis.

IV. COMPOUND DESIGN WITHIN COMBINATORIAL LIBRARIES

As the techniques associated with preparing combinatorial libraries increase in reliability, more attention is directed toward analysis of virtual libraries and compound selection. Two areas have received a lot of attention: diversity analysis and the determination of "druglikeness".

A. Library Diversity

One of the main goals of diversity analysis is to maximize the information value of each compound in a proposed library. Since it is rarely feasible to include every available synthon in a library, diversity analysis tools provide for choosing among available components to cover the broadest range of chemical space. In theory, the screening results should then allow the correlation between specific molecular descriptors and biological activity. In the past few years, a variety of descriptors and methods have been tested for their suitability in diversity assessment (82–85). Several recent reviews cover this rapidly changing field (86–90).

Because one wants to ensure the value of a library prior to construction, it has become common practice to build and evaluate a virtual library *"in silico"* prior to synthesis. Virtual libraries represent a best guess regarding the structures and precursors that can be included in the library. One can then evaluate the virtual library with respect to diversity criteria before selecting a subset for actual synthesis. Diversity analysis can be applied to the set of starting materials, as well as to the final products. While analysis of the products appears to be more effective at maximizing the diversity of a library (91), reactant analysis also has some practical applications. Since the efficiency of combinatorial chemistry relies to a large part on the fact that few starting materials give rise to many products, the list of starting materials is likely to be much smaller than a complete listing of enumerated library products. In addition, given a list of available reactants, diversity can be assessed with respect to different criteria. For example, a recent report describes the evaluation of a set of reactants in terms of reactivity by calculating the influence of molecular properties on the reactive center (92). The results of this analysis would likely be different from an assessment of the same list of reactants with respect to descriptors of biological relevance. With this in mind, it is important consider that if a compound fails to undergo the desired transformations, it really does not matter if the compound was predicted to have biological activity. Thus, unless the chemistry for the proposed library is very well developed, some amount of iteration between theoretical design and empirical testing must occur.

The difficulty with diversity analysis in general is that in spite of all the attempts to arrive at a universal definition of diversity, our understanding of the term remains subjective and highly dependent on the specific target. A central assumption is that even if the optimal substituent is not present in the library, one that is similar (according to diversity calculations) should have some observable activity. This assumption is not always correct. Most medicinal chemists have likely encountered situations where seemingly very small changes in structure—such as replacing methyl with ethyl, or changing the position of a substituent on an aromatic ring—led to very large changes in biological activity (93,94). To one target, methyl and ethyl substituents may have similar affinities; to another, there results an order-of-magnitude difference. With respect to the first target, including both methyl and ethyl in the library is redundant and a waste of effort. For the second target, the penalty for not including the correct one is, perhaps, to observe no activity at all. The complexity of this problem is illustrated with the morphine analogs in Fig. 14 (95,96). In this example, both affinity for the receptor and the type of activity (agonist, antagonist) are sensitive to subtle changes in the nitrogen substituent.

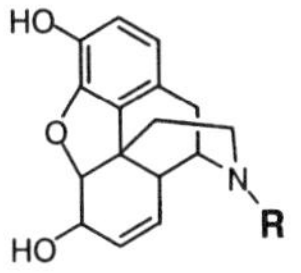

Nitrogen substituent	Activity
R = methyl	potent agonist (morphine)
R = ethyl	"inactive"
R = allyl, propyl	potent antagonist
R = butyl	"inactive"
R = pentyl, hexyl	potent agonist

Figure 14 Subtle changes can have dramatic effects on biological activity.

B. Controlling Molecular Properties

Poor biopharmaceutical properties comprise one of the main reasons for the failure of drug candidates to reach the market (3). In particular, issues such as toxicity, bioavailability, and metabolic fate become more important than affinity and selectivity as the compound progresses through the development cycle. In an effort to avoid late-stage failures, most companies have begun to emphasize ADME suitability earlier in the compound development process. Several groups have begun publishing algorithms for determining such pharmaceutical suitability (97–99). The typical approach is to choose a basis set from among known drug entities, then to identify common characteristics that can conveniently be computed. From there, the characteristics of any new compound can be compared against the basis set to determine just how "druglike" the compound may be. An example of this is the often reported "rule of five" which offers a simple method for assessing the absorption potential of an unknown compound (Fig. 15) (100). Any compound that exceeds the rule-of-five recommendations in two or more categories is predicted to have poor oral absorption.

Other algorithms, aimed at predicting aqueous solubility, metabolic suscep-

Rule of five:

1. Molecular weight< 500
2. clogP < 5
3. < 5 H-bond donors
4. < 10 H-bond acceptors

Figure 15 Characteristics of a good drug.

tibility, and blood–brain barrier penetration, are also being pursued. While progress in developing these methods is hampered by the difficulty of obtaining good data for deriving a useful basis set, advances are being made in the development of high throughput assays that model pharmacokinetic processes. High throughput screens to evaluate the metabolism by cytochrome P450, solubility, and cell permeability are all being pursued throughout the industry.

The tools described thus far offer guidance to the chemist in designing a quality library. However, several issues that influence library and compound design have not yet been captured by an expert tool or computational algorithm. For example, a chemist must use their understanding of reactant reactivity to choose fragments that are compatible with all the steps in the proposed design. Consider the hypothetical example in Fig. 16 (101). In the actual library, the substituents at R^1 were represented only by simple alkyl and benzyl derivatives. Including a tertiary amine synthon in step 1 would certainly enhance the diversity in that position. However, since the final step involves alkylation with activated halides, having nucleophiles at other positions in the molecule would likely produce by-products.

In another commonly encountered situation, portions of a library need to be steered through different paths. This is done not only to avoid reactivity problems, as in the hypothetical example above, but also to ensure that the products are balanced with respect to diversity and druglike characteristics. A simple example can be described by using the pooled synthesis technique represented in Figs. 10 and 11. According to that scheme, the products include every combination of every variable. If one synthon from each reactant list contains a carboxylic acid, then there will be at least one compound in the library that has three acidic groups. The triacid thus formed may be a small percentage of the

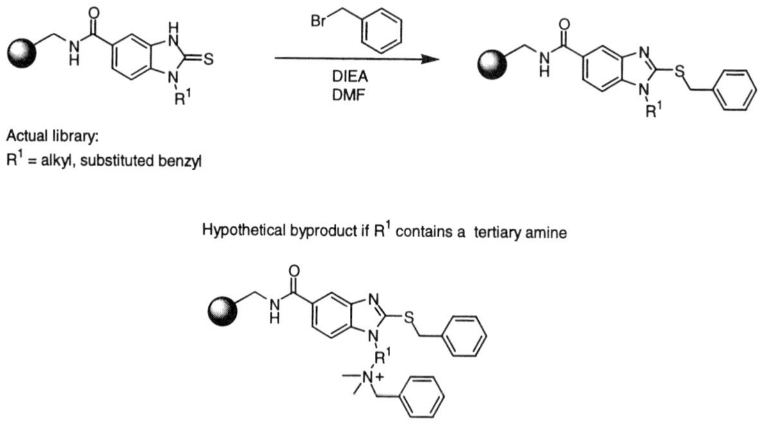

Actual library:
R^1 = alkyl, substituted benzyl

Hypothetical byproduct if R^1 contains a tertiary amine

Figure 16   Chemical compatibility must be considered in addition to diversity.

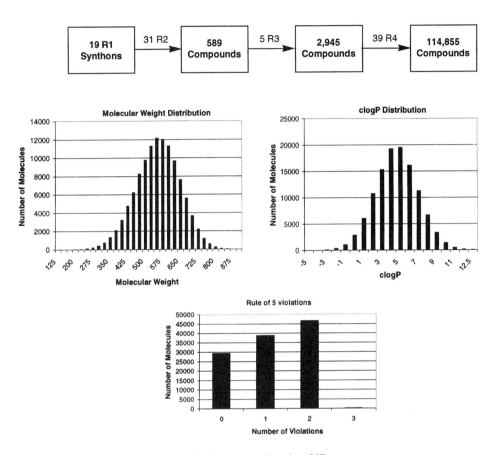

Figure 17 Property analysis of a first-generation virtual library.

total library, but it is still a poor lead. With the computational tools currently available, a chemist planning a library can repeatedly calculate properties and reorganize the course of synthetic operations until the entire library has the features and properties desired. To illustrate this concept, consider the library described in Fig. 17. The first-generation library contained all combinations of the four variables R^1–R^4, leading to a virtual library of 114,855 members. Analysis of the enumerated products that would be generated in this library reveals an excess of lipophilic and high molecular weight compounds. Examination of the rule-of-five exceptions indicate that the majority of the library has more than one exception.

Figure 18 outlines the final version of the library. As can be seen, a more elaborate splitting scheme was adopted to avoid the negative effects of combining

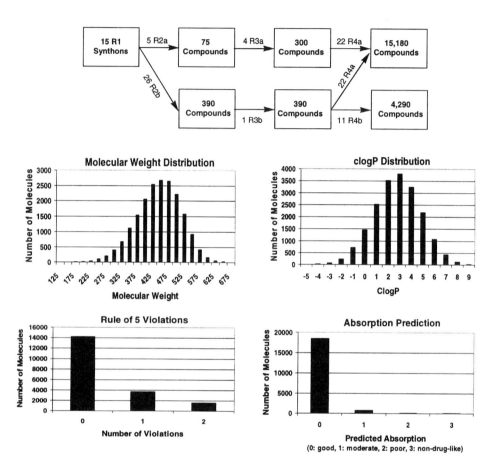

Figure 18 Property analysis of final virtual library with predicted oral absorption.

synthons with additive properties. For example, only the portion of the library with low molecular weight R^{2a} synthons is combined with the relatively high molecular weight R^{3a} synthons. Analysis of this version of the library shows substantial improvement in molecular properties and predicted bioavailability. The portion of the library that was removed is shown in Fig. 19.

In summary, there is no "one best design" that can be obtained from virtual library analysis. Instead, computational tools need to aid chemists in performing trade-off analyses so that they can better weigh the impact of each design decision. Library design involves balancing many parameters, and just as single-compound design involves a high level of creativity on the part of the chemist, so does library design, which focuses on compound design within the constraints of the high throughput process.

Sections VI–VII include examples of libraries taken from the literature that

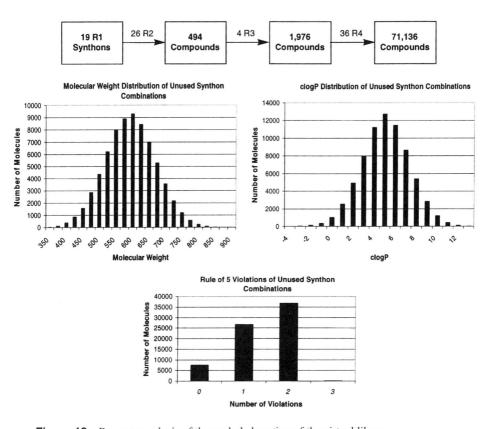

Figure 19   Property analysis of the excluded portion of the virtual library.

serve to illustrate the principles already discussed. The sections are partitioned according to whether the libraries were intended to be discovery, targeted, or optimization libraries.

V.   LOOKING FOR LEADS: DISCOVERY LIBRARIES

In contrast to targeted libraries, where there are typically specific structural features guiding the library design, the design principles that guide discovery libraries are synthetic efficiency, diversity, and lead quality.

A.   Synthesis of Oligomers

Oligomers have proven to be good synthetic targets for combinatorial synthesis. Because they are made up of repeating units, relatively few reactions need to be optimized. In addition, when the monomer units are readily available, they can rapidly produce very large and diverse libraries.

Peptides are the natural regulators of many physiological and disease processes, and peptide libraries were among the first discovery libraries to be explored using combinatorial techniques (102). While peptides are easily constructed and a wide variety of amino acids monomers are available, peptides beyond approximately four amino acids in size tend to be poor drug candidates. Specifically, degradation by proteolytic enzymes, poor gastrointestinal absorption, rapid elimination, and conformational flexibility leading to poor selectivity are problems frequently encountered with peptide leads. On the other hand, the fact that diverse libraries can be made from optimizing only one reaction is very compelling. One approach to balancing the ease of synthesis while addressing the pharmacological issues has been to devise oligomers that contain isosteric replacements for the peptide backbone. Examples are peptoids (103), oligocarbamates (104–106), oligoureas (107), azatides (108), oligothioureas (109), and oligomers of vinylogous sulfonamides (Fig. 20) (110).

A common feature of each example is the retrosynthesis of the structure to monomer units, which are easily obtained. Comparison of the peptoid syntheses outlined in Fig. 21, and entry 1 in Fig. 20, serves to illustrate the importance of this retrosynthetic principle. The first reported synthesis of peptoids (Fig. 21) used *N*-

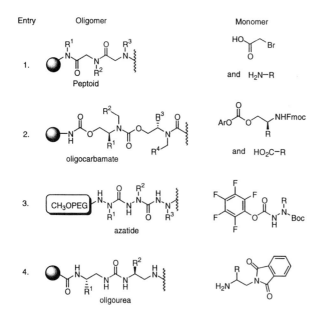

Figure 20 Oligomer libraries containing isosteric replacements for a peptide backbone.

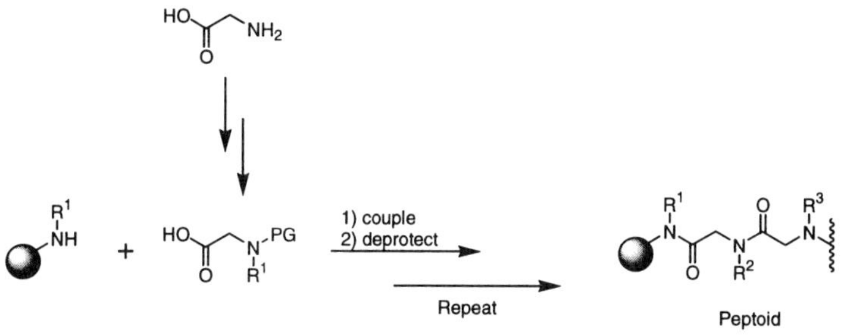

Figure 21 Peptoid synthesis using preformed N-alkylglycines: PG, protecting group.

alkylglycines as the monomer units. Each monomer had to be synthesized in solution in suitably protected form for incorporation into the growing oligomer. Modification of the original procedure so that the N-alkylglycine monomers are synthesized on resin from α-bromoacetic acid and an alkylamine afforded access to a wider range of input materials, while removing the need for solution synthesis and several protecting groups (111).

B. Efficient Construction

The fewer the linear steps in the synthesis of the library, the higher the yield. Even when the steps involve very high-yielding deprotections, it is generally better if such steps can be avoided. Libraries constructed around triazines and pyrimidines are extremely efficient in this regard. Each halide on the heteroaromatic core can be displaced with a nitrogen. However, as nitrogens are introduced, the reactivity of the remaining halides is dramatically, reduced allowing sequential elaboration of the core without the use of protecting groups (Fig. 22) (112,113).

Another method for increasing the efficiency of compound synthesis is to employ multicomponent reactions, in which several diversity elements are

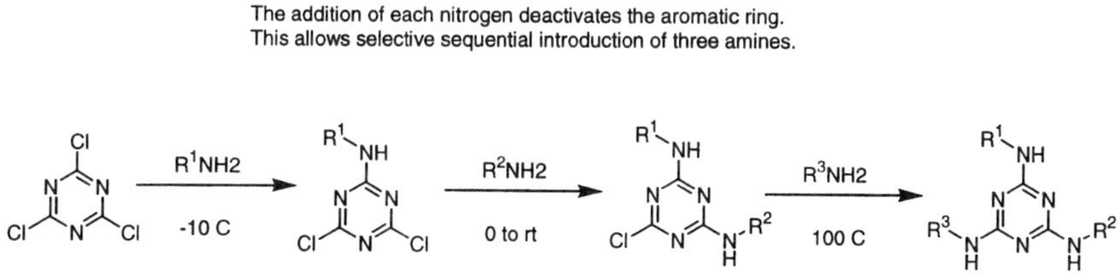

The addition of each nitrogen deactivates the aromatic ring.
This allows selective sequential introduction of three amines.

Figure 22 Triazine libraries can be constructed very efficiently because no protecting groups are required.

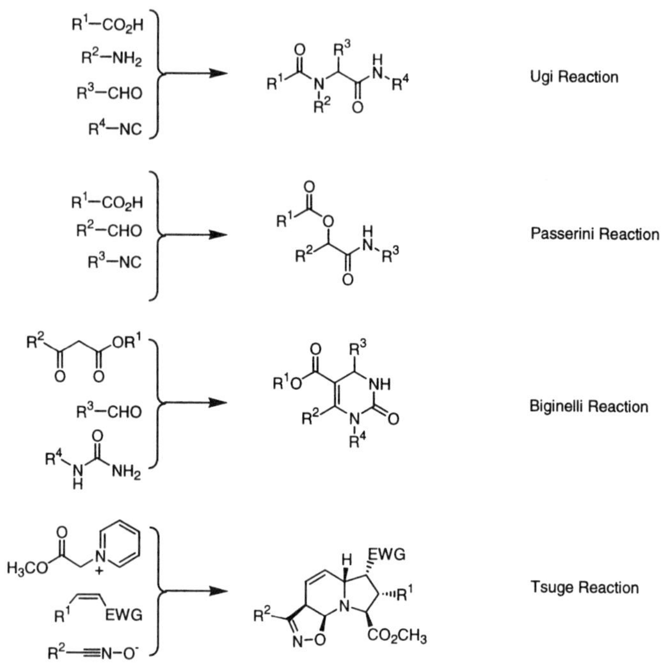

Figure 23 Multicomponent reactions introduce several variables in a single step.

brought together in a single step. Examples include the Ugi, Biginelli, Passerini, and Tsuge reactions (Fig. 23). Both solid phase and solution phase examples have been reported for several of these transformations (114–120).

C. Branching Strategies

Since developing chemistry for use with high throughput techniques requires significant time in the reaction optimization phase, it makes sense to use that initial investment in as many ways as possible. Several strategies have appeared that make use of libraries that have already been constructed to produce new libraries with different properties. Houghten and coworkers described the concept of "libraries from libraries" (121). Using this idea, a previously prepared peptide library was converted to a new library by application of an additional reaction that affects every member of the library (Fig. 24). In terms of efficiency, the one extra step doubles the output. That is, another library is produced that is the same size as the original.

Advanced intermediates can also be used as branching points to allow libraries presenting different structural motifs to be prepared. In Fig. 25 linear in-

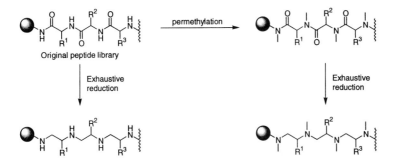

Figure 24 "Libraries from libraries" strategy.

termediates are treated with different reagents to affect different modes of cyclization (67,122).

Certain reactions or functional groups can be pivotal in enabling branching (123). For example, one of the key limitations of the Ugi reaction is the lack of easily obtainable diverse isocyanides. Keating and Armstrong (116,124) addressed this key limitation by developing 1-isocyanocyclohexene as a "universal isocyanide." Thus, the Ugi product was an activated intermediate that could be used to branch in several directions (Fig. 26).

A particularly impressive example of a single reaction that enabled substantial branching and product diversification is outlined in Fig. 27. The use of nitro-activated aryl halides to facilitate the arylation of heteroatom nucleophiles has been reported (125,126). This reaction alone allows the creation of motifs such as

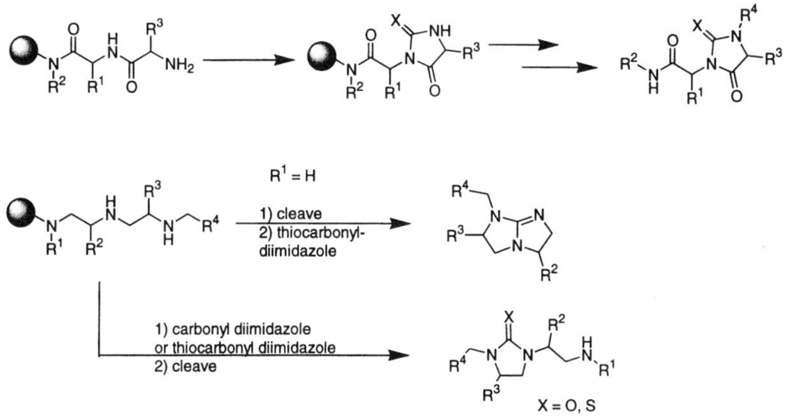

Figure 25 Branching from an advanced intermediate creates different structural classes.

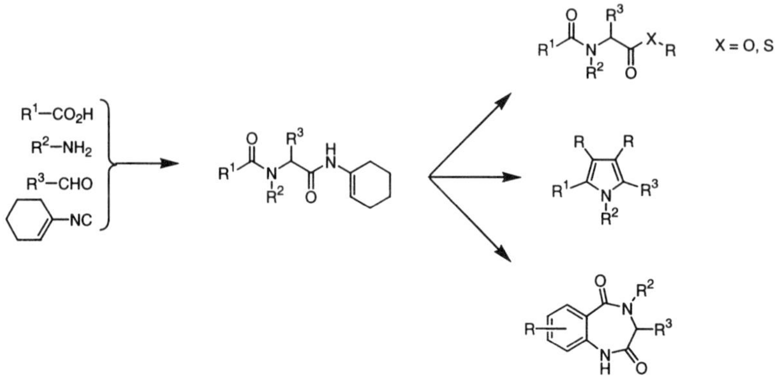

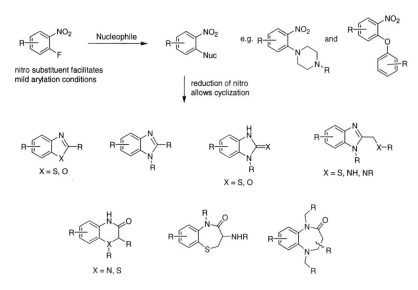

Figure 26   Different scaffolds accessible through a single Ugi intermediate.

aryl piperazines and diaryl ethers, which have substantial precedent in bioactive compounds. The aryl nitro group, however, is not typically a desirable aromatic substituent for drug development purposes. Developing chemistry for the reduction of the nitro group on solid support (127–129) not only removed a potential

Figure 27   The use of nitro-activated aryl halides facilitates the synthesis of a wide range of benzo-fused heterocycles.

metabolic liability but also created a new center for functionalization. Several libraries have taken advantage of this chemistry to prepare a wide range of benzo-fused heterocycles (130–136).

VI. LEVERAGING KNOWLEDGE: TARGETED LIBRARIES

In contrast to discovery libraries, many libraries are built with a specific target in mind. These targeted libraries are commonly constructed around motifs or functional arrangements that are already known to bind or inhibit the target. Whereas discovery libraries can be more liberal with respect to structural variation within the library, targeted libraries usually conform more closely to known templates to maximize the chance of finding biological activity. As a side benefit, the motivation to prepare specific structures in the library often fuels the development of new synthetic methods for their high throughput synthesis.

Libraries directed toward enzyme inhibitors are good examples of this type of strategy. Not only do many enzymes have X-ray structures available and well-understood mechanisms of action, but often there are functional groups known to target the active site and to act as transition state mimics. Another reason enzymes are good targets for combinatorial approaches is that there may be several distinct enzymes within a mechanistic class. Libraries with a bias toward that class of en-

Table 1 Functional Groups that Bias Compounds Toward Enzymes of Specific Classes

Enzyme class	Functional group
Metalloenzymes	Metal ligands
	Hydroxamic acid
	Thiol
	Sulfonamide
	Imidazole
	Carboxylate
Trypsin family serine proteases	Basic nitrogen for S_1' specificity pocket
	Amidines
	Guanidines
	Lysine
	Activated carbonyl
	Aldehyde
	α-Ketoester/amide
Aspartyl proteases	Hydroxyethylene isosteres
	Statine derivatives
Cysteine proteases	α-Activated ketones

zyme have improved chances of finding leads against multiple targets. Table 1 presents examples of functional groups commonly used to bias compounds toward enzymes of specific classes (137).

Carroll and coworkers at Pharmacopeia used this strategy when constructing an encoded library of statine analogs (138,139). The goal of this exercise was the identification of leads against the aspartyl protease plasmepsin II, which plays an important role in the malarial parasite *Plasmodium falciparum*. The selection of synthons for the library was guided by published information, and the library was constructed by using encoded pooled synthesis techniques. To evaluate selectivity, the human enzyme cathepsin D was chosen. Cathepsin D is an aspartyl protease 37% homologous to plasmepsin II. After library synthesis, the compounds were cleaved from the resin photochemically; then the bead eluent was split and screened simultaneously against the two related proteases (Fig. 28).

Based on 96 compounds that were decoded, interesting SARs emerged relating to the control of selectivity. Compounds of interest were then resynthesized to confirm activity. An important principle illustrated in this example is that because of the large number of data points gathered, SAR was obtained for both enzymes. In this case, it was found that cathepsin D was particularly sensitive to changes in the R^4 position, while plasmepsin II was more sensitive to changes in R^3. Figure 29 illustrates how the SARs derived from library screening can be used to rationally design compounds selective for each enzyme target.

For nonenzyme targets, bias can also be built into the library design. In this case, instead of being based on a mechanism of action, the structural features contributing to a bias are frequently derived from empirical observation. Libraries based on "privileged structures" are an appealing strategy for broad screening against related families of targets. Examples of some privileged structures for G-protein-coupled receptors (GPCRs) which have been used in library design are shown in Fig. 30 (73,137,140,141).

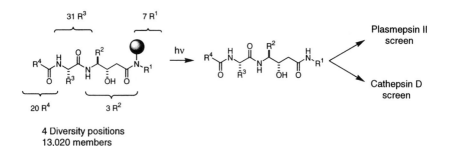

Figure 28   Using statine derivatives to target a library toward aspartyl proteases.

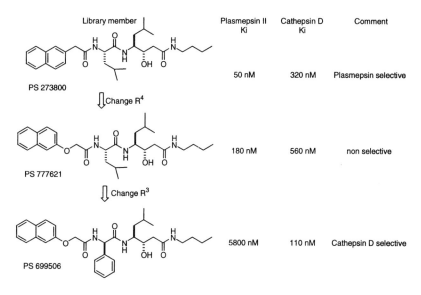

Figure 29   Selective inhibitors of two aspartyl proteases can be found in a single targeted library.

An excellent example of targeted library design comes from researchers at Merck (142–144). The objective of the research was the identification of leads for each of the five somatostatin receptor subtypes (*sstr1–sstr5*). The library design began with conformational analysis of the cyclic hexapeptide L-363,377, which

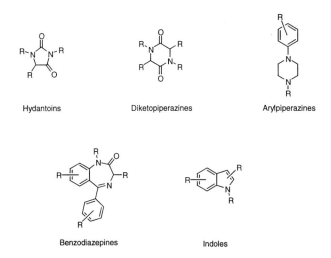

Figure 30   Examples of some GPCR privileged structures.

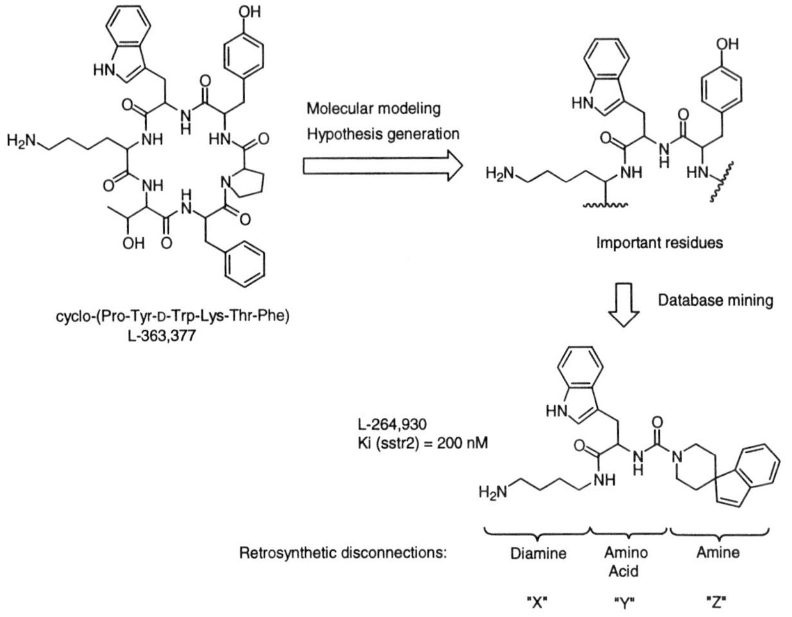

Figure 31   Molecular modeling and database mining aid library design.

was a selective antagonist for *sstr2* (Fig. 31). From the peptide, a pharmacophore model was generated, which was then used to search their internal compound collection. This resulted in the identification of the nonpeptide *sstr2* inhibitor L-264,930 which could be conveniently dissected into three portions, designated X, Y, and Z.

A 31,600-member library (accounting for uncontrolled asymmetric centers, the total number of compounds was 88,270) was constructed by means of mix and split techniques, and iterative deconvolution was used to identify the active compounds.

Because the library was designed around an *sstr2* selective inhibitor, the majority of activity found in the first library was *sstr2* activity (Fig. 32). However, enough activity could be seen in assays for the other receptors to guide the design of additional libraries. Three additional libraries were constructed, each design guided by the knowledge obtained from the preceding ones. Library 2 increased the diversity of the Z (amine) synthon and yielded selective inhibitors for *sstr1* and *sstr3,* and libraries 3 and 4 narrowed the

Variable	Design concepts
Diamine "X"	Linear - explore chain length
	Branched - bias acyclic conformation
	Cyclic - constrain conformation
Amino acid "Y"	Aromatic/Indole derivatives
Amine "Z"	Bias toward piperidine/piperazine derivatives

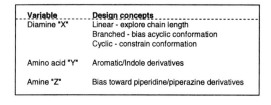

Figure 32   Merck's somatostatin library 1 yielded *sstr2* selective structures.

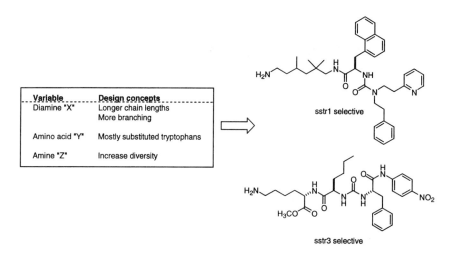

Variable	Design concepts
Diamine "X"	Longer chain lengths
	More branching
Amino acid "Y"	Mostly substituted tryptophans
Amine "Z"	Increase diversity

Figure 33   Merck's somatostatin library 2 produced selective inhibitors for *sstr1* and *sstr3*.

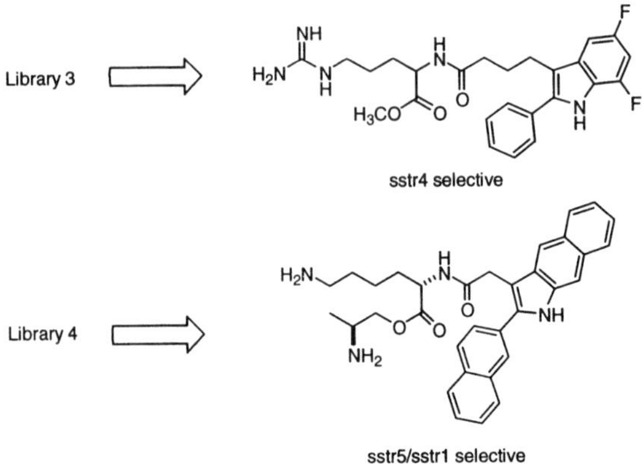

Figure 34 Merck's somatostatin libraries 3 and 4 resulted in *sstr4* and *sstr5* selective compounds.

search to aryl indoles and afforded inhibitors for *sstr4* and *sstr5* (Figs. 33 and 34).

Among the many useful concepts highlighted by this work, the use of a large library directed toward one target (*sstr2*), to identify initial leads to guide the design of follow-up libraries is particularly noteworthy.

VII. MOVING BEYOND AFFINITY: OPTIMIZATION LIBRARIES

It is relatively common in drug discovery programs for high affinity leads to fail owing to poor pharmacokinetic performance. Because of this, an increasing number of high throughput synthesis applications are surfacing in the area of lead optimization, where the modification of pharmacokinetic properties is the primary objective. Thrombin and FPT, the examples to be discussed in this section, are particularly interesting because both involve the use of combinatorial chemistry to search for alternatives to the privileged structure motifs that are highly preferred by the targets.

A. Improving Molecular Properties

1. Thrombin

Thrombin has been a target for structure-based drug design for many years. It is a classic example for targeted library design as well, since it has well-understood structural preferences. In particular, the P1 site of thrombin ('recognition pocket') contains an acidic residue (Asp189) that has been extensively exploited in compound and library design. Using concepts like those discussed in Section VI, compounds and libraries containing guanidine and amidine moieties have been designed to interact with Asp189 and orient inhibitors in the active site (20,145–148). Despite the accumulated knowledge, however, success in bringing small-molecule inhibitors into the clinic has been limited. In many cases this is due to poor oral bioavailability that is caused, in part, by the basic P1 binding moiety. It may appear somewhat ironic that the basic residues, which were designed into the structures to bias them toward thrombin, are key contributors to poor pharmacokinetic performance and are thus among the most desirable to remove (149).

A library described by scientists at Merck is a good example of the use of high throughput synthesis in conjunction with molecular modeling to address pharmacokinetic limitations of the lead structure (150). The rationale for library design came from the observation of a hydrophobic region in the P1 pocket, indicated by X-ray structures, that was not being filled by most of the known inhibitors. Based on this observation, a parallel library was constructed focussing on variation of the P1 binding unit (Fig. 35). From this library was found the 3-chlorobenzyl derivative, which was further optimized in solution to afford the 2,5-dichlorobenzyl inhibitor.

In general terms, this example illustrates a common strategy for drug discovery that is increasingly being applied to combinatorial chemistry: First, a lead

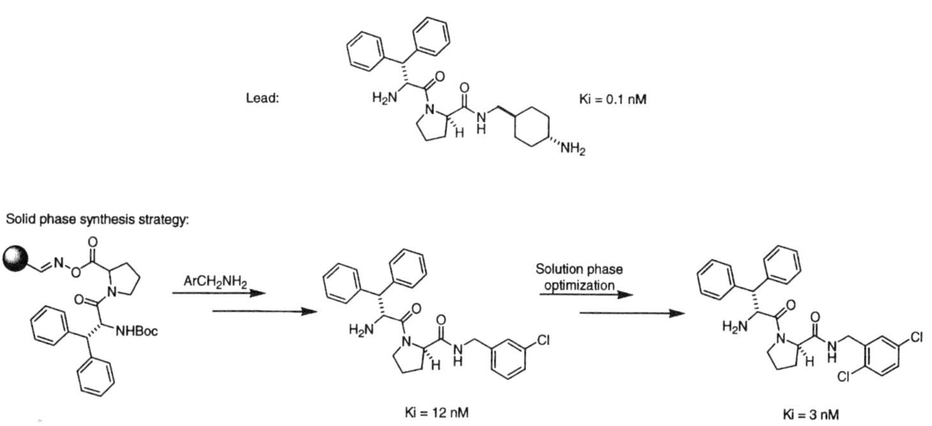

Figure 35 Rational design of a parallel library to improve physicochemical properties of thrombin inhibitors.

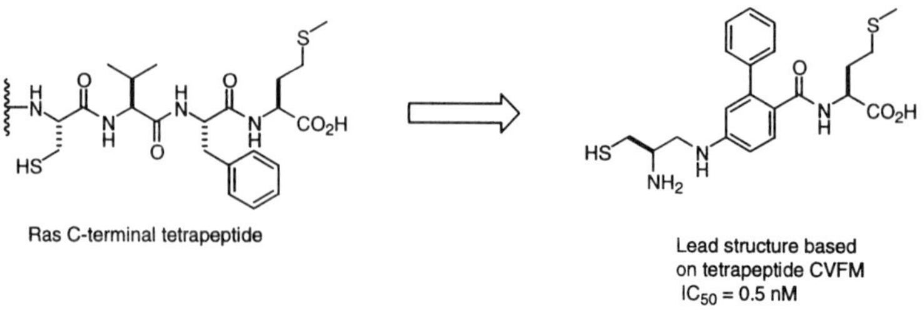

Figure 36   Structure of the Ras C-terminal peptide mimic described by Abbott.

was obtained by using privileged structure motifs, such as the tripeptide backbone and basic residues for P1 binding, to bias compounds toward thrombin and to orient the structures in the active site. Next, efforts focused on optimizing the remaining positions to improve affinity. Finally, with affinity that was more than adequate, efforts could be directed toward removing the biasing functionality, which in this case was detrimental to oral bioavailability.

2. FPT

The work from Abbott Laboratories on non-cysteine-containing Ras farnesyl-transferase (FPT) inhibitors represents another good example of the use of high-throughput techniques to search for new leads with improved molecular properties (151). This case has several similarities to the thrombin example just described. Leads for FPT have been identified by using rational drug design. One of the privileged structural arrangements that allow biasing toward FPT is a thiol, which serves to mimic a critical cysteine residue in the natural substrate. The same privileged functional group is ultimately a detriment to drug development owing to its metabolic susceptibility. Figure 36 shows the C-terminal tetrapeptide of Ras and the lead that was designed to mimic it.

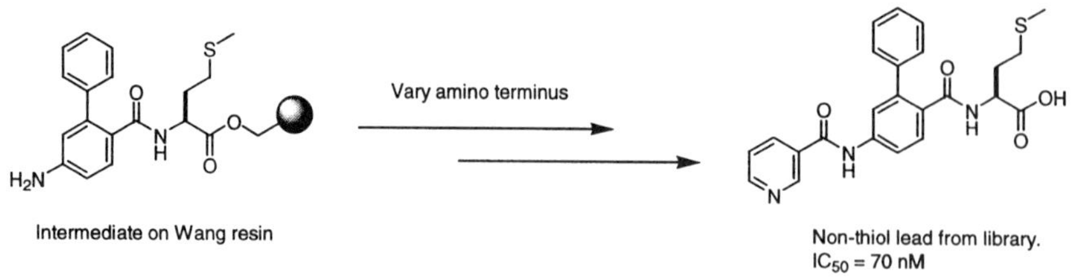

Figure 37   Parallel synthesis was used to search for a non-thiol-containing lead compound.

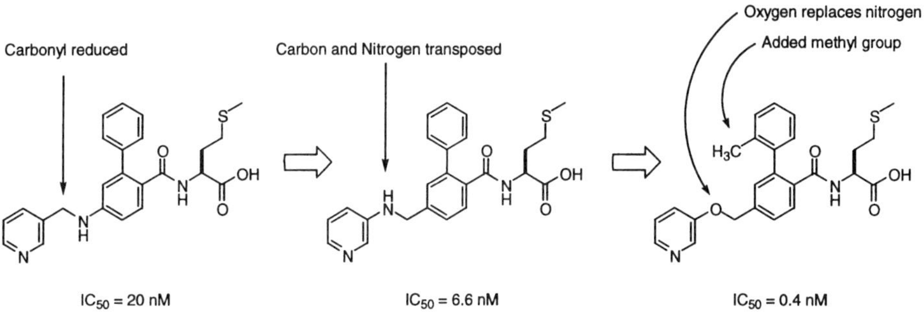

Figure 38 Further optimization of a library-generated lead through medicinal chemistry.

The objective of library synthesis was to identify a replacement for the thiol functionality in the lead. As illustrated in Fig. 37, the biaryl aniline was used as a convenient intermediate. From the library was found a nicotinic amide derivative with activity that was weaker than the original lead, but still considered promising. The nicotinamide derivative was then subjected to a variety of modifications through use of traditional medicinal chemistry. The results of medicinal chemical optimization are outlined in Fig. 38.

This example illustrates the complementarity of combinatorial and medicinal chemical approaches. Combinatorial chemistry was well suited to rapid modification of the biaryl aniline intermediate because the chemistry for derivatization of the aniline is well established and reliable, and there are an enormous number of commercially available synthons that could be examined. Other changes, such as transposing the nitrogen or adding a methyl group to the biaryl, are more easily accomplished via traditional synthetic methods.

VIII. CONCLUSION

A short history and some of the key contributions that led to modern combinatorial chemistry can be listed.

1. Solid phase peptide synthesis demonstrated the value of polymeric beads to aid in handling and purification.
2. The success of solid phase peptide synthesis provided incentive for the development of equipment that could perform automated synthesis.
3. Automated peptide synthesis demonstrated that a single reaction (amide bond formation) could be optimized to be very nearly independent of the specific substrates.

4. The split-pool concept illustrated how reactions could be performed simultaneously, amplifying one's effort manyfold.
5. High throughput screening creates a demand for more compounds.
6. Gene sequencing and functional genomics create an opportunity for new targets.

Fueled by the interest of the pharmaceutical industry, progress in developing combinatorial chemistry as a practical tool for drug discovery is moving very rapidly. Early in the development of combinatorial chemistry, the main parameters that constrained library design were the lack of versatile chemistry and tools. Today, there are ample reactions that have been proven to succeed by using high throughput techniques, and a wide variety of instruments are available for automating the process. With the implementation details reliably established, and a wealth of chemical transformations available, the focus has moved back to what has always been the medicinal chemist's main objective: how to design the best drugs. A big difference, however, is the number of compounds being designed together. Combinatorial chemistry has become a statistical science. Instead of thinking in terms of individual compounds, the emphasis is on populations, and operations that can be done on large groups have become increasingly valuable. Because of this, more chemists are looking to computational tools to aid them in making design decisions. The development of these tools is currently one of the most active areas of combinatorial chemistry and computer-aided drug design, as evidenced by the remaining chapters in this book. Thus to answer the question "How does one design a library?" one responds, "It depends." Many parameters need to be considered together, and ultimately the answer is dependent on the creativity and value judgments of the individual designer.

REFERENCES

1. Lipper RA. E pluribus product. Modern Drug Discovery 1999; Jan/Feb:55–60.
2. Johnson D. The discovery-development interface has become the new interfacial phenomenon. Drug Discovery Today 1999; 4:535–536.
3. Studt T. Drug development bottlenecks not cured by technology alone. Drug Discovery Dev 1999; Jan:40–41.
4. Spaller MR, Burger MT, Fardis M, Bartlett PA. Synthetic strategies in combinatorial chemistry. Curr. Opin Chem Biol 1997; 1:47–53.
5. Antel J. Integration of combinatorial chemistry and structure-based drug design. Curr Opin Drug Discovery Dev 1999; 2:224–233.
6. Merrifield RB. Peptide synthesis. I. The synthesis of a tetrapeptide. J Am Chem Soc 1963; 85:2149–2154.
7. Letsinger RL, Mahadevan V. Oligonucleotide synthesis on a polymer support. J Am Chem Soc 1965; 87:3526–3527.

8. Hermkens P, Ottenheijm H, Rees D. Solid-phase organic reactions: A review of recent literature. Tetrahedron 1996; 52:4527–4554.

9. Hermkens P, Ottenheijm H, Rees D. Solid-phase organic reactions. II. A review of the literature Nov 95–Nov 96. Tetrahedron 1997; 53:5643–5678.

10. Booth S, Hermkens PHH, Ottenheijm HCJ, Rees DC. Solid-phase organic reactions. III. A review of the literature Nov 96–Dec 97. Tetrahedron 1998; 54:15385–15443.

11. Früchtel JS, Jung G. Organic chemistry on solid supports. Angew Chem Int Ed Engl 1996; 35:17–41.

12. Osborn HMI, Khan TH. Recent developments in polymer supported synthesis of oligosaccharides and glycopeptides. Tetrahedron 1999; 55:1807–1850.

13. Lorsbach BA, Kurth MJ. Carbon–carbon bond forming solid-phase reactions. Chem Rev 1999; 99:1549–1581.

14. Gordon K, Balasubramanian S. Recent advances in solid-phase chemical methodologies. Curr. Opin Drug Discovery Dev 1999; 2:342–349.

15. Hudson D. Matrix assisted synthetic transformations: A mosaic of diverse contributions. I. The pattern emerges. J Comb Chem 1999; 1:333–360.

16. Hudson D. Matrix assisted synthetic transformations: A mosaic of diverse contributions. II. The pattern is completed. J Comb Chem 1999; 1:403–457.

17. Brown RCD. Recent developments in solid-phase organic synthesis. J Chem Soc Perkin Trans. 1 1998:3293–3320.

18. James IW. Linkers for solid phase organic synthesis. Tetrahedron 1999; 55: 4855–4946.

19. Floyd CD, Lewis CN, Patel SR, Whittaker M. A method for the synthesis of hydroxamic acids on solid phase. Tetrahedron Lett 1996; 37:8045–8048.

20. Kim SW, Hong CY, Koh JS, Lee EJ, Lee K. Solid phase synthesis of benzamidine-derived sulfonamide libraries. Mol Diversity 1998; 3:133–136.

21. Mohan R, Yun W, Buckman BO, Liang A, Trinh L, Morrissey MM. Solid-phase synthesis of N-substituted amidinophenoxy pyridines as factor Xa inhibitors. Bioorg Med Chem Lett 1998; 8:1877–1882.

22. Fivush AM, Wilson TM. AMEBA: An acid sensitive aldehyde resin for solid phase synthesis. Tetrahedron Lett 1997; 41:7151–7154.

23. Gayo LM, Suto MJ. Traceless linker: Oxidative activation and displacement of a sulfur-based linker. Tetrahedron Lett 1997; 38:211–214.

24. DeGrado WF, Kaiser ET. Solid-phase synthesis of protected peptides on a polymer-bound oxime: Preparation of segments comprising the sequence of a cytotoxic 26-peptide analog. J Org Chem 1982; 47:3258–3261.

25. Voyer N, Lavoie A, Pinette M, Bernier J. A convenient solid phase preparation of peptide substituted amides. Tetrahedron Lett 1994; 35:355–358.

26. Reitz AB. Recent advances in traceless linkers. Curr Opin Drug Discovery Dev 1999; 2:358–364.

27. Woolard FX, Paetsch J, Ellman JA. A silicon linker for direct loading of aromatic compounds to supports. Traceless synthesis of pyridine-based tricyclics. J Org Chem 1997; 62:6102–6103.

28. Cobb JM, Fiorini MT, Goddard CR, Theoclitou M-E, Abell C. A decarboxylative traceless linker approach for the solid phase synthesis of quinazolines. Tetrahedron Lett 1999; 40:1045–1048.

29. Bräse S, Enders D, Köbberling J, Avermaria F. A surprising solid-phase effect: Development of a recyclable "traceless" linker system for reactions on solid support. Angew Chem Int Ed Engl 1998; 37:3413–3415.
30. Vojkovsky T. Detection of secondary amines on solid phase. Pept. Res 1995; 8:236–237.
31. Krchnak V, Vagner J, Lebl M. Noninvasive continuous monitoring of solid-phase peptide synthesis by acid–base indicator. Int J Pept Protein Res 1988; 32:415–416.
32. Chu SS, Reich SH. NPIT: A new reagent for quantitatively monitoring reactions of amines in combinatorial synthesis. Bioorg Med Chem Lett 1995; 5:1053–1058.
33. Anderson RC, Stokes JP, Shapiro MJ. Structure determination in combinatorial chemistry: Utilization of magic angle spinning HMQC and TOCSY NMR spectra in the structure determination of Wang-bound lysine. Tetrahedron Lett 1995; 30: 5311–5314.
34. Shapiro MJ, Chin J, Marti RE, Jarosinski MA. Enhanced resolution in MAS NMR for combinatorial chemistry. Tetrahedron Lett 1997; 38:1333–1336.
35. Look GC, Holmes CP, Chinn JP, Gallop MA. Methods for combinatorial organic synthesis: The use of fast ^{13}C analysis for gel phase reaction monitoring. J Org Chem 1994; 59:7588–7590.
36. Yan B, Sun Q, Wareing JR, Jewell CF. Real-time monitoring of the catalytic oxidation of alcohols to aldehydes and ketones on resin support by single-bead Fourier transform infrared microspectroscopy. J Org Chem 1996; 61:8765–8770.
37. Rahman SS, Busby DJ, Lee DC. Infrared and Raman spectra of a single resin bead for analysis of solid-phase reactions and use in encoding combinatorial libraries. J Org Chem 1998; 63:6196–6199.
38. Taylor EW, Qian MG, Dollinger GD. Simultaneous on-line characterization of small organic molecules derived from combinatorial libraries for identity, quantity, and purity by reversed-phase HPLC with chemilumiscent nitrogen, UV, and mass spectrometric detection. Anal Chem 1998; 70:3339–3347.
39. Fitch WL, Szardenings AK. Chemiluminescent nitrogen detection for HPLC: An important new tool in organic analytical chemistry. Tetrahedron Lett 1997; 38: 1689–1692.
40. Kibbey CE. Quantitation of combinatorial libraries of small organic molecules by normal-phase HPLC with evaporative light-scattering detection. Mol Diversity 1995; 1:247–258.
41. Suto MJ. Developments in solution-phase combinatorial chemistry. Curr Opin Drug Discovery Dev 1999; 2:377–384.
42. Parlow JJ, Devraj RV, South MS. Solution-phase chemical library synthesis using polymer-assisted purification techniques. Curr Opin Chem Biol 1999; 3:320–336.
43. Mutter M, Hagenmaier H, Bayer E. New method of polypeptide synthesis. Angew Chem Int Ed Engl 1971; 12:811–812.
44. Bayer E, Mutter M. Liquid phase synthesis of peptides. Nature 1972; 237:512–513.
45. Han H, Wolfe MM, Brenner S, Janda KD. Liquid-phase combinatorial synthesis. Proc Natl Acad Sci USA 1995; 92:6419–6423.
46. Gravert DJ, Janda KD. Synthesis on soluble polymers: New reactions and the construction of small molecules. Curr Opin Chem Biol 1997; 1:107–113.

47. Gravert DJ, Janda KD. Organic synthesis on soluble polymer supports: Liquid-phase methodologies. Chem Rev 1997; 97:489–509.
48. Lee JK, Angulo A, Ghazal P, Janda KD. Soluble-polymer supported synthesis of a prostanoid library: Identification of antiviral activity. Org Lett 1999; 1:1859–1862.
49. Studer A, Hadida S, Ferritto R, Kim S-Y, Jeger P, Wipf P, Curran D. Fluorous synthesis: A fluorous-phase strategy for improving separation efficiency in organic synthesis. Science 1997; 275:823–826.
50. Curran DP, Hadida S. Tris(2-(perfluorohexyl)ethyl)tin hydride: A new fluorous reagent for use in traditional organic synthesis and liquid phase combinatorial synthesis. J Am Chem Soc 1996; 118:2531–2532.
51. Studer A, Jeger P, Wipf P, Curran DP. Fluorous synthesis: Fluorous protocols for the Ugi and Biginelli multicomponent condensations. J Org Chem 1997; 62: 2917–2924.
52. Flynn DL, Crich JZ, Devraj RV, Hockerman SL, Parlow JJ, South MS, Woodard S. Chemical library purification strategies based on principles of complementary molecular reactivity and molecular recognition. J Am Chem Soc 1997; 119: 4874–4881.
53. Kulkarni BA, Genesan A. Ion-exchange resins for combinatorial synthesis: 2,4-Pyrrolidinediones by Dieckmann condensation. Angew Chem Int Ed Engl 1997; 36:2454–2455.
54. Flynn DL, Devraj RV, Parlow JJ. Recent advances in polymer-assisted solution-phase chemical library synthesis and purification. Curr Opin Drug Discovery Dev 1998; 1:41–50.
55. Kaldor SW, Fritz JE, Tang J, McKinney ER. Discovery of antirhinoviral leads by screening a combinatorial library of ureas prepared using covalent scavengers. Bioorg Med Chem Lett 1996; 24:3041–3044.
56. Selway CN, Terrett NK. Parallel-compound synthesis: Methodology for accelerating drug discovery. Bioorg Med Chem 1996; 4:645–654.
57. Weller HN, Young MG, Michalczyk SJ, Reitnauer GH, Cooley RS, Rahn PC, Loyd DJ, Fiore D, Fischman SJ. High throughput analysis and purification in support of automated parallel synthesis. Mol Diversity 1997; 3:61–70.
58. Kyranos JN, Hogan JCJ. High-throughput characterization of combinatorial libraries generated by parallel synthesis. Anal Chem News Features 1998; June: 389–395.
59. Calvert S, Stewart FP, Swarna K, Wiseman JS. The use of informatics and automation to remove bottlenecks in drug discovery. Curr Opin Drug Discovery Dev 1999; 2:234–238.
60. Merrifield RB. Solid-phase synthesis. Science 1986; 232:341–347.
61. Sebestyen F, Dibo G, Kovacs A, Furka A. Chemical synthesis of peptide libraries. Bioorg Med Chem Lett 1993; 3:413–418.
62. Lam KS, Salmon SE, Hersh EM, Hruby VJ, Kazmierski WM, Knapp RJ. A new type of synthetic peptide library for identifying ligand-binding activity. Nature 1991; 354:84–86.
63. Houghten RA, Pinilla C, Blondelle SE, Appel JR, Dooley CT, Cuervo JH. Generation and use of synthetic peptide combinatorial libraries for basic research and drug discovery. Nature 1991; 354:84–86.

64. Furka A, Sebestyen F, Asgedom M, Dibo G. General method for rapid synthesis of multicomponent peptide mixtures. Int J Pept Protein Res 1991; 37:487–493.

65. Baldwin JJ, Horlbeck E. Encoded libraries may be created using split-pool or direct divide synthesis. US Patent 5,663,046, 1997.

66. Burbaum JJ, Ohlmeyer MHJ, Reader JC, Henderson I, Dillard LW, Li G, Randle TL, Sigal NH, Chelsky D, Baldwin JJ. A paradigm for drug discovery employing encoded combinatorial libraries. Proc Natl Acad Sci USA 1995; 92:6027–6031.

67. Houghten RA, Pinilla C, Appel JR, Blondelle SE, Dooley CT, Eichler J, Nefzi A, Ostresh JM. Mixture-based synthetic combinatorial libraries. J Med Chem 1999; 42:3743–3778.

68. Feng S, Kapoor TM, Shirai F, Combs AP, Schreiber SL. Molecular basis for the binding of SH3 ligands with non-peptide elements identified by combinatorial synthesis. Chem Biol 1996; 3:661–670.

69. Erb E, Janda KD, Brenner S. Recursive deconvolution of combinatorial chemical libraries. Proc Natl Acad Sci USA 1994; 91:11422–11426.

70. Berk SC, Chapman KT. Spatially arrayed mixture (SpAM) technology: Synthesis of two-dimensionally indexed orthogonal combinatorial libraries. Bioorg Med Chem Lett 1997; 7:837–842.

71. Dooley CT, Houghten RA. The use of positional scanning synthetic peptide combinatorial libraries for the rapid determination of opioid receptor ligands. Life Sci 1993; 52:1509–1517.

72. Pinilla C, Appel JR, Bondelle SE, Dooley CT, Eichler J, Ostresh JM, Houghten RA. Versatility of positional scanning synthetic combinatorial libraries for the identification of individual compounds. Drug Dev Res. 1994; 33:133–145.

73. Floyd CD, Leblanc C, Whittaker M. Combinatorial chemistry as a tool for drug discovery. In: King FD, Oxford AW, eds. Progress in Medicinal Chemistry. Vol. 36. Amsterdam: Elsevier Science, 1999, pp. 91–168.

74. Baldwin JJ, Dolle R. Deconvolution tools for solid-phase synthesis. In: Czarnik AW, DeWitt SH, eds. A Practical Guide to Combinatorial Chemistry. Washington, DC: American Chemical Society, 1997, pp 153–174.

75. Ohlmeyer MHJ, Swanson RN, Dillard LW, Reader JC, Asouline G, Kobayashi R, Wigler M, Still WC. Complex synthetic chemical libraries indexed with molecular tags. Proc Natl Acad Sci USA 1993; 90:10922–10926.

76. Nestler HP, Bartlett PA, Still WC. A general method for molecular tagging of encoded combinatorial chemistry libraries. J Org Chem 1994; 59:4723–4724.

77. Ni Z-J, Maclean D, Holmes CP, Murphy MM, Ruhland B, Jacobs JW, Gordon EM, Gallop MA. Versatile approach to encoding combinatorial organic syntheses using chemically robust secondary amine tags. J Med Chem 1996; 39:1601–1608.

78. Wagner DS, Markworth CJ, Wagner CD, Schoenen FJ, Rewerts CE, Kay BK, Geysen HM. Ratio encoding combinatorial libraries with stable isotopes and their utility in pharmaceutical research. Comb. Chem. High Throughput Screening 1998: 143–153.

79. Nicolaou KC, Xiao XY, Parandoosh Z, Senyei A, Nova MP. Radiofrequency encoded combinatorial chemistry. Angew Chem Int Ed Engl 1995; 34:2289–2291.

80. Moran EJ, Sarshar S, Cargill JF, Shahbaz MM, Lio A, Mjalli AMM, Armstrong RW. Radio frequency tag encoded combinatorial library method for the discovery

of tripeptide-substituted cinnamic acid inhibitors of the protein tyrosine phosphatase PTP1B. J Am Chem Soc 1995; 117:10787–10788.

81. Xiao X-Y, Zhao C, Polash H, Nova MP. Combinatorial chemistry with laser optical encoding. Angew. Chem Int Ed Engl 1997; 36:780–781.

82. Martin EJ, Blaney JM, Siani MA, Spellmeyer DC, Wong AK, Moos WH. Measuring diversity: Experimental design of combinatorial libraries for drug discovery. J Med Chem 1995; 38:1431–1436.

83. Pearlman RS, Smith KM. Novel software tools for chemical diversity. Perspect. Drug Discovery Design 1998; 9:339–353.

84. Brown RD, Martin YC. Use of structure–activity data to compare structure-based clustering methods and descriptors for use in compound selection. J Chem Inf Comput Sci 1996; 36:572–584.

85. Patterson DE, Cramer RD, Ferguson AM, Clark RD, Weinberger LE. Neighborhood behavior: A useful concept for validation of "molecular diversity" descriptors. J Med Chem 1996; 39:3049–3059.

86. Willett P. Using computational tools to analyze molecular diversity. In: Czarnik AW, DeWitt SH, eds. A Practical Guide to Combinatorial Chemistry. Washington DC: American Chemical Society, 1997.

87. Mason JS, Pickett SD. Partition-based selection. Persp Drug Discovery Res 1997; 7/8:85–114.

88. Mason JS, Hermsmeier MA. Diversity assessment. Curr Opin Chem Biol 1999; 3:342–349.

89. Gorse D, Rees A, Kaczorek M, Lahana R. Molecular diversity and its analysis. Drug Discovery Today 1999; 4:257–264.

90. Kauvar LM, Laborde E. The diversity challenge in combinatorial chemistry. Curr Opin Drug Discovery Dev. 1998; 1:66–70.

91. Gillet VJ, Willett P, Bradshaw J. The effectiveness of reactant pools for generating structurally diverse combinatorial libraries. J Chem Inf Comput Sci 1997; 37:731–740.

92. Braban M, Pop I, Willard X, Horvath D. Reactivity prediction models applied to the selection of novel candidate building blocks for high-throughput organic synthesis of combinatorial libraries. J Chem Inf Comput Sci 1999; 39:1119–1127.

93. Kubinyi H. Similarity and dissimilarity: A medicinal chemist's view. Perspect. Drug Discovery Design 1998; 9/10/11:225–252.

94. Roques BP, Noble F, Daugé V, Fournié-Zaluski M-C, Beaumont A. Neutral endopeptidase 24.11: Structure, inhibition, and experimental and clinical pharmacology. Pharm Rev 1993; 45:87–146.

95. Sugg EE. Nonpeptide agonists for peptide receptors: Lessons from ligands. Annu. Rep Med Chem 1997; 32:277–283.

96. Giannis A, Kolter T. Peptide mimetics for receptor ligands: Discovery, development, and medicinal perspectives. Angew Chem Int Ed Engl 1993; 32:1244–1267.

97. Koehler RT, Dixon SL, Villar HO. LASSOO: A generalized directed diversity approach to the design and enrichment of chemical libraries. J Med Chem 1999; 42:4695–4704.

98. Wang J, Ramnarayan K. Toward designing drug-like libraries: A novel computational approach for prediction of drug feasibility of compounds. J Comb Chem 1999; 1:524–533.

99. Clark DE. Rapid calculation of polar molecular surface area and its application to the prediction of transport phenomena. 1. Prediction of intestinal absorption. J Pharm Sci 1999; 88:807–814.

100. Lipinski CA, Lombardo F, Dominy BW, Feeney PJ. Experimental and computational approaches to estimate solubility and permeability in drug discovery and development settings. Adv Drug Delivery Rev 1997; 23:3–25.

101. Lee M, Nakanishi H, Kahn M. Enlistment of combinatorial techniques in drug development. Curr Opin Drug Discovery Dev 1999; 2:332–341.

102. Pavia MR, Sawyer TK, Moos WH. The generation of molecular diversity. Bioorg Med Chem Lett 1993; 3:whole issue.

103. Simon RJ, Kania RS, Zuckerman RN, Huebner VD, Jewell DA, Banville S, Ng S, Wang L, Rosenberg S, Marlowe CK, Spellmeyer DC, Yan R, Frankel AD, Santi DV, Cohen FE, Bartlett PA. Peptoids: A modular approach to drug discovery. Proc Natl Acad Sci USA 1992; 89:9367–9371.

104. Cho CY, Moran EJ, Cherry SR, Stephans JC, Fodor SPA, Adams CL, Sundaram A, Jacobs JW, Schultz PA. An unnatural biopolymer. Science 1993; 261:1303–1305.

105. Moran EJ, Wilson TE, Cho CY, Cherry SR, Schultz PG. Novel biopolymers for drug discovery. Biopolymers 1995; 37:213–219.

106. Paikoff SJ, Wilson TE, Cho CY, Schultz PG. The solid phase synthesis of N-alkyl-carbamate oligomers. Tetrahedron Lett 1996; 37:5653–5656.

107. Burgess K, Ibarzo J, Linthicum DS, Shin H, Shitangkoon A, Totani R, Zhang AJ. Solid phase syntheses of oligoureas. J Am Chem Soc 1997; 119:1556–1564.

108. Han H, Janda KD. Azatides: Solution and liquid phase syntheses of a new peptidomimetic. J Am Chem Soc 1996; 118:2539–2544.

109. Smith J, Liras JL, Schneider SE, Anslyn EV. Solid and solution phase organic syntheses of oligomeric thioureas. J Org Chem 1996; 61:8811–8818.

110. Gennari C, Nestler HP, Salom B, Still WC. Solid-phase synthesis of vinylogous sulfonyl peptides. Angew Chem Int Ed Engl 1995; 34:1763–1765.

111. Zuckerman RN, Kerr JM, Kent SBH, Moos WH. Efficient method for the preparation of peptoids [oligo(N-substituted glycines)] by submonomer solid-phase synthesis. J Am Chem Soc 1992; 114:10646–10647.

112. Hajduk PJ, Dinges J, Schkeryantz JM, Janowick D, Kaminski M, Tufano M, Augeri DJ, Petros A, Nienaber V, Zhong P, Hammond R, Coen M, Beutel B, Katz L, Fesik SW. Novel inhibitors of Erm methyltransferases from NMR and parallel synthesis. J Med Chem 1999; 42:3852–3859.

113. Chen C, Dagnino RJ, De Souza EB, Grigoriadis DE, Huang CQ, Kim KI, Liu Z, Moran T, Webb TR, Whitten JP, Xie YF, McCarthy JR. Design and synthesis of a series of non-peptide high-affinity human corticotropin-releasing factor1 receptor antagonists. J Med Chem 1996; 39:4354–4357.

114. Bicknell AJ, Hird NW, Readshaw SA. Efficient robotic synthesis. Multi-component preparation of a trycyclic template by solid phase Tsuge reaction. Tetrahedron Lett 1998; 39:5869–5872.

115. Armstrong RW, Combs AP, Tempest PA, Brown SD, Keating TA. Multiple-component condensation strategies for combinatorial library synthesis. Acc Chem Res 1996; 29:123–131.

116. Keating TA, Armstrong RW. Postcondensation modifications to Ugi four-component condensation products: 1-Isocyanocyclohexene as a convertible isocyanide.

Mechanism of conversion, synthesis of diverse structures, and demonstration of resin capture. J Am Chem Soc 1996; 118:2574–2583.

117. Wipf P, Cunningham A. A solid phase protocol of the Biginelli dihydropyrimidine synthesis suitable for combinatorial chemistry. Tetrahedron Lett 1995; 36: 7819–7822.

118. Holmes CP, Chinn JP, Look GC, Gordon EM, Gallop MA. Strategies for combinatorial organic synthesis: Solution and polymer-supported synthesis of 4-thiazolidinones and 4-metathiazanones derived from amino acids. J Org Chem 1995; 60: 7328–7333.

119. Domling A, Chi K-Z, Barrere M. A novel method to highly versatile monomeric PNA building blocks by multicomponent reactions. Bioorg Med Chem Lett 1999; 9: 2871–2874.

120. Mjalli AMM, Baiga TJ. Solid phase synthesis of pyrroles derived from a four component condensation. Tetrahedron Lett 1996; 37:2943–2946.

121. Ostresh JM, Husar GM, Blondelle SE, Dorner B, Weber PA, Houghten RA. "Libraries from libraries": Chemical transformation of combinatorial libraries to extend the range and repertoire of chemical diversity. Proc Natl Acad Sci USA 1994; 91:11138–11142.

122. Nefzi A, Dooley C, Ostresh JM, Houghten RA. Combinatorial chemistry: From peptides and peptidomimetics to small organic and heterocyclic compounds. Bioorg Med Chem Lett 1998; 8:2273–2278.

123. Hermkens PHH, Hamersma H. Functional group transformation: An efficacy-enhancing approach in combinatorial chemistry. J Comb Chem 1999; 1:307–316.

124. Keating TA, Armstrong RW. Molecular diversity via a convertible isocyanide in the Ugi four-component condensation. J Am Chem Soc 1995; 117:7842–7843.

125. Dankwardt SM, Newman SR, Krstenansky JL. Solid phase synthesis of aryl and benzylpiperazines and their application in combinatorial chemistry. Tetrahedron Lett 1995; 36:4923–4926.

126. Burgess K, Lim D, Bois-Choussy M, Zhu J. Rapid and efficient solid phase syntheses of cyclic peptides with endocyclic biaryl ether bonds. Tetrahedron Lett 1997; 38: 3345–3348.

127. Meyers HV, Dilley GJ, Durgin TL, Powers TS, Winssinger NA, Zhu H, Pavia MR. Multiple simultaneous synthesis of phenolic libraries. Mol Diversity 1995; 1:13.

128. Phillips GB, Wei GP. Solid phase synthesis of benzimidazoles. Tetrahedron Lett 1996; 37:4887–4890.

129. Hari A, Miller BL. A new method for the mild and selective reduction of aryl nitro groups on solid support. Tetrahedron Lett 1999; 40:245–248.

130. Schwarz MK, Tumelty D, Gallop MA. Solid-phase synthesis of 1,5-benzodiazepin-2-ones. Tetrahedron Lett 1998; 39:8397–8400.

131. Mayer JP, Lewis GS, McGee C, Bankaitis-Davis D. Solid-phase synthesis of benzimidazoles. Tetrahedron Lett 1998; 39:6655–6658.

132. Tumelty D, Schwarz MK, Needels MC. Solid-phase synthesis of substituted 1-phenyl-2-aminomethyl-benzimidazoles and 1-phenyl-2-thiomethyl-benzimidazoles. Tetrahedron Lett 1998; 39:7467–7470.

133. Lee J, Gauthier D, Rivero RA. Solid phase synthesis of 1-alkyl-2-alkylthio-5-carbamoylbenzimidazoles. Tetrahedron Lett 1998; 39:201–204.

134. Lee J, Murray WV, Rivero RA. Solid phase synthesis of 3,4-disubstituted-7-carbamoyl-1,2,3,4-tetrahydroquinoxalin-2-ones. J Org Chem 1997; 62:3874–3879.

135. Thomas JB, Fall MJ, Cooper JB, Burgess JP, Carroll FI. Rapid in-plate generation of benzimidazole libraries and amide formation using EEDQ. Tetrahedron Lett 1997; 38:5099–5102.

136. Wei GP, Phillips GB. Solid phase synthesis of benzimidazolones. Tetrahedron Lett. 1998; 39:179–182.

137. Dolle RE. Discovery of enzyme inhibitors through combinatorial chemistry. Mol Diversity 1997; 2:223–236.

138. Carroll CD, Johnson TO, Tao S, Lauri G, Orlowski M, Gluzman IY, Goldberg DE, Dolle RE. Evaluation of a structure-based statine cyclic diamino amide encoded combinatorial library against plasmepsin II and cathepsin D. Bioorg Med Chem Lett 1998; 8:3203–3206.

139. Carroll CD, Patel H, Johnson TO, Guo T, Orlowski M, He Z-M, Cavallaro CL, Guo J, Oksman A, Gluzman IY, Connelly J, Chelsky D, Goldberg DE, Dolle RE. Identification of potent inhibitors of *Plasmodium falciparum* plasmepsin II from an encoded statine combinatorial library. Bioorg Med Chem Lett 1998; 8:2315–2320.

140. Dolle RE. Comprehensive survey of chemical libraries yielding enzyme inhibitors, receptor agonists and antagonists, and other biologically active agents: 1992 through 1997. Annu Rep Comb Chem Mol Diversity 1999; 2:93–127.

141. Dolle RE, Nelson KHJ. Comprehensive survey of combinatorial library synthesis: 1998. J Comb Chem 1999; 1:235–282.

142. Berk SC, Rohrer SP, Degrado SJ, Birzin ET, Mosley RT, Hutchins SM, Pasternak A, Schaeffer JM, Underwood DJ, Chapman KT. A combinatorial approach toward the discovery of non-peptide, subtype-selective somatostatin receptor ligands. J Comb Chem 1999; 1:388–396.

143. Rohrer SP, Berk SC. Development of somatostatin receptor subtype selective agonists through combinatorial chemistry. Curr Opin Drug Discovery Dev. 1999; 2: 293–303.

144. Rohrer SP, Birzin ET, Mosley RT, Berk SC, Hutchins SM, Shen D-M, Xiong Y, Hayes EC, Parmar RM, Foor F, Mitra SW, Degrado SJ, Shu M, Klopp JM, Cai S-J, Blake A, Chan WWS, Pasternak A, Yang L, Patchett AA, Smith RG, Chapman KT, Schaeffer JM. Rapid identification of subtype-selective agonists of the somatostatin receptor through combinatorial chemistry. Science 1998; 282:737–740.

145. Ogbu CO, Qabar MN, Boatman PD, Urban J, Meara JP, Ferguson MD, Tulinsky J, Lum C, Babu S, Blaskovich MA, Nakanishi H, Ruan F, Cao B, Minarik R, Little T, Nelson S, Nguyen M, Gall A, Kahn M. Highly efficient and versatile synthesis of libraries of constrained β-strand mimetics. Bioorg Med Chem Lett 1998; 8: 2321–2326.

146. Kim SW, Hong CY, Lee K, Lee EJ, Koh JS, Sung J. Solid phase synthesis of benzylamine-derived sulfonamide library. Bioorg Med Chem Lett 1998; 8:735–738.

147. Illig C, Eisennagel S, Bone R, Radzicka A, Murphy L, Randle T, Spurline E, Jaeger F, Salemme R, Soll RM. Expanding the envelope of structure-based drug design using chemical libraries: Application to small-molecule inhibitors of thrombin. Med Chem Res 1998; 8:244–260.

148. Böhm H-J, Banner DW, Weber L. Combinatorial docking and combinatorial chemistry: Design of potent non-peptide thrombin inhibitors. J CAD Mol Design 1999; 13:51–56.

149. Kimball SD. Challenges in the development of orally bioavailable thrombin active site inhibitors. Blood Coagul Fibrin 1995; 6:511–519.

150. Lumma WCJ, Witherup KM, Tucker TJ, Brady SF, Sisko JT, Naylor-Olsen AM, Lewis SD, Lucas BJ, Vaca JP. Design of novel, potent, noncovalent inhibitors of thrombin with nonbasic P-1 substructures: Rapid structure–activity studies by solid-phase synthesis. J Med Chem 1998; 41:1011–1013.

151. Augeri DJ, O'Connor SJ, Janowick D, Szczepankiewicz B, Sullivan G, Larsen J, Kalvin D, Cohen J, Devine E, Zhang H, Cherian S, Saeed B, Ng S-C, Rosenberg S. Potent and selective non-cysteine-containing inhibitors of protein farnesyltransferase. J Med Chem 1998; 41:4288–4300.

2

The Fundamentals of Pharmacophore Modeling in Combinatorial Chemistry

Arup K. Ghose and Vellarkad N. Viswanadhan
Amgen Inc.
Thousand Oaks, California

John J. Wendoloski

AstraZeneca
Boston, Massachusetts

I. INTRODUCTION

The concept of pharmacophore modeling is one of the oldest yet most widely used concepts in today's drug discovery research. The essential substructural moieties of a molecule necessary for its pharmacological activity are called *pharmacophores*. This terminology was first introduced by Ehrlich (1), following the term *chromophore*, which was used to represent the functional groups responsible for the color of a compound. The interest in the idea of pharmacophores has grown enormously in recent years owing to the availability of various automated computerized software for identifying pharmacophores as well as their geometry (2–7). The pharmacophoric information as well as their three-dimensional structure can often be used to identify novel pharmacologically active lead compounds by searching various databases of known chemicals, like the Available Chemical Directory (ACD) (8). Compounds having similar pharmacophoric groups often have similar biological activity. Understandably, the existence of similar or the same pharmacophoric groups does not guarantee that biological activity will be similar. The differentiating structural moieties may cause enough repulsive interaction with the target protein/receptor to diminish or abolish its binding affinity, or its chemical or physicochemical properties may be altered enough to prevent it from reaching the binding site.

II. THE TRADITIONAL APPROACH TO IDENTIFYING A PHARMACOPHORIC GROUP

Traditionally, pharmacophoric groups have been identified from structure–activity relationship (SAR) data. When a new class of compound is identified for a particular biological activity, medicinal chemists slowly modify its structure to optimize its potency as well as to identify the important substructural moieties responsible for the biological activity. The substructure or functional group that decreases the biological activity most on removal is considered to be a pharmacophore. During SAR studies it is often found that chemically or physicochemically similar structural moieties keep the activity, although the efficacy may change. The pharmacophore will not be identified unless such groups are categorized under the same class. This makes the identification of a pharmacophore somewhat more complex than a simple search of substructural units. In other words, to correctly identify and understand pharmacophores, substructural units having comparable chemical or physicochemical properties should be classified under a broader class.

III. A GENERAL CLASSIFICATION OF PHARMACOPHORIC GROUPS

Traditionally pharmacophoric groups are broadly classified as H-bond donors, H-bond acceptors, hydrophobic groups, positively charged groups or dipoles, and negatively charged groups or dipoles. This type of classification is useful for the identification of pharmacophores, as well as for a qualitative interpretation of the biological activity. An added advantage of this qualitative representation is that it allows one to search an available chemical database leading to a substantially larger number of virtual hits. A somewhat extended list of virtual hits may increase the chance to find a better or novel lead, where reliable high throughput screening is available. A broader classification of the pharmacophoric groups may be advantageous. A hydrophobic group in a chemical structure comes in different shapes and sizes, and the strength of a hydrogen donor or acceptor may also vary considerably from one compound to other. Hence, for a particular biological activity in a specified chemical class, only a few groups may give acceptable efficacy. This may require a more precise definition of the groups, rather than just specifying a class (hydrogen donor, hydrophobic, etc.). A quantitative interpretation of biological activity may need more elaborate properties like H-bonding ability, extent of van der Waals and hydrophobic interactions, solvation effect, and partition coefficients. Please consult Chapters 3 and 4 in this volume for a more elaborate discussion of these techniques.

IV. IDENTIFICATION OF PHARMACOPHORIC GROUPS

Identification of pharmacophoric groups may not be as simple as indicated thus far when similar biological activity is exhibited by different classes of compounds and the protein–ligand complex structures for different classes of ligands are unknown. However, the situation may not always be as hopeless as we may think. The identification of pharmacophores involves the following steps:

1. Selection of a set of active compounds
2. Dissection of the molecular structure into different classes of pharmacophoric groups
3. Analysis of the maximum number of common pharmacophoric groups in all the selected active compounds
4. Deciding the geometrical feasibility in superimposing the equivalent pharmacophoric groups in different compounds
5. Testing the pharmacophoric hypothesis

One should remember a few critical aspects of these steps:

1. Selection of compounds is a compromise: structurally very different compounds may bind to different targets or locations to give the comparable biological activity. Combining different classes of compounds binding with different targets, or at different binding sites of the same target, may lead to a wrong pharmacophoric definition. However, when they bind at the same active site, ideas for getting a new lead, or for modifying existing leads by adding features from the other classes of compounds to make more potent (patentable, proprietary) compounds, are generated. This is a valuable approach in industrial drug discovery research. It is also known from X-ray crystallographic studies that very similar compounds may sometimes bind in different modes in the same binding site. Despite such occurrences, in drug discovery research, it is riskier not to rationalize the biological activities than to make a few mistakes in the process of rationalization.

2. The dissection of a molecule into substructural moieties for a pharmacophore identification problem may be best done by keeping a predefined set of substructures with their pharmacophoric class types defined. The disadvantage of this approach, however, is that deciding on a finite set of substructures that will cover all organic molecules may be difficult if not impossible.

3. Classification of the substructural moieties in a few general classes creates problems in deciding the equivalent pharmacophoric groups, especially when

the active compounds belong to many different structural types. Consideration of shape, size, and physicochemical properties along with the geometric factors may help the decision process.

4. The decision of the geometric feasibility for a superimposition of the pharmacophore is a complicated factor and is determined by two issues: finding the conformation(s) having the common geometry of the equivalent pharmacophores and evaluating the energetic accessibility of the conformation. For most organic druglike molecules the conformational space may be too large to decide these issues without ambiguity.

V. DETERMINATION OF PHARMACOPHORIC GEOMETRY

There are many different chemical and computational methods for the determination of pharmacophoric geometry. *The chemical approaches* try to constrain the conformation by fixing the torsion angles to a few estimated torsion angles. Adding rings or a double bond is common practice for constraining a molecular conformation. If the resulting compound keeps acceptable activity, we accept the geometry. Unfortunately, if the resulting compound does not show the activity, the opposite conclusion is not always true. A compound may lose activity simply because of the unfavorable interaction with the extra structural moiety added during the constraint of the geometry.

The basic ideas behind all computational methods are very similar. Locate a geometry that can be satisfied by a "physically realistic conformation" of all the active compounds. The difference comes from the search method for the conformation and the assumption of the "physically realistic conformation." The most simple conceptually, yet the most expensive search method computationally may consist of generating all combinatorial grid conformations (divide each torsion angle space, usually $0–360°$, into smaller grids of $~20°$) and comparing the low energy conformations. Distances of the pharmacophoric groups are the best coordinates for structural matching, since they are invariant to translation or rotation. For the same reason, some variations of distance features have been applied repeatedly in the description of pharmacophores (9–11).

Since 1977 when Crippen introduced the modern version of distance geometry (12), there has been considerable improvement along this line. Crippen used the common distance range information of the pharmacophoric points to deduce their geometry (13). *Distance geometry* is a mathematical procedure of generating the Cartesian coordinates of a set of points from their distance information. It can handle situations with a limited inconsistent distance information to give the best possible set of coordinates. It handles distance range information by selecting randomly a value from the specified range and setting a penalty if, during iterative improvement of the coordinates, the distance goes outside the specified range.

This approach works very effectively when there is only one structure satisfying the distance range. When a large number of conformations can be mapped from the distance range information, it does not guarantee the lowest energy conformation. To improve the situation two approaches were taken. Crippen (14) developed "energy embedding" and Ghose and Crippen (15) suggested distance mapping in the conformational space. In the latter approach multiple local distance matrices are constructed in the low energy conformational region (Fig. 1).

Marshall et al. used an alternative approach in which they mapped the torsion angles in distance space (16). The generation of a torsion angle map is computationally more efficient, since once a grid size has been defined, one can very easily, during the conformational search, assign a grid to each of the energetically allowed conformations according to its distance property. Figure 2 presents a schematic representation of torsional (orientation) mapping in the distance space. The simplest approach to testing the feasibility of superimposition of a set of equivalent atoms in two molecules is to compare their orientation maps. If a grid is occupied by both molecules, those conformations are superimposable if their

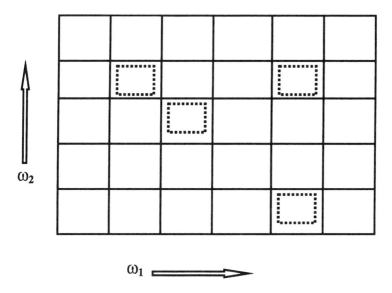

Figure 1 A schematic two-dimensional representation of distance mapping in the torsional space. Here ω_1 and ω_2 are two torsion angles of interest. The grid points represent the explored angles during torsional increment, the dotted squares represent the local distance range matrix of the pharmacophoric groups from the neighboring energetically acceptable grid points. For a feasibility of superposition of the pharmacophoric groups of two molecules, we have to compare all local distance matrices of one with that of the other.

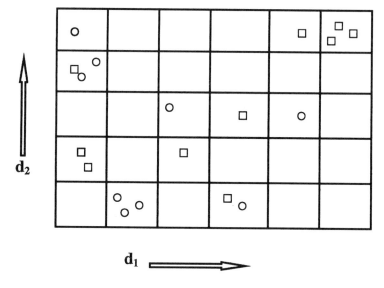

Figure 2 A schematic two-dimensional representation of torsional mapping in the distance space. Here d_1 and d_2 are two pharmacophoric distances of interest. The small squares and circles represent the energetically acceptable conformations of two molecules. For a feasibility of superposition of the pharmacophoric groups of two molecules, we have to locate a grid which is occupied by both the molecules. The distance grid size may be very important during the geometric comparison. Small grid size may create too many unoccupied grids, leading ultimately to no common grid. A large grid size may put very different conformations on the same grid.

chirality is the same (15). The ultimate result from this type of comparison often is dependent on the distance grid size.

The DHYDM (**d**istance **hy**perspace **d**istance **m**easurement) method was developed by Ghose et al. (2) to avoid the grid size problem of the orientation map approach developed by Marshall et al. (16). In DHYDM, the distance of the centroid of the grids occupied by the low energy conformations of the first molecule from those of the second molecule in the distance space are used to judge their superimposability. It is important to remember that two conformations lying near two corners of the same grid may be less similar than two conformations lying in two neighboring grids or even two conformations separated by a grid. Such artifacts can be avoided if the conformational similarity is determined by the distance in the "distance hyperspace." It was possible to predict a fairly accurate active conformation of compounds having eleven torsion angles, using this method.

Sheridan et al. (17) formulated an elegant way of using Crippen's distance geometry program for superimposing molecules. In this approach they created an ensemble distance matrix of all atoms of all the molecules that are to be superimposed. The intramolecular atomic distance ranges are given the usual distance ranges for that molecule. Many of these distances are simply the bond distances, bond angle distances, and fixed torsion distances. The intermolecular atomic distance ranges are zero to a small tolerance distance for the atoms to be superimposed. The intermolecular atomic distances of the nonsuperimposed atoms are evaluated from triangular inequality constraints by the distance geometry program. The method has some advantages over Crippen's original method, in which the constraints of the nonsuperimposed atoms of different molecules from the superimposed atoms further decreased the acceptable solution space. However, it still has most of the limitations of regular distance geometry: namely, when there are multiple solutions of conformations, it does not give any information about the other acceptable solutions, nor does it guarantee low energy conformations.

Instead of searching and comparing all low energy conformations of all the molecules of interest one may think of optimizing the fit function by modifying the torsion angles. Payne et al. (18) went along that direction and applied a *genetic algorithm* to optimize the fit function during pharmacophoric geometry determination.

VI. MOLECULAR SIMILARITY METHODS AS AN ALTERNATIVE TO PHARMACOPHORE MODELING

The objective of pharmacophore modeling often is to visualize similarities in apparently different-looking molecules with similar biological activity, or differences in molecules with apparently similar-looking molecules with different biological activity. One may think of using overall molecular similarity for this purpose. Although three-dimensional similarity is required for understanding pharmacophoric features, we will discuss briefly the concept of two-dimensional similarity as represented in the Tanimoto coefficient (19). In this method a molecular structure is represented by a set of bit strings. Each bit string represents by a substructural unit. When a substructure is present, the bit is turned on. Usually a substructural unit has six bits. Multiple bits of the same kind are turned on when there are multiple substructural units of this kind. The similarity between a query structure and a test structure is represented by the Tanimoto coefficient T as follows:

$$T = \frac{N_c}{N_q + N_t - N_c} \tag{1}$$

where N_q = number of bit screens set in the query structure,
N_t = number of bit screens set in the test structure, and
N_c = number of bit screens common to both the query and test structures.

A few interesting features of Tanimoto coefficients are as follows: (1) for a pair of molecules, it is independent of which is the query structure and which is the test structure; (2) for very similar monomeric and dimeric compounds, the coefficient will be close to 0.5; and (3) the maximum and minimum possible values of the Tanimoto coefficient are 1 and 0, respectively. Modifications of the Tanimoto coefficient are used in measurements for three-dimensional shape analysis (20).

We found that the combination of ALOGP atom type (21,22) (also known as Ghose and Crippen atom type) representation of a molecule and the Tanimoto coefficient is an excellent way to determine similarity between molecules. According to this coefficient, which can be appropriately termed GCT (Ghose–Crippen's Tanimoto coefficient), N_c is the number of common atom types, N_q is the number of atoms in the query molecule, and N_t is the number of atom in the test molecule. One major advantage here is that in the molecular fingerprint there is a limit on the number of bits for the reoccurrence of a substructure. Any substructure that occurs more frequently than the limit in the bit representation is lost in this representation, as well as in the Tanimoto coefficient of similarity. Any approach that keeps the number of counts of the occurrence does not lose the information. The trade-off here is in the computation time, since bit operation is much faster.

However, molecular similarity should be judged not only by matching atoms or substructures but also by the physicochemical properties of the atoms. Neglecting different substructures with similar physicochemical nature will decrease similarity in otherwise similar structures. To tackle this problem, Ghose et al. (23,24) proposed several functions based on physicochemical properties to estimate the similarity between two molecules. The first two functions are of the all-or-none type. Here, two atoms are assumed to be superimposed if they are within a certain preassigned distance. Otherwise, they are not. The first function assumes that the reference structure is the best possible structure, and any deviation of the physicochemical properties from the reference structure will incur a penalty:

$$F_1 = \sum_k |X_k| - |X_k - X_{j(k)}| \tag{2}$$

where X_k represents the physicochemical property of the *kth* atom of the reference structure and *j(k)* is the test atom superposed on the *kth* atom of the reference structure. The corresponding physicochemical property will be zero if no atom is superimposed. The noncovalent interaction between ligand and receptor has three major components: electrostatic, dispersive, and hydrophobic. Three physicochemical properties of atoms represent these forces, namely, atomic charge for electrostatic interaction, atomic refractivity (25) for dispersive or van der Waals interaction, and ALOGP (22) for hydrophobic interaction. When the relative importance of these interactions is not known, the atomic parameters may be scaled to give them equal weight. The foregoing function assumed that the interaction of the ligand atoms with the receptor site is quadratic, and the reference structure lies on the peak; therefore the interaction will decrease if the physicochemical property changes on either side of the ideal value. Unlike the Tanimoto coefficient, the score is not normalized, and it depends on the reference. The Tanimoto coefficient is mathematically advantageous. This scoring function represents physical status.

The interaction, however, is a linear function of these physicochemical properties. The proportionality constant from one region to another may vary depending on the receptor environment. Since the proportionality constants are not known, let us assume that the reference molecule experiences attraction from the receptor for all its atoms. In other words, the sign of the physicochemical property for attractive interaction is the sign of the corresponding physicochemical property of the reference structure. This gives a second function to express the goodness of a superposition:

$$F_2 = \sum_k \frac{X_k X_{j(k)}}{|X_k|} \tag{3}$$

One deficiency (!) of functions (2) and (3) is that they assume that the interactions are "all-or-none" type. In other words, full interaction is obtained when a function is within a preassigned distance but none otherwise. This type of function has both merits and demerits. The interaction of the ligand atoms with the neighboring receptor atoms depends not only on their types but also on their distance of separation. In the binding process, both protein and ligand relax to get the maximum overall interaction. The relaxation is some times difficult to determine even after the protein structure has been ascertained by X-ray crystallographic techniques. When the protein structure is not known, and superposition is based on grid conformations, the all-or-none hypothesis will avoid many complications. However, those who prefer to use a distance-dependent fit function for superposition can use a third function. According to this function, each atom of the test molecule will create a field on the various atoms of the reference molecule. The interaction will be maximum when it occupies the same position as the reference atom. These features are reflected in the following function (26):

$$F_3 = \sum_i \sum_j \sum_k \frac{X_j X_k}{(1 + r_{jk})^{n_i}} \tag{4}$$

where the summation over i represents the various physicochemical properties to be used to evaluate the fit, the summation over j represents the various atoms of the reference molecule, and the summation over k represents the various atoms of the test molecule; r_{jk} represents the distance between the jth atom of the reference and the kth atom of the test, and different distance dependencies n_i were used for different physicochemical properties. For electrostatic interaction the value was 1, for hydrophobic interaction the value 3 was used, and for dispersive interaction the value 6 was used for n_i. Most of these methods eventually help to decide a better superposition of the molecules. The superimposed structures can be used to identify the equivalent pharmacophores, which otherwise may not be obvious.

All these fit functions can often suggest a very plausible superposition of apparently different-looking molecules that showed comparable biological activities (27,28).

VII. PROTEIN BINDING SITE FOR THE EXPLORATION OF PHARMACOPHORIC GROUPS AND GEOMETRY

The best verification of a pharmacophore hypothesis is the solution of the ligand–protein complex structure. The equivalent pharmacophore groups will bind to the same region of the binding site. Alternatively, a protein structure and the knowledge of the binding site can be utilized to generate a pharmacophore hypothesis. Several programs can automatically dock small organic fragment at the active site and identify the preferred binding site. DOCK (29), MCSS (30), LUDI (31), and AUTODOCK (32) are a few of the ever growing list of programs (33) for this purpose. See also Chapters 6 and 17 in this book. Such a pharmacophore hypothesis based on a target protein structure can be used for searching compounds from a molecular database and for designing compounds or combinatorial libraries (34).

VIII. GEOMETRY OF NATURAL SUBSTRATES, AGONISTS, AND ANTAGONISTS AS A SOURCE OF PHARMACOPHORIC HYPOTHESIS

In the natural biological process, often protein, peptides, and other bioorganic compounds bind with the target protein of interest. Although protein–protein interaction often involves a large surface and may be difficult to inhibit by means of a small molecule (35), the structural study of such peptides and bioorganic molecules often gives a good pharmacophoric hypothesis. Such a pharmacophore hypothesis can be used to search databases of chemical compounds to generate novel leads.

IX. THE VARIOUS EXPERIMENTAL APPROACHES TO VERIFY A PHARMACOPHORE MODEL

A complete pharmacophore model has two aspects: the nature of the pharmacophores and their orientation in three-dimensional space. A complete and unambiguous verification of a pharmacophore model needs a fairly rigid compound with the required pharmacophores that showed the biological activity. If the compound is flexible, we need the X-ray or NMR structure of the ligand–protein complex. The synthesis of a fairly rigid system is often difficult, and such an approach often fails simply because the extra structural moiety necessary to fix the conformation may create a bad interaction with the binding protein, destroying its activity. Proposing a pharmacophore and later verifying it by X-ray or NMR techniques is also not very common, although there are a few such examples in the literature. Ghose et al. (2), for example, using their *DHYDM* method on a set of partially constrained matrix metalloproteinase (MMP) inhibitors (Fig. 3), pro-

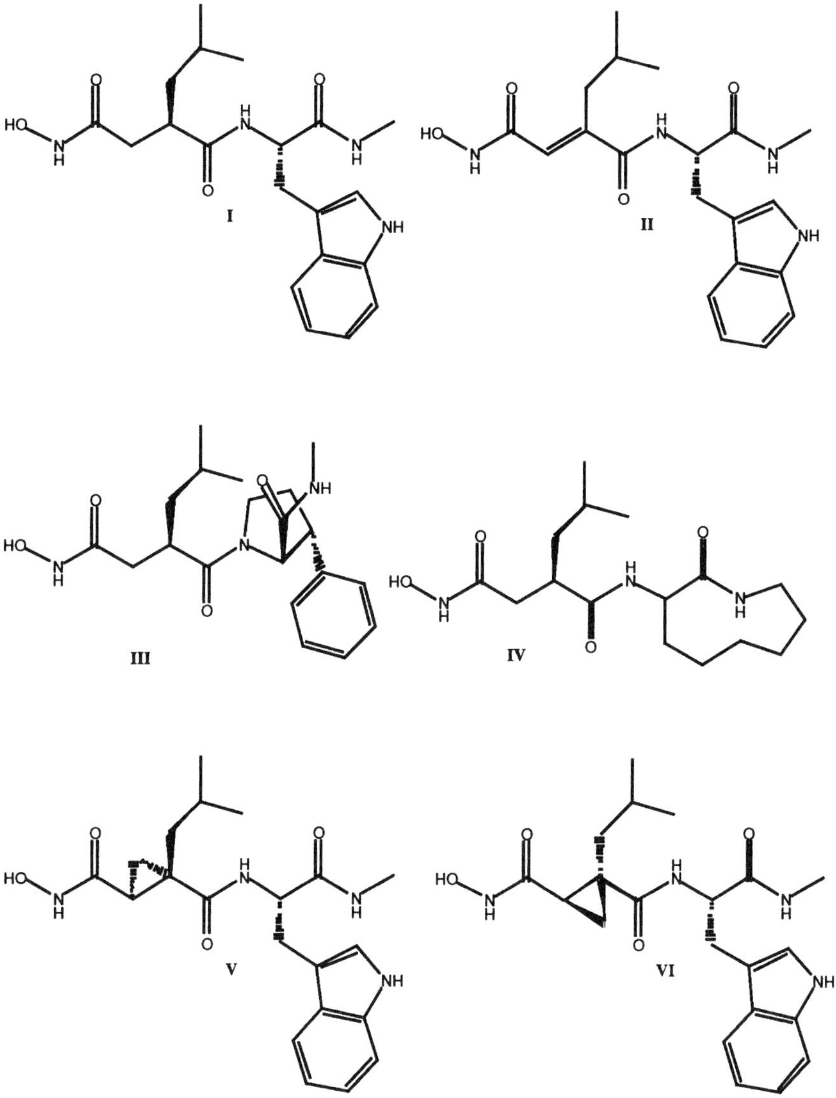

Figure 3 The various constrained and unconstrained hydroxamic acid inhibitors of collagenase, a matrix metalloproteinase, used for the evaluation of pharmacophoric geomoety.

posed a pharmacophoric geometry that was later found to be very close to the X-ray crystallographic structure of the ligand in the ligand–protein complex (Fig. 4). However, there are many different, somewhat ambiguous methods for the verification of a pharmacophore model. The study of the crystal structure of the ligand

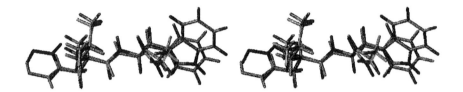

Figure 4   A comparison of the active conformation of I (dual color) as predicted by the DHYDM method with X-ray crystallographic conformation (monochrome).

alone by X-ray or the solution of conformations by NMR spectra is sometimes used to judge the validity of a pharmacophore model. However, since the active conformation often is not the minimum energy conformation, or even the crystal conformation (36), such approaches should be used with caution. Analysis of multiple active as well as inactive compounds of similar structures often throws more light on the active conformation.

X.   USE OF THE PHARMACOPHORE MODEL IN MEDICINAL CHEMISTRY

The pharmacophore model is extremely useful in medicinal chemistry during the process of a novel lead identification as well as during lead optimization. Depending on the throughput of the biological or biochemical screening, as well as the availability of the compound library, the pharmacophore model can be used in many different ways.

A.   Search for Pharmacophoric Groups

If the natural substrate or a few active compounds are known in the literature, one can dissect the structure to the standard pharmacophoric groups (as discussed in Sect. III) and simply do a pharmacophore search in the available chemical library or in a purchasable chemical directory like ACD. This approach is especially useful if one has high throughput screening (HTS) and a relaxed budget for chemicals. In absence of HTS, or if the budget for chemicals is tighter, one may try to make an educated guess regarding the geometry of the pharmacophore, as discussed next.

B.   Search for the Pharmacophores Satisfying the Geometric Requirements

Knowledge of pharmacophore geometry may be very important because it trims down the number of compounds that can satisfy the geometry. When a lot of information about the active compounds is available in the literature, one may use one of the methods discussed in Section V. Most database software as well as molecular modeling software allow searching with distance constraints. The search process is comparable to a substructure search and needs an extra step of adding the necessary distance constraints. The search process may be rigid or flexible. In a rigid search, only the conformations in the database are used for the geometric fitting. In a flexible search, the conformations are optimized for the fitting. The number of hits in a three-dimensional search is often considerably less than the substructural search. If there are not too many active conformations, one may still use a few lowest energy conformations as a possible pharmacophoric geometry.

C. Search for the Proper Scaffold to Hold the Pharmacophoric Groups

When one is interested in synthesizing a few compounds that may satisfy the pharmacophoric geometry, one can use the linking bonds of the pharmacophores to search for scaffolds (37).

XI. USE OF PHARMACOPHORES IN COMBINATORIAL LIBRARY DESIGN

There are two different types of use of pharmacophores in the combinatorial library design. A drug molecule usually contain pharmacophoric groups of various types (as discussed in Sec. IV) in a flexible or rigid scaffold. When one is making libraries for testing a large number of target proteins, the *pharmacophoric diversity* (covering geometry, number of occurrences, and nature) should be as broad as possible without sacrificing "druglikeness." Most modeling software these days gives various tools for studying the nature of pharmacophore diversity in a library. Chapter 14 gives a detailed discussion of aspects of pharmacophore diversity.

A. Generating Focused Libraries Satisfying Pharmacophore Geometry and Nature

Given a pharmacophore hypothesis, it may be a better idea to generate a library of compounds with some minor variation of geometry and nature than to synthesize a single precise compound. Such variations and the number of compounds will definitely increase the chance of finding a lead or a better compound. Although several computer programs described in the literature [NEW LEAD (38), SPROUT (39), HOOK (40)] that claim to do this job, a computational chemist often must first convince a bench chemist, who will be making the compounds. Complexity of structure and lack of a chemical synthesis plan are often obstacles. We provide here a semiautomated approach that we found very useful to generate ideas for "doable" libraries (34).

1.   Start from one of the pharmacophoric groups. If it is a common sub-structural moiety or a common functional group, use it as it is; otherwise consider a functional group from which it can be made easily. For example, a carboxylic acid may be a better choice than an amide, and so on.

2.   Now search for reagents with the first pharmacophore and one or more functional groups for combinatorial reactions to be used to attach the other pharmacophores.

3.   Now anchor the first pharmacophoric group of the reagent on the corresponding pharmacophore and orient the functional groups so that they point toward the pharmacophore that will be added in the next steps by the combinatorial reaction. The orientation process may be straightforward when the reagent is fairly rigid or protein binding site is known.

4.   Measure the approximate distance of the functional groups for combinatorial reactions to the pharmacophore to be attached.

5.   Search for the reagent with the complementary functional group and the pharmacophore group with the required distance. If all the pharmacophores are not covered, it may be necessary to keep an extra functional group in this reagent for further reaction.

The whole process can be best illustrated by a pharmacophore hypothesis generated from the SH-PTP2 phosphotyrosine binding pocket (41). Consideration of the various peptidic inhibitors or analysis of the binding pocket with probing fragments and autodocking programs like MCSS showed at least four pharmacophoric binding sites (Fig. 5).

A cationic phosphate binding pocket
A hydrophobic pocket
Two other cationic and anionic charged pockets near Lys91 and Glu17,
   respectively

The energetics of the binding groups suggests that the phosphate is the most important pharmacophore. This is consistent with the fact that dephosphorylated peptides had very low binding affinity for SH-PTP2. Phosphotyrosine-type compounds with different phosphate mimicking groups may be the first choice if we do not want to make any major change here (Fig. 6). We want to use the carboxy functionality to add reagents that can reach the hydrophobic pocket and the Lys91 pocket. This decision may sometimes be dictated by the optimal docking of the reagent or partially modified reagent (to make it similar to the final product), where the X-ray structure of comparable ligand is not available. Here we will

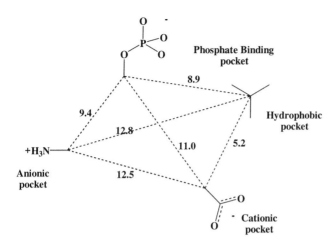

Figure 5 SHPT-P2 pharmacophores generated from the quantitative or qualitative analysis of the binding site or the various peptidic inhibitors.

search for reagents that can be coupled with the carboxyl group (e.g., amines or alcohols). Such a reagent should have a hydrogen-accepting functionality at a desired distance, orientation that will bind with the lysine side chain, and so on. If the first reagent is not very flexible, such a distance and orientation may be obtained either by model building and minimization or even from the structure of the X-ray ligand and the MCSS functional group. A few immediate hits in this search were the m-aminobenzoic acids. To add the hydrophobic groups, one can easily think of a phenolic OH group, which can be coupled with different alkyl halides.

These binding features of the protein were consistent with the relative binding affinities of several synthetic peptides, made here, whose IC_{50} values ranged between the submicromolar and micromolar levels. Interestingly enough, an independent effort by Lunney et al. (42) showed that several closely related compounds (Fig. 7) have a moderate binding affinity for the pp60 Src SH2 domain.

XII. PHARMACOPHORE MODEL AND 3D-QSAR

Pharmacophore modeling is most useful during the identification of a lead compound. However, after that quantitative structure–activity relationship (QSAR) and/or three-dimensional (3-D) QSAR becomes necessary for the lead optimization. Most 3D-QSAR techniques need to superimpose the three-dimensional structures of the inhibitors as the first step. The pharmacophore modeling may be used as the basis for such a superposition. In general there are two types of 3D-

QSAR approach. In approaches like REMOTEDISC, which are based on physicochemical properties, the atom-based local properties are clustered spatially from the superimposed three-dimensional structures of the ligands and the clustered properties are correlated with the biological activity; such methods give a physical interpretation of the nature of the hypothetical binding pocket of the protein. The second type, the field-based approaches like CoMFA (**C**omparative **M**olecular **F**ield **A**nalysis), calculate the interaction of the ligand molecules with atoms representative of the protein atoms at an arbitrarily defined set of grid points. The

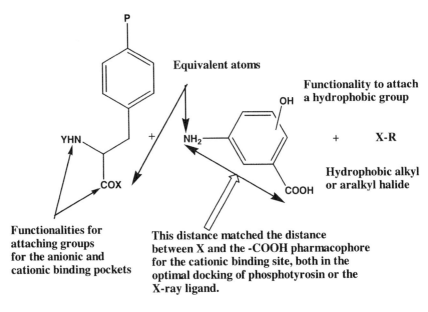

Figure 6 A hypothetical reaction scheme for a library for SH-PTP2, a cell division signaling tyrosine phosphatase protein. The first reagent is a tyrosine phosphate mimicking compound with the functionalities to attach the other pharmacophoric groups. The second reagent was searched by means of the distance of the reaction functionality of the first reagent from the cationic binding pocket. The second reagent can be any *m*-aminobenzoic acid derivative, especially the ones with functionality like—OH, which can be used for adding the hydrophobic pharmacophoric groups. In such a situation an alkyl halide satisfying the distance between the OH and the hydrophobic pharmacophore may be used as the third reagent. By taking multiple compounds with the required distances for each reagent class, one can easily make a focused library for the SH-PTP2 protein.

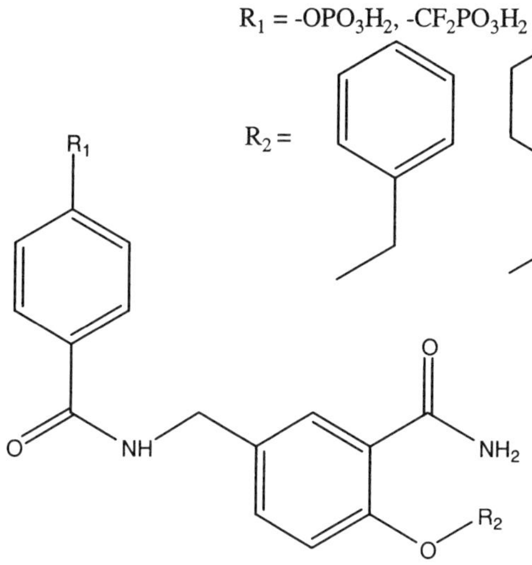

$R_1 = -OPO_3H_2, -CF_2PO_3H_2$

$R_2 =$

Figure 7 A few moderately active compounds found by Lunney et al. (42) for pp60 Src SH2 domain, a closely related protein of SH-PTP2.

interaction energies are correlated with the binding affinity to develop a comparable interpretation of the nature of the protein binding pocket. What comes out of these approaches depends on the conformation used as well as the mode of superimposition of the ligands. Since CoMFA does not supply any definite algorithm for the superimposition of ligands, the pharmacophore modeling can be a good way to initiate CoMFA. Hopfinger's molecular shape analysis and APEX-3D often used pharmacophore hypothesis to develop a 3D-QSAR analysis.

XIII. CRITICAL ASPECTS AND COMMENTS

Pharmacophore modeling is an extremely useful tool in the early stages of drug discovery research. It helps to find novel lead compounds and rationalize pharmacological activity in apparently different looking compounds. In addition, it qualitatively helps to optimize the activity of a lead compound. The popularity lies in the qualitative nature of the approach also. Unlike a QSAR approach, it does not need very precisely measured biological activity data. In general, pharmacophore modeling is more useful than QSAR approaches at the initial phase of drug discovery. The QSAR approaches become more useful during lead optimization.

One should, however, be critical while using a pharmacophore hypothesis. Constraining the conformation often leads to a loss of binding affinity, although entropy gain arising from the reduced conformational distribution should facilitate binding. Problems in such molecules may be due to constraining the conformation at an angle somewhat distant from the ideal angle or to bad interactions of the constraining structural moiety with the enzyme binding site. It is not certain how great a drop in the activity of a molecule should be acceptable during the development or validation of a model.

The idea of pharmacophoric modeling does not hold if the inhibitors show considerable activity even when one or more pharmacophoric groups do not reach the same region of the active site. Forcing the molecules to attain a conformation where equivalent groups occupy the same location in such a situation may give a distorted pharmacophoric model. However, success in the drug design process is so rare that the researchers in this area are often eager to take these risks. The only suggestion to be offered here may be to analyze the suggested (computed) conformation using existing knowledge of conformation of similar molecules.

Pharmacophore modeling is based on the idea that similar inhibitors bind in the same way at the active site. The X-ray crystallographic data of most ligand–protein complexes usually confirm this hypothesis. However, there are several exceptions to this basic idea. Multiple binding modes may often result in a binding pocket dominated by nondirectional forces (e.g., van der Waals, hydrophobic). The application of pharmacophore modeling in such a system may be risky.

Staying close to a lead compound maximizes the chance of finding an active compound. Unless it is necessary to get a very different compound (maybe for patentability) one should try to make the smallest possible change from the lead compound while searching a database.

Combinatorial chemistry is a major blessing for pharmacophore modeling. Often the fate of a single compound was used to validate a pharmacophore model. However, drug discovery is such an approximate science that one who decides from the result of a single compound risks looking like a fool. With combinatorial chemistry, we will be able to make a larger number of compounds with a specified range of pharmacophoric geometry and nature, although we do not recommend a very big library for this purpose.

REFERENCES

1. Ehrlich P. Über den jetzigen Stand der Chemotherapie. Chem Ber 1909; 42:17.
2. Ghose AK, Logan ME, Treasurywala AM, Wang H, Wahl RC, Tomczuk B, Gowravaram M, Jaeger EP, Wendoloski JJ. Determination of pharmacophoric geometry for collagenase inhibitors using a novel computational method and its verification using molecular dynamics, NMR and X-ray crystallography. J Am Chem Soc 1995; 117:4671–4682.

3. Ghose AK, Wendoloski J. Pharmacophore modeling: Methods, experimental verifications and applications. Kubinyi, H, Folker, G, Martin YC, eds., 3DQSAR in Drug Design: Theory, Methods, and Applications. Dordrecht: Kluwer, 1997.

4. Gund P. Pharmacophoric pattern searching and receptor mapping. Ann Rep Med Chem 1979; 14:299–308.

5. Humblet C, Marshall GR. Pharmacophore identification and receptor mapping. Ann Rep Med Chem 1980; 15:267–276.

6. Klebe G. Structural alignment of molecules. In: Kubinyi H, ed. 3D QSAR in Drug Design: Theory, Methods and Applications. Leiden: ESCOM, 1993, pp. 173–199.

7. Martin YC. Pharmacophore mapping. In: Martin, YC, Willett P, eds. Designing Bioactive Molecules. Washington DC: American Chemical Society, 1997, pp. 121–148.

8. MDL Information Systems. I. ISIS; v 2.0.2 ed. San Leandro, CA, 1997.

9. Pullman A. Compt Rend 1945; 221:140.

10. Coulson CA. Adv Cancer Res 1953; 1:1.

11. Kier LB. Molecular Orbital Theory in Drug Research. New York: Academic Press, 1971.

12. Crippen GM. Rapid calculation of coordinates from distance matrices. J Comput Phys 1977; 26:449–452.

13. Crippen GM. Quantitative structure–activity relationships by distance geometry: Systematic analysis of dihydrofolate reductase inhibitors. J Med Chem 1980; 23: 599–606.

14. Crippen GM. Why energy embedding works. J Phys Chem 1987; 91:6341–6343.

15. Ghose AK, Crippen GM. Geometrically feasible binding modes of the flexible ligand molecule at the receptor site. J Comput Chem 1985; 6:350–359.

16. Motoc I, Dammkoehler RA, Marshall GR. Three-dimensional structure activity relationships and biological receptor mapping. In: Trinajstic, N, ed. Mathematical and Computational Concepts in Chemistry. Chichester: Ellis Horwood, 1986, pp. 222–251.

17. Sheridan RP, Nilakantan R, Dixon JS, Venkataraghavan R. The ensemble distance geometry: Application to the nicotinic pharmacophore. J Med Chem 1986; 29: 899–906.

18. Payne AW, Glen RC. Molecular recognition using a binary genetic search algorithm. J Mol Graphics 1993; 11:74–91.

19. Willett P. Similarity and Clustering in Chemical Information Systems. Letchworth: Research Studies Press, 1986.

20. Hahn M. Three-dimensional shape-based searching of conformationally flexible compounds. J Chem Inf Comput Sci 1997; 33:80–86.

21. Ghose AK, Crippen GM. Atomic physicochemical parameters for three-dimensional structure directed quantitative structure–activity relationships I. Partition coefficients as a measure of hydrophobicity. J Comput. Chem 1986; 7:565–577.

22. Ghose AK, Viswanadhan VN, Wendoloski JJ. Prediction of hydrophobic properties of small organic molecules using fragmental methods: An analysis of ALOGP and CLOGP methods. J Phys Chem 1998; 102:3762–3772.

23. Ghose AK, Crippen GM, Revankar GR, McKernan PA, Smee DF, Robins RK. Analysis of the in vitro antiviral activity of certain ribonucleosides against parainfluenza

virus using a novel computer aided receptor modeling procedure. J Med Chem 1989; 32:746–756.

24. Viswanadhan VN, Ghose AK, Revankar GR, Robins RK. Atomic physicochemical parameters for three-dimensional structure directed quantitative structure–activity relationships. 4. Additional parameters for hydrophobic and dispersive interactions and their application for an automated superposition of certain naturally occurring nucleoside antibiotics. J Chem Inf Comput Sci 1989; 29:163–172.

25. Ghose AK, Crippen GM. Atomic physicochemical parameters for three-dimensional structure-directed quantitative structure–activity relationships. 2. Modeling dispersive and hydrophobic interactions. J Chem Inf Comp Sci 1987; 27:21–35.

26. AM-Technologies. Galaxy, v2.6, Manual Vol. 2 San Antonio, TX, 1999. www.amtech.com

27. Ghose AK, Viswanadhan VN, Sanghvi YS, Nord LD, Willis RC, Revankar GR, Robins RK. Structural mimicry of adenosine by the antitumor agents 4-methoxy- and 4-amino-8-(β-D-ribofuranosylamino)pyrimido[5,4-d]pyrimidine as viewed by a molecular modeling method. Proc Nat Acad Sci USA 1989; 86:8242–6.

28. Ghose AK, Sangvi YS, Larson SB, Revankar GR, Robins RK. Structural studies of the novel antitumor agents 4-amino and 4-methoxy-8(β-D-ribofuranosylamino) pyrimido- 5,4-d]pyrimidines and their a-anomers using X-ray, ^{1}H NMR and theoretical methods. J Am Chem Soc 1990; 112:3622–3628.

29. DesJarlais R, Sheridan RP, Seibel, GL, Dixon JS, Kuntz ID, Venkataraghavan R. Using shape complementarity as an initial screen in designing ligands for a receptor binding site of known three-dimensional structure. J Med Chem 1988; 31:722–729.

30. Miranker A, Karplus M. Functionality maps of binding sites: A multiple copy simultaneous search method. Proteins 1991; 11:29–34.

31. Bohm H-J. The development of a simple empirical scoring function to estimate the binding constant for a protein–ligand complex of known three-dimensional structure. J CAD Mol Des 1994; 8:243–256.

32. Goodsell DS, Olson AJ. Automated docking of substrates to proteins by simulated annealing. Proteins Str Func Genet 1990; 8:195–202.

33. Gillet VJ, Johnson AP. Structure generation for de novo design. In: Martin, YC, Willett, P, eds. Designing Bioactive Molecules. Washington, DC: American Chemical Society, 1997, pp. 149–174.

34. Ghose AK, Viswanadhan VN, Wendoloski, JJ. Adapting structure-based drug design in the paradigm of combinatorial chemistry and high-throughput screening: An overview and new examples with important caveats for newcomers to combinatorial library design using pharmacophore models or multiple copy simultaneous search fragments. In: Parrill, AL, Reddy, MR, eds. Rational Drug Design: Novel Methodology and Practical Applications. Washington, DC: American Chemical Society, 1999, pp. 226–238.

35. Livnah O, Stura EA, Johnson DL, Middleton SA, Mulcahy LS, Wrighton NC, Dower WJ, Joliffe LK, Wilson IA. Functional mimicry of a protein hormone by a peptide agonist: The EPO receptor complex at 2.8 Å. Science 1996; 273:464–471.

36. Nicklaus MC, Wang S, Driscol JS, Milne GW. Conformational changes of small molecules binding to proteins. Bioorg Med Chem 1995; 3:411–428.

37. Bartlett PA, Shea JT, Telfer SJ, Waterman S. Molecular recognition in chemical and biological problems. In: CAVEAT: A program to facilitate the structure-derived design of biologically active molecules. Special publication 78. Royal Chemical Society, 1989, pp. 182–196.

38. Tschinke V, Cohen NC. The NEWLEAD program: A new method for the design of candidate structures from pharmacophoric hypotheses. J Med Chem 1993; 36: 3863–3870.

39. Gillet V, Johnson AP, Mata P, Sike S, Williams P. SPROUT: A program for structure generation. J CAD Mol Des 1993; 7:127–153.

40. Eisen MB, Wiley DC, Karplus M, Hubbard RE. HOOK: A program for finding novel molecular architectures that satisfy the chemical and steric requirements of a macromolecule binding site. Proteins 1994; 19:199–221.

41. Eck MJ, Pulskey S, Trub T, Harrison SC, Shoelson SE. Spatial constraints on the recognition of phosphoproteins by the tandem SH2 domains of the phosphate SH-PTP2. Nature 1996; 379:277–280.

42. Lunney EA, Para KS, Rubin JR, Humblet C, Fergus JH, Marks JS, Sawyer TK. Structure-based design of a novel series of nonpeptide ligands that bind to the pp60 SRC SH2 domain. J Am Chem Soc 1997; 119:12471.

3

Quantitative Structure–Activity Relationship (QSAR): A Versatile Tool in Drug Design

Asim Kumar Debnath

Lindsley F. Kimball Research Institute
The New York Blood Center
New York, New York

I. INTRODUCTION

Drug design is a complex process and requires multidisciplinary approaches from concept generation to lead optimization and to eventually putting a drug on the market. As this subject is so vast, it is impossible to draw even an abbreviated picture of the whole topic in one chapter. Instead, the intention of the author is to present the importance of the one-dimensional and two-dimensional quantitative structure–activity relationship (QSAR) paradigm in drug design in as simplistic a way as possible to a readership not sufficiently familiar with the field. Three-dimensional QSAR is covered in Chapter 4.

A. Background

Before we proceed with details of QSAR and its role in drug design, it is pertinent to go over a brief background of the history of QSAR. The quest for structure–activity relationship studies probably started in 1870 when the Russian chemist Mendeleev tried to arrange elements in a periodic system in order of atomic weights and discovered that similar properties were found in each of eight groups in the matrix he drew. Though heavier elements were exceptions, the periodic table achieved its preliminary success in classifying elements. True biological structure–activity relationship studies, however, started in 1868 when Crum-Brown and Fraser (1) first observed that the physiological activity of a series of strychnine alkaloids was dependent on their quaternary substituents. They developed the following mathematical equation for the correlation.

$$\phi = f(C)$$

According to this equation, biological activity ϕ must be a function f of chemical structure C. In 1893, Richet (2) showed that the toxicity of some organic chemicals (e.g., alcohols, ethers, ketones) was inversely correlated with their solubility in water. Meyer (3) and Overton (4) in 1899 and 1901, respectively, correlated oil/water partition coefficients with the activity of narcotic substances. Overton observed that the narcotic activity progressively increased as the chain length of the substituents increased. Following Overton's work, Traube (5) in 1904 observed a linear relationship between narcotic activity and surface tension. In a different twist to a series of structure–activity studies on narcosis already mentioned, Seidell (6) in 1912 measured the solubility and partition coefficients of thymol to understand the activity of this compound against hookworms. The first thermodynamic interpretation of narcotic activity came from the work of Ferguson (7) in 1939. In 1942, Bell and Roblin (8) first showed a bilinear relationship of inhibitory activity of a large number of sulfonamides against *E. coli* with pK_a. This work later provided the basis for investigations into the mode of action of sulfonamide drugs. The major breakthrough in QSAR came indirectly through the work of Hammett, when he postulated that the electronic effect of the substituents in the benzene ring could be modeled by the ionization of simple meta- and para-substituted benzoic acids. This work led to the Hammett equation (9). Hammett defined the electronic effect σ as follows:

$$\sigma = \log K_X - \log K_H$$

where K_X and K_H are the ionization constant of the substituted benzoic acid and the parent benzoic acid, respectively. The first application of the Hammett-type relationship to a biological system was reported by Hansen, in describing the effect of electronic parameter σ on the toxicities of substituted benzoic acids (10). During the same period, Zahradnik was also trying to develop a biological Hammett equation from more fundamental principles and developed the constant β in an analogous fashion to σ (11)

$$\log\left(\frac{\gamma_i}{\gamma_{Et}}\right) = \alpha\beta$$

where γ_i is the molar concentration of the ith compound and γ_{Et} is the corresponding concentration of the ethyl analog; α depends on the sensitivity of the system, and β is the substituent constant.

Neither the Hansen nor the Zahradnik approach led to much success because each was based on single-parameter linear relationships. Nevertheless, these works motivated others to consider undertaking the development of biological correlation studies. The first application of QSAR came from Hansch et al., who in 1962 correlated the plant growth regulatory activity of phenoxyacetic acids to Hammett constants and partition coefficients (12).

The major breakthrough in QSAR resulted when Hansch and Fujita in 1964 first showed that the biological activity could be correlated linearly by free energy–related terms (different physicochemical parameters) (13). In the same year, Free and Wilson proposed a de novo method to derive the group contribution values and rationalize the data sets in quantitative terms (14). These two seminal works opened up a new field of QSAR studies in biological systems.

II. METHODOLOGIES

Before I present different QSAR methods with examples, it is pertinent to discuss briefly a few important criteria that need to be addressed before any QSAR model can be developed correctly and effectively. These are (1) the expression of biological activity and (2) the selection of parameters.

1. *Expression of biological activity.* The success of any QSAR study depends on a number of different factors, and biological activity data certainly are crucial. The quality of any QSAR model depends on the type of data used to generate the model. The activity data must be reproducible and generated from a smooth dose–response curve and multiple measurements for a particular dose with as little error as possible. QSAR studies are often performed with free energy–related descriptors. Therefore, it is important and necessary to express the biological activity in free energy–related terms (e.g., equilibrium or rate constants). The change in free energy (ΔG_0) in a biological system can be expressed as the inverse logarithm of the molar concentration C to produce a certain biological response as follows:

$$\Delta G_0 = -2.3\, RT \log K \approx \log\left(\frac{1}{C}\right)$$

The logarithmic conversion also transforms a skewed distribution, which is typically observed, of biological response to a normal distribution.

2. *Selection of parameters.* Selection of one or more appropriate parameters is another important aspect in developing a meaningful QSAR. There is no specific rule for selecting parameters; rather, the choice depends on the type of system one is using, the nature of the interactions possibly responsible for the activity, and so on. For example, if the activity depends on the passive transport of the drugs to the receptor site, then hydrophobic terms (e.g., π, $\log P$) may play an important role. Generally, a number of different and wide-ranging parameters are

selected, and an attempt is made to correlate them with the activity data. All parameters used to develop a model must be validated by a stepwise regression method. For any multiparameter model, special care must be taken to ascertain any collinearity between parameters. Collinearity, which may underestimate the importance of certain parameter(s), may skew some important information for future design processes.

Topliss and Costello suggest using at least five to six data points per parameter to avoid chance correlation and obtain meaningful statistical results with a multiple regression approach (15). A large number of independent variables can be used if the partial least-squares (PLS) technique is used (see Sect. IV. B).

A. Free–Wilson Method

Free and Wilson proposed a simplistic mathematical approach for structure–activity studies in a congeneric series (14). The underlying concept of the approach was based on the additivity principle. According to this principle, substituent groups contribute in constant amounts and in an additive manner toward biological activity and do not depend on any other structural changes in the compound. In their original publication (14), Free and Wilson put forward a generalized mathematical model, as follows:

$$A = \Sigma a_i x_i + \mu$$

where A is the activity data, a_i is the contribution of the ith substituent and x_i takes a value of 1 when the substituent is present and 0 when it is absent; μ is the average contribution of the parent molecule and is a constant term.

The mathematical model also includes symmetry equations (often called restriction equation), which assume that the summation of contributions of all substituents in a particular position is 0. According to this assumption, $\Sigma a_i x_i = 0$.

The major drawbacks of the Free–Wilson method are as follows: (1) the activity contribution of all substituents including H must be considered; (2) the summation of the group contributions at each position, the so-called symmetry assumption, must be zero; and (3) the constant term (μ) should be an overall average of the biological activity of all the compounds used to develop the QSAR model.

These limitations led to the modification of the Free–Wilson method, known as Fujita–Ban method. For an excellent review on Free–Wilson and Fujita–Ban methods, readers are referred to the discussion by Kubinyi (16) and references cited there.

B. Fujita–Ban Method

Fujita and Ban in 1971 proposed a modified mathematical equation (17), using the logarithm of activity, which is a free energy–related term and additive in nature, represented as follows:

$$\log\left(\frac{A}{A_0}\right) = \Sigma G_i X_i$$

where A and A_0 are the activity data of the substituted and unsubstituted compounds, respectively, G_i is the logarithm of the activity contribution of the ith substituent, and X_i has a value of 1 or 0 depending on the presence of the substituent or its absence, respectively. For a set of substituents, the equation takes the following form:

$$\log A = \Sigma G_i X_i + \mu$$

where μ is a constant.

The major advantages of this modified method are as follows: (1) the structural matrix does not need to be transformed; (2) no restriction equation is necessary; (3) the group contribution at each position is based on the parent compound (i.e., H); and (4) the constant term (μ) is calculated by the least-squares method and is the theoretically predicted value for the unsubstituted compound. The addition or omission of a compound does not affect markedly the value of group contributions. These advantages make the Fujita–Ban method preferable to the Free–Wilson method. A number of applications of this method have been reported (18–22).

C. Hansch–Fujita Method

The major breakthrough in defining the QSAR methodology occurred in the early 1960s when Corwin Hansch and Toshio Fujita recognized the importance of physicochemical properties of chemical compounds in determining biological activity and suggested that biological activity could be correlated with the summation of linear free energy–related terms. Thus, this approach was originally coined as linear free energy relationships (LFER). Later this approach was termed, more appropriately, as the extrathermodynamic approach and expressed by the following equation:

$$\log\left(\frac{1}{C}\right) = a\pi + b\sigma + cE_s + \cdots + \text{constant}$$

where C is the molar concentration of the compound to produce a certain biological response, π is the hydrophobic contribution of the substituent and represented by $\log P_X/P_H$, σ is the Hammett electronic descriptor of the substituent, represented by $\log K_X/K_H$, E_s is Taft's steric parameter, and a, b, and c are the appropriate coefficients. In these expressions P_X and P_H are the octanol/water partition coefficients of the substituted and unsubstituted compounds, respectively, and K_X and K_H are the ionization constants at 25°C of the meta- or para-substituted and unsubstituted benzoic acid, respectively. A detailed description of all these parameters is provided later (see Sec. III).

Using a single parameter or multiple parameters, depending on the data set, one can generate the correlation models. The parameters can be from experimental values (π, log P, σ, etc.) or from theoretically calculated values [c log P, energies of lowest unoccupied molecular orbitals (LUMO) and highest occupied molecular orbitals (HOMO), charge, etc].

1. Linear Model

As noted, biological activity can be correlated with a linear combination of physicochemical parameters. There are hundreds of such examples in the literature. A number of examples showing linear dependence of activity on a single parameter are cited later (see Sec. III: Parameters). Here we give some examples of activity that was shown to correlate with a combination of physicochemical parameters in a linear fashion. Unless otherwise mentioned specifically, statistical terms in all correlation equations are represented by n as the number of compounds used to derive the equation, r as correlation coefficient or r^2 as correlation variance, and s as standard deviation. In some cases, a cross-validated correlation coefficient (q^2 or r_{cv}^2) also is reported. Generally, whenever possible the structure of the parent compound is presented before the QSAR equation to provide information on the substituent positions.

Smith et al. reported a classic example of QSAR analysis of the hydrolysis of X-phenyl hippurates (**I**) by papain (23) as represented by the following linear equation:

I

$$\log\left(\frac{1}{K_m}\right) = 0.57(\pm 0.20)\sigma + 1.03(\pm 0.25)\,\pi_3'$$
$$- 0.61(\pm 0.29)\,\mathrm{MR}_4 + 3.80(\pm 0.17)$$
$$n = 25,\ r = 0.907,\ s = 0.208$$

In this equation K_m is the Michaelis–Menten constant at 25°C and pH 6.0, σ is the Hammett constant, π_3' is the hydrophobicity of the substituent at the meta position [when the substituent π had a negative value, one used π for H and 0 for all para substituents; when two meta substituents were present e.g., 3,5-(Cl)$_2$, only one was used; for unsymmetrical meta-disubstituted compounds a π value of 0 was assigned for the hydrophilic substituent]. The rationale was that only one meta substituent can contact the enzyme and the other metasubstituent is thus located in the surrounding aqueous phase. An earlier graphic analysis supports this rationale (24). Finally, MR_4 is the molar refractivity parameter for the para substituents; all meta substituents got a value of 0. The coefficient of π'_3 of 1 indicates that the substituents are largely desolvated.

Another example can be cited from the work of Debnath et al., concerning the mutagenicity of quinolines (**II**) in *Salmonella typhimurium* TA100 (25):

II

$$\log TA100 = 0.99(\pm 0.44) \log P - 1.48(\pm 1.19) R_8$$
$$- 2.68(\pm 2.32) R_6 - 3.19(\pm 0.98)$$

$n = 21, r = 0.842, s = 0.599$

In this expression, TA100 represents the rate of mutation in revertants/nmol, P is the experimentally determined octanol/water partition coefficient, and R is the resonance parameter. The negative coefficients indicate that electron release to the ring increases activity. The importance of this effect is more pronounced for six-substituents than for eight-substituents, thus indicating that the field/inductive effect of the nearby nitrogen atom places the eight-substituent electrons under considerable constraint.

2. Parabolic Model

Hansch and coworkers introduced the parabolic model in QSAR analysis upon realizing that biological activity of hydrophobic drugs started to level off or decrease after reaching the optimum value. This was attributed to the entrapment of the drugs in the lipid phase of the transport process (13,26). They analyzed a series of examples and proposed a second-order relationship of hydrophobicity (log P) with biological activity as follows:

$$\log\left(\frac{1}{C}\right) = a \log P + b(\log P)^2 + c$$

where a and b are the coefficients of the log P and $(\log P)^2$ terms, respectively, and c is a constant term. Since then a great number of parabolic QSAR models have been reported in the literature. The following examples illustrate the utility of parabolic relationships in delineating biological activity.

The inhibitory activity of uptake of the thyroid hormone L-triiodothyronine (T_3) by a series of phenylanthranilic acids (**III**) has been shown to correlate with calculated log P (c log P) in a parabolic fashion as follows (27):

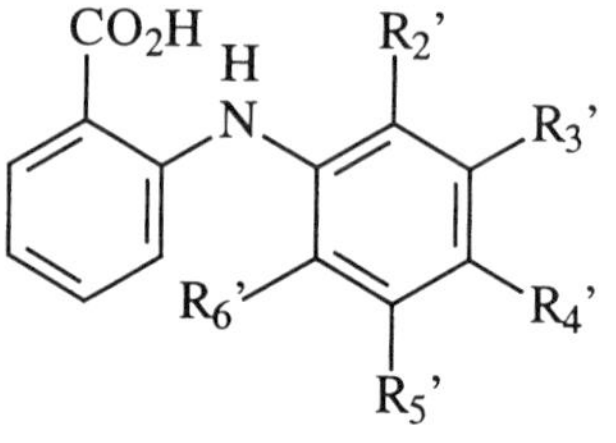

III

$$\text{logit } (\%I) = 3.92(\pm0.62) \text{ c log } P - 0.35(\pm0.06) \text{ (c log } P)^2 - 11.0 \ (\pm1.6)$$

$$n = 22, \ r^2 = 0.814, \ s = 0.164; \ c \text{ log } P_0 = 5.7$$

where logit $(\%I) = \log [\%I/(100 - \%I)]$; $\%I$, is the percentage inhibition of $[^{125}I]T_3$ uptake by H4 rat hepatocyte cells; c log P was calculated using CLOGP v3.54 (28).

Hansch et al. recently used the affinity of a series of 1-ethyl-4-phenyl-1,2,5,6-tetrahydropyridine-3-carboxylic acid derivatives (**IV**) to derive parabolic models with the calculated bulk and polarizability parameter *CMR* for the five human muscarinic receptor subtypes (Hm1–Hm5) (29). The affinity was determined by [^{3}H]NMS (tritiated normal mouse serum) binding using membranes from transfected Chinese hamster ovarian (CHO) cells. The activity, IC_{50}, was expressed as the nanomolar (nM)

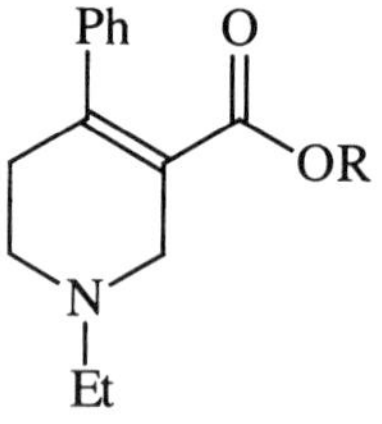

IV

concentrations of the compound required to displace [^{3}H]NMS by 50%. The equation developed using muscarinic receptor 1 (Hm1) is shown as an illustrative example:

$$\log \left(\frac{1}{C}\right) = 5.57(\pm1.27)CMR - 0.28(\pm0.07) \ CMR^2 - 19.79(\pm6.11)$$

$$n = 20, \ r = 0.914, \ s = 0.316$$

In this equation C is the concentration required to displace [^{3}H]NMS by 50%. The model shows that the affinity toward m1 increases initially as the substituent bulk increases and starts decreasing as the bulk increases, probably indicating limited bulk tolerance at the receptor site.

3. Bilinear Model

In a large number of cases, it has been found that biological activity increases with hydrophobicity linearly up to a certain point and then decreases in a linear fashion. The differences between observed and calculated biological activity were found to be high if parabolic models were used. The bilinear model to describe the nonlinear dependence of biological activity of drugs on hydrophobicity was proposed by Kubinyi (30,31) and expressed as follows:

$$\log \left(\frac{1}{C} \right) = a \log P - b \log (\beta P + 1) + c$$

The terms a, b, and c are linear and can be calculated by multiple regression analysis, whereas β is a nonlinear term and must be calculated by an iterative method.

A few examples of the application of bilinear models in QSAR will illustrate this model. Kubinyi derived a bilinear QSAR on the reported neurotoxicity data (32) [log (1/C)] of a series of alcohols with log P as follows (32):

$$\log \left(\frac{1}{C} \right) = 0.89 \ (\pm 0.05) \log P - 1.77 \ (\pm 0.10) \log (\beta P + 1) + 1.59$$

$$n = 10, r = 0.998, s = 0.041; \log \beta = -1.933, \log P_0 = 1.94$$

where C is the molar concentration causing 50% ataxia, and P is the octanol/water partition coefficient. It is interesting to note that the ideal value of log P was found to be around 2.0, an optimum value found in a number of cases, including barbiturates, and anesthetic ethers, for crossing the blood–brain barrier.

Garg et al. recently reported (33) a bilinear model for the antiviral activity data of cyclic urea (**V**) derivatives (34) against HIV-1 protease with c log P as follows:

V

$$\log \left(\frac{1}{C}\right) = 0.77 \ (\pm 0.25) \ \text{c} \log P - 1.24 \ (\pm 0.48)$$
$$\times \log(\beta.10^{\text{clog}P} + 1) + 1.05 \ (\pm 1.37)$$

$$n = 15, \ r = 0.902, \ s = 0.33; \ \log \beta = -6.84, \ \log P_0 = 6.96 \ (7.75–6.17)$$

Here C is the concentration of inhibitors producing 90% inhibition of viral RNA production in HIV-1 infected MT-2 cells. Ideal log P for this series of compounds was around 7.0.

D. Multi-CASE Method (MCASE)

Klopman and his coworkers (35,36) developed a new-generation computer auto-mated structure evaluation (CASE) program useful in drug design. The program generates fragments consisting of 2–10 heavy atoms of all possible chains from the input structure. These fragments are considered to be structural descriptors and used to derive QSAR models for prediction of biological activity. Fragments responsible for activity (biophore) and detrimental for activity (biophobe) are identified by statistical evaluation based on a binomial distribution. For the prediction of biological activity of a new compound, the program uses the information from the learning set and assigns probabilities for the compound to be active or inactive depending on the presence of biophores or biophobes.

Klopman and Ptchelintsev recently applied the Multi-CASE method to a series of 71 triazole alcohols to derive structure–antifungal, structure–teratogenicity, and structure–therapeutic index relationships (36). The purpose of the study was to identify potential biophores for antifungal activity and teratogenic activity from the learning set (69 compounds) in order to predict the potential antifungal and teratogenic activity of two compounds, ICI 153,066 from ICI and SCH 39304 from Schering-Plough and Sumitomo. The presence of certain biophores for antifungal activity, identified through the analysis, predicted a probability of 97.1% for the ICI 153,066 compound to be antifungal, but the presence of some biophores for teratogenic activity predicted a probability of 93% for the compound to be teratogenic. In the case of SCH 39304, the model did not predict any teratogenic potential, whereas it gave more than 87% probability to its having antifungal potency. The development of ICI 153,066 was aborted because of the teratogenic effects, whereas SCH 39304 went into clinical development. The successful application of the Multi-CASE program demonstrates the utility of this program for drug design.

In a more recent study, Klopman and Tu selected a set of 1819 chemicals out of 14,156 tested by the National Cancer Institute (NCI) and identified 74 fragments that could explain the anti-HIV activity of all compounds (37). Ten diverse sets of compounds were used as a test set, the activities of which were not known at the time of model development. The model correctly predicted the anti-HIV activity of 8 out of 10 chemicals. This again shows the utility of this program in predicting biological activity.

E. Other Methods

1. Artificial Neural Network (ANNs)

Neural network (NN) techniques have been used successfully in QSAR (38–51). This method is generally used when the data set size is large and the data cannot be interpreted easily by linear functions. This method is generally used in QSARs to describe a model with a nonlinear hypersurface.

Although there are different ways of constructing neural networks, the multilayer feedforward network with back-propagation is the one primarily used in drug design. In such a model the units are organized in layers starting with input units that are connected to the output unit through some layers of hidden units as shown in Fig. 1. Signals of representative input information about the drugs (parameters) are propagated forward using connecting weights, from the input layer to the output layer via the hidden units, and the output signals represent the predicted activity. Differences between the predictions and known activity are then used to adjust the weights "backward" until those differences become small. The major step in a neural network is to train the network using a representative training set. The design aspect of the neural network is very critical. If a network is trained with a large number of parameters, it may overtrain the model and generate unreliable prediction results because of overfitting. On the other hand, an undertrained neural network with too few parameters may generate poor results for new predictions. The quality of the fit can be validated by "leave-one-out" cross-

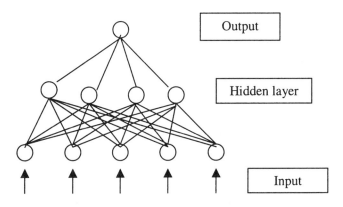

Figure 1 A prototype back-propagation neural network.

validation, whereby data are removed systematically, a neural network is trained, and a prediction of the removed data is made based on the trained network. The residual root-mean-squared error (rmse) or correlation coefficient (R) values are also calculated to validate the model. A Bayesian regularization algorithm has also been used to eliminate the need for a test model, since it minimizes a linear combination of errors and weights (52,53).

Breindl et al. recently reported the use of a back-propagation artificial neural network for predicting the octanol/water partition coefficient (log P) of a large number of organic chemicals (39). The neural net was trained with descriptors found to be important as input parameters correlating log P with these parameters by the multiple regression analysis. These parameters were calculated by means of semiempirical molecular orbital calculations using AM1 and PM3 methodology. A total of 1085 compounds were chosen for the back-propagation network; 980 compounds were used as training set and the rest as a test set. A three-layered back-propagation network was used entailing 16–25 descriptors. The network was trained until the rms error for log P fell below the 4% level. The overfitting of the training set was avoided by checking the standard deviation of the test set. A very good result was obtained. The correlation coefficient for the AM1 test set (r^2) and standard deviation (s) of the training set was 0.965 and 0.41, respectively. The data were cross-validated, and an r_{cv} of 0.93 was obtained. For the test set, the statistics were $r^2 = 0.902$, $s = 0.53$, and $r_{cv} = 0.915$. The results obtained from the PM3-generated set were slightly worse.

The major advantages of the neural networks are that they are nonparametric and nonlinear, and few statistical assumptions are required to build the model. Its major disadvantage is that the model cannot be easily interpreted, especially in physicochemical terms.

F.   New QSAR Methods

1.   HQSAR

Hologram QSAR (HQSAR) is a relatively new technique that does not require any physicochemical descriptors or three-dimensional structure to generate structure–activity models. The method is based on the input of two-dimensional structures and biological activity. The structures are converted to all possible linear and branched and overlapping fragments. The fragments are assigned a specific integer by using a cyclic redundancy check (CRC) algorithm. These integers are then hashed to a bin in an integer array of fixed length. The arrays are known as the molecular hologram, and the bin occupancies of the molecular holograms are used as the descriptors. These descriptors are expected to encode the chemical and topological information of molecules. The QSAR model is developed by using the partial least-squares (PLS) regression technique and validated by the "leave-one-out" cross-validation technique. Once the final model has been obtained, PLS yields the following equation correlating hologram bins with activity:

$$A_i = C + \sum_{l=1}^{L} X_{il} C_{il}$$

In this expression, A_i is the activity of compound i, C is a constant, X_{il} is the hologram occupancy value at position i or bin l, and C_{il} is the coefficient for the corresponding bin from the PLS run; L is the hologram length.

The fundamental difference between the HQSAR method and other fragment-based methods (e.g., the Free–Wilson method, the multi-CASE method) is that HQSAR encodes all possible fragments including overlapping fragments. The method is very fast and can also be used to predict physicochemical properties (e.g., c log P). A number of applications of the HQSAR method were reported in the late 1990s (54–57).

2. Binary QSAR

Binary QSAR is a relatively new QSAR technique, introduced to analyze large data sets, especially from high throughput screening, where the biological data are expressed in binary form (e.g., active/inactive, pass/fail). We discuss this method in more detail later (see Sec. VI.C.3).

III. PARAMETERS

The fundamental concept in QSAR analysis is to correlate activity data (biological or physical) with chemical structures having certain characteristics and physicochemical properties. These structural characteristics of the compounds as a whole or parts thereof are defined as parameters (also known as descriptors). When these parameters are used to correlate the activity data to generate models, they should help to decode the information contained in them. The most important forces responsible for drug action, starting from transport and metabolism to drug–receptor interactions, can be classified under three main categories: electronic, hydrophobic, and steric. Since the detailed applications of all the parameters are not within the scope of this chapter, we discuss only the parameters relevant to and used most frequently in QSAR.

A. Electronic Parameters

A multitude of electronic parameters (e.g., σ and modified σ values, inductive and field and resonance effects, pK_a, dipole moments μ, hydrogen-bonding parameters, quantum chemical indices) have been used in QSAR models. A brief description of some of the most widely used parameters is provided with simple examples (with only a single parameter used) of their use.

The most widely used electronic parameter is Hammett's σ, represented by the following equation: $\log K_X = \rho\sigma + \log K_H$, where K_X and K_H are the equilibrium or rate constants for the substituted X and unsubstituted analog, respectively. In this equation σ is the electronic substituent constant measured from the benzoic acid system, while ρ is a susceptibility constant signifying the measure of the importance of the electronic effect on a rate constant. The Hammett equation holds when the reaction center (Q) in a compound ($X-C_6H_4CH_2-Q$) has no effect due to the resonance interaction of the aromatic ring system. A negative σ value indicates electron release by the substituent, whereas a positive value implies electron withdrawal. The Hammett equation is an empirical equation, which has been applied to thousands of organic reaction systems. This approach also has been used to explain activity in many biological QSAR models (58–60).

The role of σ on the kinetic constant k_{cat} for the hydrolysis of a series of N-benzoylglycine pyridyl esters by papain has been demonstrated in the following equation (61):

$$\log k_{cat} = 0.45\ (\pm 0.08)\sigma + 0.17\ (\pm 0.04)$$

$$n = 23,\ r = 0.933,\ s = 0.094$$

The Hammett σ constants for substituents can be applied only when the electron-releasing or withdrawing substituents are not in conjugation with the reaction center. Modified σ constants (e.g., σ^+ and σ^-) have been derived to incorporate through-resonance effects due to the conjugation. Few applications of σ^0 in QSAR models have been reported. The electronic parameter σ^- has been derived from para-substituted phenols or anilines where the substituents can donate an electron pair to the reaction center, whereas σ^+ has been derived for substituents capable of delocalizing a $+$ charge. A number of applications of both these parameters in biological QSAR have been found.

Hansch et al. reported a correlation equation for mutagenicity of a set of 13 substituted (o-phenylenediamine)platinum dichlorides (**VI**) in the Ames test using *Salmonella typhimurium* (TA92) showing the importance of electron withdrawal by substituents R through resonance, as indicated in the following equation (62):

VI

$$\log\left(\frac{1}{C}\right) = 2.23(\pm0.32)\ \sigma^- + 5.78(\pm0.18)$$

$$n = 13, r^2 = 0.956, s = 0.260$$

In this expression C is the molar concentration of a compound producing 30 mutations/10^8 bacteria above background.

Hansch and Zhang recently described a number of QSAR models in which toxicity values for phenols and anilines were correlated with a σ^+ term of the substituents (63). Substituents that increase the electron density on the aromatic ring system were shown to potentiate toxicity. The toxicity was postulated to be mediated by a radical reaction.

Oxidation of phenols by horseradish peroxidase (compound **II**):

$$\log k = -2.48\sigma^+ + 4.68$$

$$n = 9, r = 0.972, s = 0.352$$

Oxidation of anilines by horseradish peroxidase (compound **II**):

$$\log k = -3.00\sigma^+ + 4.79$$

$$n = 7, r = 0.986, s = 0.264$$

For nonaromatic and saturated compounds, Taft introduced the inductive constant σ^* based on the rate constants of the hydrolysis of acetates (64).

The electronic effect of a substituent has been considered to be a combination of field/inductive (F) and resonance (R) effects. Swain and Lupton in 1968 proposed the following equation to define the field effect (65):

$$F = a\sigma_m + b\sigma_p + c$$

They further proposed that σ_p has both a field and a resonance component and can be represented as follows:

$$\sigma_p = \alpha F + R$$

The major drawback of Swain and Lupton's derivation of F and R values is that they were not placed on the same scale as the Hammett σ constants. Hansch et al. later modified these values by appropriately scaling F and redefining Swain and Lupton's equation as follows (60):

$$F = 1.369\ \sigma_m - 0.373\ \sigma_p - 0.009$$

$$n = 14, r = 0.992, s = 0.042$$

Efforts were also made to split electronic effect into field/inductive (σ_I) and resonance (σ_R) components as follows, and this effort was reviewed by Charton (66):

$$\sigma_p = \sigma_I + \sigma_R$$

Excellent compilations of electronic parameters (including Hammett σ as σ_m and σ_p) have been provided by the Hansch group (60,67).

Theoretical parameters based on quantum chemical calculations were applied to several QSAR models, which are especially useful when no Hammett-type electronic constants are available. Energy of lowest unoccupied orbitals (ε_{LUMO}) and energy of highest occupied orbitals (ε_{HOMO}) have been quite extensively used. According to the Koopmans theorem, ε_{LUMO} and ε_{HOMO} of a molecule can be approximated as its electron affinity and ionization, respectively. Following are some illustrative examples:

Tuppurainen et al. correlated mutagenicity of a series of lactone derivatives with ε_{LUMO} as follows (68):

$$\ln \text{TA}100 = -13.7 \, (\pm 1.0) \, \varepsilon_{LUMO} - 12.7 \, (\pm 1.0)$$

$$n = 17, r^2 = 0.924, s = 1.26$$

The AM1 semiempirical method was used to calculate ε_{LUMO} values.

The application of ε_{HOMO} can be illustrated by the following example, in which the radical scavenging activity of carbazoles was correlated with ε_{HOMO} (63):

$$\log\left(\frac{1}{C}\right) = 3.21 \, (\pm 1.9) \, \varepsilon_{HOMO} + 3.05 \, (\pm 0.16)$$

$$n = 6, r = 0.916, s = 0.017$$

In this equation, C is the concentration of carbazole required to inhibit lipid peroxidation by 50% when administered orally to mice; ε_{HOMO} was calculated by the AM1 methodology.

A multitude of electronic descriptor have found some application in QSAR analysis. Examples include dipole moment μ (69,70), ionization constants pK_a (71,72), atomic net charge q (73,74), superdelocalizability S (75,76), and molecular electrostatic potential MEP (77,78).

B. Hydrophobic Parameters

The influence of hydrophobicity of chemicals (drugs) on biological activity is probably the most widely studied area in drug design. Though Meyer and Overton in 1889–1901 first deduced the successful correlation between oil/water partition coefficients of some chemicals and their narcotic activity, it was not until 1964 that Hansch and his group first defined the hydrophobic substituent constants for the benzene system in an analogous fashion to the Hammett equation, paving the way for the hydrophobicity parameter used in QSAR. The hydrophobic substituent constant π was defined as follows:

$$\log \frac{P_X}{P_H} = \pi$$

where P_X and P_H are the partition coefficients of a compound with substituent X and the parent compound, respectively, in an octanol/water system, and π is the substituent constant for the group X. This solvent system has been accepted as a standard system to measure the partition coefficients. The octanol/water partition coefficient, designated log P, represents the overall hydrophobicity of a molecule, whereas π represents the hydrophobicity of a substituent. Both these parameters have been extensively used in QSAR models. Taylor has provided an elegant account of the history on development of hydrophobicity concept (79). An extensive compilation concerning the use of hydrophobicity can be found in a book by Hansch and Leo (80).

Since the experimental determination of the partition coefficients of a large set of compounds is a very tedious job, and often not feasible, a method of calculation of partition coefficients was proposed and pioneered by Rekker and his associates (81), using a fragmental approach. This approach had a number of drawbacks and was later extensively modified and extended by Leo et al. (67,80,82), resulting in the development of a computer program, CLOGP (28). The latest version of this CLOGP program has no missing fragment and has achieved a significant milestone, demonstrated by the following correlation between measured log P (m log P) and calculated log P (c log P).

$$\text{m log } P = 0.9 \text{ c log } P + 0.5$$

$$n = 10{,}000, \; r^2 = 0.970, \; s = 0.278, \; q^2 = 0.970$$

Several methods to calculate log P have been reported since then. For example, Ghose and Crippen's atom contribution method (83), Bodor's quantum chemical method (84), Klopman's MultiCASE method (85), and Moriguchi's method (86) have been used in QSAR applications. We will try to illustrate the utility of hydrophobic parameters π and log P through several examples.

Hansch et al. correlated the Michaelis constant K_m (reciprocal of the binding constant of substrates) for the interaction with papain of 16 hippurates of the structure $CH_3OCOCH_2NHCOC_6H_5$—Y and found a linear relationship with π as follows (24):

$$\log\left(\frac{1}{K_m}\right) = 1.01 \; \pi + 1.46$$

$$n = 16, \; r = 0.981, \; s = 0.165$$

The authors suggested that the partitioning patterns of the substrates into the enzyme–substrate (ES) complex and into the octanol system are similar and proposed that the desolvation of the substituents is the principal driving force for binding activity.

In another study, Sinclair et al. showed that the induction of cytochrome P450 by a series of barbiturates (**VII**) is dependent linearly on the hydrophobicity of the whole molecule (87), as shown in the following equation:

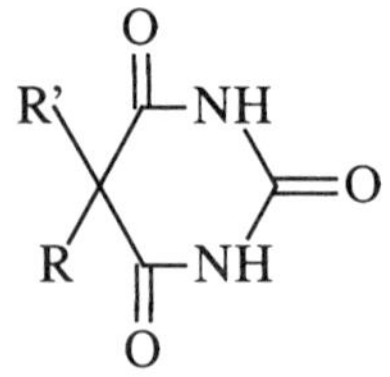

VII

$$\log \left(\frac{1}{C}\right) = 1.02 \log P + 2.75$$

$$n = 9, r = 0.984, s = 0.186$$

where C is the molar concentration of barbiturates producing a 50% increase in cytochrome P450 induction and P is the experimentally determined octanol/water partition coefficient of the barbiturates.

The two examples above illustrate the influence of hydrophobicity of aromatic and cyclic heterocyclic compounds, respectively, on biological activity. The next example correlates the hydrophobicity of short-chain aliphatic alcohols with their ability to alter plasma membrane integrity (88). The membrane integrity was quantified as a measure of lactate dehydrogenase (LDH) release by the alcohols, designated LDH_{50}:

$$\log \left(\frac{1}{LDH_{50}}\right) = 0.896 \log P + 0.117$$

$$n = 11, r = 0.993, s = 0.131$$

Again log P is the experimentally determined partition coefficient in octanol/water.

Van Steen et al. have reported an excellent correlation between the inhibition constant K_i (expressed as pK_i) of a set of eltoprazine (**VIII**) as 5-HT$_{1A}$ receptor ligands and the calculated hydrophobic parameter c log P, as shown in the following equation (89):

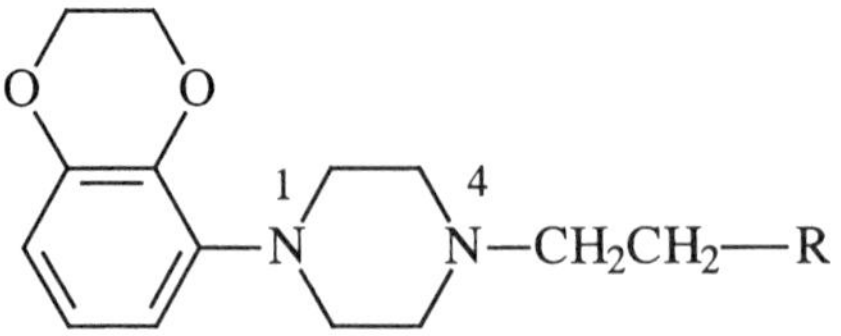

VIII

$$pK_i = 1.26 \, c \log P + 3.32$$

$$n = 12, \ r^2 = 0.83, \ s = \text{not reported}$$

For the homologous series of compounds investigated, the inhibition constant was governed by the lipophilicity of the compounds; $c \log P$ values were calculated by using the CLOGP program from Daylight C.I.S. Inc. (90).

Many studies use other hydrophobic parameters, such as R_M and $\log k'$ (known as capacity factor), derived from high performance liquid chromatography (HPLC) measurements (91–95). The R_M value is deduced from the R_f value from the following equation:

$$R_M = \log\left(\frac{1}{R_f - 1}\right)$$

Log k' has been used to calculate the octanol/water partition coefficients, since there is a linear relationship between these two parameters as outlined:

$$\log P_{oct} = a \log k' + b$$

The values of a and b depend on the mobile and stationary phases used.

For ionizable compounds, values of $\log D$ (D is the distribution coefficient of drugs in two different solvent system, e.g., octanol/water), calculated on the basis of experimental $\log P$ values at different pH are also used in QSAR (96,97).

C. Steric Parameters

Steric (or bulk) effects often play an important and critical role in QSAR. Sometimes it is difficult to separate steric from electronic and hydrophobic effects. In an attempt to distinguish between electronic and steric effects, Taft first introduced the steric term E_s based on the hydrolysis of esters of substituted acetic acids (XCH_2COOR) in acid solution and defined it as follows:

$$E_s = \log\left(\frac{k_X}{k_H}\right)_A$$

where k is the rate constant for the parent (H) and the substituted (X) acetic acid.

Hansch's group developed a QSAR equation (98) correlating the potency of seven pyridine (**IX**) derivatives in antagonizing the ability of angiotensin II to contract the rabbit aorta with the Taft's steric parameter E_s as follows:

IX

BPT =

$$pA_2 = -2.281 \ (\pm 1.086) \ E_s + 5.996 \ (\pm 1.492)$$

$$n = 7, r = 0.924, s = 0.591$$

Hancock et al. modified the E_s parameter to include the contribution of C—H hyperconjugation and proposed the following equation (99):

$$E_s^c = E_s + 0.306(n - 3)$$

where n is the number of α hydrogens and 0.306 is a constant term obtained by molecular orbital calculation.

Fujita and his group proposed that for complex biochemical steric effects of branched substituents ($-CR_1R_2R_3$), E_s^c could be calculated as a summation of steric effects of each component substituent (100). The fungicidal effects of N-substituted aminoacetonitrile ($RNHCH_2CN$) against "yellows" of the Japanese radish, expressed as the 50% preventive dose (pED_{50}), was found to be correlated well with Fujita's modified E_s^c as follows (101):

$$pED_{50} = -0.61(\pm 0.18) \ E_s^c + 1.52(\pm 0.26)$$

$$n = 16, r = 0.88, s = 0.20$$

Molar refractivity (MR) defined by the Lorentz–Lorentz equation, is used in QSAR as a bulk and polarizability parameter of a compound or a group and is scaled by 0.1 to make it equiscalar with log P, σ, and so on. In the Lorentz–Lorentz equation:

$$MR = \frac{n^2 - 1}{n^2 + 2} \frac{MW}{d}$$

where n is the refractive index, MW is the molecular weight, and d is the density of the compound. Special care should be taken when using MR because a descriptor as MR and π are often highly collinear.

Hansch and Caldwell reported (102) a correlation equation relating the inhibition of binding of tryptamine to rat cortical membranes by phenylethylamines (**X**) with the steric parameter MR.

X

$$\log\left(\frac{1}{C}\right) = 2.12 \ (\pm 0.81) \ MR_4 + 6.32 \ (\pm 0.45)$$

$$n = 7, r = 0.949, s = 0.247$$

The positive coefficient in MR_4 (MR value of substituent at the para position in the benzene ring) indicates that larger substituents may favor binding of the ligands to the receptor.

In another study, Gao et al. reported negative coefficient with MR term while correlating the relative binding affinities (RBAs) of 16α-substituted estradiol (**XI**) to the rat uterine estrogen receptor.

XI

$$\log \text{RBA} = -0.48 \; (\pm 0.10)\text{MR} + 2.08 \; (\pm 0.28)$$

$$n = 22, \; r^2 = 0.84, \; s = 0.432$$

The negative slope in MR suggests that the estrogen receptor cannot accommodate molecules with large substituents.

Verloop and his colleague recognized that the intermolecular interactions of ligands or its substituents with the receptor are important in drug action and developed an algorithm, known as STERIMOL, to calculate five directionality parameters: L, B_1, B_2, B_3, and B_4 (103). The length parameter L is defined as the length of the substituent along the axis of the bond between the first atom of the substituent and the parent molecule, and B_1-B_4, the width parameters, are determined by the rotation of the substituent around the x axis, where B_1 has the smallest value and B_4 the largest. Realizing that a large number of data are required to establish any meaningful statistics if all five parameters are to be used, Verloop et al. later modified these parameters and proposed to use only the length parameter L and two width parameters B_1 and B_5, where B_5 is the maximum width parameter (104).

Gao et al. developed an equation with a sterimol parameter L using the relative binding affinities of six 7α-undecylestradiol (**XII**) derivatives to the calf uterine estrogen receptor as follows (105):

XII

$$\log \text{RBA} = -0.74(\pm0.25)L + 2.25(\pm0.92)$$

$$n = 6, r^2 = 0.94, s = 0.155$$

The negative coefficient in L suggests that longer substituents may have detrimental effects on receptor binding.

Hansch et al. developed a correlation model for H_3–antagonist potency of a set of eight para- and meta-substituted 4(5)-phenyl-2-[[2-[4(5)-imidazolyl]ethyl]thio]imidazoles (**XIII**) using only the B_5 parameter as follows (98):

XIII

$$pA_2 = -0.403(\pm0.158) B_5\text{-}4 + 8.620 (\pm 0.448)$$

$$n = 8, r = 0.931, s = 0.242$$

The negative coefficient in B_5-4 (B_5 values at the para position) indicates that substituents beyond a permitted maximum width at the para position may have detrimental steric effects on activity.

Molecular weight (MW), molecular volume (MV), and parachor are also used as steric parameters (106–109).

D. Other Parameters

1. Hydrogen-Bonding Parameters

Hydrogen bonding is an important and critical property for biological activity of a drug depending on drug–receptor interactions. The hydrogen bond parameters are often used in correlation studies to develop QSAR models. One of the most successful hydrogen bond parameters as indicator variable was devised by Fujita et al. (110). The indicator value 1 was assigned when a molecule had the capability of forming a hydrogen bond with the receptor and 0 when it had no such capability. Charton and Charton (111) modified the scale to include the number of hydrogen bonds forming capacity of a molecule or substituent. For example, the amino group—NH_2 has the proton donor capacity of 2, whereas the acceptor capacity is 1. Similarly,—OH has a proton donor capacity of 1 but a proton acceptor capacity of 2 because it has two lone pairs. Yang et al. introduced enthalpy values of hydrogen bonds to calculate a hydrogen bond donor parameter HB_1, and a hydrogen bond acceptor parameter HB_2 (112). Abraham et al. (113) devised a group hydrogen-bonding parameter K_α for proton donors and K_β for proton acceptors from a solvatochromic approach. Dearden and Ghafourian (114) proposed that charge (Q_H) on a hydrogen atom connected to a electronegative atom and charge (Q_{MN}) on a electronegative atom should represent hydrogen bond donor ability and hydrogen bond acceptor ability, respectively.

2. Topological Descriptors

The topology of any molecule (i.e., positions of atoms and the bonds between them) determines its three-dimensional structure. Attempts have been made to generate molecular descriptors that encode the information about the structural environment of atoms, bonds, branching, unsaturation, heteroatom content, cyclic nature, and so on. These descriptors are often termed as topological descriptors and used to develop QSAR models (see Table 1 for reference). The detailed description of each of these parameters is beyond the scope of this chapter. Some of the descriptors used in QSAR studies are listed in Table 1, and their use is illustrated by selected examples.

A simple application of the connectivity index was reported by Kier et al. in their very early work on the development of this parameter (115). The local anesthetic activity of 36 distinct compounds, represented as the minimum blocking concentration (MBC) and measured experimentally on isolated nerve or muscle

Table 1 List of Some Commonly Used Topological Descriptors in QSAR Studies

Index	Description	Ref.
Wiener index	W, based on molecular graph distance	204
Randic indices	χ, connectivity index; W_n, cyclicity constant	205, 206
Kier and Hall indices		116, 146, 207–212
Valence molecular connectivity (chi) indices	$^m\chi$, $^m\chi^v$, $m = 0, 1, 2, \ldots$, indicating zero, first, second, . . . , order; v stands for valence	
Molecular shape (kappa) indices	$^1\kappa$, $^2\kappa$, etc; $^1\kappa_\alpha$, $^2\kappa_\alpha$, etc.	
Topological state indices	T_i, where i is the ith atom	
Electrotopological state (E state)	I_i, intrinsic state value of ith atom; S_i, electrotopological state of atom i; $I(H)$, E state for hydrogen atoms; atom type and bond type E-state indices	
Balaban indices	J, J_b, J_x, J_y; D_2, centric index	117, 213–216
Basak indices	IC_r, information contents; SIC_r, structural information content; CIC_r, complementary information content	216, 217

fibers, was correlated with the connectivity index χ as follows:

$$\log \text{MBC} = 3.55 - 0.762\chi$$

$$n = 36, r = 0.983, s = 0.390$$

A recent report by Gough and Hall utilized a number of different Kier and Hall indices, including electrotopological state (E state), molecular connectivity, and molecular shape, to correlate the antileukemic activity (medium effective dose, MED) of a series of 37 carboquinones (**XIV**) (116):

XIV

$$\text{pMED} = -0.208 \,(\pm 0.040)^1\chi^v + 2.112 \,(\pm 0.289) \,^4\chi^v_{PC}$$

$$- 0.338 \,(\pm 0.030)\text{SsCH}_3 - 0.128 \,(\pm 0.009)$$

$$\text{Sarom} + 5.071 \,(\pm 0.436)$$

$$n = 37, r^2 = 0.90, s = 0.21$$

In this expression $^1\chi^v$ and $^4\chi^v_{PC}$ are the valence molecular connectivity indices and SsCH$_3$ and Sarom are the atom-type E-state indices. The first-order connectivity index $^1\chi^v$ decreases with increased chain branching, and a negative coefficient indicates that the branching will enhance the activity. The $^4\chi^v_{PC}$ index increases with increased adjacency. The positive coefficient of $^4\chi^v_{PC}$ signifies that the increased adjacency will also increase the potency. The value of atom-type E state increases as the number of participating atoms increases, since this value is the summation E state of all atoms. The negative coefficient in SsCH$_3$ indicates that as the number of CH$_3$ groups increases the activity decreases. Similarly, since the coefficient in Sarom is also negative, any decrease in the value of this index will increase activity.

Recently, Balaban et al. have shown the use of several topological indices in a correlation study on the structure and the normal boiling points (NBPs) of a large number of acyclic carbonyl compounds by the following equation (117):

$$\text{NBP} = 302 \,(\pm 13) \, J_y - 372 \,(\pm 15) \, J + 223 \,(\pm 6) \, s_0$$

$$+ 116(\pm 9) \, \text{IC}_2 - 272(\pm 13) \, \text{IC}_1 - 109(\pm 11)$$

$$n = 200, r^2 = 0.964, s = 6.93°C$$

In this expression, J_y, J, IC_1, and IC_2 are defined as in Table 1; s_0 is the sum of square roots of the vertex degree (δ_i).

IV. STATISTICAL TECHNIQUES

Different statistical techniques are used in solving the correlation equation of biological activity with different parameters (physicochemical or others). Only two techniques used most often in QSAR analysis are discussed.

A. Multiple Linear Regression (MLR)

Drug action is a complex process and often cannot be explained by a simple correlation equation $y = ax_1 +$ constant. Rather, it requires a multitude of descriptors, and then the correlation equation can be represented as follows:

$$y = ax_1 + bx_2 + cx_3 + \cdots + constant$$

Multiple linear regressions can be used to solve this complex equation. A least-squares technique can be used to obtain the coefficients of each independent variable (x_1, x_2, x_3, . . .). Several statistical parameters are used to evaluate the validity of the equation. These include the correlation coefficient r (often r^2, which is a measure of the explained variance), the standard deviation s, and F statistics. The confidence intervals of all the regression terms are also calculated.

B. Partial Least Squares (PLS)

The PLS technique was introduced in QSAR to circumvent the problem of dealing with more descriptors (independent variables) than the number of data points (dependent variables) and for cases of descriptors that are not orthogonal, which cannot be handled by the MLR method (for details on methods, advantages, and disadvantages, readers are referred to Refs. 118 and 119). It is advisable to have at least five dependent variables (e.g., biological data) per descriptor for an MLR analysis. Since PLS deals with a large number of independent variables, a significantly high correlation coefficient (r^2) is often achieved. However, this statistic is not always sufficient unless validated by a cross-validation technique. In this procedure, generally the leave-one-out method is used. In other words, one compound is removed from the data set randomly or systematically as the correlation model is developed, and the excluded compound's activity is then predicted from the model. The cross-validated correlation (q^2 or $r^2_{CV)}$ is then calculated. This value is generally the indicator of the predictive value of the model. Other statistical parameters (e.g., standard deviation S_{PRESS}, SDEP) are used for statistical validation. For example, in the relations:

$$S_{\text{PRESS}} = \left(\frac{\text{PRESS}}{n - a - 1}\right)^{1/2} \quad \text{and} \quad \text{SDEP} = \left(\frac{\text{PRESS}}{n}\right)^{1/2}$$

PRESS is **p**redictive **r**esidual **s**um of **s**quares, n, the number of data points, and a is the number of PLS components (often called latent variables).

C. Other Statistical Methods

Several other statistical techniques have been used in QSAR, including discriminant analysis (74,120–125), principal component analysis (PCA) and factor analysis (74,126,127), cluster analysis (128–132), multivariate analysis (121,133–135), and adaptive least squares (136–138).

V. POTENTIAL ROLE OF QSAR IN DRUG DISCOVERY

QSAR methodologies have been applied to a wide variety of physical and biological systems, and several excellent books and review articles on this topic have been published (80,101,139–146). A review on this topic is beyond the scope of this chapter, but we provide some examples of QSAR applications to different areas important in drug design. The idea here is to give some flavor of how this important technique can be beneficial in predicting not only biological activity but also a multitude of other parameters important for the complex cycle of drug design.

A. QSAR Applications

1. Physicochemical Properties

Quantitative structure–property relationship (QSPR) (147,148) has been applied to calculate different physicochemical properties of chemicals, such as octanol/water partition coefficient (log P), pK_a, boiling point, and solubility (149–151).

  Prediction of the Octanol/Water Partition Coefficient (log P).   Moriguchi et al. developed a quantitative structure–log P relationship method for calculating log P (partition coefficient in octanol/water) based on 1230 diverse organic chemicals that included several drugs and agrochemicals using parameters for hydrophobic atoms, hydrophilic atoms, their proximity effect, unsaturated bonds, intramolecular hydrogen bonds, ring structures, amphoteric properties, and some other structural parameters (Table 2) (86). The QSAR model to calculate log P was developed using all 13 parameters described in Table 2 as follows:

$$\log P = 1.244(CX)^{0.6} + 1.017(NO)^{0.9} + 0.406PRX - 0.145(UB)^{0.8}$$
$$(t = 60.5) \qquad (t = 58.5) \qquad (t = 33.8) \qquad (t = 9.5)$$

$$+ 0.511HB + 0.268POL - 2.215AMP + 0.912ALK$$
$$(t = 5.9) \qquad (t = 19.6) \qquad (t = 19.5) \qquad (t = 9.5)$$

$$- 0.392RNG - 3.684QN + 0.474NO_2 + 1.582NCS$$
$$(t = 13.1) \qquad (t = 22.1) \qquad (t = 10.8) \qquad (t = 16.4)$$

$$+ 0.773BLM - 1.041$$
$$(t = 5.0)$$

$$n = 1230, r = 0.952, s = 0.411$$

Table 2 Parameters Used and Their Description

Parameter	Type[a]	Description
CX	N	Summation of numbers of carbon and halogen atoms weighted by C:1.0, F:0.5, Br:1.5, and I:2.0
NO	N	Total number of N and O atoms
PRX	N	Proximity effect of N/O;X—Y:2.0, X—A—Y:1.0 (X,Y:N/O, A:C,S, or P) with a correction (−1) for carboxamide/sulfonamide
UB	N	Total number of unsaturated bonds except those in NO_2
HB	D	Dummy variable for the presence of intramolecular hydrogen bonds as ortho —OH and —CO—R, —OH and —NH_2 and —COOH, or 8-OH/NH_2 in quinolines, 5- or 8-OH/NH_2 in quinoxalines, etc.
POL	N	Number of aromatic polar substituents (aromatic substituents excluding Ar—CX_2— and Ar—CX=C<, X: C or H
AMP	N	Amphoteric property; α-amino acid, 1.0, aminobenzoic acid, 0.5, pyridinecarboxylic acid, 0.5
ALK	D	Dummy variables for alkane, alkene, cycloalkane, cycloalkene (hydrocarbon with 0 or 1 double bond)
RNG	D	Dummy variable for the presence of ring structures except benzene and its condensed rings (aromatic, heteroaromatic, and hydrocarbon rings)
QN	N	Quaternary nitrogen: >N^+<, 1.0; N oxide, 0.5
NO_2	N	Number of nitro groups
NCS	N	Isothiacyanato (—N=C=S), 1.0; thiocyanato (—S—C≡N), 0.5
BLM	D	Dummy variable for the presence of β-lactam

[a] N, numerical variable; D, dummy variable; takes a value of 1 if the molecule contains a certain substructure (e.g., if a ring is present, the RNG dummy variable gets a value of 1) or a certain feature (e.g., for the presence of intramolecular hydrogen bond a value of 1 is assigned).

In this equation t is the t statistic for the coefficients, n the number of compounds, r the correlation coefficient, and s standard deviation of the estimation error. The higher t-statistic values in the hydrophobic term $(CX)^{0.6}$ and hydrophilic term $(NO)^{0.9}$ indicate that log P depends predominantly on these two nonlinear terms. The application of this method was also demonstrated. The utility of this method is illustrated by the following example, in which the approach is used to calculate log P for ampicillin (**XV**).

XV

Molecular formula: $C_{16}H_{19}N_3O_4S$
$CX = 1.0 \times 16$ (for C_{16}) $= 16.0$
$NO = 3 + 4$ (for N_3O_4) $= 7.0$
$PRX = 1.0$ (for —CO—NH—) $+ 1.0$ (for —CO—N<)
$\qquad\qquad\qquad\qquad\qquad + 2.0$ (for —CO—OH) $= 4.0$
$UB = 6.0$ (for 6 double bonds) $= 6.0$
$RNG = 1.0$ (for ring) $= 1.0$
$BLM = 1.0$ (for β-lactam) $= 1.0$
calculated log $P = 1.244 \times (16.0)^{0.6} - 1.017 \times (7.0)^{0.9} + 0.406 \times 4.0$
$\qquad\qquad\qquad - 0.145 (6.0)^{0.8} - 0.392 \times 1.0 + 0.773 \times 1.0 - 1.041$
$\qquad\qquad = 1.06$
CLOGP $= 1.35$ and experimental log $P = 1.00$

2. Prediction of pKₐ Values

A simple Hansch-type equation was reported by Mor et al. for a series of 4(5)-phenyl-2-[[2-[4(5)-imidazolyl]ethyl]thio]imidazoles (see structure **XIII**) correlating pK_a values with simple Hammett σ parameters (152):

$$pK_a = -1.17(\pm 0.15) \sigma + 4.18(\pm 0.06)$$

$$n = 11, r^2 = 0.865, s = 0.161$$

The negative coefficient of the electronic term σ indicates that electron-withdrawing groups at X increase acidity. Absence of any hydrophobic or steric terms indicates that the ionization of these compounds depends primarily on the electronic effect. An attempt to separate inductive (F) and resonance effects (R) did not produce any substantial improvement to the correlation:

$$pK_a = -1.41(\pm 0.24)\,F - 0.943(\pm 0.249)\,R + 4.27\,(\pm 0.10)$$

$$n = 11,\ r^2 = 0.885,\ s = 0.157$$

3. Prediction of Aqueous Solubility

A six-descriptor QSPR model for aqueous solubility (S_W) of a large data set (411 compounds) was proposed by Katritzky et al. (149) using the program CODESSA (153):

$$\log(S_w) = -16.1(\pm 0.7)\,Q_{min} - 0.113\,(\pm 0.005)\,N_{el} + 2.55(\pm 0.22)$$
$$\times\,\text{FHDSA(2)} + 0.781(\pm 0.064)\,\text{ABO(N)} + 0.328(\pm 0.037)$$
$$^0\text{SIC} - 0.014(\pm 0.002)\,\text{RNCS} - 0.882(\pm 0.138)$$

$$n = 411,\ r^2 = 0.879,\ s = 0.573,\ r^2_{cv} = 0.874$$

where Q_{min} (charge on the most negative atom) and RNCS (relative negative-charged surface area) are electrostatic descriptors, N_{el} (number of electrons) and ABO(N) (the average bond order of nitrogen atoms) are quantum chemical descriptors, and ^{0}SIC is a topological descriptor and represents the structural information content of zeroth order; FHDSA(2) is the fractional hydrogen donor surface area descriptor. The t-test results show that Q_{min} and N_{el} are the two most important contributing descriptors.

B. Biological Activity

1. Antiulcer Compounds

A report by Kamenska et al. (154) elegantly demonstrated the utility of QSAR models developed for the H_2-receptor antagonist data (pA_2) of 14 N-[3-[3-(1-piperidinomethyl)phenoxyl]propyl]amines (**XVI**) in predicting the activity of a series of new compounds belonging to N-[3-[3-(1-piperidinomethyl) phenoxyl]propyl]benzamides (**XVII**). The in vivo activity corresponded well to the *in vitro* activity predicted from different QSAR models. The physicochemical parameters used in the computer program OASIS were used. The best four models were as follows:

XVI   XVII

$pA_2 = 22.96(\pm 4.17) - 85.03(\pm 12.16) \, S_{11}^{E} - 1.445(\pm 0.504) \, \varepsilon_{HOMO}$

$n = 12, \, R^2 = 0.85, \, s^2 = 0.25$

$pA_2 = 34.34(\pm 4.92) - 89.06(\pm 16.91) \, S_{11}^{E} + 2.148(\pm 1.131) \, I_x$

$n = 12, \, R^2 = 0.79, \, s^2 = 0.35$

$pA_2 = -29.37(\pm 7.75) + 148.1(\pm 29.63) \, S_{11}^{N} + 3.354(\pm 1.364) \, I_x$

$n = 12, \, R^2 = 0.77, \, s^2 = 0.37$

$pA_2 = 32.12(\pm 5.49) - 72.64(\pm 12.16) \, S_{11}^{E} - 0.200(\pm 0.20) \log P$

$n = 12, \, R^2 = 0.74, \, s^2 = 0.44$

where n is the number of data points, R^2 is the coefficient of determination, and s^2 is the variance; ε_{HOMO}, S_{11}^{E} and S_{11}^{N} are electronic parameters and represent energies of the highest occupied molecular orbitals, donor and acceptor superdelocalizability at position 11 (see **XVI** and **XVII**), respectively, and log P, the hydrophobicity parameter, was calculated by the Ghose and Crippen method (155). The models showed dominance of electronic effects on pA_2 activity. The negative coefficients in ε_{HOMO} and S_{11}^{E} terms indicate that the pA_2 activity depends on the donor (acceptor) properties of the molecule, especially, at the NH reaction site at position 11. This analysis is consistent with the finding that the H_2-receptor activity depends on this NH group to form H bonds with the receptor. The dependence of activity on log P, though, reflects the receptor interactions, but a very low coefficient with poor confidence interval value makes it virtually negligible. The most notable feature of the study is that the QSAR models correctly predicted that two of the most active compounds in vitro also would be most active in the *in vivo* assay system.

2. Calcium Channel Activator

In another example for QSAR guided synthesis, a potent calcium channel activator was discovered (156). A QSAR model was developed for the inotropic activity "force EC_{50}" (concentration required to increase developed tension to 50% of the isoprenaline maximum) of a series of benzoylpyrroles using the simple hydrophobic substituent constant π and the Verloop STERIMOL parameter B_x corresponding to the width of the substituent opposite to the minimum width B_1 (104):

$$\log\left[\frac{1}{\text{force}}\right] = 0.42(\pm 0.15) \, \pi + 0.33(\pm 0.19) B_x - 1.97$$

$n = 8, \, r^2 = 0.79, \, s = 0.22$

where [force] is the "force EC_{50}" as defined before. According to the model, increased force potency is expected by simply increasing the lipophilicity and size of the substituent at position 2. (All the substituents in the data set were at position 2.) The authors selected the benzyl group for possible synthesis because it has large π (2.01) and B_x (6.02) values. According to the equation, the predicted "force EC_{50}" value was 0.14 μM. After synthesis and testing, the compound showed a "force EC_{50}" value of 0.049 μM, an 80% increase over the most potent compound tested in the original data set.

These examples show the utility of QSAR methods in drug design.

C. Toxicity

1. Mutagenicity and Carcinogenicity

Evaluation of mutagenicity (carcinogenicity) is one of the important steps before compounds can be marketed as drugs. The establishment of the test system by Ames (157) opened up a great opportunity to assess the potential toxicity of thousands of industrial pollutants, chemical, and drugs. Several studies on QSAR of mutagens and carcinogens have been reported. With the advent of combinatorial synthesis and high throughput screening (HTS), millions of compounds are now considered for drug discovery. Therefore, predictive QSAR models for toxicity (including mutagenicity and carcinogenicity) will be more important for the drug design process than ever before.

Hansch's group reported a QSAR study of a series of triazenes **(XVIII)** tested against *Salmonella typhimurium* TA92 with S9 activation (158). The triazenes have been used for the treatment of melanoma. The equation shows that the mutagenicity of this class of compounds can be explained by hydrophobicity (log P) and the through-resonance electronic term σ^+ as follows:

XVIII

$$\log\left(\frac{1}{C}\right) = 1.04(\pm 0.17) \log P - 1.63(\pm 0.34)\, \sigma^+ + 3.06(\pm 0.43)$$

$$n = 17,\ r^2 = 0.949,\ s = 0.315$$

where C is the molar concentration of triazenes producing 30 mutations above background in 10^8 bacteria.

Because of the unavailability of σ^+ parameters, Shusterman et al. later used MO-generated parameters (ε_{LUMO} and q_N HOMO) to accommodate more complex triazene molecules for QSAR studies (159).

The mutagenicity of aromatic nitro and amino compounds has been studied extensively for QSAR model development (68,73,160–169). These groups are present in many chemicals, pharmaceutical compounds, and environmental pollutants and in compounds from grilled meat. In a study with a large number of data sets on aromatic nitro compounds, Debnath et al. reported the following equation (160):

$$\log TA98 = 0.65(\pm0.16) \log P - 2.90(\pm0.59) \log (\beta \times 10^{\log P} + 1)$$
$$- 1.38(\pm0.25) \, \varepsilon_{LUMO} + 1.88(\pm0.39) \, I_1 - 2.89(\pm0.81) \, I_a$$
$$- 4.15(\pm0.58)$$

$$n = 188, \, r^2 = 0.810, \, s = 0.886, \, \log P_0 = 4.93(\pm0.35), \, \log \beta = -5.48$$

Here TA98 is the number of revertants per nanomole of nitro compounds, ε_{LUMO} is the energy of the lowest unoccupied molecular orbital as calculated by AM1, I_1 is an indicator variable with a value of one for all compounds with more than two fused rings, and I_a is an indicator variable for five compounds containing acenthrylene rings.

The ε_{LUMO} term indicates that lower the energy of the lowest unoccupied MO, the better that LUMO can accept electrons and facilitate the reduction process of the nitro group, which was thought to be important for the mutagenic activity of this class of compounds. The bilinear relationship with log P shows that the mutagenic potential is also enhanced with hydrophobicity up to a certain limit and falls sharply over a value of around 5.0.

2. Phenol Toxicity

Selassie et al. reported an elegant study delineating the mechanism of toxicity of phenols in leukemia cells (170). In a large series of simple substituted phenols, two "novel and unusual" QSARs emerged. When only electron-releasing substituents were considered, both hydrophobicity (log P) and the electronic parameter σ^+ were found to be important:

$$\log\left(\frac{1}{IC_{50}}\right) = -1.58(\pm0.26) \, \sigma^+ + 0.21(\pm0.06) \log P + 3.10 \, (\pm0.24)$$

$$n = 23, \, r^2 = 0.898, \, s = 0.191$$

In this equation, IC_{50} is the concentration of phenols required to reduce inhibition of growth by 50%, log P is the hydrophobicity parameter, and σ^+ is Brown's modified Hammett electronic parameter. Two substituents, 3-NH_2 and 4-$NHCONH_2$, had to be omitted to derive the model.

The QSAR on electron-attracting substituents did not show any dependence on the electronic parameter. A simple equation with the log P term could explain most of the variance:

$$\log\left(\frac{1}{IC_{50}}\right) = 0.62(\pm0.16) \log P + 2.35(\pm0.31)$$

$$n = 15, \, r^2 = 0.845, \, s = 0.232$$

One compound (3-OH) had to be omitted. The study clearly demonstrated that for two different categories of substituted phenols, two types of mechanism were operative for toxicity against leukemia cells. For electron-releasing substituents, formation of radicals probably dominates in the toxicity process, whereas for electron-attracting substituents the process is probably governed by nonspecific toxicity as manifested by dependence on a sole hydrophobicity term.

Selassie's group has reported an improved QSAR model on a much larger data set of phenols; homolytic bond dissociation energy (BDE) served as the electronic parameter, and log P was used (171). BDE was found by this group to be a superior parameter to σ^+ and it directly measures the thermodynamics of phenoxyl radical formation. The model is represented by the following equation:

$$\log\left(\frac{1}{C}\right) = -0.19(\pm0.02)\text{BDE} + 0.21(\pm0.03)\log P + 3.11(\pm0.10)$$

$$n = 52, \, r^2 = 0.920, \, s = 0.202; \, q^2 = 0.909$$

As in the preceding study, only electron-releasing substituents were used to derive the equation. The coefficient in log P has identical values in both the equations. The introduction of BDE parameter helped the inclusion of a large variety of phenols, including naphthols, in the equation.

D. Pharmacokinetics

Absorption, distribution, metabolism, and excretion (ADME) play an important role in designing drugs with the most desirable properties. Often, a candidate drug that exhibits potent activity in in vitro systems fails to show the desired clinical effects in later stages of development and must be abandoned. To avoid this situation, application of ADME in the early stage of drug development is getting considerable attention now a days (172,173). QSAR has been shown to be a useful tool in using pharmacokinetics data to develop predictive ADME models (for an excellent compilation of QSAR models, see Ref. 174) and can be of great use in decision making on the potential of a compound to have ideal pharmacokinetic profiles. This is a vast subject, which is not reviewed here. Instead, a few examples are presented to introduce the new practitioner of QSAR to the use of these methods in structure–pharmacokinetics relationship studies. A number of excellent articles have been published on this topic (174–179).

Recently, Clark reported a simple correlation model for the prediction of blood–brain barrier (BBB) penetration from a large set of organic molecules using polar surface area (PSA) and calculated log P (180). The degree of BBB penetration is usually expressed as the logarithm of the ratio of concentrations of compounds in the brain and in the blood as log ($C_{\text{brain}}/C_{\text{blood}}$) or simply as log BB. The model was developed with training set of 57 compounds and validated with two different test sets. The following equation shows the QSAR model:

$$\log \text{BB} = -0.0148\,(\pm 0.001)\,\text{PSA} + 0.152$$
$$\times\,(\pm 0.036)\,\text{c}\,\log P + 0.139\,(\pm 0.073)$$

$$n = 55,\, r = 0.887,\, s = 0.354$$

Two compounds were dropped to derive the model. The negative coefficient in PSA (PSA is an indicator of compound's ability to form hydrogen bonds) and the positive coefficient in c log P indicate that compounds that form fewer hydrogen bonds with the solvent and are more lipophilic will be most likely to penetrate to the brain.

Seydel et al. described studies on the relationship between the rate of elimination (K_e) from the body of a series of ortho-, meta-, para- and ortho, para-substituted sulfapyridine (**XIX**) derivatives with physicochemical parameters (181). The elimination of the compounds was measured in rats by analyzing blood samples after intravenous injection of 50 mg of compounds per kilogram of body weight. The lipophilicity parameter ΔR_m, the acid dissociation constant pK_a, the indicator variable I, and the Taft E_s parameters were used in the correlation studies.

XIX

The correlation study for para- and meta-substituted derivatives provided the following equation:

$$\log K_e = -0.58(\pm 0.1)\Delta R_m + 0.31(\pm 0.03)\,pK_a - 3.03$$

$$n = 9,\, r = 0.98,\, s = 0.12$$

When ortho- and ortho, para-substituted derivatives were considered, the following QSAR was obtained:

$$\log K_e = -0.83(\pm 0.12)\Delta R_m + 0.16(\pm 0.05)\,pK_a - 1.53$$

$$n = 10,\, r = 0.94,\, s = 0.19$$

The combined data of these two series of compounds yielded the following equation:

$$\log K_e = -0.77(\pm 0.09)\Delta R_m + 0.23(\pm 0.03)\,pK_a + 0.26(\pm 0.096)I - 2.8$$

$$n = 19,\, r = 0.95,\, s = 0.18$$

In this equation, I is the indicator variable and has a value of 1 for all ortho-substituted compounds and 0 for meta- or para-substituted derivatives. The indicator I could be replaced by the E_s parameter and probably represents the steric effect of ortho-substituted compounds. About 90% of the variance could be explained just by the lipophilic and electronic terms.

VI. ROLE OF DIFFERENT DESCRIPTORS (PARAMETERS) AND QSAR IN LIBRARY DESIGN

The introduction of combinatorial chemistry and high throughput screening technology in drug discovery has enabled the generation of enormous amounts of data in a very short period of time. If the database is not selected properly, these random screening techniques may generate redundant information, and a huge number of compounds may be produced that do not represent a sufficiently diverse set of unique compounds. Eventually, the entire process may not be cost-effective. To circumvent these problems, several useful and important techniques have been developed to optimize the diversity and to select druglike compounds with desirable physicochemical properties from a large database or combinatorial library. Though these methods cover a diverse array of techniques, we will concentrate on the use of some of the descriptors and QSAR methods for designing combinatorial libraries and for analyzing their molecular diversity.

A. Separation of Druglike Molecules and Nondrugs from Large Database

For the last few years, it has been realized that understanding the "druglikeness" of chemical compounds with appropriate physicochemical properties in a database is important and can be of great use for designing combinatorial libraries or for selecting compounds from a large database for screening (43,182–184).

Lipinski et al. at Pfizer formulated the "rule of five" as a guiding tool to select drugs that should have a high probability of being orally available (185). This rule was established after analysis of a set of calculated properties of 2245 drugs in the World Drug Index (WDI) database. According to this rule, a compound is likely to have poor absorption and permeation properties when number of hydrogen bond donors (any N—H, O—H) exceeds 5, the number of hydrogen bond acceptors (any N, O) exceeds 10, the molecular weight (MW) exceeds 500, and the calculated octanol/water partition coefficient (c log P) exceeds 5. This general rule applies to the majority of the drugs except for substrates for biological transporters. Though this rule is not directly used to classify drugs and nondrugs, it can be used to design combinatorial libraries or select compounds with ideal properties for screening.

Ajay et al. used a Bayesian neural network (BNN) to distinguish drugs from nondrugs in the Comprehensive Medicinal Chemistry (CMC) database, containing mostly druglike molecules, and in the Available Chemical Directory (ACD), containing mostly nondruglike molecules (43). They have used seven one-dimensional descriptors, namely log P, molecular weight (MW), number of hydrogen bond donors (ND), number of hydrogen bond acceptors (NA), number of rotatable bonds (NR), aromatic density (AR), kappa index ($^2\kappa_\alpha$, representing the degree of branching), and 166 two-dimensional descriptors (keys) based on ISIS (Integrated Scientific Information System) fingerprints, to train a set of 3500 compounds selected by random partitioning from both the CMC and ACD databases. The seven-descriptor set predicted over 80% of the CMC compounds correctly as potential drugs, while a large number (30%) of compounds from ACD were classified as druglike. The performance with two-dimensional descriptors was better, but the best prediction was obtained when investigators uses a combination of one- and two-dimensional descriptors. The BNN method correctly classified 90% of the compounds from the CMC database, while it misclassified only 10% of the compounds in the ACD database. The extrapolation ability of the model was also demonstrated by the fact that it classified 80% of the molecules in the MDDR (MACCS-II Drug Data Report) database as druglike.

Atom descriptors (186) were used by Sadowski et al. (182) in a feedforward neural network method to develop a scoring scheme. These scoring functions were used to distinguish drugs from nondrugs in two large databases, ACD (containing 169,331 compounds) and World Drug Index (WDI), containing 38,416 compounds. This method successfully classified 77% of the compounds in the WDI database as drugs and 83% of the compounds in the ACD database as non-drugs. The robustness of the method was further ascertained by omitting from the training set four categories of drugs in the WDI database (hormones and antagonists, and drugs affecting the nervous system, blood and cardiovascular, and respiratory system). The model developed with the rest of the compounds mostly correctly classified all these compounds as drugs.

Gillet et al. used the calculated property profiles (e.g., molecular weight, the $^2\kappa_\alpha$ shape indices, number of aromatic rings, rotatable bonds, hydrogen bond donors, and hydrogen bond acceptors) as a tool to differentiate between druglike molecules in the WDI database (14,861 active molecules) from nondruglike molecules in the SPRESI database (16,807 inactive molecules) (187). A genetic algorithm approach for generating optimal weighted scores of the profiles was used to rank unknown libraries of molecules for the probability of having biological activity. The authors have applied these druglike profiles to unknown sets consisting of NCI AIDS antiviral compounds, a dictionary of natural products, and the Glaxo Wellcome proprietary database, and they report three to five times enhancement over the random selection methods in finding the active molecules from inactive ones.

Ghose et al. recently described a knowledge-based approach for designing combinatorial (or large) libraries containing "druglike" molecules (184). They analyzed the physicochemical properties and structural characteristics of the molecules in a commercially available database, the Comprehensive Medicinal Chemistry (CMC) database, and developed a consensus definition of "druglike" properties of a molecule. A total of 6304 different drugs that belong to seven different groups (based on diseases they affect) were analyzed for their calculated log P, molecular refractivity (MR), molecular weight (MW), and number of atoms. The log P and MR values were calculated using the ALOGP and AMR method, respectively, of Ghose et al. (83,188). Two sets of ranges for the physicochemical properties were determined based on drug molecules available in the CMC database: (1) qualifying ranges (80% of the drugs in the database belonged to this group) and (2) preferred ranges (50% of the drugs belonged to this group). The qualifying range for calculated log P of the druglike molecules in the CMC database was -0.4 to 5.6, whereas the preferred range was 1.3–4.1. The ranges for other physicochemical properties were also determined. The ranges can be used as a guiding tool to select druglike molecules from other databases. The ranges determined for seven most important categories of drugs can be useful when one is selecting druglike molecules for those particular targets.

The authors provided a succinct picture of the physicochemical characteristics of different classes of drugs and described the characteristics of the drugs that did not fall within these values. The authors also evaluated the presence or absence of rings, special functional groups, or any particular substructure in a specific class of drugs. This information is expected to be useful in designing combinatorial libraries or selecting compounds from libraries with appropriate physicochemical properties.

A recent study (123) reported classification methods to separate active and inactive compounds in a diverse large database. A number of classification technique, (e.g., discriminant analysis, recursive partitioning and hierarchical agglomerative clustering, and standard topology-based descriptors generated using the Molconn-X program and binary structural keys from the ISIS programs) were used to identify ACE inhibitors, β-adrenergic antagonists, and H_2-receptor antagonists from the CMC database. The discriminant-based classification approach showed better results in correctly identifying active compounds, while the binary structural keys from the ISIS program showed better performance in classifying inactive compounds.

B. Molecular Diversity Analysis

Physicochemical parameters have been used to assess the diversity of combinatorial libraries (131,189). An oligo(N-substituted)glycine (NSG) peptoid combinatorial library based on the tyramine submonomer was reported by Martin et al.

(189). A set of 15–20 descriptors representing lipophilicity, shape and branching, chemical functionality, and receptor recognition descriptors, such as hydrogen bond donors, hydrogen bond acceptors, and aromatics, were used for diversity analysis of a pool of 721 primary amines and 1133 carboxylates. The D-optimal design technique (190) was used to design the library with optimal diversity.

Shemetulskis et al. used calculated physicochemical properties of chemicals, such as CLOGP, CMR, and CDM (electronic dipole moment) for two commercial databases, Chemical Abstracts structural database (CAST-3D) and Maybridge (MAY), to compare the diversity space in their corporate database (CBI) (131). A clustering technique was used to analyze the diversity. This technique, which is useful in helping to select compounds from different databases to enhance the diversity space in an existing database, may increase the chances of finding novel lead compounds.

Lewis et al. reported the use of molecular descriptors to generate a novel diversity property descriptor (DPD) code for the selection of a diverse set of compounds from combinatorial libraries for biological screening. Six descriptors (H acceptor, H donor, flexibility, electrotopological index, c log *P,* and aromatic density) were used (191).

C. Application of QSAR in Library Design

1. Inverse QSAR

An inverse QSAR method implemented in a new library design technique, known as Focus-2D, has been recently reported by Cho et al. to rationally design a virtual peptide combinatorial library (192,193). A preconstructed QSAR served as one of the methods to select compounds with high predicted activity in a virtual library. The method was validated by developing a QSAR equation using the GA-PLS (genetic algorithm–partial least-squares) method from a training set of 28 bradykinin-potentiating (BK) pentapeptides and predicting the activity of the two most active peptides from the equation. These researchers used topological descriptors, calculated by the Molconn-X program, as well as several amino acid based descriptors (Z_1, Z_2, and Z_3) (194) related to hydrophobicity; bulk, and electronic properties, respectively, and isotropic surface area (ISA) and electronic charge index (ECI) (195). Significant cross-validated correlation coefficients and low standard error of predictions were achieved with both studies (i.e., The one based on topology and the one based on amino acid descriptors). The method suggested a number of amino acids as the preferred building blocks. These amino acids were also present most frequently in the known active BK peptides. The results obtained from the training set of 28 pentapeptides were used to extrapolate on a theoretically possible (3.2 million) pentapetides and comparable results were obtained. Since the training set population was very small compared to the theoretically possible peptides, a modified "degree of fit" condition was used to control the degree of extrapolation by excluding peptides that are structurally too distant from the training peptides.

2. Artificial Neural Network (ANN)

Burden and Winkler (52) recently reported the use of a multilayered feedforward network trained by back-propagation methods to develop a benzodiazepine virtual receptor for screening a 40,000-compounds database from Maybridge. A diverse set of 321 compounds belonging to benzodiazepines, arylpyrazoloquinolines, β-carbolines, imidazopyridazines, and cyclopyrrolones with reported biological activity (pI_{50}) was separated into a training set (270 compounds), a validation set (30 compounds), and a test set (21 compounds). The input parameters for the neural nets were based on simple atomistic and functional group representations of the molecules. The 21 input parameters were $C_{(aromatic)}$, C-4, C-3, C-2, $N_{(aromatic)}$, N-3, N-2, N-1, O-2, O-1, S, P, Cl, F, Br, I, seven-membered rings, six-membered rings, five-membered rings, four-membered rings, and three-membered rings. The best results were obtained with a five-layered (21:8:5:3:1) network with a low rmse (root mean square error) value and a correlation coefficient r of 0.794.

3. Binary QSAR

Introduction of combinatorial chemistry for designing large libraries compelled researchers to discover rapid robotic methods for assaying literally millions of compounds in a short period of time. This rapid method is referred to as high throughput screening (HTS). Often, this method just generates yes/no (active/inactive; pass/fail) data, and the results are prone to error. Current QSAR methodologies require less heterogeneous compounds with continuous activity data and lower error margins to have any predictive value. To overcome methodological problems in current QSAR techniques and to handle such a huge amount of binary data from HTS, Labute (196) introduced a method termed "binary QSAR" to handle binary measurement data from HTS. Two very recent reports described the successful use of this method for analyzing large sets of binary data (197,198). This method is expected to help in extracting important structural information required for biological activity and to design more focused libraries for drug discovery. For methodological details, readers are referred to recent articles by Labute's (196) and Gao et al. (198). The flowchart of the method is shown in Fig. 2.

The performance of the QSAR model is measured by evaluating three levels of prediction from the model. If m_0 is the number of inactive compounds, m_1 is the number of active compounds, c_0 is the number of correctly predicted inac-

Structure:

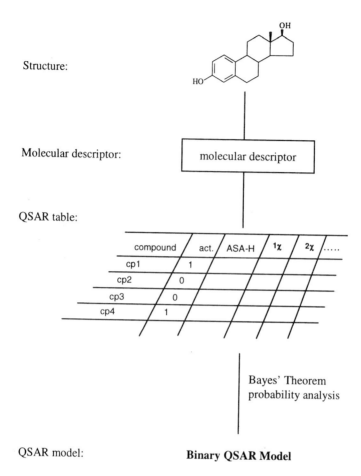

Molecular descriptor:

QSAR table:

Bayes' Theorem
probability analysis

QSAR model: **Binary QSAR Model**

Figure 2   The flowchart of binary QSAR analysis. (From Ref. 198.)

tive compounds, and c_1 is the number of correctly predicted active compounds by the model, then:

$100(c_0/m_0)$ represents the percentage accuracy for active compounds
$100(c_1/m_1)$ represents the percentage accuracy for inactive compounds
$100(c_0 + c_1)/(m_0 + m_1)$ represents the percentage of overall accuracy for all
    compounds

The first report concerning the application of binary QSAR to a drug discovery problem involving the QSAR analysis of estrogen receptor ligands was published by Gao et al. (198) in 1999. The binding data, log RBA (relative binding affinity) for 463 different estrogen analogs, were selected from the literature and converted to binary data based on certain threshold criteria of activity. Specifically, if the activity of certain compounds was equal to or greater than the threshold value, any such compound was considered to be active; the rest were classified as inactive. Four hundred and ten compounds were selected for the training set and 53 were used as a test set to validate the model. Thirteen different molecular descriptors were used: four Kier and Hall connectivity indices (χ), three Kier shape indices (κ), one flexibility index (Φ), one total hydrophobic surface area (ASA-H), one aromatic bond (b-Ar) descriptor, one charge descriptor (peoe-PC$^+$), and two indicator variables. Principal component analysis (PCA) was used with a smoothing function to derive the model. The best non-cross-validated model accurately predicted 85% of the active and 93% of the inactive, whereas the cross-validated model predicted 76% of active and 93% of inactive compounds. The model was validated on the test set, and 78% of the active compounds and 98% of the inactive compounds were correctly predicted with an overall accuracy of 94%.

The advantages and disadvantages of the binary QSAR method also were addressed by Labute (196). The method should be useful for selecting (prioritizing) compounds for HTS, for designing focused combinatorial libraries, and for screening and synthesizing virtual libraries. One major drawback of this method is the difficulty of interpreting the importance of descriptors in developing the model.

VII. SOME EXAMPLES OF SUCCESSFUL QSAR APPLICATIONS

The success of QSAR methodologies should not be judged by counting the number of drugs or pesticides that it helped to put on the market. QSAR plays a multifaceted role in a number of stages in the drug design process, and thus any contribution of QSAR along this path should be considered a success. Moreover, drug discovery usually takes place in pharmaceutical companies, which rarely report successes in this area due to confidentiality restrictions. Nevertheless, it is worthwhile to briefly list (Table 3) some of the successful applications of QSAR known from published data, to give additional background to readers new to this field. For detailed accounts of successful applications of QSAR methodologies, readers may consult Refs. 101 and 199–203.

VIII. LIST OF SOFTWARE/DATABASES RELEVANT FOR QSAR

Table 4 lists software relevant to QSAR studies, along with capabilities and sources.

Table 3 Successful Applications of QSAR in Developing Drugs and Herbicides

Type of agent (proprietary name)	QSAR method used	Stage of development reached
Antiallergic: purinones (Zaprinast)	Hansch type	Clinical trial
Antiallergic: pyranemaines	Hansch type	Clinical trial
Cerebral vasodialator: benzyldiphenylmethylpiperazines	Hansch type	The QSAR study helped to avoid redundant synthesis (the best candidate designed by QSAR had to be dropped because of toxicity)
Antibacterial: quinolone carboxylic acid (Norfloxacin)	Hansch type	Marketed drug
Antihypertensive: quinazolines (MY-5561)	Hansch type	Preclinical
Anti-inflammatory: furoindole carboxamides	Hansch type	Lead optimization
Anticancer: acridines (*m*-AMSA)	Hansch type	Marketed drug
Antiulcer: (Cimetidine)	Quantitative comparison of physicochemical properties and bioisosterism	Marketed drug
Diuretics: 3-amino-1-benzylpyrazolin-5-one (Muzolimine)	Topliss procedure (nonstatistical)	Marketed drug
Cardioselective β-adrenergic blocker: (RS-51288)	Hansch type	Clinical trial
Cardiotonic: (AR-L115)	Cluster analysis	Advanced clinical trial
Alzheimer disease: donezepil hydrochloride (Aricept)	QSAR, molecular modeling, molecular shape analysis, docking	Marketed drug
Antimigraine: (Lomerizine)	QSAR	Marketed drug
Herbicide: N-benzylacylamide (Bromobutide)	Hansch type	Marketed herbicide
Herbicide: triazinone (Metamitron)	Hansch type	Marketed herbicide

Table 4 List of QSAR-Related Software, Capabilities, and Sources

Software/database	QSAR-related capabilities	Source (URL)
ACD/Physicochemical Laboratory	Calculates pK_a, log P, log D, solubility, boiling point and vapor pressure, Hammett electronic constants, liquid properties (refraction, density, molar volume, etc.)	Advanced Chemistry Development, Inc. (www.acdlabs.co.uk)
ADAPT	Calculates descriptors, develops QSAR	Academic; available from Dr. Peter C. Jurs (zeus.chem.psu.edu)
C²*QSAR	Creation of QSAR	Molecular Simulations, Inc. (www.msi.com)
ChemPlus	QSAR properties	Hypercube, Inc. (www.hyper.com)
CLOGP C-QSAR	Calculates log P Database of QSAR models, develops QSAR	Biobyte (www.biobyte.com)
CODESSA	Descriptor generator, QSAR	Semicon (www.semichem.com)
Galaxy	Property calculation and QSAR	AM Technologies, Inc. (www.am-tech.com)
HINT Molconn-Z	Calculates log P, generates molecular connectivity, shape, and information indices for QSAR	eduSoft, LC (www.eslc.vabiotech.com)
Kow Win	Calculates log P	Environmental Research Center of Syracuse Research Corporation (esc_plaza.syrres.com)
M-CASE, KLOGP	Develops QSAR, calculates log P	Multicase, Inc. (www.multicase.com)
Molecular Analysis Pro	Calculates molecular properties; QSAR	ChemSW, Inc. (www.chemsw.com)
pKalc, PrologP, PrologD	Calculates pK_a, log P, and log D	Compudrug International, Inc. (www.compudrug.com)
PCModels	log P and polarizability prediction	Daylight Chemical Information Systems, Inc. (www.daylight.com)

continues

Table 4 *Continued*

Software/database	QSAR-related capabilities	Source (URL)
POLLY	Calculates molecular indices	Contact Dr. Subhash C. Basak, University of Minnesota[a]
ProChemist	Calculates log P, pKa, solubility; performs QSAR	Cadcom (hom.worldnet.fr/cadcom)
QlogP	Calculates log P	Contact Nicholas Bodor, University of Florida[b]
QuaSAR-Binary	Binary QSAR	Chemical Computing Group, Inc. (www.chemcomp.com)
ScilogP	Calculates log P	SciVision
SciQSAR	Creation of QSAR/QSPR	(www.scivision.com)
SIMCA	Multivariate data analysis	Umetrics (www.umetrics.com)
Sybyl QSAR	Creation of QSAR	Tripos, Inc.
HQSAR	Hologram of QSAR	(www.tripos.com)
TOPIX, CLUSTER	Calculates topological descriptors; performs cluster analysis	Software Development Lohninger (SDL) (www.lohninger.com)
Tsar	Creation of QSAR	Oxford Molecular Group (www.oxmol.com)
XlogP	Calculates log P	Academic (www.ipc.pku.edu.cn)

[a] e-mail address: slasak@wyle.nrri.vmn.edu
[b] e-mail address: bodorn@nervn.nerdc.ufl.edu

IX. CONCLUSIONS

The QSAR paradigm that started about 40 years ago continues to make its mark in almost every area of the drug design process involving chemical library design and the understanding of physicochemical properties of drugs and their pharmacokinetics, drug actions and mechanisms, or toxicity. Since the introduction of hydrophobicity in drug research by Hansch, we have learned to consider hydrophobicity in the transport of drugs and their bioavailability, in drug–receptor interactions, and in many other aspects of drug actions. Lipinski's "rule of five" is now in everybody's mind during the design of a drug or the selection of a drug from a database. We have learned the concept of optimum values for hydrophobicity in certain drug actions (e.g., drugs acting on the central nervous system). Hammett's σ parameter in organic chemistry encouraged us to think about biological actions in terms of electronic effect and facilitated drug design. The concept of topology has been widely used in drug design. All these and many other principles described in this chapter contributed to the concept of rational design of drugs. Therefore, the success of QSAR should not and must not be judged by its predictive potential for designing new drugs. The true success of QSAR lies in the fact that the method fundamentally helped to motivate researchers to think rationally and to apply knowledge from multidisciplinary areas to design effective drugs. As we move through the new millennium, and the Human Genome Project matures, we will see the emergence of new targets for drug discovery. There will be a pressing need to discover new and more selective drugs. QSAR methods, if applied judiciously, will play an even more important and critical role in the drug discovery cycle.

ACKNOWLEDGMENTS

I wish to thank Drs. A. Robert Neurath, Cathy Falk, Cynthia Selassie, and Corwin Hansch for thoroughly reading the manuscript and their critical comments and Jean Ng. Pack for preparing the figures. This work was supported by grants from Phillip Morris, Inc., Hugoton Foundation, and Johnson & Johnson.

REFERENCES

1. Crum-Brown A, Fraser TR. On the connection between chemical constitution and physiological action. I. On the physiological action of the salts of the ammonia bases, derived from strychnia, brucia, thebaia, codeia, morphia, and nocotia. Trans R Soc Edinburgh 1868; 25:151–203.
2. Richet C. On the relationship between the toxicity and the physical properties of substances. CR Seances Soc Biol Ses Fil 1893; 9:775–776.
3. Meyer H. On the theory of alcohol narcosis. I. Which property of anaesthetics gives them their narcotic activity? Arch Exp Path Pharmakol 1899; 42:109–118.
4. Overton E. Studien über die Narkose, zugleich ein Beitrag zur allgemeinen Pharmakologie. Jena: 1901.
5. Traube J. Arch Physio 1904; 105:541.
6. Seidell A. A new bromine method for the determination of thymol, salicylates, and similar compounds. Am Chem J 1912; 47:508–526.
7. Ferguson J. The use of chemical potentials as indices of toxicity. Proc R Soc London Ser B 1939; 127:387–403.
8. Bell PH, Roblin RO. Studies in chemotherapy. VII. A theory of the relation of structure to activity of sulfanilamide type compounds. J Am Chem Soc 1942; 64:2905–2917.
9. Hammett LP. Physical Organic Chemistry. 2nd ed. New York: McGraw-Hill, 1970.
10. Hansen OR. Hammett series with biological activity. Acta Chem Scand 1962; 16:1593–1600.

11. Zahradnik R. Influence of the structure of aliphatic substituents on the magnitude of the biological effect of substances. Arch Int Pharmacodyn Ther 1962; 135:311–329.

12. Hansch C, Maloney PP, Fujita T, Muir RM. Correlation of biological activity of phenoxyacetic acids with Hammett substituent constants and partition coefficients. Nature 1962; 194:178–180.

13. Hansch C, Fujita T. ρ-σ-π Analysis. A method for the correlation of biological activity and chemical structure. J Am Chem Soc 1964; 86:1616–1626.

14. Free SM, Wilson JW. A mathematical contribution to structure–activity studies. J Med Chem 1964; 7:395–399.

15. Topliss JG, Costello RJ. Chance correlations in structure–activity studies using multiple regression analysis. J Med Chem 1972; 15:1066–1068.

16. Kubinyi H. The Free–Wilson method and its relationship to the extrathermodynamic approach. In: Hansch C, Sammes PG, Taylor JB, Ramsden CA, eds. Comprehensive Medicinal Chemistry. Vol 4: Quantitative Drug Design. Oxford: Pergamon Press, 1990, pp. 589–643.

17. Fujita T, Ban T. Structure–activity study of phenethylamines as substrates of biosynthetic enzymes of sympathetic transmitters. J Med Chem 1971; 14:148–152.

18. Singh P, Kumar R. 1,2-Diarylimidazoles as inhibitors of cyclooxygenase-2: A quantitative structure–activity relationship study. J Enzyme Inhib 1999; 14:277–288.

19. Gupta SP, Paleti A. Quantitative structure–activity relationship studies on some nonbenzodiazepine series of compounds acting at the benzodiazepine receptor. Bioorg Med Chem 1998; 6:2213–2218.

20. Hernández-Gallegos Z, Lehmann PA. A Free–Wilson/Fujita–Ban analysis and prediction of the analgesic potency of some 3-hydroxy- and 3-methoxy-N-alkylmorphinan-6-one opioids. J Med Chem 1990; 33:2813–2817.

21. Higginbottom M, Kneen C, Ratcliffe GS. Rationally designed dipeptoid analogues of CCK. A Free–Wilson/Fujita–Ban analysis of some α-methyltryptophan derivatives as CCK-B antagonists. J Med Chem 1992; 35:1572–1577.

22. Kubinyi H. Quantitative structure–activity relationships. V. A simple algorithm for Fujita–Ban and Free–Wilson analyses. Arzneim Forsch 1977; 27:750–758.

23. Smith RN, Hansch C, Kim KH, Omiya B, Fukumura G, Selassie CD, Jow PY, Blaney JM, Langridge R. The use of crystallography, graphics, and quantitative structure–activity relationships in the analysis of the papain hydrolysis of X-phenyl hippurates. Arch Biochem Biophys 1982; 215:319–328.

24. Hansch C, Smith RN, Rockoff A, Calef DF, Jow PY, Fukunaga JY. Structure–activity relationships in papain and bromelain ligand interactions. Arch Biochem Biophys 1977; 183:383–392.

25. Debnath AK, de Compadre RL, Hansch C. Mutagenicity of quinolines in *Salmonella typhimurium* TA100. A QSAR study based on hydrophobicity and molecular orbital determinants. Mutat Res 1992; 280:55–65.

26. Hansch C, Clayton JM. Lipophilic character and biological activity of drugs. II. The parabolic case. J Pharm Sci 1973; 62:1–21.

27. Chalmers DK, Scholz GH, Topliss DJ, Kolliniatis E, Munro SL, Craik DJ, Iskander MN, Stockigt JR. Thyroid hormone uptake by hepatocytes: Structure–activity relationships of phenylanthranilic acids with inhibitory activity. J Med Chem 1993; 36:1272–1277.

28. CLOGP. V. 4.0. Biobyte Corporation, California, 1999.

29. QSAR Database. Pomona College, California, 1999.

30. Kubinyi H. Quantitative structure–activity relationships. 7. The bilinear model, a new model for nonlinear dependence of biological activity on hydrophobic character. J Med Chem 1977; 20:625–629.

31. Kubinyi H. Nonlinear dependence of biological activity on hydrophobic character: The bilinear model. Farmaco [Sci] 1979; 34:248–276.

32. Kubinyi H. Lipophilicity and biological activity. Drug transport and drug distribution in model systems and in biological systems. Arzneim Forsch 1979; 29: 1067–1080.

33. Garg R, Gupta SP, Gao H, Babu MS, Debnath AK, Hansch C. Comparative quantitative structure–activity relationship studies on anti-HIV drugs. Chem Rev 1999; 99:3525–3602.

34. Lam PY, Ru Y, Jadhav PK, Aldrich PE, DeLucca GV, Eyermann CJ, Chang CH, Emmett G, Holler ER, Daneker WF, Li L, Confalone PN, McHugh RJ, Han Q, Li R, Markwalder JA, Seitz SP, Sharpe TR, Bacheler LT, Rayner MM, Klabe RM, Shum L, Winslow DL, Kornhauser DM, Hodge CN. Cyclic HIV protease inhibitors: Synthesis, conformational analysis, P2/P2′ structure–activity relationship, and molecular recognition of cyclic ureas. J Med Chem 1996; 39:3514–3525.

35. Klopman G. MULTICASE 1. A hierarchical computer automated structure evaluation program. Quant Struct-Act Relat 1992; 11:176–184.

36. Klopman G, Ptchelintsev D. Antifungal triazole alcohols: A comparative analysis of structure–activity, structure–teratogenicity and structure–therapeutic index relationships using the Multiple Computer-Automated Structure Evaluation (Multi-CASE) methodology. J Comput Aided Mol Design 1993; 7:349–362.

37. Klopman G, Tu M. Diversity analysis of 14,156 molecules tested by the National Cancer Institute for anti-HIV activity using the quantitative structure–activity relational expert system MCASE. J Med Chem 1999; 42:992–998.

38. Manallack DT, Livingstone DJ. Neural network in drug discovery: Have they lived up to their promise? Eur J Med Chem 1999; 34:195–208.

39. Breindl A, Beck B, Clark T, Glen RC. Prediction of the N-octanol/water partition coefficient, log P, using a combination of semiempirical MO-calculation and a neural network. J Mol Model 1997; 3:142–155.

40. de Gregorio C, Kier LB, Hall LH. QSAR modeling with the electrotopological state indices: Corticosteroids. J Comput Aided Mol Design 1998; 12:557–561.

41. King RD, Hirst JD, Sternberg MJE. New approaches to QSAR: Neural networks and machine learning. Perspect Drug Discovery Design 1993; 1:279–290.

42. Ajay. A unified framework for using neural networks to build QSARs. J Med Chem 1993; 36:3565–3571.

43. Ajay, Walters WP, Murcko MA. Can we learn to distinguish between drug-like and nondrug-like molecules? J Med Chem 1998; 41:3314–3324.

44. Brinn M, Walsh PT, Payne MP, Bott B. Neural network classification of mutagens using structural fragment data. SAR QSAR Environ Res 1993; 1:169–210.

45. Duprat AF, Huynh T, Dreyfus G. Toward a principled methodology for neural network design and performance evaluation in QSAR. Application to the prediction of log P. J Chem Inf Comput Sci 1998; 38:586–594.

46. Hosseini M, Madalena DJ, Spence I. Using artificial neural networks to classify the activity of capsaicin and its analogues. J Chem Inf Comput Sci 1997; 37:1129–1137.
47. Manallack DT, Ellis DD, Livingstone DJ. Analysis of linear and nonlinear QSAR data using neural networks. J Med Chem 1994; 37:3758–3767.
48. So SS, Karplus M. Evolutionary optimization in quantitative structure–activity relationship: An application of genetic neural networks. J Med Chem 1996; 39: 1521–1530.
49. Tang Y, Wang HW, Chen KX, Ji RY. QSAR of 3-methylfentanyl derivatives studied with neural networks method. Chung Kuo Yao Li Hsueh Pao 1995; 16:26–32.
50. Tetko IV, Villa AE, Livingstone DJ. Neural network studies. 2. Variable selection. J Chem Inf Comput Sci 1996; 36:794–803.
51. Good AC, So SS, Richards WG. Structure–activity relationships from molecular similarity matrices. J Med Chem 1993; 36:433–438.
52. Burden FR, Winkler DA. New QSAR methods applied to structure–activity mapping and combinatorial chemistry. J Chem Inf Comput Sci 1999; 39:236–242.
53. Burden FR, Winkler DA. Robust QSAR models using Bayesian regularized neural networks. J Med Chem 1999; 42:3183–3187.
54. Naumann T, Lowis DR. A new, highly predictive QSAR technique. First International Electronic Conference on Synthetic Organic Chemistry (ECSOC-1) 1997.
55. Tong W, Lowis DR, Perkins R, Chen Y, Welsh WJ, Goddette DW, Heritage TW, Sheehan DM. Evaluation of quantitative structure–activity relationship methods for large-scale prediction of chemicals binding to the estrogen receptor. J Chem Inf Comput Sci 1998; 38:669–677.
56. Winkler DA, Bureau R. Holographic QSAR of benzodiazepines. Quant Struct-Act Relat 1998; 17:224–231.
57. Heritage TW, Lowis DR. Molecular hologram QSAR. In: Parrill AL, Reddy MR, eds. Rational Drug Design Novel Methodology and Practical Applications. Washington, DC: American Chemical Society, 1999, pp 212–225.
58. Hansch C, Hoekman D, Gao H. Comparative QSAR: Toward a deeper understanding of chemicobiological interactions. Chem Rev 1999; 96:1045–1075.
59. Hansch C. The use of sigma in structure–activity correlations. J Med Chem 1970; 13:964–966.
60. Hansch C, Leo A, Taft RW. A survey of Hammett substituent constants and resonance and field parameters. Chem Rev 1999; 91:165–195.
61. Compadre CM, Hansch C, Klein TE, Petridou-Fischer J, Selassie CD, Smith RN, Steinmetz W, Yang CZ, Yang GZ. Separation of electronic and hydrophobic effects for the papain hydrolysis of substituted N-benzoylglycine esters. Biochim Biophys Acta 1991; 1079:43–52.
62. Hansch C, Venger BH, Panthananickal A. Mutagenicity of substituted (o-phenylenediamine)platinum dichloride in the Ames test. A quantitative structure– activity analysis. J Med Chem 1980; 23:459–461.
63. Hansch C, Zhang L. Comparative QSAR: Radical toxicity and scavenging. Two different sides of the same coin. SAR QSAR Environ Res 1995; 4:73–82.
64. Taft RW. Steric Effects in Organic Chemistry. New York: Wiley, 1956.
65. Swain CG, Lupton EC. Field and resonance components of substituent effects. J Am Chem Soc 1968; 90:4328–4337.

66. Charton M. Electrical effect substituent constants analysis. Prog Phys Org Chem 1981; 13:119–251.

67. Hansch C, Leo A, Hoekman D. Exploring QSAR. Vol. 2. Hydrophobic, Electronic, and Steric Constants. Washington, DC: American Chemical Society, 1995.

68. Tuppurainen K, Lötjönen S, Laatikainen R, Vartiainen T. Structural and electronic properties of MX compounds related to TA100 mutagenicity. A semi-empirical molecular orbital QSAR study. Mutat Res 1992; 266:181–188.

69. Lien EJ, Kumler WD. Dipole moments and pharmacological activity of cyclic ureas, cyclic thioureas, and the N, N'-dimethylated compounds. J Med Chem 1968; 11:214–219.

70. Lien EJ, Kumler WD. Dipole moment and structure of thiophene derivatives and benzene analogs. J Pharm Sci 1970; 59:1685–1688.

71. Inoue S. On the prediction of pK_a values of amino sugars. Chem Pharm Bull (Tokyo) 1968; 16:1134–1137.

72. Neiman Z, Quinn FR. Quantitative structure–activity relationships of purines. I. Choice of parameters and prediction of pK_a values. J Pharm Sci 1981; 70:425–430.

73. Debnath AK, Hansch C, Kim KH, Martin YC. Mechanistic interpretation of the genotoxicity of nitrofurans (antibacterial agents) using quantitative structure–activity relationships and comparative molecular field analysis. J Med Chem 1993; 36:1007–1016.

74. Blin N, Federici C, Koscielniak T, Strosberg AD. Predictive quantitative structure–activity relationships (QSAR) analysis of β-3-adrenergic ligands. Drug Design Discovery 1995; 12:297–311.

75. Bearden AP, Schultz TW. Comparison of *Tetrahymena* and *Pimephales* toxicity based on mechanism of action. SAR QSAR Environ Res 1998; 9:127–153.

76. Mekenyan OG, Veith GD. The electronic factor in QSAR: MO-parameters, competing interactions, reactivity and toxicity. SAR QSAR Environ Res 1994; 2:129–143.

77. Bikker JA, Kubanek J, Weaver DF. Quantum pharmacologic studies applicable to the design of anticonvulsants: Theoretical conformational analysis and structure–activity studies of barbiturates. Epilepsia 1994; 35:411–425.

78. Kwiatkowski W, Karolak-Wojciechowska J. Use of molecular electrostatic potentials for analysis of anticonvulsant activities of phenylsuccinimides. SAR QSAR Environ Res 1993; 1:233–244.

79. Taylor PJ. Hydrophobic properties of drugs. In: Hansch C, Sammes PG, Taylor JB, Ramsden CA, eds. Comprehensive Medicinal Chemistry. Vol. 4. Quantitative Drug Design. 1st ed. Oxford: Pergamon Press, 1990, pp 241–294.

80. Hansch C, Leo A. Exploring QSAR-Fundamentals and Applications in Chemistry and Biology. Washington, DC: American Chemical Society, 1995.

81. Rekker RF. The Hydrophobic Fragment Constants. Its Derivation and Application. A Means of Characterizing Membrane Systems. Amsterdam: Elsevier, 1977.

82. el Tayar N, Tsai RS, Testa B, Carrupt PA, Leo A. Partitioning of solutes in different solvent systems: The contribution of hydrogen-bonding capacity and polarity. J Pharm Sci 1991; 80:590–598.

83. Ghose AK, Viswanadhan VN, Wendoloski JJ. Prediction of hydrophobic properties of small organic molecules using fragmental methods: An analysis of ALOGP and CLOGP methods. J Phys Chem 1998; 102:3762–3772.

84. Bodor N, Gabanyi Z, Wong C-K. A new method for the estimation of partition coefficient. J Am Chem Soc 1989; 111:3783–3786.
85. Klopman G, Wang S. A computer automated structure evaluation (CASE) approach to the calculation of partition coefficients. J Comput Chem 1991; 12:1025–1032.
86. Moriguchi I, Hirono S, Liu Q, Nakagome I, Matsushita Y. Simple method of calculating octanol/water partition coefficient. Chem Pharm Bull (Tokyo) 1992; 40:127–130.
87. Sinclair J, Cornell NW, Zaitlin L, Hansch C. Induction of cytochrome P-450 by alcohols and 4-substituted pyrazoles. Comparison of structure–activity relationships. Biochem Pharmacol 1986; 35:707–710.
88. McKarns SC, Hansch C, Caldwell WS, Morgan WT, Moore SK, Doolittle DJ. Correlation between hydrophobicity of short-chain aliphatic alcohols and their ability to alter plasma membrane integrity. Fundam Appl Toxicol 1997; 36:62–70.
89. van Steen BJ, van Wijngaarden I, Tulp MT, Soudijn W. A series of N4-imidoethyl derivatives of 1-(2,3-dihydro-1,4-benzodioxin-5-yl)piperazine as 5-HT1A receptor ligands: Synthesis and structure–affinity relationships. J Med Chem 1995; 38:4303–4308.
90. CLOGP. Daylight C.I.S. Inc., California, 1999.
91. Yamagami C, Takami H, Yamamoto K, Miyoshi K, Takao N. Hydrophobic properties of anticonvulsant phenylacetanilides. Relationship between octanol–water partition coefficient and capacity factor determined by reversed-phase liquid chromatography. Chem Pharm Bull (Tokyo) 1984; 32:4994–5002.
92. Yamagami C, Fujita T. Hydrophobicity parameters of heteroaromatic compounds derived from various partitioning systems. In: Hansch C, Fujita T, eds. Classical and Three-Dimensional QSAR in Agrochemistry. Washington, DC: American Chemical Society, 1995, pp 36–47.
93. Altomare C, Tsai RS, el Tayar N, Testa B, Carotti A, Cellamare S, De Benedetti PG. Determination of lipophilicity and hydrogen-bond donor acidity of bioactive sulphonyl-containing compounds by reversed-phase HPLC and centrifugal partition chromatography and their application to structure–activity relations. J Pharm Pharmacol 1991; 43:191–197.
94. Altomare C, Carotti A, Cellamare S, Carrieri A, Ciabatti R, Malabarba A. Lipophilicity of teicoplanin antibiotics as assessed by reversed phase high-performance liquid chromatography: Quantitative structure–property and structure–activity relationships. J Pharm Pharmacol 1994; 46:994–999.
95. Biagi GL, Barbaro AM, Guerra MC, Babbini M, Gaiardi M, Bartoletti M, Borea PA. R_m values and structure–activity relationship of benzodiazepines. J Med Chem 1980; 23:193–201.
96. McFarland JW, Berger CM, Froshauer SA, Hayashi SF, Hecker SJ, Jaynes BH, Jefson MR, Kamicker BJ, Lipinski CA, Lundy KM, Reese CP, Vu CB. Quantitative structure–activity relationships among macrolide antibacterial agents: In vitro and in vivo potency against Pasteurella multocida. J Med Chem 1997; 40:1340–1346.
97. Franke R, Gruska A, Presber W. Combined factor and QSAR analysis for antibacterial and pharmacokinetic data from parallel biological measurements. Pharmazie 1994; 49:600–605.
98. Medchem/Biobyte QSAR Database. Pomona College, California, 1999.

99. Hancock CK, Meyer EA, Yager BJ. Quantitative separation of hyperconjugation effects from steric substituent constants. J Am Chem Soc 1961; 83:4211–4213.

100. Fujita T, Iwamura H. Applications of various steric constants to quantitative analysis of structure–activity relationships. Topics Curr Chem 1983; 114:119–157.

101. Fujita T. The extrathermodynamic approach to drug design. In: Hansch C, Sammes PG, Taylor JB, Ramsden CA, eds. Comprehensive Medicinal Chemistry. Vol. 4. Quantitative Drug Design. 1st ed. Oxford: Pergamon Press, 1990, pp 497–560.

102. Hansch C, Caldwell J. The structure–activity relationship of inhibitors of serotonin uptake and receptor binding. J Comput Aided Mol Design 1991; 5:441–453.

103. Verloop A, Hoogenstraaten W, Tipker J. Development and application of new steric substituent parameters in drug design. In: Ariens EJ, ed. Drug Design. New York: Academic Press, 1976, pp 165–207.

104. Verloop A. The STERIMOL Approach to Drug Design. New York: Dekker, 1987.

105. Gao H, Katzenellenbogen JA, Garg R, Hansch C. Comparative QSAR analysis of estrogen receptor ligands. Chem Rev 1999; 99:723–744.

106. Lien EJ, Gao H. QSAR analysis of skin permeability of various drugs in man as compared to in vivo and in vitro studies in rodents. Pharm Res 1995; 12:583–587.

107. Ren S, Das A, Lien EJ. QSAR analysis of membrane permeability to organic compounds. J Drug Target 1996; 4:103–107.

108. Selassie CD, Hansch C, Khwaja TA. Structure–activity relationships of antineoplastic agents in multidrug resistance. J Med Chem 1990; 33:1914–1919.

109. Lewis DF, Ioannides C, Parke DV. A quantitative structure–activity relationship study on a series of 10 para-substituted toluenes binding to cytochrome P4502B4 (CYP2B4), and their hydroxylation rates. Biochem Pharmacol 1995; 50:619–625.

110. Fujita T, Nishioka T, Nakajima M. Hydrogen-bonding parameter and its significance in quantitative structure–activity studies. J Med Chem 1977; 20:1071–1081.

111. Charton M, Charton BI. The structural dependence of amino acid hydrophobicity parameters. J Theor Biol 1982; 99:629–644.

112. Yang G, Lien EJ, Guo Z. Physical factors contributing to hydrophobic constant π. Quant Struct-Act Relat 1986; 5:12–18.

113. Abraham MH, Duce PP, Prior DV, Barratt DG, Morris JJ, Taylor PJ. Hydrogen bonding. 9. Solute proton donor and proton acceptor scales for use in drug design. J Chem Soc Perkin Trans II 1989; 1355–1375.

114. Dearden JC, Ghafourian T. Hydrogen bonding parameter for QSAR: Comparison of indicator variables, hydrogen bond counts, molecular orbital and other parameters. J Chem Inf Comput Sci 1999; 39:231–235.

115. Kier LB, Hall LH, Murray WJ, Randic M. Molecular connectivity. I. Relationship to nonspecific local anesthesia. J Pharm Sci 1975; 64:1971–1974.

116. Gough JD, Hall LH. Modeling antileukemic activity of carboquinones with electrotopological state and chi indices. J Chem Inf Comput Sci 1999; 39:356–361.

117. Balaban AT, Mills D, Basak SC. Correlation between structure and normal boiling points of acyclic carbonyl compounds. J Chem Inf Comput Sci 1999; 39:758–764.

118. Cramer RD. Partial least squares (PLS): Its strength and limitations. Perspect Drug Discovery Design 1993; 1:269–278.

119. Wold S, Johansson E, Cocchi M. PLS—Partial least-squares projections to latent structures. In: Kubinyi H, ed. 3D QSAR in Drug Design—Theory, Methods, and Applications. Leiden: ESCOM, 1993, pp 523–550.

120. Chen BK, Horváth C, Bertino JR. Multivariate analysis and quantitative structure–activity relationships. Inhibition of dihydrofolate reductase and thymidylate synthetase by quinazolines. J Med Chem 1979; 22:483–491.

121. Cronin MT, Basketter DA. Multivariate QSAR analysis of a skin sensitization database. SAR QSAR Environ Res 1994; 2:159–179.

122. Dearden JC. Descriptors and techniques for quantitative structure–biodegradability studies. SAR QSAR Environ Res 1996; 5:17–26.

123. Dixon SL, Villar HO. Investigation of classification methods for the prediction of activity in diverse chemical libraries. J Comput Aided Mol Design 1999; 13:533–545.

124. Liu Q, Hirono S, Moriguchi I. Quantitative structure–activity relationships for calmodulin inhibitors. Chem Pharm Bull (Tokyo) 1990; 38:2184–2189.

125. Sugai S, Murata K, Kitagaki T, Tomita I. Studies on eye irritation caused by chemicals in rabbits. 1. A quantitative structure–activity relationships approach to primary eye irritation of chemicals in rabbits. J Toxicol Sci 1990; 15:245–262.

126. Knight JL, Weaver DF. A computational quantitative structure–activity relationship study of carbamate anticonvulsants using quantum pharmacological methods. Seizure 1998; 7:347–354.

127. Lindgren F, Sjöström M, Berglind R, Nyberg B. Modelling of the biological activity for a set of ceramic fibre materials: A QSAR study. SAR QSAR Environ Res 1996; 5:299–310.

128. Brown RD, Martin YC. An evaluation of structural descriptors and clustering methods for use in diversity selection. SAR QSAR Environ Res 1998; 8:23–39.

129. Campbell P, Srinivasan R, Knoell T, Phipps D, Ishida K, Safarik J, Cormack T, Ridgway H. Quantitative structure–activity relationship (QSAR) analysis of surfactants influencing attachment of a *Mycobacterium* sp. to cellulose acetate and aromatic polyamide reverse osmosis membranes. Biotechnol Bioeng 1999; 64: 527–544.

130. Cronin MT. The use of cluster significance analysis to identify asymmetric QSAR data sets in toxicology. An example with eye irritation data. SAR QSAR Environ Res 1996; 5:167–175.

131. Shemetulskis NE, Dunbar JBJ, Dunbar BW, Moreland DW, Humblet C. Enhancing the diversity of a corporate database using chemical database clustering and analysis. J Comput Aided Mol Design 1995; 9:407–416.

132. van de Waterbeemd H, Testa B, Marrel C, Cooper DR, Jenner P, Marsden CD. Multivariate statistical analysis of L-dopa esters as potential anti-parkinsonian prodrugs. Drug Design Delivery 1987; 2:135–143.

133. Devillers J, Domine D, Bintein S. Multivariate analysis of the first 10 MEIC chemicals. SAR QSAR Environ Res 1994; 2:261–270.

134. Mannhold R, Cruciani G, Dross K, Rekker R. Multivariate analysis of experimental and computational descriptors of molecular lipophilicity. J Comput Aided Mol Design 1998; 12:573–581.

135. Van der Graaf PH, Nilsson J, Van Schaick EA, Danhof M. Multivariate quantitative structure–pharmacokinetic relationships (QSPKR) analysis of adenosine A1 receptor agonists in rat. J Pharm Sci 1999; 88:306–312.

136. Hirono S, Nakagome I, Hirano H, Matsushita Y, Yoshii F, Moriguchi I. Non-congeneric structure–pharmacokinetic property correlation studies using fuzzy adaptive least-squares: Oral bioavailability. Biol Pharm Bull 1994; 17:306–309.

137. Hirono S, Nakagome I, Hirano H, Yoshii F, Moriguchi I. Non-congeneric structure–pharmacokinetic property correlation studies using fuzzy adaptive least-squares: Volume of distribution. Biol Pharm Bull 1994; 17:686–690.

138. Wiese M, Schaper KJ. Application of neural networks in the QSAR analysis of percent effect biological data: Comparison with adaptive least squares and nonlinear regression analysis. SAR QSAR Environ Res 1993; 1:137–152.

139. Topliss JG, ed. Quantitative Structure–Activity Relationships of Drugs. New York: Academic Press, 1983.

140. Martin YC, Kutter E, Austel V, eds. Modern Drug Research. Path to Better and Safer Drugs. New York: Dekker, 1989.

141. Hansch C, Fujita T, eds. Classical and Three-Dimensional QSAR in Agrochemistry. Washington, DC: American Chemical Society, 1994.

142. Fujita T, ed. QSAR and Drug Design: New Developments and Applications. Amsterdam: Elsevier, 1995.

143. Devillers J, ed. Comparative QSAR. Washington, DC: Taylor & Francis, 1997.

144. Parrill AL, Reddy MR, eds. · Rational Drug Design—Novel Methods and Practical Applications. Washington, DC: American Chemical Society, 1999.

145. Kubinyi H. QSAR: Hansch Analysis and Related Approaches. In: Mannhold R, Krogsgaard-Larsen P, Timmerman H, eds. Methods and Principles in Medicinal Chemistry. Vol. 1. Weinheim: VCH, 1993.

146. Kier LB, Hall LH. Molecular connectivity in structure–activity analysis. New York: Wiley, 1986.

147. Katritzky AR, Lobanov VS, Karelson M. QSPR: The correlation and quantitative prediction of chemical and physical properties from structure. Chem Soc Rev 1995; 279–287.

148. Katritzky AR, Karelson M, Lobanov VS. QSPR as a means of predicting and understanding chemical and physical properties in terms of structure. Pure Appl Chem 1997; 69:245–248.

149. Katritzky AR, Wang Y, Sild S, Tamm T. QSPR studies on vapor pressure, aqueous solubity, and the prediction of water–air partition coefficients. J Chem Inf Comput Sci 1998; 38:720–725.

150. Huuskonen J, Salo M, Taskinen J. Aqueous solubility prediction of drugs based on molecular topology and neural network modeling. J Chem Inf Comput Sci 1998; 38:450–456.

151. Suttter JM, Jurs PC. Prediction of aqueous solubility for a diverse set of heteroatom-containing organic compounds using a quantitative structure–property relationship. J Chem Inf Comput Sci 1996; 36:100–107.

152. Mor M, Bordi F, Silva C, Rivara S, Crivori P, Plazzi PV, Ballabeni V, Caretta A, Barocelli E, Impicciatore M, Carrupt PA, Testa B. H3-receptor antagonists: synthesis and structure–activity relationships of para- and meta-substituted 4(5)-phenyl-2-[[2-[4(5)-imidazolyl]ethyl]thio]imidazoles. J Med Chem 1997; 40:2571–2578.

153. Katritzky AR, Lobanov VS, Karelson M. CODESSA, (2.0) Semichem, Kansas City, MO, 1994.

154. Kamenska V, Ivanov T, Nedyalkova Z, Petkov O, Lutekov G, Taskov M, Nikolov G, Mekenyan O. Computer design and syntheses of antiulcer compounds. 1st communication: *N*-[3-[3-(1-piperidinomethyl)phenoxy]propyl]amines and benzamides. Arzneim Forsch 1996; 46:1090–1095.

155. Ghose AK, Crippen GM. Atomic physicochemical parameters for three-dimensional-structure-directed quantitative structure–activity relationships. 2. Modeling dispersive and hydrophobic interactions. J Chem Inf Comput Sci 1987; 27:21–35.

156. Baxter AJG, Dixon J, Ince F, Manners CN, Teague SJ. Discovery and synthesis of methyl 2,5-dimethyl-4-[2-(phenylmethyl)benzoyl]-1*H*-pyrrole-3-carboxy late (FPL 64176) and analogues: The first examples of a new class of calcium channel activator. J Med Chem 1993; 36:2739–2744.

157. Ames BN, Durston WE, Yamasaki E, Lee FD. Carcinogens are mutagens: A simple test system combining liver homogenates for activation and bacteria for detection. Proc Natl Acad Sci USA 1973; 70:2281–2285.

158. Venger BH, Hansch C, Hatheway GJ, Amrein YU. Ames test of 1-(X-phenyl)-3,3-dialkyltriazenes. A quantitative structure–activity study. J Med Chem 1979; 22: 473–476.

159. Shusterman AJ, Johnson AS, Hansch C. Correlation of mutagenicity of 1,1-dimethyl-3(X-phenyl)-triazenes with molecular orbital energies and hydrophobicity. Int J Quantum Chem 1989; 36:19–33.

160. Debnath AK, Lopez DC, Debnath G, Shusterman AJ, Hansch C. Structure–activity relationship of mutagenic aromatic and heteroaromatic nitro compounds. Correlation with molecular orbital energies and hydrophobicity. J Med Chem 1991; 34:786–797.

161. Debnath AK, Debnath G, Shusterman AJ, Hansch C. A QSAR investigation of the role of hydrophobicity in regulating mutagenicity in the Ames test. 1. Mutagenicity of aromatic and heteroaromatic amines in *Salmonella typhimurium* TA98 and TA100. Environ Mol Mutagen 1992; 19:37–52.

162. Debnath AK, Lopez DC, Shusterman AJ, Hansch C. Quantitative structure–activity relationship investigation of the role of hydrophobicity in regulating mutagenicity in the Ames test. 2. Mutagenicity of aromatic and heteroaromatic nitro compounds in *Salmonella typhimurium* TA100. Environ Mol Mutagen 1992; 19:53–70.

163. Debnath AK, Hansch C. Structure–activity relationship of genotoxic polycyclic aromatic nitro compounds: Further evidence for the importance of hydrophobicity and molecular orbital energies in genetic toxicity. Environ Mol Mutagen 1992; 20:140–144.

164. Debnath AK, Shusterman AJ, Lopez DC, Hansch C. International Commission for Protection Against Environmental Mutagens and Carcinogens. The importance of the hydrophobic interaction in the mutagenicity of organic compounds. Mutat Res 1994; 305:63–72.

165. Klopman G, Rosenkranz HS. International Commission for Protection Against Environmental Mutagens and Carcinogens. Approaches to SAR in carcinogenesis and mutagenesis. Prediction of carcinogenicity/mutagenicity using MULTI-CASE. Mutat Res 1994; 305:33–46.

166. Mersch-Sundermann V, Rosenkranz HS, Klopman G. The structural basis of the genotoxicity of nitroarenofurans and related compounds. Mutat Res 1994; 304:271–284.

167. Smith C, Payne V, Doolittle DJ, Debnath AK, Lawlor T, Hansch C. Mutagenic activity of a series of synthetic and naturally occurring heterocyclic amines in *Salmonella.* Mutat Res 1992; 279:61–73.

168. Basak SC, Grunwald GD. Predicting mutagenicity of chemicals using topological and quantum chemical parameters: a similarity based study. Chemosphere 1995; 31:2529–2546.

169. Benigni R, Andreoli C, Giuliani A. QSAR models for both mutagenic potency nd activity: Application to nitroarenes and aromatic amines. Environ Mol Mutagen 1994; 24:208–219.

170. Selassie CD, DeSoyza TV, Rosario M, Gao H, Hansch C. Phenol toxicity in leukemia cells: A radical process? Chem Biol Interact 1998; 113:175–190.

171. Selassie CD, Shusterman AJ, Kapur S, Verma RP, Zhang L, Hansch C. On the toxicity of phenols to fast-growing cells. A QSAR model for a radical-based toxicity. J Chem Soc Perkin Trans 2 1999; 2729–2733.

172. Smith DA, van de Waterbeemd H. Pharmacokinetics and metabolism in early drug discovery. Curr Opin Chem Biol 1999; 3:373–378.

173. Tarbit MH, Berman J. High-throughput approaches for evaluating absorption, distribution, metabolism and excretion properties of lead compounds. Curr Opin Chem Biol 1998; 2:411–416.

174. Austel V, Kutter E. Absorption, distribution, and metabolism of drugs. In: Topliss JG, ed. Quantitative Structure–Activity Relationships of Drugs. New York: Academic Press, pp 1983:437–496.

175. Mayer JM, van de Waterbeemd H. Development of quantitative structure–pharmacokinetic relationships. Environ Health Perspect 1985; 61:295–306.

176. Seydel JK. Pharmacokinetics in drug design. In: Dearden JC, ed. Quantitative Approaches to Drug Design. Amsterdam: Elsevier, 1983, pp 163–181.

177. Rowland M. Pharmaco-Kinetics-QSAR: Definitions, concepts and models. In: Dearden JC, ed. Quantitative Approaches to Drug Design. Amsterdam: Elsevier, 1983, pp 155–161.

178. Lien EJ. Structure–absorption–distribution relationships: Significance for drug design. In: Ariens EJ, ed. Drug Design. Vol V. New York: Academic Press, 1975, pp 81–132.

179. Luco JM. Prediction of the brain–blood distribution of a large set of drugs from structurally derived descriptors using partial least-squares (PLS) modeling. J Chem Inf Comput Sci 1999; 39:396–404.

180. Clark DE. Rapid calculation of polar molecular surface area and its application to the prediction of transport phenomena. 2. Prediction of blood–brain barrier penetration. J Pharm Sci 1999; 88:815–821.

181. Seydel JK, Trettin D, Cordes HP, Wassermann O, Malyusz M. Quantitative structure–pharmacokinetic relationships derived on antibacterial sulfonamides in rats and its comparison to quantitative structure–activity relationships. J Med Chem 1980; 23:607–613.

182. Sadowski J, Kubinyi H. A scoring scheme for discriminating between drugs and nondrugs. J Med Chem 1998; 41:3325–3329.
183. Walters WP, Ajay, Murcko MA. Recognizing molecules with drug-like properties. Curr Opin Chem Biol 1999; 3:384–387.
184. Ghose AK, Viswanadhan VN, Wendoloski JJ. A knowledge-based approach in designing combinatorial or medicinal chemistry libraries for drug discovery. 1. A qualitative and quantitative characterization of known drug databases. J Comb Chem 1999; 1:55–68.
185. Lipinski CA, Lombardo F, Dominy BW, Feeney PJ. Experimental and computational approaches to estimate solubility and permeability in drug discovery and development settings. Adv Drug Delivery Rev 1997; 23:3–25.
186. Viswanadhan VN, Ghose AK, Revankar GR, Robins RK. Atomic physicochemical parameters for three-dimensional structure-directed quantitative structure–activity relationships. IV. Additional parameters for hydrophobic and dispersive interactions and their application for and automated superposition of certain naturally occurring nucleoside antibiotics. J Chem Inf Comput Sci 1989; 29:163–172.
187. Gillet VJ, Willett P, Bradshaw J. Identification of biological activity profiles using substructural analysis and genetic algorithms. J Chem Inf Comput Sci 1998; 38: 165–179.
188. Ghose AK, Crippen GM. Atomic physicochemical parameters for three-dimensional structure directed quantitative structure activity relationships. I. Partition coefficients as a measure of hydrophobicity. J Comput Chem 1986; 7:565–577.
189. Martin EJ, Blaney JM, Siani MA, Spellmeyer DC, Wong AK, Moos WH. Measuring diversity: Experimental design of combinatorial libraries for drug discovery. J Med Chem 1995; 38:1431–1436.
190. Federov VV. Theory of optimal experiments. New York: Academic Press, 1972.
191. Lewis RA, Mason JS, McLay IM. Similarity measures for rational set selection and analysis of combinatorial libraries: The diverse property-derived (DPD) approach. J Chem Inf Comput Sci 1997; 37:599–614.
192. Cho SJ, Zheng W, Tropsha A. Focus-2D: A new approach to the design of targeted combinatorial chemical libraries. Pac Symp Biocomput 1998; 305–316.
193. Cho SJ, Zheng W, Tropsha A. Rational combinatorial library design. 2. Rational design of targeted combinatorial peptide libraries using chemical similarity probe and the inverse QSAR approaches. J Chem Inf Comput Sci 1998; 38:259–268.
194. Hellberg S, Sjöström M, Skagerberg B, Wold S. Peptide quantitative structure–activity relationships, A multivariate approach. J Med Chem 1987; 30:1126–1135.
195. Collantes ER, Dunn WJ. Amino acid side chain descriptors for quantitative structure–activity relationship studies of peptide analogues. J Med Chem 1995; 38: 2705–2713.
196. Labute P. Binary QSAR: A new method for the determination of quantitative structure activity relationships. Pac Symp Biocomput 1999; 444–455.
197. Gao H, Bajorath J. Comparison of binary and 2D QSAR analyses using inhibition of human carbonic anhydrase II as a test case. Mol Diversity 1999; 4:115–130.
198. Gao H, Williams C, Labute P, Bajorath J. Binary quantitative structure–activity relationship (QSAR) analysis of estrogen receptor ligands. J Chem Inf Comput Sci 1999; 39:164–168.

199. Fujita T. Drug Design: Fact or Fantasy? New York: Academic Press, 1984.
200. Topliss JG. Some observations on classical QSAR. Perspect Drug Discovery Design 1993; 1:253–268.
201. Boyd DB. Progress in rational design of therapeutically interesting compounds. In: Liljefors T, Jørgensen FS, Krogsgaard-Larsen P, eds. Rational Molecular Design in Drug Research. Copenhagen: Munksgaard, 1998, pp 15–23.
202. Boyd DB. Innovation and the rational designs of drugs. CHEMTECH 1998; 28:19–23.
203. Boyd DB. Is rational design good for anything? In: Parrill AL, Reddy MR, eds. Rational Drug Design: Novel Methodology and Practical Applications. Washington, DC: American Chemical Society, 1999 pp 346–356.
204. Wiener H. Structural determination of paraffin boiling points. J Am Chem Soc 1947; 69:17–20.
205. Pisanski T, Plavsic D, Randic M. On numerical characterization of cyclicity. J Chem Inf Comput Sci 2000; 40:520–523.
206. Randic M. On characterization of molecular branching. J Am Chem Soc 1975; 97:6609–6615.
207. Hall LH, Kier LB. The molecular connectivity chi indices and kappa shape indices in structure–property modeling. In: Boyd DB, Lipkowitz K, eds. Reviews of Computational Chemistry. Vol. 2. New York: VCH, 1991, pp 367–422.
208. Kier LB. An index of flexibility from molecular shape descriptors. Prog Clin Biol Res 1989; 291:105–109.
209. Hall LH. Computational aspects of molecular connectivity and its role in structure–activity modeling. In: Rouvray DH, ed. Computational Chemical Graph Theory. New York: Nova Press, 1990, pp 202–233.
210. Kier LB, Hall LH. An electrotopological-state index for atoms in molecules. Pharm Res 1990; 7:801–807.
211. Kier LB, Hall LH. The electrotopological state index: An atom-centered index for QSAR. In: Testa B, ed. Advances in Drug Design. Vol. 22. New York: Academic Press, 1992
212. Kellogg GE, Kier LB, Gaillard P, Hall LH. E-state fields: Applications to 3D QSAR. J Comput Aided Mol Design 1996; 10:513–520.
213. Balaban AT. Highly discriminating distance-based topological index. Chem Phys Lett 1982; 89:399–404.
214. Balaban AT, Catana C, Dawson M, Niculescu-Duvaz I. Application of weighted topological index J for QSAR of carcinogenesis inhibitors (retinoic acid derivatives). Rev Roum Chim 1990; 35:997–1003.
215. Bonchev D, Seitz WA, Mountain CF, Balaban AT. Modeling the anticarcinogenic action of retinoids by making use of the OASIS method. 3. Inhibition of the induction of ornithine decarboxylase by arotinoids. J Med Chem 1994; 37:2300–2307.
216. Basak SC, Gute BD. Characterization of molecular structure using topological indices. SAR QSAR Environ Res 1997; 7:1–21.
217. Basak SC, Grunwald GD, Niemi GJ. Use of graph theoretic and geometrical molecular descriptors in structure-activity relationships. In: Balaban AT, ed. From Chemical Topology to Three-Dimensional Geometry. New York: Plenum Press, 1997, pp. 73–116.

4

Quantitative Structure–Activity Relationships (QSAR):

A Review of 3D QSAR

Gordon M. Crippen and Scott A. Wildman

*University of Michigan
Ann Arbor, Michigan*

I. INTRODUCTION

Here we consider a narrowly defined but still frequently occurring problem in drug design that one might refer to as the unknown receptor 3D QSAR problem. Given are the experimentally determined biological activity measurements for a set of compounds and their chemical structures. These measured activities for all the compounds, which are presumed to be the result of reversible binding to some common site on some macromolecule, such as a protein, are expressed in terms of ΔG_{bind} or some roughly equivalent scale, such as log K_i, or log IC_{50}. While the three-dimensional structures of these small ligand molecules may be estimated with good molecular mechanics potentials, the three-dimensional structure of the receptor site is unknown. (Otherwise, we are dealing with structure-based binding modeling, quite a different problem, addressed by quite different methods: Good-sell and Olson, 1990; Jones et al., 1997; Mankino and Kuntz, 1997; Sandak et al., 1998). The objective is to use this given information to construct some sort of correlation between ligand structure, including three-dimensional features, and the observed activities. This relationship should map the three-dimensional structure of a ligand into a quantitative estimate of activity that agrees well in some sense with the observations for the training set of compounds, and it should produce useful predictions for test compounds. If the only ligand features employed in the relationship come from the covalent bonding of the ligand, the problem belongs to traditional QSAR (Martin, 1978; Hansch and Leo, 1995).

The focus of this chapter is on the methodology employed in a bewildering array of approaches going back at least 25 years (see Table 1). They share the common motivation that since the real receptor physically interacts with a three-dimensional ligand molecule in sometimes very specific ways that dramatically distinguish subtle isomers, the appropriate relation between structure and activity must involve more three-dimensional features than physicochemical parameters of substituents and indicator variables. From that common beginning, however, the overall strategic goal splits between those who would construct a relation in terms of such a realistic model of the receptor that prediction uses structure-based modeling, versus those who seek a robust, predictive correlation that may in fact bear little resemblance to the receptor–ligand complex when it is later determined ex-

Table 1　Overview of 3D QSAR Methods

Paradigm Method	Realism	Molecular alignment	Conformational flexibility	Stratagem
Receptor				
Pharmacophore	+	User	None	Correlate residue–ligand interaction to activity
YAK	+ + + +	User	None	Deduce receptor structure
Distance geometry	+	User modes	Implicit	Distance interval matrix
Voronoi	+ +	Automatic	Implicit	Search over interaction energy
Egsets	+ +	Automatic	Implicit	Search over partition modes
Egsite	+	Automatic	Implicit	Search over site geometry
EGSITE2	+	Automatic	Sampled	Mixed integer program
Ligand				
MSA	− −	User reference	Sampled	Overlap on reference
MTD	− −	User reference	None	Hypermolecule vertex activity
REMOTEDISC	−	Automatic	Sampled	Primary site point occupancy
CoMFA	− −	User	None	PLS and grid field
Molecular similarity	− −	User	None	PLS and grid field
Compass	−	Automatic	Sampled	Neural net and grid field

perimentally. As we will see, the degree of realism in a receptor model is not necessarily reflected in the predictive power. These differences in overall goal may affect the paradigm underlying a method. In the first extreme, the idea is to construct a model of the receptor site at some level of realism by deductions made from the given ligand structures and activities. The other extreme concentrates on the ligand structures, reasoning that there must be common features found in the active molecules and other features found only in the inactive ones. While the receptor recognizes these features, they are properties of the ligands, not the receptor site.

Whatever the goal and paradigm a method subscribes to, there is a broad range along the subjective/objective axis. At the subjective extreme, the user may have to guess practically all features of the QSAR, and the computer algorithm may do little more than quantitatively assess the quality of the guess. At the objective extreme, the user inputs the observed activities and covalent structures of the ligands, and the algorithm deduces the receptor structure in full detail, including the positioning of each ligand in the site, as well as a complete model for the energetics of their interaction. Typically, the earlier methods have many subjective features, and even the most recent ones are not completely objective. Greater objectivity requires more CPU time, so a key feature of many methods is the particular stratagem used to make the approach computationally feasible.

In what follows, we will first cover some of the receptor paradigm methods, and then some of the ligand paradigm. In either case, important features are the subjectivity/objectivity level and the stratagems employed.

II. RECEPTOR PARADIGM

A.   Realistic Receptors

1. Pharmacophore

Before 3D QSAR there was the concept of the pharmacophore, a constellation of a few special atomic groups in fixed relative positions, such that possessing the pharmacophore is a requirement for high activity (Golender and Vorpagel, 1993; Wermuth and Langer, 1993). Höltje and Kier (1974) started from a three-point pharmacophore associated with sweet taste and a homologous series of compounds sharing two of the required pharmacophore groups and varying the structure of the third. They found a particular positioning of a particular amino acid sidechain, tyrosine, such that the observed activity correlated well with the calculated enthalpy of interaction between the sidechain and the third pharmacophoric group. While this approach appears to be motivated by a desire to move away from a completely empirical pharmacophore toward at least a sketchy view of part of the receptor, the adjustment goal was clearly centered on good correlation, rather than realistic portrayal of the binding site.

2. YAK

The logical extrapolation of placing one amino acid sidechain of a receptor model is to place many, preferably connected to some extent with peptide bonds (Höltje et al., 1993). One method of constructing such minireceptors or pseudoreceptors (Snyder et al., 1993) is YAK, which begins with a user-supplied superposition of active ligands and uses a combination of forcefields and knowledge extracted from the Protein Data Bank to place residues of likely types around the superimposed ligands at likely positions. The general goal of such a study is to deduce the structure of the real receptor, rather than produce a correlation between activity and calculated interaction between a ligand and the site model. Furthermore, the result may well depend strongly on the superposition supplied to YAK. While that need not be a pure guess on the part of the user, the superposition step is external to the method and thus counts as a subjective factor. In one successful validation study (Vendani et al., 1993), the superposition was determined from the crystal structure of a protein–ligand complex, and then YAK reproduced many features of that same protein's binding site.

B. Abstract Receptors

1. Distance Geometry

For n atoms there are $3n$ coordinates and $n(n-1)/2$ interatomic distances, yet distances are sometimes more convenient to work with. Suppose the receptor site is represented as a few "site points" described in terms of their interpoint distances plus or minus some flexibility factor δ, and a conformationally flexible molecule is described as interatomic distance intervals, $[l_{ij}, u_{ij}]$, taken over a survey of low energy conformations. Then if atom i is supposed to lie at site point I and atom j at site point J, a necessary condition is simply that the intervals $[l_{ij}, u_{ij}]$ and $[d_{IJ} - \delta, d_{IJ} + \delta]$ overlap. Let a binding mode denote which atom is supposed to be in contact with which site point, allowing for unoccupied site points and unused atoms, and a maximum occupancy of one atom at a site point (Fig. 1). Although no elaborate superposition calculation is required to check the necessary geometric conditions for a mode, these distance conditions are not sufficient. It may be impossible to find site point coordinates satisfying a given set of $d_{IJ} \pm \delta$, and there may be no single conformation having interatomic distances in all the required overlap intervals. In practice, however, standard distance geometry methods (Crippen and Havel, 1988) are usually successful at producing coordinates of the site points for visual display.

The calculated binding or activity is assumed to be a sum of interactions between the atoms matched with site points and those site points. To speed subsequent calculations and to keep down the number of adjustable parameters, atoms in the ligands are grouped together into a few united atoms (common ring struc-

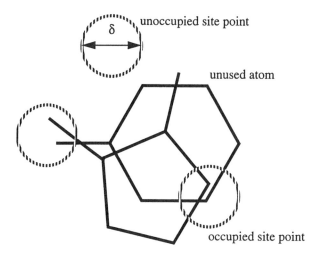

unoccupied site point

unused atom

occupied site point

Figure 1 Schematic example of a distance geometry site with two molecules superimposed.

tures, functional groups, etc.). Likewise, the user assigns site points to a few different types. Thus the adjustable parameters are a table of interaction energies depending on the united atom type and on the site point type. The objective is a least-squares fit of observed and calculated activities as a function of the adjustable interaction parameters. A touch of realism is added by permitting the model ligands to try all the different geometrically allowed binding modes, just as the real ligand at equilibrium occupies the real receptor site predominantly in the orientation and internal conformation that is energetically most favored.

The original distance geometry 3D QSAR (Crippen, 1979) was very subjective in that the user had to choose a number site points of specified types in a certain geometry, group ligand atoms into a few united atoms, and propose optimal binding modes for each ligand. Only then did the algorithm take over, adjusting the energy parameters for a least-squares fit and exhaustively searching all geometrically allowed modes to determine what binding mode of each ligand was actually optimal, given the current energy parameter values.

There followed a long series of steps toward greater objectivity. Congeneric series of ligands could be easily decomposed automatically into a small set of shared united atoms (Crippen, 1980). Adjustment of the energy parameters became much simpler with the introduction of quadratic programming, which is a general procedure for optimizing a quadratic function subject to linear inequality constraints. Here the optimization is still the least-squares fit of observed versus calculated activities, given the user's proposed optimal binding modes. The constraints are that the proposed optimal modes are indeed better than other geometrically allowed modes. Next (Crippen, 1981), interatomic distance intervals were automatically determined by a coarse systematic search over rotatable bonds with steric bump checking. This step also aids the user's choice of united atoms. Automatic recognition of conformationally independent chiral quartets of atoms permitted construction of stereospecific site models in terms of intersite point distance matrices and noting chirality of some quartets of site points.

In fact, the user's suggested binding modes can be guided in part by the same decomposition algorithm applied to even sets of chemically diverse ligands (Crippen, 1982), and then these modes imply intersite point distance constraints, so that the site geometry can be determined from the proposed modes. Refinement of this procedure (Ghose and Crippen, 1982) produces more clearly defined site geometries. The initial dependency on suggested binding modes was relaxed subsequently by an automatic procedure for revising the proposed modes so as to improve the fit (Ghose and Crippen, 1983a, b).

Up to this point, predictions were limited to molecules that could be viewed as constructed from the same set of united atoms found in the training set. On the basis of covalent structure, one can assign atomic physicochemical parameters such that the sum of the atomic hydrophobicities accurately match experimental values for the octanol/water partitioning coefficient $\log P$, and summed atomic polarizability properties match experimental molar refractivities (Ghose et al., 1987). Thus instead of modeling the calculated activity as $G_{calc} = \Sigma_{contacts} e(t_A, t_s)$, where t_A is the type of the united atom positioned at the contacted site point having type t_s, it can be expressed as still a linear combination of atomic hydrophobicities h_a, atomic refractivities m_a, and partial charges q_a for those atoms constituting the united atoms in the contacts for that binding mode, and the corresponding three adjustable parameters (e_h, e_m, e_q) for site points of those types.

$$G_{calc} = \sum_{contacts} \sum_{a \,\varepsilon\, A} [h_a e_h (t_s) + m_a e_m(t_s) + q_a e_q(t_s)] \qquad (1)$$

That way, the model can be tested with novel compounds as long as their constituent atoms can be assigned physicochemical parameters (Ghose and Crippen, 1985a).

The distance interval matrix stratagem certainly saved a lot of computational expense in trying out different binding modes, but the conditions checked were only necessary, not completely sufficient. For example, there may be an energetically reasonable conformation of a molecule such that a particular pair of atoms are at their distance of closest approach, l_{ij}, and some other conformation gives their greatest separation, u_{ij}; but there may well be intermediate distances in the interval $[l_{ij}, u_{ij}]$ that are found in no allowed conformation. An improvement on this (Ghose and Crippen, 1985b) represents a conformationally flexible ligand by a set of distance interval matrices, each covering a smaller patch of energetically allowed conformation space without including so many disallowed conformations. Checking the geometric feasibility of important binding modes can also be further verified by determining site point coordinates and positioning the ligand onto them in three-dimensional space (Ghose and Crippen, 1984). The approach lacks realism also because it neglects activity contributions from atoms that fall between the small site point spheres and because of the difficulty of modeling steric effects by surrounding a site point available to atoms with a cluster of "filled" site points that must be avoided by any atom (Crippen, 1979).

2. Voronoi Models

The advent of linearized embedding (Crippen, 1989) suggested a stratagem switch. Here the ligand molecule is represented in the rigid valence geometry approximation as a collection of rigid rings or other groups connected by rotatable bonds. Each rigid group has a local coordinate system specified by up to three unit vectors, and the position of any atom in a particular conformation is a linear combination of an overall molecular translation vector and these various unit vectors. As conformations change, the relative orientations of some unit vectors change, so conformational flexibility can be summarized as a matrix of intervals of inner products on all pairs of unit vectors. On the other side (Crippen, 1987), a receptor site can be abstractly modeled as a Voronoi complex, which consists of one or more polyhedra or regions that divide up all space without overlaps. Each region represents some pocket for binding possibly many atoms or some place that might be sterically forbidden because it is where the polypeptide chain of the receptor lies. The geometry of the regions are determined by the coordinates of one so-called generating point per region, and each polyhedral region is just the set of all points closer to its generating point than to any other (Fig. 2). It turns out that demanding that a given ligand bind in a particular mode (i.e., assignment of now each atom to one and only one region) implies that a corresponding set of linear inequalities has a solution. By discretizing unit vector orientations, an exhaustive search for all possible binding modes is feasible for moderately flexible ligands and site models built out of few regions.

The initial Voronoi method (Crippen, 1987) requires of the user the coordinates of the generating points and choices of united atoms. Every atom in a ligand contributes additively to the calculated activity, depending on its atom type and the type of the region to which the particular mode assigns it, regardless of exactly where in the region that atom would lie. The variables in the problem are these interaction parameters. From there, the algorithm first enumerates all possible binding modes and then searches for a set of interaction parameter values such that the optimal binding mode for each ligand agrees with the observed activity. Multiple solutions are possible, so this search is carried out as a rough global search from many random starts, locally minimizing by subgradient optimization a penalty

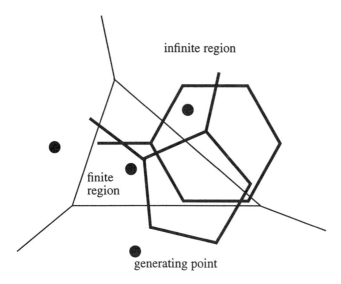

infinite region

finite
region

generating point

Figure 2   A schematic Voronoi site consisting of three infinite regions surrounding one finite one.

function that is everywhere continuous but only piecewise differentiable. This represents a significant step toward objectivity because the user no longer supplies suggested optimal modes, which is the equivalent in other methods of the user-supplied superposition hypothesis. It also leads to producing multiple site models for a given training set, as a way of expressing the limitations of the training set and estimated error in subsequent predictions.

Another innovation of Crippen (1987) is that the observed activities are no longer taken to be single values, but intervals on the binding strength scale, like error bars. This permits fitting imprecise data using given error estimates, and it also implies that the fit is no longer a least-squares procedure. The penalty function used in parameter adjustment becomes zero whenever the calculated activity for the optimal binding mode of each ligand lies anywhere within the corresponding observed interval. Although the observed intervals may be broad in some cases, no outliers beyond that are permitted for a solution. This implies that standard deviations, correlation coefficients, and even leave-one-out cross-validation measures of the fit to the training set are not applicable any more.

Contrasting the key problems in the explicit site versus the ligand paradigms led to a useful concept. In the site paradigm, there is an explicitly represented model of the receptor site, and much of the computational effort goes into optimally positioning a ligand in the site. In the ligand paradigm, the crucial step is to superimpose particularly the active ligands in such a way as to demonstrate maximal molecular similarity. The intuitive choice in a congeneric series is to super-impose the common ring system typically and then discuss differences in activity as a function of the substituents at different positions, where now the positions in different molecules having the same label in organic nomenclature coincide in space. This idea has been generalized several different ways (Good et al., 1993a and b; Klebe, 1993; Klebe et al., 1994; Carbó-Dorca et al., 1995; Castro and Richards, 1998; Robert et al., 1999) to produce definitions of abstract molecular similarity, such as the Carbo index, comparing electron density, and the Meyer in-dex, which is based on the shape of the van der Waals surface. On the other hand, one real enzyme might recognize two different ligands as equally good inhibitors while another enzyme might find one to be greatly superior to the other. This translates in the site paradigm to defining molecular similarity *with respect to a given site model,* be it an approximation to a real receptor site or an artificial con-struct. Bradley et al. (1993) gave such a definition of molecular similarity in terms of simple Voronoi site models, showing how different models did or did not dis-tinguish between certain kinds of isomers and so on. Site models involving large regions might determine two ligands to be very similar because they always gave similar calculated binding activities over a wide range of interaction energy pa-rameters, even though their optimal binding modes for some particular choice of parameters might not closely superimpose them!

The original Voronoi modeling programs were improved by relaxing the discretization approximation in the search over all possible binding modes (Boulu and Crippen, 1989). This required the numerical solution of a set of linear in-equalities and quadratic equalities at each mode being tested. The search was nev-ertheless made faster by branch-and-bound stratagems that eliminated classes of modes on the basis of lack of overlap between interatomic distance intervals and the matching interregion distance intervals. (For example, the minimal distance between two polyhedral regions is the shortest distance between one point in the first region and some point in the other, taken over all such pairs of points.) An-other stratagem was that Voronoi regions are always convex, so the set of atoms supposed to bind in one region must also be convex. Typically broad error bars on the observed activities permit solutions involving few regions, and more precise input data demanded more geometric detail in the site models. In any case, the user still has to supply subjective guesses for generating point coordinates, usually starting with few regions and proceeding to more complicated models until the ob-served activities can be fit (Boulu et al., 1990).

3. Egsets

Site models are adjusted in the Voronoi approach primarily in terms of interaction parameters. A particular choice of parameters implies that certain binding modes of the ligands are optimal, and the corresponding calculated activities may or may not agree with the observed intervals. An alternative search strategy (Bradley and Crippen, 1993; Srivastava and Crippen, 1993; Srivastava et al., 1993) is to systematically try different choices of proposed binding modes (as was done manually in the distance geometry approach). Such a choice puts constraints on the interregion distance intervals and sometimes the chirality of quartets of regions. If these constraints can be satisfied, the mode choice further puts constraints on the interaction parameters to ensure that the proposed modes are actually optimal and give calculated activities in agreement with the observed intervals. If no solution can be found for a small number of regions, try more regions. If a solution is found, its geometry is given in terms of interregion distance intervals and possibly some chirality relations. Distance geometry embedding can then produce coordinates of the centers of these regions, although they may not be necessarily convex.

While Egsets relieves the user of supplying Voronoi generating point coordinates, to keep the combinatorial search over proposed modes feasible, the user still must greatly simplify the ligands into a few united atoms apiece. United atoms are, however, required to be convex sets, even though eventually the regions may not be convex. Convexity is an important stratagem because there are relatively few partitions of a ligand into mutually exclusive and exhaustive convex sets of superatoms. The binding modes in the search are then just assignments of each convex set in a partition to a different region.

The adjustable interaction parameters are three physicochemical parameters for each region, as in Eq. (1). These are further constrained by introducing an artificial ligand into the training set, namely, a very large, very hydrophobic one: graphite (Bradley and Crippen, 1993). This ligand's modest activity forces any hydrophobic regions to be limited in size, but one with weak interactions could be large and would represent the bulk solvent outside the receptor.

4. Egsite

The search for site models can be carried out in terms of the interaction energy parameters (Voronoi), the suggested optimal binding modes (Egsets), or the site geometry (Egsite). Choosing particular values for one aspect of the problem at least constrains what the other aspects can be, and then it comes down to a quantitative assessment of how efficient the search is in realistic problems. Egsite is centered around a branch-and-bound search over possible site geometries for a specified number of regions, beginning with the vaguest possible description and gradually splitting up into mutually exclusive and more precisely specified alternative geometries. As an analogy, the interval $[-1000, 1000]$ is a vague representation of the number $5 =$ the interval $[5,5]$, and $[0, 6]$ is a more precise approximation. Since a precise description of a site in this approach is expressed as a matrix of intervals of interregion distances plus some possibly empty set of chirality designations, then an imprecise description employs intervals of intervals of interregion distances plus fewer chiral quartets. The mathematics of dealing with intervals and intervals of intervals is all summarized by Crippen (1995).

Without getting into a lot of detail, the original Egsite algorithm starts with initial conformational searches for the ligands summarized as interatomic distance intervals and possibly some chiral quartets of atoms. The user specifies convex sets of atoms to become united atoms. As before, modes amount to relatively few partitions of the united atoms into convex subsets and assignment of each subset to a different site region. Site regions are assumed to be convex, and the number of regions is chosen by the user, typically starting with only a few. Interaction parameters amount to a hydrophobicity and a molar refractivity factor for each region, corresponding to the atomic hydrophobicities and refractivities. For a given indefinite site description, modes are classed as certainly geometrically allowed ("sure"), possibly but not certainly allowed depending on how subsequent refinement proceeds ("may"), and certainly disallowed by this site and any subsequent refinement of it ("bad"). Starting with the vaguest site description, the algorithm carries out a depth-first search of increasing the precision of the description, which tends to move "may" modes into either the "sure" or the "bad" category. At each stage, a linear program is solved to determine whether there is any set of interaction parameters such that some sure/may mode is optimal for each ligand and that the corresponding calculated activity falls within the given observed interval.

For a sufficiently large number of regions, one or more solutions may be found at different branches of the search tree. The geometric part of the solution consists of more or less narrow intervals of interregion distance intervals plus possibly some chiral quartets of regions when stereospecific binding is important. The energetic part really consists of not just a particular set of values for the interaction parameters, but rather a feasible region in the multidimensional parameter space, as the linear programming terminology would call it. A subsequent prediction of activity for some test ligand interacting with just one of the solution site models yields now an interval of calculated activity, determined by linear programming where the predicted activity is first maximized and then minimized within the parameter feasible region.

Another innovation in a subsequent refinement of the approach (Schnitker et al., 1997) is the constraint that one of the regions must represent the bulk solvent by having fixed infinite diameter and zero interactions with atoms. The latter is appropriate when the activity data compare the free solvated state of a ligand with its bound state, as in ΔG_{bind}.

5. EGSITE2

The latest step (Crippen, 1997) on the road toward complete objectivity finally eliminates the user's subjective choice of united atoms. That means that it is no

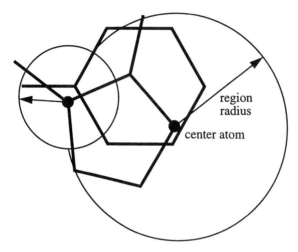

Figure 3 A schematic EGSITE2 site consisting of two explicit regions in addition to the surrounding solvent region. The larger explicit region is not convex.

longer feasible to repeatedly, exhaustively examine all possible modes for each ligand. Instead, a so-called basis set of modes is developed during the search for site geometries whenever new, currently optimal modes are discovered. The convexity stratagem used in previous approaches to drastically reduce the number of modes is also abandoned. Instead, the site is described in terms of interregion distance intervals, which includes a putative diameter for each region. Then a mode amounts to a choice of one "center" atom for some of the regions. Since conformational flexibility is now represented by a small set of low energy conformers of each ligand, we have interatomic distances for each conformer. Given the center atoms specified by the mode, another atom falls into a particular region if it is closer to that region's center atom than any other region's, and if it is within the region radius of the center atom (Fig. 3). The resulting partitioning of the ligand's atoms among regions is more stable to shifts of center atom choices to nearby atoms than in the analogous Voronoi region definition. Thus the mode definition stratagem permits a reliable search for optimal binding modes, given some site geometry and the physicochemical property-related interaction parameters.

The search for solutions is carried out in the combined space of interaction energy parameters and interregion distance bounds. Finding one solution is a moderately complicated procedure involving solving a series of "mixed integer programs" (Crippen, 1999), which determine values for the continuous geometric and energetic parameters while automatically making decisions about which basis modes are allowed or disallowed for what reasons. If a solution is found, it is automatically biased toward the least restrictive geometry possible (i.e., large regions lying near one another). Subsequent solutions can be found by adding a cutting plane to the problem that excludes the preceding solution.

For all its strange formulation, EGSITE2 performs well on standard test cases (Crippen, 1997), sometimes finding solutions with remarkably small training sets that nonetheless show respectable predictive power.

III. LIGAND PARADIGM

A. MSA

Molecular Shape Analysis (MSA) provides a variety of descriptors to be used in conjunction with conventional QSAR descriptors in a linear regression, resulting in a least-squares fit between observed and calculated activities (e.g., Hopfinger, 1980; Tokarsky and Hopfinger, 1994). Let a particular, low energy conformation of an active ligand be designated as the "reference structure." Then for any other ligand, find the one low energy conformation out of several alternatives that can be superimposed on the reference so as to maximize the overlap volume between the two van der Waals surfaces. Descriptors for that ligand include the overlap volume itself, the remaining volume outside the overlap, and related measures (Fig. 4). The choice of which conformation of which ligand is to be the reference structure is either left up to the user, or a wide variety of possibilities are tried, and

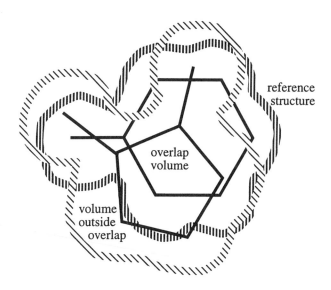

Figure 4   A schematic MSA superposition onto the reference molecule such that there is one contiguous overlap volume and three pieces of volume outside the overlap.

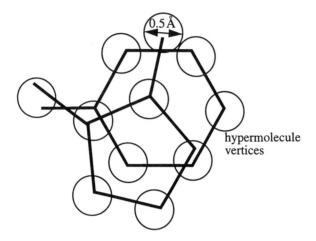

Figure 5 A schematic MTD superposition resulting in 11 sufficiently distinct hyper-molecule vertices.

the one giving the best fit is chosen. Clearly the key issue here is the reference structure and the degrees of freedom that are hidden in the process of choosing which structures to superimpose.

B. MTD

The Minimal Topological Difference (MTD) approach (e.g., Simon et al., 1976; Simon, 1993; Simon et al., 1994) resembles MSA in that the user chooses some conformation of a preferably rigid and highly active ligand and then superimposes a selected conformation of each of the other ligands onto it so as to maximize the number of equivalent atoms superimposed, or some similar objective. Considering all molecules together in their superposition, atoms in this "hypermolecule" are clustered together to form a "vertex" if they lie within 0.5 Å of each other (Fig. 5). Any one ligand occupies some subset of these vertices. Each vertex is declared to be favorable, unfavorable, or irrelevant toward activity, and one can use in a standard QSAR regression the MTD of a ligand, which is the number of unfavorable vertices occupied minus the number of favorable ones. As in MSA, there are extra degrees of fitting freedom available by searching for the favorable/unfavorable/irrelevant labelling of vertices that gives the best fit to the training set.

C. REMOTEDISC

REMOTEDISC (**re**ceptor **mo**deling from **t**hree-**di**mensional **s**tructure and physicochemical properties: Ghose et al., 1989) resembles MTD in that low energy conformations of active ligands found by systematic search over rotatable bonds are superimposed upon the user's chosen reference conformer of the selected reference compound. The alignment algorithm is specified to begin by matching triples of atoms in the ligand with triples in the reference having similar interatomic distances and atom types, distinguishing seven broad categories. Such a three-point match implies a unique rigid translation and rotation that in general brings many of the ligand's atoms near other reference atoms, so the proposed superposition is scored on the basis of similarity between the physicochemical parameters of the atom pairs. When each compound in the training set has been optimally superimposed in this way on the reference, each sufficiently distinct atom position (equivalent to an MTD vertex) is said to be a site point, but the user then selects a few highly occupied site points as having distinct types, and all other site points assume the type of the nearest selected one (Fig. 6). The calculated activity of a compound, given its optimal superposition and assignment of atoms to site points, is just as in Eq. (1), except that atoms are not grouped, and the weighted conformational internal energy of the optimally superimposed conformer is also added in. The adjustable parameters are the three interaction energy parameters (hydrophobicity, molar refractivity, and partial charge) associated with each distinct site point type, and these are adjusted to agree with the observed activities in a least squares fit.

If the user has chosen few primary site points, there will be only three times that number of adjustable parameters in the fit (Ghose and Crippen, 1990). Some variables can be eliminated by reverse stepwise regression.

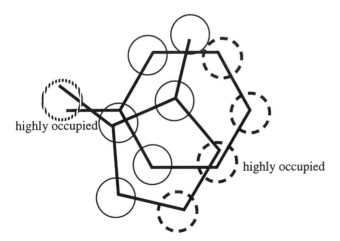

Figure 6 A schematic REMOTEDISC superposition showing three types of site points based on three highly occupied site points.

D. CoMFA

The most commonly used 3D-QSAR method today is CoMFa (comparative molecular field analysis: Cramer et al., 1988, 1993). Since development by Cramer and coworkers in 1988, CoMFA has been used in countless drug discovery projects and benchmark computational studies and has become the standard 3D-QSAR method that results of new methods are compared to. This is in part due not only to the ease of application and availability in the SYBYL molecular modeling package (Tripos), but also because it is clearly capable of producing reliable results. It can provide the user with reasonable predictions of activity and a graphical depiction of areas around a representative molecule that may be suitable for specific kinds of modification. Many reviews of CoMFA analyses, results, strengths and weaknesses are available (Folkers et al., 1993; Kubinyi, 1993; Greco et al., 1997; Oprea and Waller, 1997).

As usual for the ligand paradigm, the key step is the alignment of one chosen conformation of each ligand, but the alignment algorithm is external to CoMFA. Specific results do depend on the type and quality of the alignment. Then instead of what might be called the adaptive grid of site points in MTD and RE-MOTEDISC, the aligned structures are placed inside a cubic lattice (Fig. 7). The grid size of the lattice can affect results, but is commonly accepted at 2 Å spacing. The exact position of the ligand set within the grid is also not supposed to be crit-

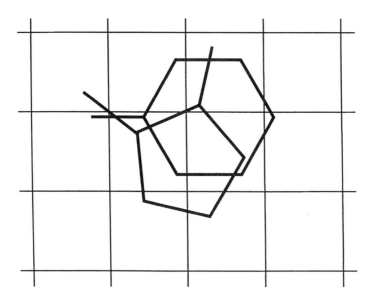

Figure 7   A schematic CoMFA, molecular similarity, or Compass superposition.

ical, especially if the grid size is small (Cramer et al., 1988). Once the structures are aligned and placed in the grid, energy values for steric and electrostatic interactions for each ligand are calculated for a probe atom at each point on the grid, and stored in a massive table of one row for each ligand and one column for each grid point. The steric energy term is calculated by a standard Lennard-Jones potential of the form

$$E_{vdw} = 4\varepsilon \left[\left(\frac{\alpha}{r_{pt}} \right)^{12} - 2 \left(\frac{\alpha}{r_{pt}} \right)^{6} \right]$$

and the electrostatic term

$$E_{el} = \frac{q_p q_t}{4\pi\varepsilon_0\varepsilon_r r_{pt}}$$

as a Coulombic interaction. More recently a third term, representing a hydrophobic field calculated by HINT (Abraham and Kellogg, 1993), has become available and is seen to improve results in some cases, while causing difficulties in others. Additional terms have also been suggested. The energy data generated then are fit to the experimentally determined biological data to generate the QSAR model. Such an overdetermined system, with at least two energy terms for each of thousands of grid points, can be a challenge to analyze in a robust way. CoMFA uses partial least squares (PLS) (Wold et al., 1984, 1993) to accomplish this. PLS identifies which terms contribute most significantly to the fit through analysis of the variance of the terms, usually finding a significant fit (if one exists) with fewer than ten terms. It is common for CoMFA models to use five or fewer terms.

While it is important to fit the existing data, the real value of CoMFA, or any QSAR method, is predictive ability. CoMFA measures predictive ability by calculating q^2 (cross-validated r^2 or predictive r^2) (Cramer et al., 1988; Kubinyi and Abraham, 1993). This is accomplished by leaving a compound out of the fit and then predicting the activity of that molecule for each compound, one at a time, until all molecules have been predicted exactly once. The q^2 value is then calculated as follows:

$$q^2 = \frac{SD - press}{SD} \tag{2}$$

where SD is the sum of the squared deviations of each biological activity value from their mean, and "press" is the predictive sum of squares, the sum over all compounds of the squared differences between the actual and predicted activity values. A q^2 in the range 0.5–0.6 indicates moderate predictive power, while a high value for q^2, a value near the fit r^2, shows excellent predictive ability of the model. This statistical analysis of the model is considered a necessary step in CoMFA and most other QSAR methods.

Even though CoMFA is widely used and accepted, it is not without problems (Greco et al., 1997; Oprea and Waller, 1997). One set of CoMFA limitations comes in the calculation of steric and electrostatic interactions. The Lennard-Jones 6–12 potential becomes very steep near the van der Waals surface of the molecule. Therefore, small differences in position of the molecules due to alignment or position in the grid can dramatically change the potential energy at grid points close to the surface. These differences may then be treated as significant descriptors in the PLS analysis, whereas they would be dismissed as "nearly identical" by visual inspection. Additionally, the steric and electrostatic potentials used by CoMFA evaluate as a singularity at atomic positions and are therefore normally limited to calculation only outside the volume of the molecule, and are cut off at some arbitrary fixed value. Since the slopes of the Lennard-Jones and Coulomb potentials differ, the cutoff values are exceeded at different distances from the molecules, and while a smoothing procedure has been offered (Cramer et al., 1993), adjusting the fields for simultaneous evaluation can result in the loss of information for one of the fields.

It has also been noted that ligand–receptor interactions in nature may not be accurately represented by including only steric and electrostatic terms. Significant work has gone into including supplemental fields in CoMFA (Oprea and Waller, 1997), particularly the hydrophobic field calculated by HINT (Kellogg et al., 1991; Abraham and Kellogg, 1993). Simply put, HINT uses the octanol/water partition coefficient ($\log P$) as atomic contributions generated from the fragment constants of Hansch, to generate a map of the hydrophobic "field" of the molecule. Inclusion of this field has led to modestly improved results in some cases and slightly worse results in others (Kellogg et al., 1991). The trend does not seem to be predictable. It is likely that there are additional fields that could be included in CoMFA to improve fit; however, it is unclear whether this would be a result of additional variables for PLS or if inclusion of these fields would reliably enhance predictive ability (Oprea and Waller, 1997).

However, the two main difficulties associated with CoMFA are the choice of a correct active ligand structure and the molecular alignment step. Almost all ligands are significantly flexible, particularly in the environmental influence of a receptor, and often ligands will not bind in that conformation determined to be the "lowest energy conformation" by *in vacuo* computation (Jain et al., 1994). CoMFA will use only a single conformation for each ligand in the data set and does not treat ligand flexibility. Consequently, the molecular structure as it is evaluated for alignment and for steric and electrostatic interactions may not be the active conformation. It has been argued that a series of similar molecules will all adopt a similar binding mode and therefore CoMFA is robust even when the binding mode is incorrect, with serious limitations on the interpretation of the graphical output (Cramer et al., 1988). While CoMFA does appear to produce a reasonable prediction of activity in these situations, the explanation is lacking.

The second main concern with the implementation of CoMFA is the necessary molecular alignment (Jain et al., 1994; Greco et al., 1997; Schnitker et al.,

1997). This difficulty is not limited to CoMFA; it appears in all 3D-QSAR methods that involve an explicit alignment step. There are several computational methods available to determine molecular alignment, including field-fit and active analog methods. The active analog approach (Marshall et al., 1979, 1993) consists of identifying a few (usually at least three) pharmacophoric groups in the set of molecules, which are then superimposed by least-squares fit, usually resulting in overlapping common internal molecular structure. The field-fit (Cramer et al., 1991) techniques involve calculating or identifying interaction fields or areas around a molecule which are then superimposed without considering the internal framework of the molecules. Each of these general techniques has been implemented in a variety of ways, some of which include a degree of conformational flexibility. In nature, however, what we would call molecular alignment is really the determination of a binding mode and is therefore the responsibility of the receptor and the ligand together (Bradley et al., 1993; Crippen, 1995). In the absence of a receptor environment, molecular alignment remains, at best, ambiguous. Even if they have very similar structures, two molecules may bind in similar spatial or in completely reversed orientations experimentally. Algorithms for molecular alignment neglect this physical reality, and therefore none of these methods can always reproduce the natural molecular alignment that would be necessary for a truly realistic CoMFA. Instead, choosing an alignment in advance corresponds to choosing only one possible binding mode, which may or may not agree with the positioning of these same ligands in the real receptor. However, this may not always have serious consequences for the fitting of the training set and even subsequent predictions.

E. Molecular Similarity

A number of researchers have been working to establish a QSAR using molecular similarity measures. In general, molecular similarity indices (Good et al., 1993a, b; Klebe et al., 1994; Carbó-Dorca et al., 1995; Castro and Richards, 1998; Robert et al., 1999) provide a quantitative measure of how similar two molecules are. There are several different measures used (Good et al., 1993b), the most common being the Carbo index, comparing electron density, and the Meyer index, which is based on the shape of the van der Waals surface. Each of these indices has also been modified to be calculated using electrostatic potential, electric field, and combinations of all these measures. In 3D-QSAR applications, the index used is calculated for the entire set of molecular structures on a three-dimensional grid and the resulting similarity matrix is analyzed, often by PLS, in a manner similar to that of CoMFA. As the similarity information is collected in a three-dimensional manner, it can indeed provide details about the local geometry or property differences, as is needed for 3D-QSAR. The different measures have each been developed into their own method, such as CoMSIA (Klebe et al., 1994), and each shows results that are roughly comparable to CoMFA.

These techniques are also subject to the same pitfalls as CoMFA. Again, there is likely some error caused by positioning molecules in the grid, the degree of which will be determined by the similarity index used. However, these errors are likely small and can be overcome by changing grid size or by attempting several models, positioning ligands differently in the grid each time. Here too, the largest problems are conformational flexibility and molecular alignment. In fact, similarity indices may be even more sensitive to incorrect or poor alignment than CoMFA.

F. Compass

An alternative to the single conformation and molecular alignment required by other methods, Compass (Jain et al., 1994, 1995) introduces an iterative alignment mechanism allowing conformational flexibility. In this regard, it is more objective than MTD, MSA, REMOTEDISC, or CoMFA. The method starts with a user-defined pharmacophore or common substructure, but then performs a conformation search to generate a list of alternative structures for each molecule. An initial set of conformations is chosen and aligned to form a first estimation of the model. Each aligned conformation, called a pose, is represented as a set of "sampling points" located outside but near the van der Waals surface of the molecule. Steric and hydrogen bond donor/acceptor interactions are determined at each of these points in a distance-dependent manner, rather than on a three-dimensional grid, and the model is generated with a neural net instead of PLS. Compass then chooses the best-fit pose of a representative molecule and develops a model based now on poses of other molecules that match the best pose, a process that involves selection of different conformations of molecules in the data set. After 50–100 iterations, the current set of poses are subjected to a new molecular alignment, creating yet another new set of poses, which will undergo the iterative model building process. After 0–5 iterations of the realignment loop, a final, self-consistent model is developed, which includes predictions of activity and active conformation as well as a graphical display of results.

Compass has several advantages over CoMFA (Jain et al., 1994; Greco et al., 1997). First, since the interactions are calculated at sampling points instead of grid points, there is no dependence on positioning in the grid and none of the difficulties associated with the CoMFA interaction cutoff values. Moreover, sampling points allow for the calculation of only a few hundred interactions rather than the thousands associated with grid methods. The descriptors used are also designed to be less sensitive to slight steric misalignment than those used in many grid-based methods. Second, Compass includes ligand flexibility in an efficient manner. As poses are selected and interactions are calculated, the information on previous poses is stored for possible future use, thereby speeding up the process, as any conformation may be chosen as the new pose at any step. Third, the molecular alignment, which is vitally important to all these methods, evolves as the model is generated, thus compensating for any errors in the initial alignment. This alone is a substantial and necessary improvement over previous methods. However, the iterative process that provides conformational and alignment flexibility is subject to local minima traps (Greco et al., 1997). If the initial set of poses is not diverse enough, it is likely that only still similar poses will be chosen in subsequent iterations, causing quick and false acceptance of the model. While methods may be introduced to identify this predicament, it seems that this situation cannot entirely be avoided.

REFERENCES

Abraham DJ, Kellogg GE. (1993) Hydrophobic Fields. In 3D *QSAR in Drug Design: Theory, Methods and Applications.* Kubinyi H, ed. Leiden: ESCOM, pp. 506–522.

Boulu LG, Crippen GM. (1989) Voronoi Binding Site Models: Calculation of Binding Modes and Influence of Drug Binding Data Accuracy. J Comput Chem 10, 673–682.

Boulu LG, Crippen GM, Barton HA, Kwon H, Marletta, M. (1990) Voronoi Binding Site Model of a Polycyclic Aromatic Hydrocarbon Binding Protein. J Med Chem 33, 771.

Bradley MP, Crippen GM. (1993) Voronoi Modeling: The Binding of Triazines and Pyrimidines to *L. casei* Dihydrofolate Reductase. J Med Chem 36, 3171–3177.

Bradley M, Richardson W, Crippen GM. (1993) Deducing Molecular Similarity Using Voronoi Binding Sites. J Chem Inf Comp Sci 33, 750–755.

Carbó-Dorca R, Besalú E, Amat Ll, Fradera X. (1995) Quantum Molecular Similarity Measures (QMSM) as a Natural Way, Leading Towards a Theoretical Foundation of Quantitative Structure Properties Relationships (QSPR). J Math Chem 18, 237–246.

Castro A, Richards WG. (1998) The Use of Molecular Similarity Indices in the Determination of a Bioactive Conformation. Eur J Med Chem 33, 617–623.

Cramer RD III, Patterson DE, Bunce JD. (1988) Comparative Molecular Field Analysis (CoMFA). 1. Effect of Shape on Binding of Steroids to Carrier Proteins. J Am Chem Soc 110, 5959–5967.

Cramer RD III, Clark M, Simeroth P, Patterson DE. (1991) Recent Developments in Comparative Molecular Field Analysis (CoMFA). In *QSAR: Rational Approaches to the Design of Bioactive Compounds.* Silipo C, Vittoria A, eds. Amsterdam: Elsevier Science, pp. 239–242.

Cramer RD III, DePriest SA, Patterson DE, Hecht P. (1993) The Developing Practice of Comparative Molecular Field Analysis. In *3D QSAR in Drug Design: Theory, Methods and Applications.* Kubinyi H, ed. Leiden: ESCOM, pp, 443–485.

Crippen GM. (1979) Distance Geometry Approach to Rationalizing Binding Data. J Med Chem 22, 988–997.

Crippen GM. (1980) Quantitative Structure–Activity Relationships by Distance Geometry: Systematic Analysis of Dihydrofolate Reductase Inhibitors. J Med Chem 23, 599–606.

Crippen GM. (1981) Quantitative Structure–Activity Relationships by Distance Geometry: Thyroxine Binding Site. J Med Chem 24, 198–203.

Crippen GM. (1982) Distance Geometry Analysis of the Benzodiazepine Binding Site. Mol Pharmacol 22, 11–19.

Crippen GM. (1987) Voronoi Binding Site Models. J Comp Chem 8, 943–955.

Crippen GM. (1989) Linearized Embedding: A New Metric Matrix Algorithm for Calculating Molecular Conformations Subject to Geometric Constraints, J Comput Chem 10, 896–902.

Crippen GM. (1995) Intervals and the Deduction of Drug Binding Site Models. J Comp Chem 16, 486–500.

Crippen GM. (1997) Validation of EGSITE2, a Mixed Integer Program for Deducing Objective Binding Site Models from Experimental Binding Data. J Med Chem 40, 3161–3172.

Crippen GM. (1999) Deducing Objective Site Models by Mixed Integer Programming. In *Rational Drug Design.* Truhlar DG, Howe WJ, Hopfinger AJ, Blaney J, and Dammkoehler RA, eds. *The IMA Volumes in Mathematics and Its Applications.* Vol. 108. New York: Springer Verlag, pp. 115–126.

Crippen GM, Havel TF. (1988) *Distance Geometry and Molecular Conformation.* New York: Research Studies Press (Wiley).

Folkers G, Merz A, Rognan D. (1993) CoMFA: Scope and Limitations. In *3D QSAR in Drug Design: Theory, Methods and Applications.* Kubinyi H, ed. Leiden: ESCOM, pp. 583–618.

Ghose AK, Crippen GM. (1982) Quantitative Structure–Activity Relationship by Distance Geometry: Quinazolines as Dihydrofolate Reductase Inhibitors. J Med Chem 25, 892–899.

Ghose AK, Crippen GM. (1983a) Essential Features of the Distance Geometry Model for the Inhibition of Dihydrofolate Reductase by Quinazolines. In *Proceedings of the Fourth European Symposium on Chemical Structure–Biological Activity: Quantitative Approaches.* In *Quantitative Approaches to Drug Design.* Dearden JC, ed. Amsterdam: Elsevier, pp. 99–108.

Ghose AK, Crippen GM. (1983b) Combined Distance Geometry Analysis of Dihydrofolate Reductase Inhibition by Quinazolines and Triazines. J Med Chem 26, 996–1010.

Ghose AK, Crippen GM. (1984) General Distance Geometry Three-Dimensional Receptor Model for Diverse Dihydrofolate Reductase Inhibitors. J Med Chem 27, 901–914.

Ghose AK, Crippen GM. (1985a) Use of Physicochemical Parameters in Distance Geometry and Related Three-Dimensional Quantitative Structure–Activity Relationships: A Demonstration Using *Escherichia coli* Dihydrofolate Reductase Inhibitors. J Med Chem 28, 333–346.

Ghose AK, Crippen GM. (1985b) Geometrically Feasible Binding Modes of a Flexible Ligand Molecule at the Receptor Site. J Comput Chem 6, 350–359.

Ghose AK, Crippen GM. (1990) Modeling the Benzodiazepine Receptor Binding Site by the General Three-Dimensional Structure-Directed Quantitative Structure–Activity Relationship Method REMOTEDISC. Mol Pharma 37, 725–734.

Ghose AK, Pritchett A, Crippen GM. (1987) Atomic Physicochemical Parameters for Three-Dimensional Structure Directed Quantitative Structure–Activity Relationships. III. Modeling Hydrophobic Interactions. J Comp Chem 9, 80–90.

Ghose AK, Crippen GM, Revankar GR, McKernan PA, Smee DF, Robins RK. (1989) Analysis of the In Vitro Antiviral Activity of Certain Ribonucleosides Against Parainfluenza Virus Using a Novel Computer Aided Receptor Modeling Procedure, J Med Chem 32, 746–756.

Golender VE, Vorpagel ER. (1993) Computer-Assisted Pharmacophore Identification. In *3D QSAR in Drug Design: Theory, Methods and Applications.* Kubinyi H, ed., Leiden: ESCOM, pp. 137–149.

Good AC, So S, Richards WG. (1993a) Structure–Activity Relationships from Similarity Matrices. J Med Chem 36, 433–438.

Good AC, Peterson SJ, Richards WG. (1993b) QSARs from Similarity Matrices. Technique Validation and the Application in the Comparison of Different Similarity Evaluation Techniques. J Med Chem 36, 2929–2937.

Goodsell DS, Olson AJ. (1990) Automated Docking of Substrates to Proteins by Simulated Annealing. Proteins, 8, 195–202.

Greco G, Novellino E, Martin YC. (1997) Approaches to Three-Dimensional Quantitative Activity Relationships. In *Reviews in Computational Chemistry,* Vol. 11. Lipkowitz KB, Boyd DB, eds. New York: Wiley-VCH, pp. 183–240.

Hansch C, Leo A. (1995) *Exploring QSAR: Fundamentals and Applications in Chemistry and Biology.* Washington, DC: American Chemical Society.

Höltje H-D, Kier LB. (1974) Sweet Taste Receptor Studies Using Model Interaction Energy Calculations. J Pharm Sci 63, 1722–1725.

Höltje, H-D; Anzali, S, Dall, N, Höltje, M. (1993) Binding Site Models. In *3D QSAR in Drug Design: Theory, Methods and Applications.* Kubinyi H, ed. Leiden: ESCOM, pp. 320–335.

Hopfinger AJ. (1980) A QSAR Investigation of Dihydrofolate Reductase Inhibition by Baker Triazines Based upon Molecular Shape Analysis. J Am Chem Soc 102, 7196.

Jain AN, Koile K, Chapman D. (1994) Compass: Predicting Biological Activities from Molecular Surface Properties. Performance Comparisons on a Steroid Benchmark. J Med Chem 37, 2315–2327.

Jain AN, Harris NL, Park JY. (1995) Quantitative Binding Site Model Generation: Compass Applied to Multiple Chemotypes Targeting the $5HT_{1A}$ Receptor. J Med Chem 37, 1295.

Jones G, Willett P, Glen RC, Leach AR, Taylor R. (1997) Development and Validation of a Genetic Algorithm for Flexible Docking. J Mol Biol 267, 727–748.

Kellogg GE, Semus SF, Abraham DJ. (1991) HINT: A New Method of Empirical Hydrophobic Field Calculation for CoMFA. J CAD Mol Des 5, 545–552.

Klebe G. (1993) Structural Alignment of Molecules. In *3D QSAR in Drug Design: Theory, Methods and Applications.* Kubinyi, H., ed., Leiden: ESCOM, pp. 173–199.

Klebe G, Abraham, U.; Mietzner, T. (1994) Molecular Similarity Indices in a Comparative Analysis (CoMSIA) of Drug Molecules to Correlate and Predict Their Biological Activity. J Med Chem 37, 4130–4146.

Kubinyi H, ed. (1993) *3D QSAR in Drug Design: Theory, Methods and Applications.* Leiden: ESCOM.

Kubinyi H, Abraham U. (1993) Practical Problems in PLS Analyses. In *3D QSAR in Drug Design: Theory, Methods and Applications.* Kubinyi H, ed. Leiden: ESCOM, pp. 717–728.

Mankino S, Kuntz ID. (1997) Automated Flexible Ligand Docking and Its Application for Database Search. J Comp Chem 18, 1812–1825.

Marshall GR. (1993) Binding-Site Modeling of Unknown Receptors. In *3D QSAR in Drug Design: Theory, Methods and Applications.* Kubinyi H, ed. Leiden: ESCOM, pp. 80–116.

Marshall GR, Barry CD, Bosshard HE, Dammkoehler RA, Dunn DA. (1979) The Conformational Parameter in Drug Design: The Active Analog Approach. In *Computer-Assisted Drug Design.* Olson EC, Christoffersen RE, eds. Washington, DC: American Chemical Society, pp. 205–226.

Martin YC. (1978) *Quantitative Drug Design.* New York: Dekker.

Oprea TI, Waller CL. (1997) Theoretical and Practical Aspects of Three-Dimensional Quantitative Activity Relationships. In *Reviews in Computational Chemistry,* Vol. 11. Lipkowitz KB, Boyd DB, eds. New York: Wiley-VCH, pp. 127–182.

Robert D, Amat LI, Carbó-Dorca R. (1999) Three-Dimensional Quantitative Activity Relationships from Tuned Molecular Quantum Similarity Measures: Prediction of the Corticosteroid-Binding Globulin Binding Affinity for a Steroid Family. J Chem Inf Comp Sci 39, 333–344.

Sandak B, Wolfson HJ, Nussinov R. (1998) Flexible Docking Allowing Induced Fit in Proteins: Insights from an Open to Closed Conformational Isomer. Proteins, 32, 159–174.

Schnitker J, Gopalaswamy R, Crippen GM. (1997) Objective Models for Steroid Binding Sites of Human Globulins. J CAD Mol Des 11, 93–110.

Simon Z. (1993) MTD and Hyperstructure Approaches. In *3D QSAR in Drug Design: Theory, Methods and Applications.* Kubinyi H, ed. Leiden: ESCOM, pp. 307–319.

Simon Z, Chiriac A, Motoc I, Holban S, Ciubotariu D, Szabadai Z. (1976) Receptor Site Mapping. Search Strategy of Standard for Correlations with Minimal Steric Differences. Stud Biophys 55, 217.

Simon Z, Chiriac A, Holban S, Ciubotariu D, Mihalas GI. (1994) *Minimum Steric Difference. The MTD-Method for QSAR Studies.* Letchworth: Research Studies Press.

Snyder JP, Rao SN, Koehler KF, Vendani A. (1993) Minireceptors and Pseudoreceptors. In *3D QSAR in Drug Design: Theory, Methods and Applications.* Kubinyi H, ed. Leiden: ESCOM, pp. 336–354.

Srivastava S, Crippen GM. (1993) Analysis of Cocaine Receptor Site Ligand Binding by Three-Dimensional Voronoi Site Modeling Approach, J Med Chem 36, 3572–3579.

Srivastava S, Richardson WW, Bradley MP, Crippen GM. (1993) Three-Dimensional Receptor Modeling Using Distance Geometry and Voronoi Polyhedra. In *3D QSAR in Drug Design: Theory, Methods and Applications.* Kubinyi H, ed. Leiden: ESCOM, pp. 409–430.

Tokarsky JS, Hopfinger AJ. (1994) Three-Dimensional Molecular Shape Analysis: Quantitative Structure–Activity Relationship of a Series of Cholecystokinin-A Receptor Antagonists, J Med Chem 37, 3639.

Vendani A, Zbinden P, Snyder JP. (1993) Pseudoreceptor Modeling: A New Concept for the Three-Dimensional Construction of Receptor Binding Sites, J Recept Res 13, 163–177.

Wermuth C, Langer T. (1993) Pharmacophore Identification. In *3D QSAR in Drug Design: Theory, Methods and Applications*. Kubinyi H, ed. Leiden: ESCOM, pp. 117–136.

Wold S, Ruhe A, Wold H, Dunn WJ. (1984) The Colinearity Problem in Linear Regression. The Partial Least Squares (PLS) Approach to Generalized Inverses. SIAM J Sci Stat Comput 5, 735.

Wold S, Johansson E, Cocchi M. (1993) PLS—Partial Least Squares Projections to Latent Structures. In *3D QSAR in Drug Design: Theory, Methods and Applications*. Kubinyi H, ed., Leiden: ESCOM, pp. 523–550.

5

Binding Energy Landscapes of Ligand–Protein Complexes and Molecular Docking

Principles, Methods, and Validation Experiments

Gennady M. Verkhivker, Djamal Bouzida, Daniel K. Gehlhaar, Paul A. Rejto, Sandra Arthurs, Anthony B. Colson, Stephan T. Freer, Veda Larson, Brock A. Luty, Tami Marrone, and Peter W. Rose

Agouron Pharmaceuticals, Inc., A Pfizer Company
San Diego, California

I. INTRODUCTION

Deciphering of the common successes and failures in ligand–protein docking, revealed by using different docking strategies and energy functions, allows one to establish fundamental connections between the topology of binding energy landscapes and the results of docking simulations. Binding energy landscape analysis is presented for the methotrexate–dihydrofolate reductase system, which represents a common success in molecular docking of ligand–protein complexes, and for the cyclodextrin glycosyltransferase–maltose complex, which is an example of a common failure in molecular docking simulations. The successes and failures in docking simulations are explained based on the thermodynamic properties determined from equilibrium simulations and the shape of the underlying binding energy landscape. The robust topology of the native structure is a decisive factor contributing to the thermodynamics and dynamics of the methotrexate–dihydrofolate reductase (MTX–DHFR) system that appear to be robust to structural perturbations, variations in the ligand composition, and accuracy of the energetic model. A hierarchical approach that involves a hierarchy of energy functions is proposed in the analysis of a common failure in molecular docking. A protocol of identifying clusters of structurally similar low-energy conformations, generated in equilibrium simulations with the simplified energy function, and subsequent energy minimization with the molecular mechanics force field, resolves a typical common failure in molecular docking.

II. PHYSICAL PRINCIPLES OF MOLECULAR RECOGNITION

Computer simulations of ligand–protein interactions have become a valuable tool in understanding the molecular recognition process on an atomic level and facilitating structure-based ligand discovery (1–12). It has been recognized that mechanisms of protein folding and ligand–protein binding are similar and can be elucidated by studying the nature of the underlying energy landscape, which describes the free energy of the system as a function of its coordinates (13–18). Common aspects of protein folding and ligand–protein binding phenomena include the existence of a thermodynamically stable native structure, a large number of conformational states available to the system, and the complex nature of interactions, which in the case of molecular recognition results in a frustrated binding energy landscape with a multitude of energetically similar but structurally different ligand–protein binding modes (16–23). It has been widely accepted that the underlying physical forces that govern the process of ligand–protein binding are the same as in protein folding (24–26). The major components of protein stabilization are hydrophobic interactions and hydrogen bonds, with the hydrophobic effect representing the dominant force in stabilizing the protein structure and defined as the combined effect of protein internal van der Waals interactions and hydration of nonpolar groups (24,27). Theoretical studies have used experimental results on the types of stabilizing forces in folding and binding to establish the appropriate energy models. Subsequently, analysis of the resulting energy landscape has been used to examine the molecular basis of protein folding and ligand–protein binding phenomena and to elucidate the thermodynamic stability and kinetic accessibility of the native structure, the nature and origins of meta-stable folding intermediates, and ligand–protein intermediate complexes (16–18,21–23,28–33).

A. Computational Models of Ligand–Protein Binding Thermodynamics

The development of structure-based drug design strategies requires the ability to combine accurate computational free energy models and elaborate calorimetric thermodynamic measurements to accurately interpret and predict binding affinity of ligand–protein complexes from structural considerations. A detailed description of ligand–protein association implies a delicate balance between van der Waals and electrostatic interactions, hydrogen bonding, solvation effects, and conformational entropy, all of which are difficult to compute accurately (2,5,25,34). In general, the first principles of statistical mechanics do not allow decomposition of binding free energy into separate terms (35). However, complex biophysical phenomena such as molecular recognition can be analyzed with empirical free energy models that postulate a small number of terms contributing to the binding process (36–39). Theoretical analysis of the thermodynamic aspects of molecular recognition has led to a number of approaches (36–51) that describe the essential steps of the binding process. Empirical binding free energy is typically comprised of enthalpy contributions resulting from the van der Waals and electrostatic ligand–protein interactions, and entropic contributions resulting from the release of water molecules from the ligand–protein interface during complex formation (43,44). Knowledge-based statistical ligand–protein interaction potentials derived from a database of inhibitor–enzyme complexes have been combined with the solvation parameters, hydrophobicity, and conformational entropy scales developed in protein folding and binding (37,39,48,51).

A structure-based thermodynamic analysis approach has been recently introduced that is based upon structural parameterization of folding and binding energetics of proteins, peptides, and synthetic ligands (52–58), as well as on the formalism that identifies the probability of different protein regions and individual amino acids being folded in equilibrium intermediates (59–64). The resulting binding free energy model includes the generic portion of the Gibbs free energy, calculated separately for the enthalpy and entropy components, and the electrostatic and ionization effects, as well as the contribution due to the change in translational degrees of freedom. The enthalpy contribution of the free energy results from the formation of van der Waals interactions, hydrogen bonding, and concomitant desolvation of the interacting groups. This free energy component is parameterized in terms of changes in apolar and polar solvent-accessible surface areas. The entropy contribution is composed of solvation component and changes in conformational degrees of freedom. Conformational entropy upon folding or binding is evaluated by considering the following three contributions for each amino acid: the entropy change associated with the transfer of a side chain from the interior of the protein to its surface; the entropy change gained by a surface-exposed side chain when the backbone changes from a unique folded conformation to an unfolded conformation; and the entropy change gained by the backbone upon unfolding from a unique native conformation. The magnitude of conformational entropy contributions for each amino acid has been estimated by computing the probability profiles of different conformational states as a function of dihedral angles (52,56,64).

A detailed structural mapping of the binding energetics, which has been performed for a number of peptidic and synthetic HIV-1 protease inhibitors (57), has shown an excellent agreement with the experimental data. The major contribution to the Gibbs free energy was determined by the hydrophobic effect resulting from the favorable entropy of water molecules released from ligand and protein groups. The enthalpy contributions were unfavorable at room temperature and were dominated by the positive enthalpy of desolvating hydrophobic groups. The unfavorable desolvation enthalpy of hydrophobic groups dominated the overall enthalpy component by offsetting the favorable enthalpy of forming inhibitor–HIV-1 protease interactions. These data are consistent with the analysis of hydration enthalpy and entropy contributions to protein folding, derived earlier in calorimetric studies (65,66), and they support the notions that stabilizing forces in protein folding and ligand–protein binding are rather similar and that appropriately derived energetic models can adequately describe both folding and binding phenomena.

A binding free energy model that includes a molecular mechanics interaction energy term, empirical solvation, and conformational entropy terms was compared with the calorimetric data on binding and protein unfolding (46,47). This energy function was parameterized in terms of changes in the polar and nonpolar solvent-accessible surface areas upon binding for a diverse set of ligand–protein complexes. It has been found not only that there is good agreement between this simple theoretical model and calorimetric binding data, but also that the binding free energy model can adequately describe the thermodynamics of protein unfolding (47). Correlation between experimental binding free energies and theoretical models that describe binding in terms of changes in contributions from the nonpolar and polar components at the ligand–protein interfaces has been found in some complexes that associate as rigid bodies (25,67). It has been also suggested that not only the hydration contributions but also the van der Waals and electrostatic intermolecular interactions may be proportional to the size of the ligand–protein interface (25). However, the presence of packing defects, coupling between local folding and binding, and the presence of strong interactions between a few residues in the ligand–protein interface (68) may complicate binding affinity analysis. Consequently, even if general contributing forces and interactions in ligand–protein binding are well established, the reasonably accurate prediction of binding free energies requires a combination of rigorous thermodynamic and structural information coupled with accurate representation of the energetics of both unbound and bound states in the ligand–protein association reaction.

B. Computational Models of Ligand–Protein Dynamics and Molecular Docking

Computational studies of molecular recognition usually require the consistent and rapid determination of the global energy minimum of a ligand–protein complex, which must correspond to the experimentally solved X-ray structure (1,4,11). Recent advances in computational structure prediction of ligand–protein complexes utilize a diverse range of energetic models, based on either surface complementarity (69–76) or atom–atom representations of the intermolecular interactions (21,77–81). A variety of optimization docking techniques include Monte Carlo methods (82–84), molecular dynamics (85,86), genetic algorithms (87–89), and tabu searching algorithm (90) and are focused primarily on molecular docking of flexible ligands into proteins that are held fixed in a bound conformation, while the internal degrees of freedom of the ligand and its rigid-body variables are optimized. Combined flexible ligand docking and protein side chain optimization techniques have been proposed in molecular recognition studies (91–93). A variant of the dead-end elimination (DEE) algorithm has been used to avoid a combinatorial explosion by restricting both the ligand and the side chains of the receptor residues to a limited number of discrete low energy conformations (91). The combinatorial problem in flexible peptide docking with major histocompatibility complexes receptors was also approached by utilizing the DEE algorithm to optimize protein side chains that adopt to the docked peptide conformations (92). A hierarchical computational approach, introduced for predicting structures of ligand–protein complexes and analyzing binding energy landscapes, combines Monte Carlo simulated annealing technique to determine the ligand-bound conformation with the DEE algorithm for side chain optimization of the protein active site residues (93). In this method, each of the docked ligand conformations is used to generate the template for a subsequent step of protein side chain optimization with the DEE procedure. Local minimizations and energy evaluations of the generated DEE solutions are performed at the final stage of this protocol. Limited protein side chain flexibility has been employed in the GOLD program, which takes into account rotational flexibility of hydrogens (89). Other approaches incorporate protein flexibility by using rotamer libraries of side chains (76,91–93), Monte Carlo simulations combined with minimization in flexible binding sites (84), or molecular dynamics docking simulations (86). A combination of energetic models with stochastic optimization techniques have led to a number of powerful strategies for computational structure prediction of ligand–protein complexes and docking of flexible ligands to a protein with a rigid backbone and flexible side chains has now become more feasible (91–95). A simplified energy function in combination with evolutionary sampling technique was developed to satisfy both thermodynamic and kinetic requirements in docking simulations by reducing frustration of the underlying binding energy landscape

(21,22,78,96). Robust structure prediction of bound ligands given a fixed conformation of the native protein was achieved with this family of energy functions by generating binding energy landscapes with coexisting correlated, funnel-like (97–99) and uncorrelated, rugged features. While adequate for nonpolar and hydrogen bonds patterns, this simplified energy model does not include a direct electrostatic component and therefore may be expected to fail when extensive networks of electrostatic interactions are present in the crystal structures. By contrast, the GOLD algorithm employs a template of protein hydrogen bond donors and acceptors, and it uses a genetic algorithm to sample intermolecular hydrogen bond networks and ligand conformations (89). This approach lacks a desolvation component and was found to be less suitable in finding hydrophobic interactions. Docking methodologies implemented in such programs as Flex X (79,80), Hammerhead (81), and GOLD (89) have been validated on a large number of ligand–protein complexes with known crystal structures to test robustness of the method. There have been also studies that employed explicit protein flexibility (100,101). However, the results of flexible ligand docking with a receptor in the absence of any experimentally known protein-bound conformation are considerably less reliable (102). Applications of flexible ligand docking techniques range from the analysis of the binding energy landscapes (103,104) to lead discovery (17,105), database mining (106), and structure-based combinatorial ligand design (107) and include simulations with ensembles of multiple ligands (108) and ensembles of multiple protein conformations (109,110).

Docking simulations usually determine a single structure of the complex with the lowest energy and postulate that the lowest energy conformation corresponds to the native structure. The number of low energy structures is usually very large, and a computationally demanding task of finding the lowest energy structure does not imply its thermodynamic stability. Nevertheless, the structure prediction problem implies determination of the ensemble of many similar conformations that describe the thermodynamically stable native basin of the global energy minimum rather than a single structure. The conjecture that there are more low energy conformations surrounding the native state than nonnative local minima was used to recognize near-native protein structures in ensembles of misfolded decoys (111). It was suggested that uniform sampling of the conformational space with the low resolution energy function followed by identification of the largest cluster of structurally related low energy conformations may be more efficient in finding the region of the conformational space that contains the native structure than the energy-based criteria (111).

The NP hardness of the ligand–protein recognition problem, as in protein folding, implies that for a given protein there may be ligands that do not find the global free energy minimum on the binding energy landscape in a reasonable amount of computer time, given a high degree of complexity and frustration of the underlying binding energy landscape. Nevertheless, ligand–protein complexes with experimentally determined X-ray structures must recognize their global free energy minimum rapidly and consistently. It was suggested that crystallographically solved structures of ligand–protein complexes have unfrustrated molecular recognition landscapes (16,21–23) and thereby provide a framework for exploring the relationship between the shape of these landscapes and the results of docking simulations.

We analyze the topological features of the native complexes that are critical for robust structure prediction and thermodynamic stability and are determined by early ordering of the recognition ligand motif in its native conformation. Structural stability of these motifs contributes decisively to the topology and thermodynamic stability of the native ligand–protein complex (112). These molecular fragments, termed recognition anchors, exhibit a high structural consensus or accessibility of the dominant native binding mode in docking simulations. In addition, when these molecular fragments are embedded in larger molecules, they maintain structural stability of the bound conformation, a property that we termed structural harmony (17,96). We have established that the results of kinetic docking simulations can be rationalized based on the thermodynamic properties of ligand–protein binding determined from equilibrium simulations and the analysis of the binding energy landscape (103,104). We present the thermodynamic analysis of the binding energy landscape for the methotrexate–dihydrofolate reductase ligand–protein complex, which is primarily determined by the topology of the native binding mode, appears to be robust to structural perturbations and variations in the ligand composition, and is relatively insensitive to the accuracy of the energetic model describing ligand–protein interactions (103,104,113).

Comparing the results of validation docking experiments on a large number of Protein Data Bank (PDB) ligand–protein complexes with the GOLD program (89) and with our docking strategy (78), we have detected a number of complexes for which both methods fail to predict the crystal structures. Misdocked predictions in ligand–protein docking can be categorized as soft and hard failures. In soft failure, the energy of the crystal structure, after minimization with the chosen force field, is lower than the energy of the lowest energy conformation found in docking simulations. A soft failure is due to a flaw in the search algorithm, which is unable to find the global energy minimum. Hard failures are more difficult; they arise when the energy of a misdocked structure is lower than the energy of the minimized crystal structure. Hard failures result from an inability to accurately reproduce subtle differences in the relative energies of alternate binding modes, a problem that compounded by competing electrostatic and van der Waals interactions, which results in a frustrated binding energy landscape.

Deciphering of the common hard failures in ligand–protein docking, revealed by using different docking strategies and energy functions, allows one to establish connections between topology of the binding energy landscapes and the results of docking simulations. Following the notion that important aspects of ligand–protein binding such as the thermodynamic stability and kinetic accessibility of the native structure can be rationalized based on similarities with protein folding, we hypothesize that the native structure of the complexes that are hard failures in docking do not provide structurally robust and thermodynamically stable topology, and misprediction of the crystal structure may result from high kinetic accessibility of the misdocked binding mode that is only a meta-stable local minimum. We find that hard failures in ligand–protein docking can arise when the underlying energy landscape is rugged, with a number of misdocked frustrated binding modes separated by high energy barriers, or when the landscape is shallow and flat with small barriers and marginally stable binding modes. The results of docking and equilibrium simulations can be used not only to generate binding energy landscapes but also to identify clusters of structurally similar conformations that can define the diversity of binding funnels (32,33,114) leading to different local minima. We demonstrate the feasibility of predicting the crystal structure of a ligand–protein complex that belongs to the class of common hard failures in molecular docking by identifying clusters of structurally similar conformations followed by the energy refinement of cluster representatives. We show how the thermodynamic analysis of the binding energy landscapes complements the results of kinetic docking simulations in resolving a typical hard failure in molecular docking and detecting the native binding mode that is represented by an isolated island on a flat energy landscape with a narrow conformational funnel leading to the crystal structure.

III. ENERGETIC ASPECTS OF LIGAND–PROTEIN BINDING: HIERARCHICAL MODELS IN MOLECULAR DOCKING

A. Low Resolution, Knowledge-Based Energy Models

We have pursued a "plug-and-play" strategy with two different energy functions, a molecular mechanics force field and a simplified energy function, along with two different sampling techniques, evolutionary programming (78) and Monte Carlo simulations (103,104,110). The knowledge-based simplified energetic model includes intramolecular energy terms for the ligand, given by torsional and nonbonded contributions of the DREIDING force field (115), and intermolecular ligand–protein steric and hydrogen bond interaction terms calculated from a piecewise linear potential summed over all protein and ligand heavy atoms (Fig. 1). The parameters of the pairwise potential depend on the six different atom types: hydrogen bond donor, hydrogen bond acceptor, both donor and acceptor, carbon-sized nonpolar, fluorine-sized nonpolar, and sulfur-sized nonpolar. Primary and secondary amines are defined to be donors, while oxygen and nitrogen atoms with no bound hydrogens are defined to be acceptors. Sulfur is modeled as being capable of making long-range, weak hydrogen bonds, which allows for sul-

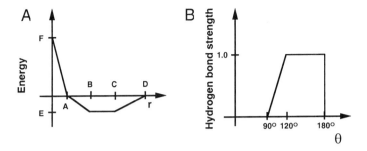

Figure 1 (A) The functional form of the ligand–protein interaction energy. For steric interactions, $A = 0.93B$, $C = 1.25B$, $D = 1.5B$, $E = -0.4$, $F = 15.0$, and $B = r_l + r_p$ is the sum of the atomic radii for the ligand and protein atoms. The atomic radius for carbon, oxygen, nitrogen atoms is 1.8 Å, 1.8 Å for fluorine and 2.2 Å for sulfur. Chlorine and bromide atoms are modeled as sulfur atoms. For hydrogen bond interactions, $A = 2.3$, $B = 2.6$, $C = 3.1$, $D = 3.4$, $E = -4.0$, $F = 15.0$. For sulfur–hydrogen bond interactions, $A = 2.7$, $B = 30.0$, $C = 3.5$, $D = 3.8$, $E = -2.0$, $F = 15.0$. For chelating interactions with the metals $A = 1.5$, $B = 1.7$, $C = 2.5$, $D = 3.0$, $E = -10.0$, $F = 15.0$. For repulsive interactions, $A = 3.2$, $E = 0.1$, $F = 15.0$, and B, C, and D are not relevant. A multiplicative desolvation penalty of 1.0 is applied to the attractive portion of the interaction between nonpolar and polar atoms. The units of A, B, C, and D are angstroms; for E and F the units are kilocalories per mole. (B) The hydrogen bond interaction energy is multiplied by the hydrogen bond strength term, which is a function of the angle θ determined by the relative orientation of the protein and ligand atoms. We defined θ as the angle between two vectors, one of which points from the protein atom to the ligand atom. For protein atoms with a single heavy-atom neighbor, the second vector connects the protein atom with its heavy-atom neighbor, while for protein atoms with two heavy-atom neighbors, it is the bisector of the vectors connecting the protein atom with its two neighbors. The long-range component of the repulsive term used for donor–donor, acceptor–acceptor, and donor–metal close contacts is scaled according to the relative positioning of the two atoms. The scaling used is equivalent to that used for hydrogen bonding. That is, the penalty is greatest when the angle θ is 180°, fading to zero at 90° and below.

fur donor closer contacts that are seen in some of the crystal structures. Crystallographic water molecules and hydroxyl groups are defined to be both donor and acceptor, and carbon atoms are defined to be nonpolar. The steric and hydrogen bond–like potentials have the same functional form, with an additional three-body contribution to the hydrogen bond term. The parameters were refined to yield the experimental crystallographic structure of a set of ligand–protein complexes as the global energy minimum (21,78). No assumptions regarding either favorable ligand conformations or any specific ligand–protein interactions were made, and all buried crystallographic water molecules are included in the simulations as part of the protein structure.

B. All-Atom-Based Energy Models

The all-atom-based energy function employed in this study contains an intramolecular term for the ligand, which consists of the van der Waals and torsional strain contributions of the DREIDING force field (115) and an intermolecular energy term that describes interactions between the ligand and the protein. The short-ranged repulsive interactions present in many molecular force fields such as AMBER (116,117) lead to rough energy surfaces with high energy barriers separating local minima. In this force field, small changes in position can lead to significant energy changes. For molecular docking simulations, it has been shown that the energy surface must be smooth for robust structure prediction of ligand–protein complexes (21); softening the potentials is a way to smooth the force field and enhance sampling of the conformational space while retaining adequate description of the binding energy landscape (103,118). We have shown that the modified AMBER force field and the simplified piecewise linear (PL) energy function produce comparable results during docking simulations in predicting crystal structures of ligand–protein complexes (103,104,110,113). Both the modified AMBER energy function and the PL energy function lack singularities at interatomic distances; they also do not effectively explore accessible ligand binding modes or sample a large fraction of conformational space, particularly at high temperature. These energy models have been adequate in structural and thermodynamic analysis of MTX–DHFR and biotin–streptavidin complexes (103). The molecular recognition energy function employed in the thermodynamic analysis of the MTX–DHFR binding energy landscape, presented in this study, employs the modified AMBER force field to describe ligand–protein interactions with a soft core smoothing component used to soften the repulsive part of the potential and includes an implicit solvation correction. A solvation term was added to the interaction potential to account for the free energy of interactions between the explicitly modeled atoms of the ligand–protein system and the implicitly modeled solvent. The term was derived by considering the transfer of an atom from an environment where it is completely surrounded by solvent to an environment in which it has explicit atomic neighbors (119). The solvation term reflects the driving force in ligand–protein binding that favors nonpolar groups to be buried at the interface, while this term imposes a penalty on dehydration of polar groups that needs to be compensated by formation of specific interactions. The proposed molecular recognition energy model is believed not only to predict crystal structures of the ligand–protein complexes as the global free energy minimum but also to faithfully reproduce the energetics of transitions on the ligand–protein association pathway.

   Although the standard AMBER force field is less amenable to searching, in principle it should describe more adequately the energetics of ligand–protein interactions, which is critical in recognizing the native binding mode for systems that belong to the category of hard failures in molecular docking. In this case, we use the standard molecular mechanics AMBER force field in conjunction with a desolvation correction (103,110,118) and an efficient parallel Monte Carlo dynamics using simulated tempering approach (120–125) to investigate the thermodynamics of the underlying binding energy landscape. This type of generalized-ensemble Monte Carlo methods employs global updates that significantly enhance thermalization of the system and overcome slow dynamics at low temperatures on rough energy landscapes, thereby permitting regions with a small density of states to be sampled accurately.

IV. COMPUTER SIMULATIONS OF LIGAND–PROTEIN BINDING

In simulations of ligand–protein interactions, the protein is held fixed in its bound conformation, while rigid-body degrees of freedom and rotatable angles of the ligand are treated as independent variables. Ligand conformations and orientations are sampled in a parallelepiped that encompasses the binding site obtained from the crystallographic structure of the corresponding complex, with a 10 Å cushion added to every side of this box in simulations of the MTX–DHFR system and 5 Å cushion in simulations with the hard failure ligand–protein complexes. Bonds allowed to rotate include those linking sp^3-hybridized atoms to either sp^3- or sp^2-hybridized atoms and single bonds linking two sp^2-hybridized atoms. The ligand bond lengths, the bond angles, and the torsional angles of the unrotated bonds were obtained from the crystal structures of the bound ligand–protein complexes. Crystallographic buried water molecules are included in the simulations as part of the protein structure.

A. Monte Carlo Simulations of Ligand–Protein Binding

In Monte Carlo simulations of ligand–protein binding, we employed the dynamically optimized acceptance ratio method whereby the maximum step sizes at each temperature are dynamically chosen to optimize the acceptance ratio, which is the ratio of accepted conformations to the total number of trial conformations (126). At a given cycle of the simulation, each degree of freedom can change randomly throughout some prespecified range determined by the acceptance ratio obtained during the preceding cycle. This range varies from one degree of freedom to another because of the complex nature of the energy landscape. At the end of each cycle, the maximum step size is updated and used during the next cycle. The moves are chosen as follows: a variable is selected at random, and then a uniform random displacement is given along each rigid-body degree of freedom or a randomly chosen dihedral angle of the ligand is rotated by a random angle along a randomly chosen axis.

In this approach, the simulations are arranged in cycles, and after a given cycle i, where the average acceptance ratio for each degree of freedom j is $\langle P_j \rangle^i$, the step sizes σ_j^i for each degree of freedom are updated for cycle $i + 1$ according to the formula

$$\sigma_j^{i+1} = \sigma_j^i \frac{\ln[a\langle P_{\text{ideal}} \rangle + b]}{}\tag{1}$$

where $\langle P_{\text{ideal}} \rangle$ is the desired acceptance ratio, chosen to be 0.5. The parameters a and b are used to ensure that the step sizes remain well behaved when the acceptance ratio approaches 0 or 1. They are assigned so that the ratio σ^{i+1}/σ^i is scaled up by a constant value s for $\langle P_j \rangle^i = 0$, and down by the same constant for $\langle P_j \rangle^i = 1$. Solving the equations

$$s^{-1} = \frac{\ln[a\langle P_{\text{ideal}} \rangle + b]}{\ln[b]}\tag{2}$$

$$s = \frac{\ln[a\langle P_{\text{ideal}} \rangle + b]}{\ln[a + b]}\tag{3}$$

with $s = 3$ yields $a = 0.673$ and $b = 0.065$.

We have performed equilibrium Monte Carlo simulations for the MTX–DHFR wild-type complex and the mutated system with a fixed DHFR protein conformation at $T = 100, 200, 300, 400, 500, 600, 700, 800, 900, 1000, 1500, 2000, 2500, 3000, 3500, 4000, 4500,$ and 5000 K. We updated the maximum step sizes using the acceptance ratio method every cycle of 1000 sweeps, and stored both the energy and the coordinates of the system at the end of each cycle. For all these simulations, we equilibrated the system for 1000 cycles (or a million sweeps), and collected data during 10,000 cycles (or 10 million sweeps) resulting in 10,000 samples at each temperature. A sweep is defined as a single trial move for each degree of freedom of the system.

An evolutionary algorithm, a stochastic optimization technique based on the ideas of natural selection (78), was used in ligand–protein docking simulations of the system that presents a hard failure in molecular docking. In this case, we have carried out equilibrium simulations by using parallel simulated tempering dynamics (120–125) with 50 replicas of the ligand–protein system attributed respectively to 50 different temperature levels that are uniformly distributed in the range between 5300 and 300 K. Independent local Monte Carlo moves are performed independently for each replica at the corresponding temperature level, but after a simulation cycle has been completed for all replicas, configuration exchanges for every pair of adjacent replicas are introduced. The mth and nth replicas, described by a common Hamiltonian $H(X)$, are associated with the inverse temperatures β_m and β_n, and the corresponding conformations X_m and X_n. The exchange of conformations between adjacent replicas m and n is accepted or rejected according to the Metropolis criterion with the probability

$$p = \min(1, \exp[-\delta])$$

where $\delta = [\beta_n - \beta_m][H(X_m) - H(X_n)]$. Starting with the highest temperature, every pair of adjacent temperature configurations is tested for swapping until the final lowest value of temperature is reached. This process of swapping configurations is repeated 50 times after each simulation cycle for all replicas, whereby the exchange of conformations presents an improved global update that increases thermalization of the canonical simulation for each replica. During the course of the simulation, each replica has a nonnegligible probability of moving through the entire temperature range and the detailed balance is never violated, two conditions that guarantee each replica of the system to be equilibrated in the canonical distribution with its own temperature (120–125).

B. The Weighted Histogram Analysis Method

The energy landscape approach (28–31) can elucidate such general properties of molecular recognition as the nature of the thermodynamic phases and barriers on the ligand–protein association pathway (103,104,110,113). This method evaluates equilibrium thermodynamic properties of the system from Monte Carlo simulations of the system at a broad temperature range with the aid of the optimized data analysis and the weighted histogram analysis technique (126–132).

Monte Carlo simulations can be used to calculate equilibrium averages of any quantity of interest, but in general computing these averages at different temperatures requires independent simulations at each temperature. With the single-histogram method, thermodynamic properties can be calculated at temperatures other than the simulation temperature provided there is accurate sampling of the density of states in the relevant range of energies (127,128,133–135). In practice, this requirement limits the applicability of the single-histogram method to temperatures near the simulation temperature. The multiple-histogram method (128,129) optimally combines simulation data obtained at many discrete temperatures to provide an improved estimate of the density of states, which can then be used over a range of continuous temperatures. A generalization of the multiple-histogram method, the weighted histogram analysis method (WHAM), estimates the density of states from data collected using umbrella sampling (128–130). All these histogram methods have been applied to simulations of biomolecules. In lattice models of protein folding, histograms have been used to calculate the native state probability density as a function of temperature (133), as well as the potential of mean force (PMF) as a function of the number of native contacts (134,135). Histograms have also been used to compute the PMF for both one-dimensional and multidimensional reaction coordinates at constant temperature (130–132). While alternate methods such as free energy perturbation and various weighting schemes are sufficient to compute one-dimensional PMFs, WHAM has been shown to be preferable for computing two-dimensional PMFs (136).

In this work, we apply the weighted histogram analysis method to compute ligand–protein binding energy landscapes, $F(R, T)$, as a continuous function of temperature and reaction coordinate. They are determined by first tabulating two-dimensional histograms $H_i(E, R)$ from the various constant temperature equilibrium simulations i, and then solving the self-consistent multiple-histogram equations (128) to yield the density of states:

$$W(E, R) = \frac{\sum_{i=1}^{M} g_i^{-1} H_i(E, R)}{\sum_{j=1}^{M} g_j^{-1} n_j \exp[-(E - F_j)/k_B T_j]} \quad (4)$$

where

$$\exp\left[\frac{-F_j}{k_B T_j}\right] = \sum_E W(E) \exp\left[\frac{-E}{k_B T_j}\right]$$

$$W(E) = \sum_R W(E, R) \quad (5)$$

and g_j depends on the correlation time τ_j as $g_j = 1 + 2\tau_j$, n_j is the number of samples at the temperature T_j.

Although these equations are expressions for the density of states as a function of both energy and reaction coordinate, the free energies are identical to those obtained from the standard one-dimensional multiple histogram equation

$$W(E) = \sum_R W(E, R) = \frac{\sum_{i=1}^{M} g_i^{-1} H_i(E)}{\sum_{j=1}^{M} g_j^{-1} n_j \exp[-(E - F_j)/k_B T_j]} \quad (6)$$

where

$$H_i(E) = \sum_R H_i(E, R) \quad (7)$$

and $H_i(E)$ is the standard one-dimensional histogram as a function of energy. These equations are precisely the self-consistent equations for the free energies in the one-dimensional multiple-histogram equations. Hence, the one-dimensional equations can be used to determine the free energies F_j and then to compute the multidimensional density of states $W(E, R)$. In this way, calculating the multidimensional density of states as a function of E and R requires no additional computational effort beyond tabulating the simulation data as a function of reaction coordinate as well as energy; the only difficulty is that more sampling is required to ensure adequate statistics.

From the probability density $W(E, R)$, the potential of mean force $F(R, T)$ at arbitrary temperature relative to a reference position R_c can be computed from the probability density $P(R, T)$ as follows:

$$F(R, T) = -k_B T \ln\left[\frac{P(R, T)}{P(R_c, T)}\right] \tag{8}$$

where

$$P(R, T) = \sum_E P_T(E, R) \tag{9}$$

$$P_T(E, R) = W(E, R) \exp\left[\frac{-E}{k_B T}\right] \tag{10}$$

We define R to be the root-mean-square deviation (rmsd) of the ligand coordinates from the native state, and the native state is chosen to be the reference state, so $R_c = 0.0$.

The ligand–protein binding reaction is studied as a sequence of transitions between different thermodynamic phases, where many conformations correspond to each phase. The ligand–protein association pathways and ligand–protein states are interpreted in terms of ensembles of conformations, where each step in the molecular recognition mechanism is regarded as a transitions between ensembles of rapidly interconverting conformations. Transitions between different phases of the ligand–protein system are analyzed by defining two characteristic equilibrium temperatures, T_{bound} and T_{native}. Temperature is a convenient control thermodynamics parameter that enables one to examine the cooperativity effects in ligand–protein binding, the relative effect of enthalpy and entropy, and mechanisms of transitions between a free energy minimum corresponding to the native structure of the ligand–protein complex and a free energy minimum corresponding to the entropically favorable phase of unbound ligand conformations. As entropy favors the random coil phase in protein folding at high temperature, so are unbound ligand and protein favored at high temperature in the binding process. At lower temperature, there may be nonspecific aggregation of the ligand either in the active site or in an alternative binding site of the protein, which corresponds to the collapsed globule state for proteins. At temperature T_{bound}, the ligand–protein system undergoes a transition to the bound phase, which is either a nativelike complex analogous to the molten globule state or a nonnative misdocked intermediate, reminiscent of the frozen collapsed state in protein folding. At temperature T_{native}, the transition of the ligand–protein system to the native complex is complete. To analyze the stage of the binding process when a transition to the native structure is complete, we define all structures within $R_{native} = 1.5$ Å from the native conformation to constitute the native binding domain or basin. From the temperature profile of the probability of being in the native domain, one can define the transition temperature T_{native} by $P(R_{\text{native}}, T_{\text{native}}) = 0.5$ (133), where the probability of being in the native binding domain is given by

$$P(R_{\text{native}}, T) = \frac{\sum_{R < R_{native}} P(R, T)}{\sum_R P(R, T)}$$

Cooperative binding is characterized by a narrow temperature interval between the transition temperatures Tbound and Tnative that guarantees the thermodynamic stability of the native complex at temperatures high enough to be kinetically reachable.

C. Similarity Clustering

Three-dimensional similarity calculations are based on the spatial proximity of atoms in a binding site and the atom type. We distinguish four types of atom: hydrogen bond donors, hydrogen bond acceptors, hydrogen bond donors and acceptors, and nonpolar atoms. The atom type compatibility $a(i, j)$ is assigned a value between 0.0 and 1.0, with the compatibility between two atoms of the same type defined as 1.0 and that between donor and acceptor atom as 0.0; other combinations of atoms have compatibilities between 0.0 and 1.0.

The spatial proximity between two atoms i and j is evaluated with a Gaussian function $p(i, j) = 10^{(-r_{ij}^2/\sigma^2)}$, where r_{ij} is the distance between atoms i and j, and $\sigma = -c^2/\log_{10}(p)$, where c and p denote the cutoff distance and the proximity threshold, respectively. Both the cutoff distance and the proximity threshold determine the shape of the Gaussian function to evaluate spatial proximity of two atoms, with $c = 3.0$ Å and $p = 0.000032$.

We calculate a descriptor $d(i,j)$ from the spatial proximity and the atom type compatibility:

$$d(i, j) = p(i, j) * a(i, j) \qquad \text{if } r(i, j) \le c$$

$$d(i, j) = 0 \qquad \text{if } r(i, j) > c$$

An atom descriptor $D_m^n(i)$ for atom i in molecule m is then calculated by summation over all N atoms in molecule n, $D_m^n(i) = \sum_{j=1}^{N} d_m^n(i, j)$. The intermolecular similarity between molecules m and n is given by the Tanimoto coefficient (137):

$$S(m, n) =$$
$$\frac{\sum_{i=1}^{M} D_m^m(i) D_m^n(i) + \sum_{j=1}^{N} D_n^m(j) D_n^n(j)}{\sum_{i=1}^{M} D_m^m(i)^2 + \sum_{j=1}^{N} D_n^n(j)^2 - \sum_{i=1}^{M} D_m^m(i) D_m^n(i) - \sum_{j=1}^{N} D_n^m(j) D_n^n(j)}$$

Molecules are grouped into clusters by comparing the intermolecular similarity coefficient. The first molecule is assigned to the first cluster. The next molecule is assigned to the cluster in which a cluster member has the highest similarity to the next molecule, if the similarity is above a threshold, chosen to be 0.85. Otherwise, the next molecule is assigned to a new cluster. The first member of a cluster is called the cluster leader. After all molecules have been assigned to clusters, the molecules are arranged in new order, starting with the largest cluster and proceeding to the smallest cluster. The reordered set of molecules is subjected to the same clustering procedure. This procedure is iterated until the information entropy converges to a minimum. We analyze clusters with at least 100 members. Since conformations that belong to the same cluster are equivalent with 85% structural similarity, different clusters are compared by analyzing cluster leaders.

V. MOLECULAR RECOGNITION IN THE MTX–DHFR COMPLEX: BINDING ENERGY LANDSCAPE ANALYSIS OF A COMMON SUCCESS IN MOLECULAR DOCKING

By studying the binding energy landscape of the MTX–DHFR complex, which presents a common success in molecular docking, we analyze the structural and thermodynamic origins of the robustness of the native structure topology to structural perturbations and modifications of the ligand composition. We have recently investigated the impact of ligand modifications on the thermodynamics of molecular recognition of the MTX–DHFR system and analyzed a relationship between binding mechanisms and the degree of ligand optimization (113) by applying the concept of hot and cold errors developed originally in protein folding studies (138). Mutations in protein sequences that are classified as cold errors affect the dynamics of the folding process but retain the native structure. By contrast, hot error mutations have more dramatic consequences and lead to protein misfolding. Native and near-native conformations are most affected by mutations, while their effect on the ensemble of unfolded conformations is less pronounced (138). Hot errors in the ligand are defined as those that destroy the thermodynamic stability and kinetic accessibility of the native binding mode, leading to a more rugged binding energy landscape with a misdocked lowest energy structure. The native binding mode of the complex is still maintained when the cold errors are introduced in the ligand, but a different shape of the energy landscape and more complicated kinetics of binding may result (113).

We have established (112) the binding modes for three major components of the MTX–DHFR complex in the complex with the *Lactobacillus casei* DHFR (139). MTX consists of three major components: a pteridine ring, a *p*-aminobenzoyl group, and a glutamic acid moiety portion (Fig. 2). While the pteridine ring is deeply buried in the active site and forms more than half of the hydrogen bonds with DHFR, the glutamate conformation is determined by specific electrostatic interactions of the glutamyl α-carboxylate with the guanidinium group of an argi-

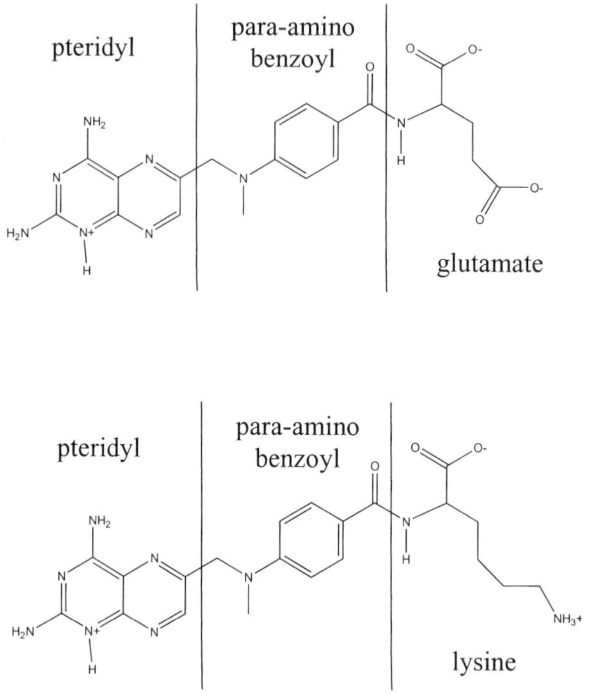

Figure 2 Breakdown of the MTX ligand into three components: pteridine ring, *p*-aminobenzoyl, and glutamyl side chain. In the mutant form of MTX, the glutamyl side chain is replaced by lysine.

nine residue. We found that both the glutamate portion and the pteridine ring are predicted consistently to bind in their crystallographic positions of the *L. casei* DHFR complex, while the *p*-aminobenzoyl group is predicted to bind in the same location that the pteridine ring binds in the MTX–DHFR complex (112). Hence, MTX is an example of a high affinity inhibitor containing two anchoring substructures that fulfill the kinetic requirements of structural consensus and structural harmony from docking simulations, but with the pteridine fragment presumably more thermodynamically stable than the glutamate moiety. In addition, we have shown that modifications in the pteridine ring of MTX belong in the category of hot errors that destroy the native binding mode (112). By analyzing the binding energy landscape of an all-carbon variant derivative of the pteridine fragment, we found that the crystallographic binding mode becomes considerably less stable relative to its alternate bound conformations. Furthermore, in a mutant of the pteridine fragment where the pteridine ring hydrogen bond acceptors are replaced with donors and donors replaced with acceptors, the crystallographic binding mode is never thermodynamically stable even at low temperatures, although it is still a local minimum (112).

The impact of ligand modifications on thermodynamic stability of the MTX–DHFR complex was studied with lysine substituted for the glutamate side chain of the MTX ligand (Fig. 2). In the native complex, each of the three moieties of the MTX ligand makes favorable interactions, and most substitutions are likely to replace some of these interactions by less favorable ones, thereby affecting the energy of the native state. The replacement of the acidic Gln by the basic Lys, which appeared to fall into this category, generates a chimera that consists of only one pteridine anchor fragment coupled with the Lys side chain. We have shown elsewhere that the predicted structure of the wild-type MTX–DHFR complex lies within 1.0 Å rmsd of the crystallographic structure (Fig. 3a); and the predicted structure of the mutant complex (Fig. 3b) is similar to the structure of the wild-type

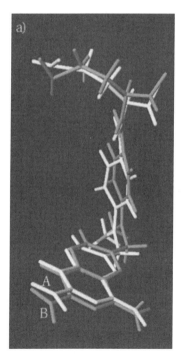
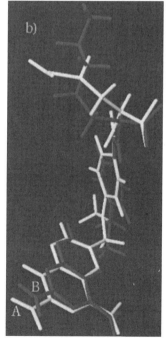

Figure 3 The crystal structure of MTX in the MTX–DHFR complex with the lowest energy conformation (red) obtained from 200 docking simulations, which defines its predicted structure (a), and the crystal structure of MTX in the MTX–DHFR complex with the lowest energy conformation of the Lys mutant of MTX obtained from docking simulations, which defines its predicted structure (b).

complex, with the exception of the portion of the ligand that was mutated (113). The fact that the structure for the mutant complex is so close to the wild-type structure suggests that the Lys mutation does not destroy the native binding mode and therefore may be regarded as a cold error.

Here, we focus on equilibrium aspects of the MTX–DHFR binding process such as the number and nature of thermodynamic phases available to the ligand–protein system and the relative stability of thermodynamically stable states; we also examine the difference between hot errors in the pteridine fragment of the MTX ligand and the potentially less severe cold error generated by the Lys substitution of the Glu side chain. The binding energy landscape for the MTX–DHFR complex at the transition temperatures is characterized by two major domains that represent the unbound and bound phases separated by a pronounced free energy barrier (Fig. 4a). The native binding domain of the MTX–DHFR system extends to nearly 5.0 Å rmsd from the native structure: there are no significant free energy barriers in this region. While there are local minima of the MTX–DHFR system between 8 and 10 Å from the native structure, there is a broad funnel of conformations that extends approximately to 7.0 Å and leads to the native binding mode with moderate barriers. At high temperatures the unbound states are a stable thermodynamic phase of the ligand–protein system. The native binding domain of the MTX–DHFR system dominates the thermodynamic equilibrium in the bound phase even at high temperatures and bound states that form alternative local minima are never thermodynamically stable. Because the crystal structure of the MTX–DHFR complex is thermodynamically stable at higher temperature than that of the mutated system (Fig. 4b), it promotes the consistent acquisition of the native state and a direct transition from random conformations in the unbound phase to the native binding domain for MTX–DHFR complex that forms the broad basin of attraction. By contrast, a lower temperature is required to stabilize the native state for the mutated system. In the computationally engineered Lys mutant of MTX, there is an additional, meta-stable, local minimum located approximately 3.0 Å rmsd from the low-energy structure (Fig. 4b). This meta-stable intermediate has lower free energy than the native state at high temperature, and only at lower temperature does the native state dominate the equilibrium. However, this intermediate is not a consequence of trapping in a specific misdocked binding mode but rather represents an ensemble of native like states. In this meta-stable intermediate, the pteridine ring is located in its native binding mode, with occasional fluctuations of the four-amino group of the pteridine ring between native and near-native conformations. However, the Lys side chain fluctuates significantly in this binding domain. The transition to the native like bound conformations for the computationally engineered system occurs at lower temperatures than for the "optimal" MTX–DHFR system, so the system may get trapped in misdocked or near-native-like conformations.

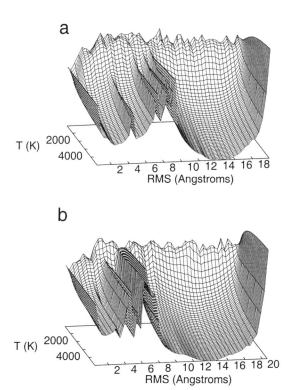

a

T (K) 2000
       4000

2 4 6 8 10 12 14 16 18
RMS (Angstroms)

b

T (K) 2000
       4000

2 4 6 8 10 12 14 16 18 20
RMS (Angstroms)

Figure 4   The binding free energy as a function of rmsd from the crystallographic conformation for the wild-type MTX–DHFR complex at the transition temperatures T_{bound} = 1500 K and T_{native} = 1000 K. The temperature-dependent binding energy profiles for (a), the MTX–DHFR complex from T = 100–5000 K and (b) the Lys mutant MTX–DHFR complex from T = 100–5000 K. In the binding energy profile, the reference energy $F(R = 0, T)$ is defined to be zero for each two-dimensional temperature slice.

The probability of being in the native structure for the pteridine anchor moiety and the entire MTX ligand coincides at the equilibrium temperature T_{bound}, when the MTX–DHFR system undergoes a transition to the bound phase (Fig. 5a). The wild-type MTX ligand has a single-step transition to the native structure of the complex. For the mutated ligand, by contrast, the pteridine ring adopts the native conformation first, at higher temperature. Only at somewhat lower temperature does the remainder of the ligand acquire the native conformation (Fig. 5b). We have determined that for the MTX–DHFR complex, the pteridine recognition anchor and the Glu side chain achieve their native conformations concurrently, and thereby the pronounced thermodynamic stability of the native structure leads to a single dominant funnel and eliminates not only

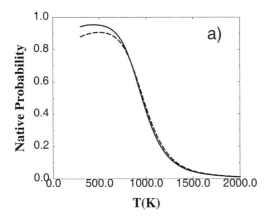

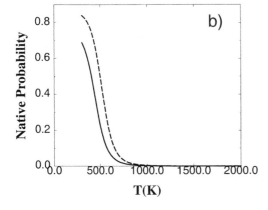

Figure 5 The native state probability as a function of temperature for (a) wild-type MTX (solid line) and the pteridine fragment in MTX (dashed line) and (b) the Lys mutant MTX ligand (solid line) and the pteridine fragment in the Lys derivative of MTX (dashed line). The native state for the pteridine fragment in wild-type MTX is defined as the crystallographic conformation of the fragment in the complex with DHFR. The native state for the pteridine fragment in the Lys mutant MTX ligand is defined as the conformation of the fragment in the lowest energy conformation of the Lys mutant of MTX.

frustrated regions, but also the necessity to form meta-stable, nativelike binding domains (103,104,113). The Lys side chain acquires its lowest energy conformation more slowly than the recognition pteridine anchor that results in the formation of the meta-stable intermediate complex, coherent with the native structure topology. The shape of the binding energy landscape for the mutant system dictates a more complicated scenario of binding, but the overall topology of the favored MTX binding mode is protected after this modification. The thermodynamic analysis has shown that Lys substitution in MTX is an example of a cold error in ligand design, since the native structure is preserved even though the dynamics of the binding process is affected. We find that structural stability of the crystal structure of the native MTX–DHFR complex results from the native interactions being significantly stronger on average than nonnative interactions, which leads to a gradual decrease of the energy of the native domain and a single, dominant conformational funnel directed to the crystal structure of the complex. Although the binding mechanism is generally determined by the overall shape of the binding energy landscape, the robust topology of the native structure contributes primarily to the thermodynamics and dynamics of the MTX–DHFR system and explains the success of molecular docking simulations in predicting the crystal structure of the complex.

V. MOLECULAR RECOGNITION OF THE CYCLODEXTRIN GLYCOSYLTRANSFERASE–MALTOSE COMPLEX: BINDING ENERGY LANDSCAPE ANALYSIS OF A COMMON FAILURE IN MOLECULAR DOCKING

The hard failure we shall examined is a complex of cyclodextrin glycosyltransferase with maltose-bound molecule (Fig. 6) in domain C (pdb entry 1cdg) (140). Three maltose binding sites have been observed on the surface of the enzyme, two in domain E, and one in domain C. The maltose molecules bound in the E domain interact with the protein residues that are implicated in a raw starch binding motif conserved among a diverse group of starch-converting enzymes (140). In the third carbohydrate binding site, located in domain C, the complex with the maltose

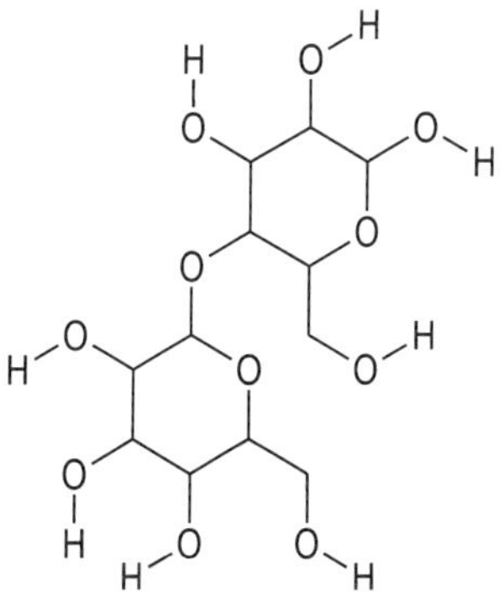

Figure 6   The structure of maltose molecule.

molecule has been determined crystallographically. Docking simulations using either the PL or the AMBER energy functions fail to predict the crystallographic binding mode of the complex as the lowest energy structure. The native binding mode was located in only a small fraction of docking simulations (Fig. 7a, c). The lowest energy structure determined from docking simulations with the PL energy function is located at rmsd = 6.1 Å from the crystal structure (Fig. 7b). The spectrum of low energy docking solutions consists primarily of the conformations that belong to the misdocked binding mode and the conformations with the nativelike binding mode are congregated at the tail of the spectrum (Fig. 7b). The lowest energy structure predicted with the AMBER energy function belongs to the same

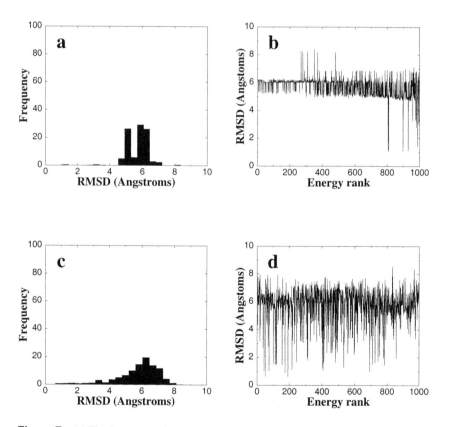

Figure 7 (a) The frequency of predicting binding modes of the 1cdg ligand–protein complex with the piecewise linear energy function and (b) the rmsd of the docked conformations from the crystal structure ranked by energy. (c) The frequency of predicting binding modes of the 1cdg ligand–protein complex with the AMBER force field. (d) The docked conformations as a function of RMSD from the crystal structure ranked by energy.

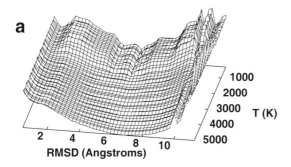

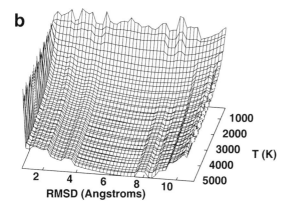

Figure 8 The binding free energy landscape for the 1cdg complex with (a) the piecewise linear energy function and (b) the AMBER force field. For each two—dimensional temperature slice, the reference energy $F(R = 0, T)$ is defined to be zero.

misdocked binding mode identified earlier by means of the PL energy function and is located at rmsd = 6.28 Å from the crystal structure (Fig. 7d). However, the corresponding spectrum of low energy docking solutions is more rugged, with three structurally different binding modes.

By performing equilibrium simulations, we generated the corresponding binding free energy profiles and registered the lowest energy structures using both energy functions. The predicted structure with the PL energy function is located at rmsd = 6.19 Å (Fig. 8a) from the native state and the lowest energy conformation obtained in simulations with the AMBER force field resides at rmsd = 7.04 Å from the crystal structure (Fig. 8b). The energy values of the predicted conformations, which belong to the same misdocked binding modes found in docking, are lower not only than the energy of the crystal structure but also than the energies of the corresponding predicted structures in the docking simulations. The binding free energy profile constructed with the PL energy function has at least four different binding domains, one of which is within 2.0 Å rmsd from the crystal structure (Fig. 8a). Nevertheless, the misdocked binding mode at rmsd = 6 Å from the crystal structure dominates thermodynamic equilibrium at the entire temperature range and is more favorable than the crystallographic conformation. The binding free energy profile is characterized by a broad basin of alternative conformations, which represents the major funnel on the energy landscape leading at lower temperatures to a family of misdocked binding domains. At high temperature, when the system can explore efficiently the conformational space, a broad basin in the region between 4 and 6 Å rmsd from the crystal structure is more stable and contributes to the thermodynamic equilibrium (Fig. 8a). The transition to the well-defined misdocked binding mode occurs only at lower temperatures. The funnel leading to the native structure narrows as temperature decreases, and the native binding mode is represented by an isolated island of conformations in the close vicinity of the crystal structure. Moreover, a significant energy barrier, which emerges at lower temperature, separates this region from the rest of the conformational space (Fig. 8a). While there is a narrow, though well-defined, funnel of conformations that leads to the native structure, the alternative binding modes are thermodynamically stable at all temperatures, and this promotes the consistent acquisition of the misdocked binding mode in docking simulations.

The true binding free energy landscape is multidimensional, whereas the binding free energy profile that we analyze is only a one-dimensional projection of this surface onto a single coordinate. One of the disadvantages of using rmsd as a reaction coordinate for the ligand–protein binding process is that it is difficult to distinguish energy minima that are distant from the crystal structure. For example, it is not possible to unambiguously conclude whether a single, broad low energy basin is located in the domain between 4 and 6 Å rmsd from the crystal structure, or whether many different local ligand–protein binding modes exist at this region. A convenient method to resolve this complication is to generate clusters of structurally similar conformations, since two distinct binding modes will result in two different conformational clusters. Structural clustering of the conformations generated from equilibrium simulations with the PL energy function produced the largest size clusters located at rmsd = 5 and 6 Å from the crystal structure (Fig. 9a). Only a relatively small cluster of native conformations was detected at rmsd = 1.0 Å from the native state, reflecting a narrow funnel in the proximity of the crystal structure. However, energy minimization with the AMBER energy function of the cluster leaders yields the lowest energy structure, located at rmsd = 0.78 Å from the crystal structure (Fig. 9b). Hence, using a two-step protocol of first identifying clusters of structurally similar conformations generated from equilibrium sampling in the uniform temperature range from 300 to 5000 K with the PL energy function, followed by minimization of the corresponding cluster

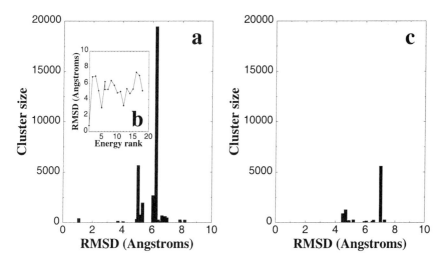

Figure 9 (a) The size of structurally similar clusters for the 1cdg ligand–protein complex with the piecewise linear energy function as a function of rmsd from the crystal. (b) The RMSD of the conformations minimized with the AMBER energy function, ranked by energy. (c) The size of structurally similar clusters for the 1cdg ligand–protein complex with the AMBER energy function as a function of rmsd from the crystal.

leaders with the AMBER force field, resolves the hard failure in ligand–protein docking and results in the structure prediction consistent with the crystallographic binding mode.

The binding energy landscape generated with the AMBER energy function is rather flat, with a number of shallow meta-stable local minima separated by small barriers. At higher temperatures a broad funnel of conformations in the region between rmsd = 4 and 8 Å from the native structure contributes significantly to the thermodynamic equilibrium (Fig. 8b). However, the relative stability of the binding modes changes as the temperature lowers, and the binding modes at rmsd = 3 and 7 Å from the crystal structure come to dominate the thermodynamic equilibrium at lower temperatures. Interestingly, most of the low docking solutions determined with the AMBER energy function belong to the region of the conformational space that is more stable at higher temperatures, and only a very small fraction of the predicted conformations populate regions that become more stable at lower temperatures. There are no clusters of structurally similar conformations sampled with AMBER that are centered closer than 4.0 Å rmsd from the crystal structure (Fig. 9c). The hard failure in docking with the AMBER energy function for the 1cdg ligand–protein system results from a number of misdocked frustrated binding modes on the flat binding energy landscape with no significant barriers.

The bound conformation of the ligand in the crystal structure interacts by its aromatic portion of the reducing sugar with the aromatic side chain of Trp413. The bound maltose molecule makes five direct hydrogen bonds to the protein. The carboxyl oxygen and NH group of Ile414 form hydrogen bonds with O-2′ and O-3′ atoms of the ligand, respectively. In addition, the carboxyl oxygen of Glu411 interacts with O-2 and the carboxyl oxygen of Gly446 forms a hydrogen bond with O-6 of the ligand (Fig. 10). The most populated binding mode, the largest cluster of structurally similar conformations generated with PL energy function, interacts with the same key protein residues, including Glu411 and Ile414, as the crystal structure. This binding mode has different networks of hydrogen bonds and, moreover, forms more hydrogen bonds than are formed in the crystal structure. The O-3 and O-3′ atoms of the maltose bound molecule form two hydrogen bonds with OE2 of Glu411 and O-2′ interacts with the carboxyl oxygen of Glu411. Other hydrogen bonds are formed with the carboxyl oxygen of Ile414, nitrogens of the Trp413, and Arg412 side chains (Fig. 10).

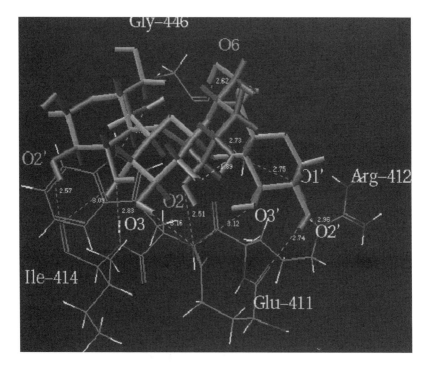

Figure 10   The crystal structure of the 1cdg ligand–protein complex with the lowest energy conformation obtained from docking simulations, which define its predicted structure.

VI. ANALYSIS OF THE COMMON HARD FAILURES IN MOLECULAR DOCKING: LESSONS FROM VALIDATION EXPERIMENTS

We discovered that for the 1cdg ligand–protein system the following two-step protocol led to predictions in better agreement with experiment than using either energy function by itself: identification of clusters of structurally similar low energy conformations generated in equilibrium simulations with the simplified PL energy function, followed by energy minimization with the AMBER energy function. This suggests that in some cases the PL energy function provides a better representation of the binding energy landscape. The robustness of simplified energy functions in structure prediction of ligand–protein complexes can generally be increased by generating more adequate energy landscapes, and regions with a high density of low energy states that describe the multitude of the binding modes must be detectable during simulations. In contrast, a more detailed energy function can produce kinetic traps because of high energy barriers and a rougher energy landscape susceptible to the precise geometry of the binding modes. Nevertheless, the more complete AMBER energy function is required to correctly rank the resulting conformations.

A simple energy function was sufficient to describe for a given complex the available binding modes that interact with the catalytic protein residues. A more detailed energy function discriminated between the relative stabilization energy of the binding modes determined by the precise balance of electrostatic and van der Waals interactions. Generally, a hierarchical process of similarity clustering of conformations generated from equilibrium simulations with a simplified energy function followed by energy refinement with a molecular mechanics force field may provide a useful approach in resolving a typical common failure in ligand–protein docking. These results, which show that a multistep protocol improves predictions of the crystal structure of the complex, may reflect hierarchical features of the binding process itself.

In light of similarities between ligand–protein binding and protein folding mechanisms, our hypothesis is related to a recent evolutionary protein folding model (141). This model has shown that thermodynamic stability and kinetic accessibility of the native structure could have emerged during the evolution of the protein function, namely, as a result of functional selection and structural stabilization of active site residues sufficient for the protein to fold to its native state. The stabilization of the protein structure was proposed to be hierarchical, where the active site residues first fold to their native conformation, followed by so-called coherent regions that are relatively unfrustrated and reach their folded conformation more slowly than the active site residues but faster than other, frustrated regions. Following this evolutionary model, it is tempting to speculate that certain conformations of a particular ligand that fulfill functional requirements, such as interactions with the catalytic protein residues, would have increased structural stability. In the case of hard failure complexes, multiple binding modes have been detected with the simplified energy function, each of which forms a stable network of interactions with the same key catalytic residues, which results in formation of multiple funnels on the binding energy landscape. Subsequent energetic stabilization of these binding modes determines the global free energy minimum corresponding to the crystal structure of the complex. It is possible that for hard failure complexes, the native binding modes with the lowest stabilization energy may have developed from narrow conformational funnels. As a result, the crystal structure could not be predicted as a single structure with the lowest energy, or by identifying the largest cluster of structurally similar conformations, but rather by energy refinement of all clusters.

VII. CONCLUSIONS

We have shown that well-optimized ligand–protein complexes, such as the methotrexate–dihydrofolate reductase system, have a single dominant binding funnel directed to the unique crystallographic binding mode, which is the thermodynamically stable global free energy minimum, robust against various structural perturbations. The results suggest that the important factor behind the consistent and rapid determination of the crystal structure of a ligand–protein complex is the shape of the binding energy landscape. Although the binding mechanism is generally determined by the shape of the entire binding energy landscape, the topology of the native structure is a major factor contributing to the thermodynamics and dynamics of the MTX–DHFR system. The pronounced thermodynamic stability of the native structure and the robust topology of the native binding mode, determined by the recognition pteridine anchor of the MTX ligand, rationalize a common success in molecular docking simulations of the MTX–DHFR complex obtained with a broad variety of energy functions. In contrast, we find that the stability of the native binding modes for the hard failures determined with different energy models is marginal, and minor structural changes will lead to changes in the relative stability of the native binding mode. The interactions that stabilize the crystal structure for these complexes may not stabilize near-native conformations that lead to a diversity of binding funnels. This marginal stability of the binding modes for hard failure complexes coupled with narrow regions in the conformational space that correspond to the nativelike conformations could make these systems highly sensitive to protein conformational fluctuations, and reliable structure prediction for the ligand–protein complexes under study becomes even more problematic with a flexible protein model. Subsequently, a very sensitive and highly optimized energy function is required to rank correctly the relative stabilization energy of the crystallographic binding mode of the hard failure complex.

This explains why common failures in molecular docking have been detected when different energy functions and searching methods were used.

Common failures in predicting crystal structures of ligand–protein complexes have been investigated for ligand–protein systems by a combined thermodynamic and kinetic analysis of the binding energy landscapes. The topography of the native binding funnels for hard failure complexes can be steep and narrow, thereby precluding robust structure prediction of the crystal structure. We have found that structural similarity can detect narrow funnels on a rugged surface and isolated regions of conformational space that may contain crystal structure conformation. The topology of the binding modes is less sensitive to the details of specific interactions, and a more adequate sampling of the multitude of the binding modes can be achieved with the PL energy function. The energy landscapes generated with the AMBER force field generally either are more rugged, with a number of misdocked meta-stable binding modes, or are characterized by shallow and flat binding domains corresponding to the marginally stable local minima. We have found that neither determining a single structure with the lowest energy nor finding the most common binding mode topology is sufficient to predict crystal structures of the complexes that belong to the category of hard failures. A comparison of a simplified, knowledge—based energy function and the molecular mechanics force field suggests that the more adequate sampling of the multitude of the binding modes is achieved with the simplified energy function. A hierarchical approach is proposed whereby structural similarity clustering of the conformations generated from equilibrium simulations with a simplified energy function is followed by energy refinement with the AMBER molecular mechanics force field. This protocol, which involves a hierarchy of energy functions (100,142), helps to resolve some common failures in ligand–protein docking. Analysis of other cases that represent hard failures in ligand–protein structure prediction involving both a rigid and a flexible protein model provides a next step in developing improved energy functions and docking strategies.

REFERENCES

1. Kuntz ID. Structure-based strategies for drug design and discovery. Science 1992; 257:1078–1082.
2. Straatsma TP, McCammon JA. Computational alchemy. Annu. Rev. Phys. Chem. 1992; 43:407–435.
3. Schoichet BK, Stroud RM, Santi DV, Kuntz ID, Perry KM. Structure-based discovery of inhibitors of thymidylate synthase. Science 1993; 259:1445–1450.
4. Cherfils J, Janin J. Protein docking algorithms, simulating molecular recognition. Curr. Opin. Struct. Biol. 1993; 3:265–269.
5. Kollman P. Free energy calculations—Applications to chemical and biological phenomena. Chem. Rev. 1993; 93:2395–2417.

6. Kuntz ID, Meng EC, Shoichet BK. Structure-based molecular design. Acc. Chem. Res. 1994; 27:117–123.

7. Ajay, Murcko MA. Computational methods to predict binding free energy in ligand–receptor complexes. J. Med. Chem. 1995; 38:4953–4967.

8. Lybrand TP. Ligand–protein docking and rational drug design. Curr. Opin. Struct. Biol. 1995; 5:224–228.

9. Rosenfeld R, Vajda S, DeLisi C. Flexible docking and design. Annu. Rev. Biophys. Biomol. Struct. 1995; 24:677–700.

10. Jones G, Willett P. Docking small-molecule ligands into active sites. Curr. Opin. Biotechnol. 1995; 6:652–656.

11. Gschwend DA, Good AC, Kuntz ID. Molecular docking towards drug discovery. J. Mol. Recognition 1996; 9:175–186.

12. Shoichet BK, Leach AR, Kuntz ID. Ligand solvation in molecular docking. Proteins Struct. Funct. Genet. 1999; 34:4–16.

13. Frauenfelder H, Sliger SG, Wolynes PG. The energy landscapes and motions of proteins. Science 1991; 254:1598–1603.

14. Frauenfelder H, Wolynes PG. Biomolecules, where the physics of complexity and simplicity meet. Phys. Today 1994; 47:58–64.

15. Frauenfelder H. Complexity in proteins. Nature Struct. Biol. 1995; 2:821–823.

16. Janin J. Quantifying biological specificity: The statistical mechanics of molecular recognition. Proteins Struct. Funct. Genet. 1996; 25:438–445.

17. Rejto PA, Verkhivker GM. Unraveling principles of lead discovery: From unfrustrated energy landscapes to novel molecular anchors. Proc. Natl. Acad. Sci. USA 1996; 93:8945–8950.

18. Verkhivker GM, Rejto PA. A mean field model of ligand–protein interactions. Implications for the structural assessment of human immunodeficiency virus type 1 protease complexes and receptor-specific binding. Proc. Natl. Acad. Sci. USA 1996; 93:60–64.

19. Mattos C, Ringe D. Multiple binding modes. In: Kubinyi H, ed. 3D QSAR in Drug Design: Theory, Methods, and Applications. Leiden: ESCOM, 1993, pp 226–254.

20. Mattos C, Rasmussen B, Ding X, Petsko GA, Ringe D. Analogous inhibitors of elastase do not always bind analogously. Nature Struct. Biol. 1994; 1:55–58.

21. Verkhivker GM, Rejto PA, Gehlhaar DK, Freer ST. Exploring energy landscapes of molecular recognition by a genetic algorithm, analysis of the requirements for robust docking of HIV-1 protease and FKBP-12 complexes. Proteins Struct. Funct. Genet. 1996; 25:342–353.

22. Rejto PA, Verkhivker GM, Gehlhaar DK, Freer ST. New trends in computational structure prediction of ligand–protein complexes for receptor-based drug design. In: van Gunsteren W, Weiner P, Wilkinson AJ, eds. Computational Simulation of Biomolecular Systems. Leiden: ESCOM, 1997, pp 451–465.

23. Verkhivker GM, Rejto PA. Mean field analysis of FKBP-12 complexes with FK-506 and rapamycin: Implications for a role of crystallographic water molecules in molecular recognition and specificity. Proteins Struct. Funct. Genet. 1997; 28:313–324.

24. Privalov PL. Physical basis of the stability of the folded conformations of proteins. In: Creighton T, ed. Protein Folding. New York: Freeman, 1992, pp 83–126.

25. Janin, J. Elusive affinities. Proteins Struct. Funct. Genet. 1995; 21:30–39.
26. Makhatadze GI, Privalov PL. Hydration effects in protein unfolding. Biophys. Chem. 1994; 51:291–309.
27. Makhatadze GI, Privalov PL. Energetic of protein structure. Adv. Protein Chem. 1995; 47:307–425.
28. Bryngelson JD, Onuchic JN, Socci ND, Wolynes PG. Funnels, pathways, and the energy landscape of protein folding, a synthesis. Proteins Struct. Funct. Genet. 1995; 21:167–195.
29. Dill KA, Bromberg S, Yue K, Fiebig KM, Yee DP, Thomas PD, Chan HS. Principle of protein folding—A perspective from simple exact models. Protein Sci. 1995; 4:561–602.
30. Dill KA, Chan HS. From Levinthal to pathways to funnels. Nature Struct. Biol. 1997; 4:10–19.
31. Shakhnovich EI. Theoretical studies of protein-folding thermodynamics and kinetics. Curr. Opin. Struct. Biol. 1997; 7:29–40.
32. Tsai C-J, Xu D, Nussinov R. Protein folding via binding and vice versa. Curr. Biol. 1998; 3:R71–R80.
33. Tsai C-J, Kumar S, Ma B, Nussinov R. Folding funnels, binding funnels and protein function. Protein Sci. 1999; 8:1181–1190.
34. Gilson MK, Given JA, Bush BL, McCammon JA. The statistical–thermodynamic basis for computation of binding affinities: A critical review. Biophys. J. 1997; 72:1047–1069.
35. Mark AE, van Gunsteren W. Decomposition of the free energy of a system in terms of specific interactions. Implications for theoretical and experimental studies. J. Mol. Biol. 1994; 240:167–176.
36. Krystek S, Stouch T, Novotny J. Affinity and specificity of serine endopeptidase–protein inhibitor interactions. Empirical free energy calculations based on X-ray crystallographic structures. J. Mol. Biol. 1993; 234:661–679.
37. Verkhivker GM, Appelt K, Freer ST, Villafranca JE. Empirical free energy calculations of ligand–protein crystallographic complexes. I. Knowledge-based ligand–protein interaction potentials applied to the prediction of HIV-1 protease binding affinity. Protein Eng. 1995; 8:677–691.
38. Wallqvist A, Jernigan RL, Covell DG. A preference—based free energy parametrization of enzyme–inhibitor binding. Applications to HIV-1 protease inhibitor design. Protein Sci. 1995; 4:1881–1903.
39. Vajda S, Sippl M, Novotny J. Empirical potentials and functions for protein folding and binding. Curr. Opin. Struct. Biol. 1997; 7:222–228.
40. Searle MS, Williams DH. The cost of conformational order: Entropy changes in molecular association. J. Am. Chem. Soc. 1992; 114:10690–10697.
41. Searle MS, Williams DH, Gerhard U. Partitioning of free energy contributions in the estimation of binding constants: Residual motions and consequences for amide–amide hydrogen bond strengths. J. Am. Chem. Soc. 1992; 114:10697–10704.
42. Williams DH, Cox JPL, Doig AJ, Gardner M, Gerhard U, Kaye PT, Lal AR, Nicholls IA, Salter CJ, Mitchell RC. Toward the semiquantitative estimation of binding constants. Guides for peptide–peptide binding in aqueous solution. J. Am. Chem. Soc. 1991; 113:7020–7030.

43. Horton N, Lewis M. Calculation of the free energy of association for protein complexes. Protein Sci. 1992; 1:169–181.
44. Böhm H-J. The development of a simple empirical scoring function to estimate the binding constant for a protein–ligand complex of known three-dimensional structure. J. Comput.-Aided Mol. Design 1994; 8:243–256.
45. Jain AN. Scoring noncovalent protein–ligand interactions, a continuous differentiable function tuned to compute binding affinities. J. Comput.-Aided Mol. Design 1996; 10:427–440.
46. Weng Z, Vajda S, DeLisi C. Prediction of protein complexes using empirical free energy function. Protein Sci. 1996; 5:614–626.
47. Weng Z, Delisi C, Vajda S. Empirical free energy calculation: Comparison to calorimetric data. Protein Sci. 1997; 6:1976–1984.
48. Zhang C, Vasmatzis G, Cornette JL, DeLisi C. Determination of atomic desolvation energies from the structures of crystallized proteins. J. Mol. Biol. 1997; 267:707–726.
49. Böhm H-J. Prediction of binding constants of protein ligands: A fast method for the prioritization of hits obtained from de novo design or 3D database search programs. J. Comput. Aided Mol. Design 1998; 12:309–323.
50. Takamatsu Y, Itai A. A new method for predicting binding free energy between receptor and ligand. Proteins Struct. Funct. Genet. 1998; 33:62–73.
51. Verkhivker GM. Empirical free energy calculations of ligand–protein crystallographic complexes. II. Knowledge-based ligand–protein interaction potentials applied to the thermodynamic analysis of hydrophobic mutations. In: Hunter L, Klein T, eds. Pacific Symposium on Biocomputing-96. Singapore: World Scientific, 1996, pp 638–652.
52. D'Aquino JA, Gomez J, Hilser VJ, Lee KH, Amzel LM, Freire E. The magnitude of the backbone conformational entropy change in protein folding. Proteins Struct. Funct. Genet. 1996; 25:143–156.
53. Gomez J, Freire E. Thermodynamic mapping of the inhibitor site of the aspartic protease endothiapepsin. J. Mol. Biol. 1995; 252:337–350.
54. Gomez J, Hilser VJ, Xie D, Freire E. The heat capacity of proteins. Proteins Struct. Funct. Genet. 1995; 22:404–412.
55. Hilser VJ, Gomez J, Freire E. The enthalpy change in protein folding and binding: Refinement of parameters for structure-based calculations. Proteins Struct. Funct. Genet. 1996; 26:123–133.
56. Luque I, Mayorga O, Freire E. Structure-based thermodynamic scale of α-helix propensities in amino acids. Biochemistry 1996; 35:13681–13688.
57. Bardi JS, Luque I, Freire E. Structure-based thermodynamic analysis of HIV-1 protease inhibitors. Biochemistry 1997; 36:6588–6596.
58. Luque I, Todd MJ, Gomez J, Semo N, Freire E. The molecular basis of resistance to HIV-1 protease inhibition: A plausible hypothesis. Biochemistry 1998; 37:5791–5797.
59. Xie D, Freire E. Molecular basis of cooperativity in protein folding. V. Thermodynamic and structural conditions for the stabilization of compact denatured states. Proteins Struct. Funct. Genet. 1994; 19:291–301.

60. Xie D, Freire E. Structure-based prediction of protein folding intermediates. J. Mol. Biol. 1994; 242:62–80.

61. Xie D, Fox R, Freire E. Thermodynamic characterization of an equilibrium folding intermediate of staphylococcal nuclease. Protein Sci. 1994; 3:2175–2184.

62. Hilser VJ, Freire E. Structure-based calculations of the equilibrium folding pathway of proteins. Correlation with hydrogen exchange protection factors. J. Mol. Biol. 1996; 262:756–772.

63. Hilser VJ, Freire E. Predicting the equilibrium protein folding pathway: Structure-based analysis of staphylococcal nuclease. Proteins Struct. Funct. Genet. 1997; 27:171–183.

64. Lee KH, Xie D, Freire E, Amzel M. Estimation of changes in side chain configurational entropy in binding and folding: General methods and application to helix formation. Proteins Struct. Funct. Genet. 1994; 20:68–84.

65. Makhatadze GI, Privalov PL. Contribution of hydration to protein folding thermodynamics. I. The enthalpy of hydration. J. Mol. Biol. 1993; 232:639–659.

66. Privalov PL, Makhatadze GI. Contribution of hydration to protein folding thermodynamics. II. The entropy and Gibbs energy of hydration. J. Mol. Biol. 1993; 232:660–679.

67. Spolar RS, Record MT Jr. Coupling of local folding to site—specific binding of proteins to DNA. Science 1994; 263:777–784.

68. Clackson T, Wells J. A hot spot of binding energy in a hormone–receptor interface. Science 1995; 267:383–386.

69. Shoichet BK, Kuntz ID. Protein docking and complementarity. J. Mol. Biol. 1991; 221:327–346.

70. Walls PH, Sternberg MJE. New algorithm to model protein–protein recognition based on surface complementarity. Applications to antibody–antigen docking. J. Mol. Biol. 1992; 228:277–297.

71. Vakser IA, Aflalo C. Hydrophobic docking, a proposed enhancement to molecular recognition techniques. Proteins Struct. Funct. Genet. 1994; 20:320–329.

72. Jackson RM, Sternberg MJE. A continuum model for protein–protein interactions: Application to the docking problem. J. Mol. Biol. 1995; 250:258–275.

73. Fisher D, Lin SL, Wolfson HJ, Nussinov R. A geometry—based suite of molecular docking processes. J. Mol. Biol. 1995; 248:459–477.

74. Norel R, Lin SL, Wolfson HJ, Nussinov R. Molecular surface complementarity at protein–protein interfaces, the critical role played by surface normals at well placed, sparse, points in docking. J. Mol. Biol. 1995; 252:263–273.

75. Gabb HA, Jackson RM, Sternberg MJ. Modeling protein docking using shape complementarity, electrostatics and biochemical information. J. Mol. Biol. 1997; 272:106–120.

76. Jackson RM, Gabb HA, Sternberg MJE. Rapid refinement of protein interfaces incorporating solvation: Application to the docking problem. J. Mol. Biol. 1998; 276:265–285.

77. Friedman AR, Roberts VA, Tainer JA. Predicting molecular interaction and inducible complementarity, fragment docking of Fab–peptide complexes. Proteins Struct. Funct. Genet. 1994; 20:15–24.

78. Gehlhaar DK, Verkhivker GM, Rejto PA, Sherman CJ, Fogel DB, Fogel L.J., Freer ST. Molecular recognition of the inhibitor AG-1343 by HIV-1 protease: Conformationally flexible docking by evolutionary programming. Chem. Biol. 1995; 2:317–324.

79. Rarey M, Kramer B, Lengauer T, Klebe G. A fast flexible docking method using an incremental construction algorithm. J. Mol. Biol. 1996; 261:470–489.

80. Rarey M, Kramer B, Lengauer T. Multiple automatic base selection, protein–ligand docking based on incremental construction without manual intervention. J. Comput. Aided. Mol. Design 1997; 11:369–384.

81. Welch W, Ruppert J, Jain AN. Hammerhead: Fast, fully automated docking of flexible ligands to protein binding sites. Chem. Biol. 1996; 3:449–462.

82. Caflisch A, Niederer P, Anliker M. Monte Carlo docking of oligopeptides to proteins. Proteins Struct. Funct. Genet. 1992; 13:223–230.

83. Hart TN, Read RJ. A multiple—start Monte Carlo docking method. Proteins Struct. Funct. Genet. 1992; 13:206–222.

84. Apostolakis J, Pluckthun A, Caflisch A. Docking small ligands in flexible binding sites. J. Comput. Chem. 1998; 19:21–37.

85. Di Nola A, Roccatano D, Berendsen HJC. Molecular dynamics simulation of the docking of substrates to proteins. Proteins Struct. Funct. Genet. 1994; 19:174–182.

86. Wasserman ZR, Hodge CN. Fitting an inhibitor into the active site of thermolysin: A molecular dynamics study. Proteins Struct. Funct. Genet. 1996; 24:227–237.

87. Clark KP, Ajay. Flexible ligand docking without parameter adjustment across four ligand–receptor complexes. J. Comput. Chem. 1995; 16:1210–1226.

88. Oshiro CM, Kuntz ID, Dixon JS. Flexible ligand docking using a genetic algorithm. J. Comput.-Aided Mol. Design 1995; 9:113–130.

89. Jones G, Willett P, Glen RC, Leach AR, Taylor R. Development and validation of a genetic algorithm for flexible docking. J. Mol. Biol. 1997; 267:727–748.

90. Baxter CA, Murray CW, Clarak DE, Westhead DR, Eldridge MD. Flexible docking using a tabu search and an empirical estimate of binding affinity. Proteins Struct. Funct. Genet. 1998; 33:367–382.

91. Leach AR. Ligand docking to proteins with discrete side-chain flexibility. J. Mol. Biol. 1994; 235:345–356.

92. Desmet J, Wilson IA, Joniau M, De Mayer M, Lasters I. Computation of the binding of fully flexible peptides to proteins with flexible side—chains. FASEB J 1997; 11:164–172.

93. Schaffer L, Verkhivker GM. Predicting structural effects in HIV-1 protease mutant complexes with flexible ligand docking and protein sidechain optimization. Proteins Struct. Funct. Genet. 1998; 33:295–310.

94. Westhead DR, Clark DE, Murray CW. A comparison of heuristic search algorithms for molecular docking. J. Comput.-Aided Mol. Design 1997; 11:209–228.

95. Totrov M, Abagyan R. Detailed *ab initio* prediction of lysozyme–antibody complex with 1.6 Å accuracy. Nature Struct. Biol. 1994; 1:259–263.

96. Shah N, Rejto PA, Verkhivker GM. Structural consensus in ligand–protein docking identifies recognition peptide motifs that bind streptavidin. Proteins Struct. Funct. Genet. 1997; 28:421–433.

97. Leopold PE, Montal M, Onuchic JN. Protein folding funnels: A kinetic approach to the sequence–structure relationship. Proc. Natl. Acad. Sci. USA 1992; 89:8721–8725.

98. Onuchic JN, Wolynes PG, Luthey-Schulten Z, Socci ND. Toward an outline of the topography of a realistic protein-folding funnel. Proc. Natl. Acad. Sci. USA 1995; 92:3626–3630.

99. Dill KA. Polymer principles and protein folding. Protein Sci. 1999; 8:1166–1180.

100. Totrov M, Abagyan R. Flexible protein–ligand docking by global energy optimization in internal coordinates. Proteins Struct. Funct. Genet. 1997; suppl 1:215–220.

101. Sandak B, Wolfson HJ, Nussinov R. Flexible docking allowing induced fit in proteins: Insights from an open and closed conformational isomers. Proteins Struct. Funct. Genet. 1998; 32:159–174.

102. Dixon JS. Evaluation of the CASP2 docking section. Proteins Struct. Funct. Genet. 1997; suppl 1:198–204.

103. Bouzida D, Arthurs S, Colson AB, Freer ST, Gehlhaar DK, Larson V, Luty BA, Rejto PA, Rose PW, Verkhivker GM. Thermodynamics and kinetics of ligand–protein binding studied with the weighted histogram analysis method and simulated annealing. In: Altman RB, Dunker AK, Hunter L, Klein T, Lauderdale K, eds. Pacific Symposium on Biocomputing-99. Singapore: World Scientific, 1999, pp: 426–437.

104. Bouzida D, Rejto PA, Verkhivker GM. Monte Carlo simulations of ligand–protein binding energy landscapes with the weighted histogram analysis method. Int. J. Quantum Chem. 1999; 73:113–121.

105. Oshiro CM, Kuntz ID. Characterization of receptors with a new negative image: Use in molecular docking and lead optimization. Proteins Struct. Funct. Genet. 1998; 30:321–336.

106. Ewing TJA, Kuntz ID. Critical evaluation of search algorithms for automated molecular docking and database searching. J. Comput. Chem. 1997; 18:1175–1189.

107. Sun Y, Ewing TJA, Skillman AG, Kuntz ID. CombiDOCK: structure-based combinatorial docking and library design. J. Comput.-Aided Mol. Design 1998; 12:597–604.

108. Lorber DM, Shoichet BK. Flexible ligand docking using conformational ensembles. Protein Sci. 1998; 7:938–950.

109. Knegtel RM, Kuntz ID, Oshiro CM. Molecular docking to ensembles of protein structures. J. Mol. Biol. 1997; 266:424–440.

110. Bouzida D, Rejto PA, Arthurs S, Colson AB, Freer ST, Gehlhaar DK, Larson V, Luty BA, Rose PW, Verkhivker GM. Computer simulations of ligand–protein binding with ensembles of protein conformations: A Monte Carlo study of HIV-1 protease binding energy landscapes. Int. J. Quantum Chem. 1999; 72:73–84.

111. Shortle D, Simons KT, Baker D. Clustering of low-energy conformations near native structures of small proteins. Proc. Natl. Acad. Sci. USA 1998; 95:11158–11162.

112. Rejto PA, Bouzida D, Verkhivker GM. Examining ligand–protein interactions with binding energy landscapes. Theor. Chem. Acc. 1999; 101:138–142.

113. Verkhivker GM, Rejto PA, Bouzida D, Arthurs S, Colson AB, Freer ST, Gehlhaar DK, Larson V, Luty BA, Marrone T, Rose PW. Towards understanding the mechanisms of molecular recognition by computer simulations of ligand–protein interactions. J. Mol. Recognition 1999; 12:371–389.

114. Zhang C, Chen J, DeLisi C. Protein–protein recognition: Exploring the energy funnels near the binding sites. Proteins Struct. Funct. Genet. 1999; 34:255–267.

115. Mayo SL, Olafson BD, Goddard III WA. DREIDING: A generic force field for molecular simulation. J. Phys. Chem. 1990; 94:8897–8909.

116. Weiner SJ, Kollman PA, Case DA, Singh UC, Chio C, Alagona G, Profeta S, Weiner P. A new force field for molecular mechanical simulation of nucleic acids and proteins. J. Am. Chem. Soc. 1984; 106:765–784.

117. Jorgensen WL, Tirado-Rives J. The OPLS potential functions for peptides. Energy minimizations for crystals of cyclic peptides and crambin. J. Am. Chem. Soc. 1988; 110:1657–1666.

118. Beutler TC, Mark AE, van Schaik RC, Gerber PR, van Gunsteren W. Avoiding singularities and numerical instabilities in free energy calculations based on molecular simulations. Chem. Phys. Lett. 1994; 222:529–539.

119. Marinari E, Parisi G. Simulated tempering: A new Monte Carlo scheme. Europhys. Lett. 1992; 19:451–458.

120. Hukushima K, Nemoto K. Exchange Monte Carlo method and application to spin glass simulations. J. Phys. Soc. (Jpn) 1996; 65:1604–1607.

121. Hansmann UHE, Okamoto Y. Monte Carlo simulations in generalized ensemble: Multicanonical algorithm versus simulated tempering. Phys. Rev. E. 1996; 54: 5863–5865.

122. Hansmann UHE, Okamoto Y. Generalized-ensemble Monte Carlo method for systems with rough energy landscape. Phys. Rev. E. 1997; 56:2228–2233.

123. Hansmann UHE, Okamoto Y. Numerical comparisons of three recently proposed algorithms in the protein folding problem. J. Comput. Chem. 1997; 18:920–933.

124. Hansmann UHE. Parallel tempering algorithm for conformational studies of biological molecules. Chem. Phys. Lett. 1997; 281:140–150.

125. Stouten PFW, Frömmel C, Nakamura H, Sander C. An effective solvation term based on atomic occupancies for use in protein simulations. Mol. Simul. 1993; 10:97–120.

126. Bouzida D, Kumar S, Swendsen RH. Efficient Monte Carlo methods for the computer simulation of biological molecules. Phys. Rev. A. 1992; 45:8894–8901.

127. Ferrenberg AM, Swendsen RH. New Monte Carlo technique for studying phase transitions. Phys. Rev. Lett. 1988; 61:2635–2638.

128. Ferrenberg AM, Swendsen RH. Optimized Monte Carlo data analysis. Phys. Rev. Lett. 1989; 63:1195–1198.

129. Boczko EM, Brooks III CL. Constant-temperature free energy surfaces for physical and chemical processes. J. Phys. Chem. 1993; 97:4509–4513.

130. Kumar S, Bouzida D, Swendsen RH, Kollman PA, Rosenberg JM. The weighted histogram analysis method for free energy calculations on biomolecules. I. The method. J. Comput Chem. 1992; 13:1011–1021.

131. Kumar S, Rosenberg JM, Bouzida D, Swendsen RH, Kollman PA. Multidimensional free energy calculations using the weighted histogram analysis method. J. Comput Chem. 1995; 16:1339–1350.

132. Kumar S, Paybe PW, Vasquez M. Methods for free energy calculations using iterative techniques. J. Comput Chem. 1996; 17:1269–1275.

133. Socci ND, Onuchic JN. Kinetic and thermodynamic analysis of protein like heteropolymers: Monte Carlo histogram technique. J. Chem. Phys. 1995; 103: 4732–4744.

134. Sali A, Shakhnovich EI, Karplus M. How does a protein fold? Nature 1994; 369:248–251.

135. Mirny L, Abkevich VI, Shakhnovich EI. Universality and diversity of the protein folding scenarios: A comprehensive analysis with the aid of lattice model. Fold. Design 1996; 1:103–116.

136. Roux B. The calculation of the potential of mean force using computer simulations. Comput Phys. Commun 1995; 91:275–282.

137. Bawden D. Browsing and clustering of chemical structures. In: Warr WA, ed. Chemical Structures: The International Language of Chemistry. Berlin, Heidelberg: Springer-Verlag, 1988, pp 145–150.

138. Tiana G, Broglia RA, Roman HE, Vigezzi E, Shakhnovich EI. Folding and misfolding of designed proteinlike chains with mutations. J. Chem. Phys. 1998; 108:757–761.

139. Bolin JT, Filman DJ, Matthews DA, Hamlin RC, Kraut J. Crystal structures of *Escherichia coli* and *Lactobacillus casei* dihydrofolate reductase refined at 1.7 Å resolution. I. General features and binding of methotrexate. J. Biol. Chem. 1982; 22:13650–13662.

140. Lawson CL, van Montfort R, Strokopytov B, Rozeboom HJ, Kalk KH, de Vries GE, Penninga D, Dijkhuizen L, Dijkstra BW. Nucleotide sequence and X-ray structure of cyclodextrin glycosyltransferase from *Bacillus circulanas* strain 251 in a maltose—dependent crystal form. J. Mol. Biol. 1994; 236:590–600.

141. Saito S, Sasai M, Yomo T. Evolution of the folding ability of proteins through functional selection. Proc Natl Acad Sci USA 1997; 94:11324–11328.

142. Given JA, Gilson MK. A hierarchical method for generating low-energy conformers of a protein–ligand complex. Proteins Struct Funct Genet 1998; 33:475–495.

6

Fast Continuum Electrostatics Methods for Structure-Based Ligand Docking

Catherine Tenette-Souaille, Nicolas Budin, Nicolas Majeux, and Amedeo Caflisch

University of Zurich
Zurich, Switzerland

I. INTRODUCTION

To discover drugs against human diseases, a plethora of methodologies have been developed. These involve many research fields and activities ranging from purely theoretical to experimental. Soon after computers became available at reasonable prices, many investigators in pharmaceutical companies and nonprofit research institutions realized the almost unlimited opportunities that the computational and data handling power of computers can offer for the very difficult task of drug discovery. After the information technology revolution of the last three decades, the coming years will be characterized by major successes in genomics and proteomics. The cloning and sequencing of the human genome as well as progress in high throughput approaches to solve protein structures will generate very valuable information on an ever increasing number of potential drug targets. Drug design is significantly facilitated if the three-dimensional conformation of the protein target is known at the atomic level of detail. The large efforts invested in the determination of protein structures by pharmaceutical and biotechnology companies are practical proof that knowledge of the three-dimensional conformation of protein targets is of paramount importance.

In this chapter we first review a number of approaches for structure-based, computer-aided design. A detailed description of the continuum electrostatic approach developed in our research group for docking library of small to medium-sized fragments is then presented. Finally, an application to the p38 mitogen-activated protein (MAP) kinase is discussed, and a brief outlook ends the chapter.

II. STRUCTURE-BASED LIGAND DESIGN APPROACHES

Computational tools that exploit knowledge of the three-dimensional structure of a protein target are used for de novo design (1,2), improving lead compounds, and helping in the selection of monomers to focus combinatorial libraries (3). Prioritization is done by empirical and knowledge-based scoring functions or force field energy functions (4). Ligands are built by connecting small molecular fragments or functional groups, often rigid, or even atoms. In the latter case, the methods have shown significant flexibility with respect to the structures that can be obtained (5–7). The main disadvantage of compounds generated by atom-based approaches is that they often have complicated structures and are in most cases very difficult to synthesize. Hence, methods that build new compounds by combining predefined fragments are more popular. The number of newly created bonds is small, and therefore it is easier to control the chemistry (i.e., the synthesizability and the chemical stability) of the designed molecules. Furthermore, fragments are easily modeled, since such model parameters as partial charges, periodicity, and the force constant of torsion angles are assumed to depend mainly on the fragment and only to a lesser extent on the rest of the structure.

Fragment-based ligand design may be achieved in two ways. In the first, small fragments are docked in the active site. The best positions of each fragment type are retained and connected to generate candidate ligands. Alternatively, an anchor fragment is docked in the binding site and the ligand is grown starting from it. These approaches should not be considered to be mutually exclusive, but rather as complementary, since they are useful for generating candidate ligands with different physicochemical characteristics and structural properties.

A. Methods Based on the Connection of Docked Fragments

This approach has the advantage that the fragments occupy optimal positions and are oriented such that their interaction with the protein is favorable. On the other hand, the geometry of the bonds connecting the fragments to each other or to a central template is not optimal and must be accepted with a certain tolerance initially. The mapping of a binding site–fragment assembly into complete ligands can be performed by separate programs (8,9) or integrated in a single computational tool (10). For site mapping, two main approaches have emerged: the first is based on binding site shape descriptors and the second on multiple-copy techniques.

With the program GRID (11), Goodford pioneered the use of molecular probes to explore the surface of a protein and search for energetically favorable positions. The interaction energies are then mapped onto a grid that describes the regions of attraction between a probe and the protein. The surface descriptors thus obtained can be used to screen a three-dimensional database of small compounds. This task can be performed by the program CLIX (12). For each molecule, CLIX attempts to make a pair of substituents spatially coincide with a pair of favorable interaction sites proposed by GRID.

Several docking programs can map a protein binding site using small to medium-sized molecular fragments, either rigid or partly flexible. The program DOCK, which was first based on rigid docking and the use of geometrical criteria to judge the complementarity between receptor and ligand, was therefore fast enough to screen databases for leads (13). DOCK uses spheres complementarity to the receptor molecular surface to create a space-filling negative image of the receptor site. Several atoms of the ligand are matched with spheres that define the binding site. Flexibility (14) and a force field, such as energy function for scoring (15), were included in later development of the program. DOCK has been used to find novel micromolar inhibitors of enzymes (16,17).

FlexX (18) is a program for the fast docking of medium-sized flexible ligands. It first positions a fragment of the ligand by mapping three interaction centers of the fragment onto three interaction points of the receptor. The ligand is then constructed in an incremental way, and Böhm's empirical function is used for scoring (19). FlexX is fast enough to allow screening of small databases of ligands. The docking of hydrophobic fragments has been slightly improved (20), and the algorithm has been extended to predict the location of water molecules in the binding site (21).

A number of genetic algorithms have been suggested for docking (22–24). They combine speed with simplicity of concept. For example, GOLD (22) is based on a genetic algorithm that encodes in the chromosomes the values of the dihedral angles around rotatable bonds and positions the ligand in the binding site by means of a simple least-squares fitting that maximizes the number of intermolecular hydrogen bonds. It also allows flexibility around bonds to hydrogen bond donors and acceptors in the receptor. GOLD uses a force field with a simple approximation of solvation consisting of precalculated atom type–based hydrogen bond energies. The method has been tested on 100 complexes, leading to a success rate, defined by the authors in a rather subjective way, of about 70% for redocking into the complexed conformation (25). Docking into the unbound conformation was tested on only three examples and gave mixed results (25). An accurate treatment of solvation is essential for docking into a flexible binding site (26).

Multiple-copy techniques use numerous fragment replicas, each transparent to the others but subject to the full force field of the receptor, to determine energetically favorable positions and orientations (functionality maps) of small molecules or functional groups in the binding site of a protein (27–29) or RNA (30). Although the multiple-copy, simultaneous search (MCSS) method was originally proposed in the context of a rigid receptor (28), it was extended to allow for ligand and receptor flexibility (29,31,32).

We have developed a new continuum solvation approach called SEED, which can be used for efficiently docking fragments into a rigid receptor (8). It combines the advantages of shape descriptors and multiple-copy methods. Polar and apolar vectors are distributed on the surfaces of the receptor binding site and the fragments, and matched with each other, allowing exhaustive docking on a discrete space. The main advantage of SEED over other docking programs is the comprehensive treatment of electrostatic solvation effects in an efficient and accurate manner. SEED is described in further detail in Section III.

The large amount of structural information in the functionality maps can be exploited by other programs. The docked fragments can be linked together with smaller (9,33) or larger (34–36) linkers. The program CAVEAT (34,35) was designed to do interactive searches of three-dimensional databases to find molecular frameworks that can position functional groups in specific relative orientations. CAVEAT focuses on relationships between bonds; methods are implemented to identify and classify structural frameworks. The HOOK algorithm (36) uses "skeletons" from a database, on which "hooks" are defined, to connect a set of functional groups previously docked in a binding site. The linkage is accomplished by fusing the hooks with two or more methyl groups from the functional groups. Computational combinatorial ligand design (CCLD) (9) is also based on docking of functional groups with MCSS or SEED (8). The fragments are ranked according to an approximated binding free energy. After classifying positions into overlapping (i.e., mutually excluding) and bonding (i.e., possibly bound by small linkers) pairs, CCLD creates ligands by linking the docked fragments with the most favorable of the small linkers. To avoid combinatorial explosion, growing is discontinued when the average binding free energy of the fragments in the new ligand exceeds a user-specified threshold.

Some programs integrate site mapping and fragment assembly. LUDI (10,37) makes extensive use of empirical information derived from structural databases. Interaction sites that indicate possible positions for functionalities complementary to the receptor are defined and used to dock fragments from a library. Alternatively, the output of GRID can be used for the definition of interaction sites. The fragments are fitted on the interaction sites with the algorithm published by Kabsch (38) and are connected with small linkers. Interaction geometries were derived from structural data on small organic molecules (39,40). The scoring function used in LUDI is empirical (19). The program was extended to take into account the synthetic accessibility of the constructed molecules (41).

The program SPROUT (42) can deal either with a three-dimensional experimental receptor structure or with a pharmacophore model derived from known inhibitors. Target sites in the binding pocket are identified and labeled by type. Fragments, from a library presorted according to atomic and molecular properties, are selected and overlaid on a target site. Once fragments have been docked into all the target sites, the linking procedure is performed, taking into account the identity of the fragments. In the second phase, atom types are exchanged with others of the same hybridization state to find a combination exhibiting optimal interactions with the binding site.

In an effort to remain close to progress in modern chemistry, a number of computational tools were further developed to facilitate the design of combinatorial libraries. CombiDOCK is a modified version of DOCK to efficiently screen a large combinatorial library for a receptor (43). CombiDOCK first positions the scaffolds in the binding site and, for each scaffold orientation, all potential fragments are attached. The interactions between substituents and receptor are individually scored, and factorial combinations of fragments are suggested. In LUDI, a new procedure has been implemented to focus the design on a chemical reaction, amenable to parallel chemistry (41,44).

B. Methods Based on the Progressive Buildup of Ligands

Ligand buildup is a powerful stepwise strategy for de novo ligand design. It starts with a seed fragment placed in an appropriate region of the binding site. New ligands are then grown by sequentially appending building blocks (fragments or atoms). To avoid combinatorial explosion, a large fraction of all building blocks is discarded at every step according to some heuristic scoring. This method has the advantage that the newly formed chemical bonds have a correct geometry and that the intraligand interactions can be taken into account during the design. On the other hand, buildup approaches have difficulties in generating ligands that bind to different pockets if these are separated by gap regions that do not allow specific interactions. Moreover, the success of the growing procedure and therefore the quality of the designed molecules depend dramatically on the position of the seed, since the latter is usually kept fixed. The seed position(s) can be determined from X-ray or NMR structures of ligand–protein complexes. If no structure is available, seeds must be obtained by manual or computer-aided docking. Many programs that implement the buildup strategy have been described in the literature, and the following list is not exhaustive.

GenStar (45) and LEGEND (46,47) use single atoms as building blocks. GenStar grows sequentially structures that are entirely composed of sp^3 carbons. It allows branching and ring formation. For each new atom generated, several hundred candidate positions with acceptable bond geometries are generated. Each position is scored based on a simple binding site contact model, and the selected position is chosen at random among the highest scoring cases. LEGEND works in a similar way but uses the MM2 force field (48). The choice of the atom type is driven by the protein electrostatic potential value at the atom position.

Both GroupBuild (49) and SMoG (50,51) use libraries of organic compounds to design ligands. In these programs, each candidate fragment is attached to the growing structure and rotated around the new bond in fixed increments. In GroupBuild a standard molecular mechanics potential function is used to rank the candidates. The chosen fragment rotamer is randomly selected among the top 25% of fragment positions. SMoG uses a knowledge-based potential for the ranking. The lowest energy rotamer's acceptance is determined by a Metropolis Monte Carlo criterion, which compares the new energy per atom with and without the candidate fragment.

The buildup strategy has also been implemented in the programs GROW (52), LUDI (41), and PRO_LIGAND (53). Their library of fragments is however restricted to amino acids and amino acid derivatives. This has the advantage that the designed ligands are synthetically accessible, but the explored chemical space is relatively small. Moreover the energetics of the ligand can be studied by well-parameterized force fields. On the other hand, peptides, besides their poor pharmacological properties, represent special problems stemming from their great conformational flexibility. This latter property is taken into account by using multiple conformers for each amino acid. The main differences between these three programs lie in the scoring functions used to rank the ligands and in the way the conformation libraries for the amino acids are generated. GROW's scoring function is based on the AMBER force field (54) supplemented by a solvent-accessible surface approximation of solvation (55). LUDI and PRO_LIGAND use empirical scoring functions combined with a rule-based interaction site strategy (19,56). The GROW and PRO_LIGAND libraries contain low energy conformations, whereas LUDI uses conformations extracted from high resolution protein structures.

C. Binding Energy

Ligand design involves the extension of the docking problem into chemical space. The degrees of freedom to be optimized are not only the positional and conformational variables of a particular compound, but, additionally, its chemical identity. This point of view makes one important problem in the field of ligand design particularly clear: the quality of the scoring or energy function used to evaluate the different solutions. When the search space is very limited, as it was, for example in the first programs that performed rigid docking (13), a very simple energy function based on geometrical criteria suffices to recognize the correctly docked structures. When flexibility in the ligand (and the protein) is allowed, the effect of solvation must be taken into account (26) to avoid sampling irrelevant parts of the conformational space.

A simple example shows the higher quality requirements on the scoring function for design purposes: assuming that the charge on an atom in a designed ligand is a (continuous or discrete) variable of the optimization, any simple force field–based energy function would tend to maximize the total charge (57). This is however in disagreement with empirical data. Although sometimes high affinity may be due to ionic interactions, often the desolvation penalty of full charges on the ligand and the protein is stronger than the direct interaction.

Desolvation is the change in the solvation energy of the ligand and the receptor due to the displacement of high dielectric solvent by low dielectric solute upon complex formation. This further indicates that the scoring function should correspond to a difference between the free and the complexed states. The calculation of such differences is not necessary in docking because the term corresponding to the free state identically cancels. Accurate and reliable prediction of the absolute binding free energy for a medium-sized flexible ligand is currently beyond the limits of routine calculations, since the most probable conformations in water must be found, as well, and averaging with the correct thermodynamic weights must be performed. Furthermore, in ligand design free energies are assumed to be additive, although it is clear that this is only a crude approximation (58).

The main task for a scoring function in a ligand design program is to find the conformations with the lowest energies for every chemical species (be it atom, fragment, or complete ligand) and, in the case of different chemical entities (e.g., a benzene and a guanidinium docking in the same binding pocket), to decide which yields the lowest binding free energy. Both tasks and especially the latter are not straightforward and will most probably have to be addressed at different levels of accuracy during different stages of the design process. Recently, methods based on the combination of several models (multilayered scoring system, consensus scoring) have been shown to increase the predictivity (59) and to reduce the number of false positives suggested by individual scoring functions (60).

III. THE SEED APPROACH

Figure 1 shows a flowchart of the library docking program SEED. A brief explanation is given here; further details of the method (e.g., the clustering procedure and evaluation of the van der Waals interactions) can be found in the original paper (8). Different fragment types are docked in the order specified by the user. After each fragment placement, the binding energy is estimated. The binding energy is the sum of the van der Waals interaction and the electrostatic energy with

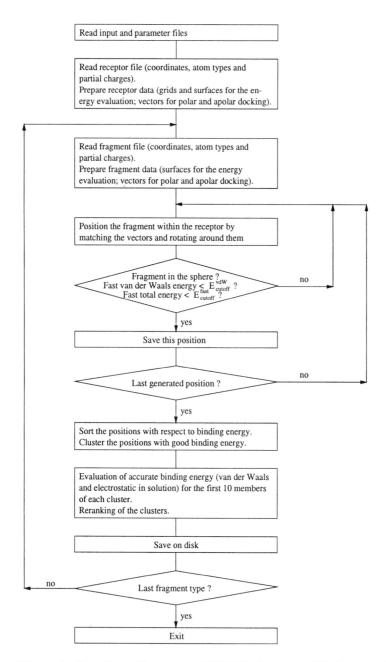

Figure 1 Flowchart of the program SEED. The fast van der Waals energy is evaluated by trilinear interpolation (114) from a grid (lookup table) using the geometric mean approximation (115–118). The fast total energy includes solvation and will be presented elsewhere (Majeux et al., Proteins, 2001. In press).

continuum solvation. Each fragment type is docked after all placement–energy evaluations of the preceding fragment type have been made. The fragment docking procedure and electrostatic energy evaluation are outlined in this Section III.A and III.B. For the docking of a library of 100 fragments into a rigid binding site of about 25 residues, the latest version of SEED requires about 5 hours of CPU time on a single processor (195 MHz R10000 or PentiumIII 550 MHz). For more than one processor, the speedup is linear, and the docking of a library of 1000 fragments would require about 6 hours on a cluster of eight commodity-price processors.

A. Fragment Docking

The binding site of the receptor is defined by a list of residues selected by the user. Fragments that have at least one H-bond donor or acceptor are docked by using the polar vectors. Given this definition, some "polar" fragments can have considerable hydrophobic character (e.g., diphenyl ether). Therefore they are also docked by the procedure for nonpolar fragments unless otherwise specified by the user.

1. Docking of Polar Fragments

Polar fragments are docked to ensure the formation of one or more hydrogen bonds with the receptor. The fragment is then rotated around the H-bond axis to increase sampling. Figure 2 shows the sampling of docked positions for pyrrole

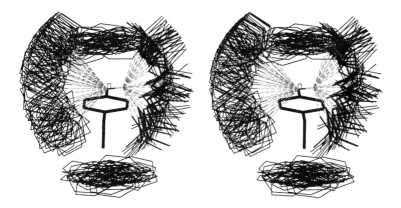

Figure 2   Relaxed-eyes stereoview of benzene, acetone, and pyrrole docked by SEED around a tyrosine side chain. The benzene groups are distributed above and below the plane of the tyrosine phenolic ring, while pyrrole and acetone fragments are involved in hydrogen bonds with the tyrosine hydroxyl as donors and acceptors, respectively. Carbon atoms are black, oxygen and nitrogen atoms dark gray, and hydrogen atoms light gray. Hydrogen bonds are drawn with dashed lines.

and acetone around a tyrosine side chain. Ideal and close-to-ideal hydrogen bond geometries are sampled in a discrete but exhaustive way.

2.   Docking of Nonpolar Fragments

Hydrophobicity maps (61) are used to dock nonpolar fragments. The points on the receptor SAS are ranked according to the sum of van der Waals interaction and receptor desolvation (61,62), and the *n* most hydrophobic points (where *n* is an input parameter) are selected for docking. As an illustrative example, Fig. 3 shows the most hydrophobic points on the ATP binding site of the p38 MAP kinase. For both the fragment and the receptor, apolar vectors are defined by joining each point on the SAS with the corresponding atom. Finally, nonpolar fragments are docked by matching an apolar vector of the fragment with an apolar vector of the receptor at the optimal van der Waals distance. To improve sampling, additional rotations of the fragment are performed around the axis joining the receptor atom and fragment atom (Fig. 2). To increase efficiency, nonpolar fragments are discarded without calculation of the electrostatic energy if the van der Waals interaction is less favorable than a threshold value.

For both polar and nonpolar fragments, the docking is exhaustive on a discrete space. The discretization originates from the limited number of preferred directions and rotations around them. Fragment symmetries are checked only once for every fragment type and are exploited to increase the efficiency in docking.

B.   Electrostatic Energy with Continuum Solvation

The main assumption underlying the evaluation of the electrostatic energy of a fragment–receptor complex is the description of the solvent effects by continuum electrostatics (62–72). The system is partitioned into solvent and solute regions, and appropriate values of the dielectric constant are assigned to each region. In this approximation only the intrasolute electrostatic interactions need to be evaluated. This strongly reduces the number of interactions with respect to an explicit treatment of the solvent. Moreover it makes feasible the inclusion of solvent effects in docking studies where the equilibration of explicit water molecules would be a major difficulty. In docking and even more in ligand design, the electrostatic effects of the solvent must be modeled accurately; it has been shown that the continuum dielectric model provides an efficient and useful approximation of molecules and molecular complexes in solution (62,63,73).

The difference in electrostatic energy in solution upon binding of a fragment to a receptor can be calculated as the sum of the following three terms (9,65):

> *Partial desolvation of the receptor:* electrostatic energy difference due to the displacement of high dielectric solvent by the fragment volume

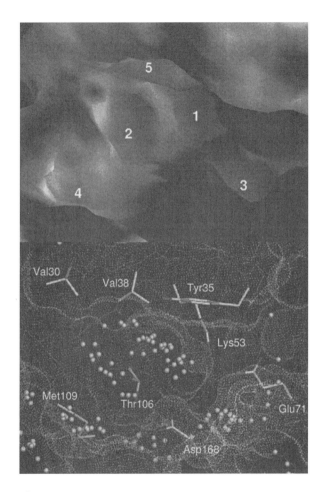

Figure 3 ATP binding site in the p38 MAP kinase (PDB file 1a9u, [94]). (Top) The hydrophobic pockets are colored in green using the hydrophobicity map approach (61). They are referenced by numbers. The figure was made with GRASP (119). (Bottom) The yellow spheres represent the most hydrophobic points (61) in the binding site. The residues lining the binding site are displayed with cylinders. The molecular surface, calculated with the Connolly algorithm (120), is displayed by white dots.

Screened receptor–fragment interaction: intermolecular electrostatic energy between the fragment and the receptor in the solvent

Partial desolvation of the fragment: electrostatic energy difference due to the displacement of high dielectric solvent by the receptor volume

The definition of the solute volume (i.e., the low dielectric volume) is central in the evaluation of these energy terms with a continuum model. The solute–solvent dielectric boundary is described by the molecular surface (MS) of the solute (74). A grid covering the receptor is utilized to identify the low dielectric volume. In a first step, the volume occupied by the isolated receptor is defined on the grid. Subsequently, for every position of a docked fragment, the volume enclosed by the MS of the fragment–receptor complex is identified.

The screened fragment–receptor interaction and the fragment desolvation are evaluated with a grid-based implementation (62,63) of the generalized Born (GB) approximation (68–72). The GB approach would be too time-consuming for the evaluation of the desolvation of the receptor, which is calculated as described next.

1. Receptor Desolvation

The electrostatic desolvation of the receptor accounts for the loss of favorable receptor–solvent electrostatic interactions due to the removal of part of the highly polarizable solvent to accommodate the volume of the fragment. This contribution always disfavors association and can be calculated within the assumption of continuum electrostatics (62–64,68–72). The electrostatic energy E of the receptor in solution can be expressed in terms of the electric displacement vector $\boldsymbol{D}(\boldsymbol{x})$ and of a location-dependent dielectric constant $\epsilon(\boldsymbol{x})$ as an integral over the three-dimensional (3D) space R^3 (75):

$$E = \frac{1}{8\pi} \int_{R^3} \frac{\boldsymbol{D}^2(\boldsymbol{x})}{\epsilon(\boldsymbol{x})} d^3x \qquad (1)$$

Since $\boldsymbol{D}(\boldsymbol{x})$ is additive, for point charges it can be rewritten as a sum over all charges i of the receptor:

$$\boldsymbol{D}(\boldsymbol{x}) = \sum_i \boldsymbol{D}_i(\boldsymbol{x}) \qquad (2)$$

As for the electrostatics, the displacement of solvent by the fragment volume at the surface of the receptor has the sole effect of modifying the dielectric properties in the space occupied by the fragment. Over this volume, the dielectric constant changes from the solvent value (ϵ_w) to the solute value (ϵ_p). Usually, ϵ_w is set to 78.5, which is the value of water at room temperature, while the value of ϵ_p can range from 1 to 4. In the limit in which $\boldsymbol{D}(\boldsymbol{x})$ does not change significantly upon docking of the ligand, the variation of the electrostatic energy of the receptor (i.e., the desolvation) can be written according to Eq. (1) as an integral over the volume occupied by the fragment (V_{fragment}):

$$\Delta E^{\text{desolv}} = \frac{\tau}{8\pi} \int_{V_{\text{fragment}}} \boldsymbol{D}^2(\boldsymbol{x}) d^3x \qquad (3)$$

where $\tau = 1/\epsilon_p - 1/\epsilon_w$. A 3D grid is built around the receptor and Eq. (3) becomes:

$$\Delta E^{desolv} = \frac{\tau}{8\pi} \sum_{k \in V_{fragment}} D^2(x_k) \Delta V_k \tag{4}$$

where the index k runs over the grid points occupied by the fragment. The grid spacing is usually 0.5 Å. The electric displacement of every charge of the receptor can be approximated by the Coulomb field (62,71,76):

$$D(x) = \sum_i q_i \frac{(x - x_i)}{|x - x_i|^3} \tag{5}$$

where x_i is the position of the receptor atom i and q_i its partial charge. Equation (5) is an analytical approximation of the total electric displacement and fulfills the condition of validity of Eqs. (3) and (4): that is, $D(x)$ is independent of the dielectric environment. The receptor desolvation in the Coulomb field approximation results from Eq. (4) together with Eq. (5):

$$\Delta E^{desolv} = \frac{\tau}{8\pi} \sum_{k \in V_{fragment}} \left(\sum_i q_i \frac{(x_k - x_i)}{|x_k - x_i|^3} \right)^2 \Delta V_k \tag{6}$$

The volume occupied by a docked fragment is the part of the volume enclosed by the MS of the complex that was not occupied by the isolated receptor. It consists of the actual volume of the fragment and the interstitial volume enclosed by the reentrant surface between fragment and receptor.

It is important to note that the desolvation of a charged ion by a small nonpolar sphere at a distance r from the ion varies approximately as $1/r^4$ (Eq. 6). This is a very short-range effect compared with the ion electrostatic potential, which varies as $1/r$.

2. Screened Fragment–Receptor Interaction

The fragment–receptor interaction in solution is calculated via the GB approximation (68). In a solvent of dielectric constant ϵ_w, the interaction energy between two charges embedded in a solute of dielectric constant ϵ_p is

$$E_{ij}^{int} = \frac{q_i q_j}{\epsilon_p r_{ij}} - \frac{q_i q_j \tau}{R_{ij}^{GB}} \tag{7}$$

where $\tau = 1/\epsilon_p - 1/\epsilon_w$,

$$R_{ij}^{GB} = \sqrt{r_{ij}^2 + R_i^{eff} R_j^{eff} \exp \left(\frac{-r_{ij}^2}{4R_i^{eff} R_j^{eff}} \right)} \tag{8}$$

and q_i is the value of the partial charge i, while r_{ij} is the distance between charges i and j, R_i^{eff}, the effective radius of charge i, is evaluated numerically on a three-dimensional grid covering the solute as described elsewhere (62). It is a quantity depending only on the solute geometry and represents an estimate of the average distance of a charge from the solvent.

The intermolecular interaction energy is calculated as follows:

$$E^{int} = \sum_{\substack{i \in \text{fragment} \\ j \in \text{list}_i}} E_{ij}^{int} \tag{9}$$

where list_i contains the receptor atoms belonging to the neighbor list of atom i. The electrostatic neighbor list includes all the receptor atoms of the van der Waals neighbor list and one atom for every charged residue whose charge center is within 13 Å of the closest binding site residue. Supplementing the van der Waals neighbor list with a monopole approximation of distant charged residues dramatically reduces the error originating from the long-range electrostatic interactions.

3. Fragment Desolvation

The fragment intramolecular energy in solution is calculated with the GB formula as described in Ref. 62:

$$E = \sum_{\substack{i \in \text{fragment} \\ j \in \text{list}_i}} E_i^{self} + \sum_{\substack{i > j \\ i,j \in \text{fragment}}} \left(\frac{q_i q_j}{\epsilon_p r_{ij}} - \frac{q_i q_j \tau}{R_{ij}^{GB}} \right) \tag{10}$$

where the two sums run over the partial charges of the fragment. Equation (10) differs from Eq. (9) by virtue of the presence of the *self-energy* term $\sum_i E_i^{self}$. This term is not zero only in the case of intramolecular energies; E_i^{self} is the *self-energy* of charge i and represents the interaction between the charge itself and the solvent. It is calculated (62,71) as follows:

$$E_i^{self} = \frac{q_i^2}{2R_i^{vdW} \epsilon_p} - \frac{q_i^2 \tau}{2R_i^{eff}} \tag{11}$$

where R_i^{vdW} is the van der Waals radius of charge i.

The difference in the intramolecular fragment energy upon binding to an uncharged receptor in solution is

$$\Delta E = E^{docked} - E^{free} \tag{12}$$

where E^{docked} and E^{free} are the energies in solution of the fragment bound and unbound to the receptor, respectively. They are evaluated according to Eq. (10). For the unbound fragment (E^{free}), the effective radii are calculated considering the volume enclosed by the molecular surface of the fragment to be the solute. For the bound fragment (E^{docked}), the solute is the volume enclosed by the molecular surface of the receptor–fragment complex. E^{free} is evaluated only once per fragment type, while E^{docked} is recalculated for every fragment position in the binding site.

C. Validation

The approximations inherent to our continuum electrostatic approach were validated by comparison with finite difference solutions of the Poisson equation (8). For this purpose, the three electrostatic energy terms were calculated with SEED and simulations from the University of Houston Browning Dynamics (UHBD) program (67,77), for a set of small molecules and ions distributed over the binding site of thrombin and at the dimerization interface of the HIV-1 aspartic protease monomer. The molecule set included acetate ion, benzoate ion, methylsulfonate ion, methylammonium ion, methylguanidinium ion, 2,5-diketopiperazine, and benzene. There were 1025 fragment–receptor complexes analyzed for thrombin and 1490 for the HIV-1 protease monomer. The agreement between the two methods is very good, and it is better for a solute dielectric constant of 4.0 than 1.0 (see Table II of Ref. 8). It was also shown that systematic errors (slope $\neq$ 1) are independent of the receptor and the solute dielectric constant and consequently can be corrected by the use of appropriate scaling factors for the different energy terms (62).

IV. ILLUSTRATIVE APPLICATION

A. The MAP Kinase Family

Mitogen-activated protein (MAP) kinases are essential enzymes for intracellular signaling cascades because they phosphorylate several regulatory proteins. They are responsive to hormones, cytokines, environmental stresses, and other extracellular stimuli and are activated by a dual phosphorylation of a threonine and a tyrosine residues in a TXY motif in the so-called phosphorylation lip. A MAP kinase is characterized by its downstream substrates and by the kinases by which it is preferentially activated. The best-characterized MAP kinases are the extracellular signal–regulated kinases ERKs (TEY activation motif) (78,79), the c-Jun N-terminal kinases JNKs (TPY motif) (80–82), and p38 (TGY motif) (83–86); p38 MAP kinase (also called CSBP2) plays a role in processes as diverse as transcriptional regulation, production of interleukins, and apoptosis of neuronal cells. Inhibitors of p38 activity could therefore be useful as a treatment strategy for inflammatory and neurodegenerative diseases.

All unphosphorylated forms of MAP kinases are similar in their topology, which consists of two domains separated by a substrate binding cleft (87). The N-terminal domain incorporates the glycine-rich loop, which contains the ATP binding motif GXGXXG, while the C-terminal domain contains the magnesium sites,

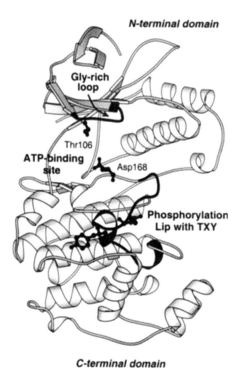

Figure 4   Molscript picture (121) of the p38 MAP kinase (PDB file 1a9u, Ref 94). Biologically relevant regions of p38 are in dark grey.

the catalytic base, and the phosphorylation "lip" with the TXY motif (Fig. 4). Yet, some noticeable differences affect the geometry of the ATP binding site: (1) the phosphorylation lips vary in sequence and structure; (2) the relative orientation of the two domains is different, leading to different domain interface shapes; and (3) some residues in the ATP binding site are different (e.g., at position 106—sequence number according to p38—there is a Thr in p38, a Gln in the ERKs, and a Met in the JNKs).

B.   Specific Inhibitors of p38

The CSAID™ (cytokine suppressive anti-inflammatory drugs) class of anti-inflammatory compounds inhibits the synthesis of cytokines, such as interleukin 1 and tumor necrosis factor, by specific inhibition of the p38 MAP kinase (83,88,89). They have a common chemical pattern: a central five-membered ring, either imidazole or pyrrole, substituted by a pyridine or a pyrimidine ring at C-5, a fluorinated or iodinated phenyl ring at C-4, and a third substituent at position N-1 or C-2 (Fig. 5). These low molecular weight inhibitors and their analogs bind to the ATP binding cleft of the unactivated form of p38 and are competitive with respect to ATP. They are potent inhibitors, with IC_{50} in the nanomolar range, and highly selective for p38 compared with the other MAP kinases.

Both biochemical and structural data suggest that the specificity toward p38 kinases is determined by differences in nonconserved regions within or near the ATP binding site. In particular, a site-directed mutagenesis study demonstrated the crucial role of Thr106 (90). The determination of the three-dimensional structures of the apo, unphosphorylated human, and murine p38 MAP kinases, free and bound to various small-sized inhibitors, allowed investigators to gain more insights into the structural affinity and specificity determinants of the p38 binding site (91–94). The binding modes of the inhibitors share common properties: the pyridine or pyrimidine nitrogen acts as hydrogen bond acceptor for the Met109 backbone NH, and the phenyl ring is inserted into a mainly hydrophobic pocket, delimited by the Lys53 and Thr106 side chains. The former feature is analogous to what is observed in complexes with ATP and with other protein kinase inhibitors (95–98). Conversely, the latter is unique to p38. The third substituent of the central scaffold may also be involved in inhibitor selectivity by interacting with various residues of the glycine-rich loop and occasionally with the Asp168 side chain. The imidazole scaffold is hydrogen-bonded to the Lys53 side chain. Despite these common properties of inhibitor binding, the detailed organization of the binding cleft differs in the structures of the complexes between p38 and two

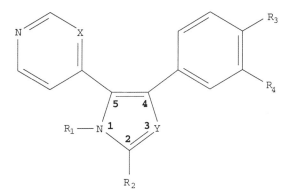

Figure 5 Chemical structure of a series of potent and selective p38 inhibitors: [94]. X = N, CH; Y = N, CH; R_1 = H, methylcyclopropyl, piperidinyl; R_2 = H, 4-methylsulfinylphenyl, piperidinyl; R_3 = H, F; R_4 = H, I, CF_3. Compound **1** of the SEED fragment library is 4-(4-fluorophenyl)-1-methyl-5-(4-pyridyl)-imidazole (i.e., R_1 = CH_3, R_2 = H, R_3 = F, R_4 = H, X = CH, Y = N). (From Ref. 94.)

series of inhibitors (93,94). The loop containing the ATP binding motif adopts different orientations, leading to a more or less open site, and the side chain orientation of some critical residues, (e.g., Lys53, Thr106, Asp168) depends on the presence or absence of the inhibitor in the site, as well as on the type of inhibitor (93,94). Overall, the ATP binding site of p38 displays a remarkable flexibility.

C. Docking of Molecular Fragments with SEED

1. Fragment Library

A library of 70 mainly rigid fragments ranging in size from 7 to 31 atoms was used in this study. It contains 17 apolar fragments (no hydrogen bond donors or acceptors), 39 polar and neutral compounds, and 14 fragments with one or two formal charges (Table 1). Many of the molecular frameworks found frequently in known drugs (99) are included (e.g., benzene, pyridine, naphthalene, 5-phenyl-1,4-benzodiazepine), and some of them can be used for the synthesis of combinatorial libraries in the solid phase (100) or by portioning and mixing (101). Fragment structures were generated with the molecular modeling program WITNOTP (A. Widmer, Novartis Pharma Basel, unpublished). For each fragment type, all the low energy conformations are included in the library (e.g., cis and trans for 2-butene). Partial charges were assigned with an iterative method based on the partial equalization of orbital electronegativities (MPEOE) (102,103). The MPEOE implementation in WITNOTP reproduces the all-hydrogen CHARMm22 parameter set (Molecular Simulations Inc.) for proteins and proteinaceous fragments. Fragment coordinates were minimized with the program CHARMM (104), and the CHARMm22 parameter was set to an average value of the energy gradient of 0.01 kcal/(mol Å) using a linear distance-dependent dielectric function.

2. SEED Input Parameters

The same input parameters as in the original SEED paper (Table I in Ref. 8) were used, except for the following ones. The interior dielectric constant was set to 4 to model the electronic polarizability and dipolar reorientation effects of the solute (105). The number of apolar vectors was increased from 100 to 150 because of the very large binding site, discussed shortly. Finally, the radius of the probe sphere for the definition of the SAS (used for the selection of the apolar vectors) was set to 1.4 Å (instead of 1.8 Å) to better define microcavities and small crevices. The positions of the fragments were preselected by discarding those whose geometric center was outside a sphere of 10 Å radius centered on the center of the binding site. All polar fragments were docked using both polar and apolar surface points.

Table 1   Fragment Library

	Number of atoms		Number of configurations	H-bond		Formal charges
Fragment	Heavy	Total		Acceptors	Donors	
Apolar fragments						
Ethane	2	8	1	—	—	0
Propane	3	11	1	—	—	0
Cyclopropane	3	9	1	—	—	0
2-Methylpropane	4	14	1	—	—	0
1-Butene	4	12	4	—	—	0
2-Butene	4	12	2	—	—	0
2-Methyl-2-butene	5	15	1	—	—	0
2,2-Dimethylpropane	5	17	1	—	—	0
Cyclopentane	5	15	1	—	—	0
Benzene	6	12	1	—	—	0
Cyclohexane	6	18	1	—	—	0
Adamantane	10	26	1	—	—	0
Dekaline	10	28	1	—	—	0
Naphthalene	10	18	1	—	—	0
N-Methylindole	10	19	1	—	—	0
Tetraline	10	22	1	—	—	0
Dibenzocyclohexane	14	26	1	—	—	0
Neutral fragments with one or more hydrogen bond groups						
Dimethyl sulfoxide	4	10	1	1	—	0
Isopropanol	4	12	2	1	1	0
Imidazole	5	9	1	1	1	0
N-Methylacetamide	5	12	1	1	1	0
Pyrrole	5	10	1	—	1	0
N-methyl-methylsulfonamide	6	13	1	2	1	0
Oxazolidinone	6	11	1	2	1	0
Pyridine	6	11	1	1	—	0
Pyrimidine	6	10	1	2	—	0
2-Pyrrolidinone	6	13	1	1	1	0
4-Thiazolidinone	6	11	1	1	1	0
δ-Valero lactam	7	16	2	1	1	0
3,4-Dihydroxytetrahydrofuran	7	15	4	3	2	0
Phenol	7	13	1	1	1	0
Tetrahydro-2-pyrimidinone	7	15	1	1	2	0
Cytosine	8	13	1	2	2	0
1,2-Dihydroxy benzene	8	14	1	2	2	0
1,2-Dihydroxy cyclohexane	8	20	4	2	2	0
2,5-Diketo-1,4-piperazine	8	14	2	2	2	0
Uracil	8	12	1	2	2	0
Indole	9	16	1	—	1	0
2-Methyl-3-amino N-methylbutanamide	9	23	18	1	2	0

(*continued*)

Table 1 *Continued.*

Fragment	Number of atoms Heavy	Number of atoms Total	Number of configurations	H-bond Acceptors	H-bond Donors	Formal charges
Adenine	10	15	1	3	2	0
3,6-Dimethyl-2,5-diketo-1,4-piperazine	10	20	3	2	2	0
Isoquinoline	10	17	1	1	—	0
N-Formyl-L-proline	20	19	8	3	1	0
Quinazoline	10	16	1	2	—	0
Quinoline	10	17	1	1	—	0
Tetrahydroquinoline	10	21	1	—	1	0
Guanine	11	16	1	3	3	0
Meso-Inositol	12	24	1	6	6	0
α-Carboline	13	21	1	1	1	0
β-Carboline	13	21	1	1	1	0
Diphenyl ether	13	23	1	1	—	0
5-Methyl-3-methylsulfox-deacetophenone	13	25	8	2	—	0
2,3,4-Furantricarboxylic acid	14	18	8	7	3	0
5-Phenyl-1,4-benzodiazepine	17	29	2	2	—	0
5-Phenyl-1,4-benzo-diazepine-2-one	18	30	2	2	1	0
4-(4-Fluorophenyl)-1-methyl 5-(4-pyridyl)-imidazole (1)	19	31	2	2	—	0
Charged fragments						
Methylammonium	2	8	1	—	1	+
Methylamidine	4	11	1	—	2	+
Methylguanidine	5	13	1	—	3	+
Tetrahydropyrrole	5	15	1	—	1	+
Piperidine	6	18	1	—	1	+
Benzamidine	9	18	2	—	2	+
5-Amidineindole	12	22	2	—	3	+
Acetate	4	7	1	2	—	−
Methylsulfonate	5	8	1	3	—	−
Benzoic acid	9	14	1	2	—	−
L-Proline	8	17	2	2	1	+−
D-Proline	8	17	2	2	1	+−
Piperazine	6	18	1	—	2	++
Methylphosphate	6	9	1	4	—	−−

3. System Setup

The structure of the human p38 with the inhibitor SB203580 inside the ATP binding site (94) was downloaded from the Protein Data Bank (code 1a9u) (106). The water molecules and the inhibitor were removed. Hydrogen atoms were added with the program WITNOTP. Partial charges were assigned to p38 with the

MPEOE method (102,103) implemented in WITNOTP which, as mentioned earlier, reproduces the all-hydrogen MSI CHARMm22 parameter set. The following 32 residues of the ATP binding site were used by SEED to position the apolar and polar surface points for docking: Val30, Gly31, Ser32, Gly33, Ala34, Tyr35, Gly36, Ser37, Val38, Cys39, Val50, Ala51, Val52, Lys53, Glu71, Leu75, Ile84, Gly85, Leu86, Leu104, Val105, Thr106, His107, Leu108, Met109, Gly110, Ala111, Asp112, Asn155, Leu167, Asp168, and Arg173.

4. Description of the Binding Site

Figure 3 displays the most hydrophobic regions of the ATP binding site together with the most hydrophobic points. The hydrophobic regions were determined by means of the method developed by Scarsi et al. (61). Five hydrophobic regions of concave shape were found. They are designated as pockets 1–5 henceforth. Region 5 is almost flat, and pockets 2 and 4 are less concave than pockets 1 and 3. Pocket 1, located between the Thr106 and Lys53 side chains, is occupied by the phenyl group of the known inhibitors, while pocket 2, lined by the Thr106 and Met109 side chains, is occupied by the pyridine or pyrimidine cycle. The N-substituent of the central imidazol or pyrrole groups is in contact with pocket 5, close to the Val30 and Val38 side chains. Surprisingly, pockets 3 and 4 are empty in the available crystal structures of the MAP kinase p38–inhibitor complexes.

The backbone NH of Met109 and the side chains of Lys53, Tyr35, Arg67, and Arg173 are hydrogen bond donors in the protein binding site that can interact with candidate ligands. Interesting hydrogen bond acceptors are in the side chains of Asp168, Glu71, and Tyr35.

B. p38 Functionality Maps

The library used in this study contains 4-(4-fluorophenyl)-1-methyl-5-(4-pyridyl)imidazole (compound 1, Fig. 5), which is a close analog of a class of potent CSAID inhibitors of p38 (83,88,89). Compound 1 has three of the four rings of the known inhibitors but lacks the C-2 substituent of the imidazole. Figure 6 shows that SEED docks compound 1 in the right orientation [heavy atom root-mean-square deviation of 0.9 Å from the position of the inhibitor SB203580, PDB code 1a9u (94)] and ranks it as best, among the 70 fragments of the library, with a very favorable van der Waals energy (Table 2). Furthermore, the aromatic rings of the second and third best fragments, diphenyl ether and dibenzocyclohexane, overlap the pyridine and phenyl rings of the inhibitor SB203580. To better describe the SEED results, the functionality map discussion is divided into three subsections according to the polar character of the fragments.

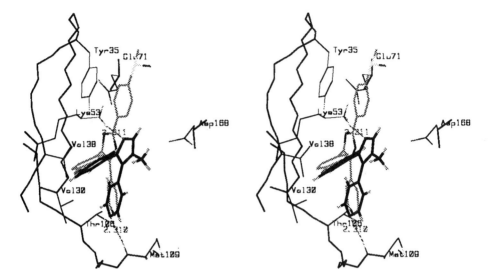

Figure 6   Relaxed-eyes stereoview of the p38 MAP kinase binding site (thin lines) with the SB203580 inhibitor [94] (medium lines, green carbon atoms), and a close analog—that is compound **1** (thick lines)—docked by SEED. Oxygen atoms are colored in red, nitrogen in blue, sulfur in yellow, and hydrogen in cyan. Hydrogen bonds are shown by a red dashed line.

1.   Fragments Docked Using the Apolar Vectors

Common trends are observed for most of the apolar fragments containing a phenyl ring, on one hand, and for most of the alkyl groups, on the other hand. Generally, all hydrophobic pockets accommodate apolar fragments, but some preference is observed for pockets 1–4, while pocket 5 is less favorable because it is almost flat. As expected, the electrostatic intermolecular term and the desolvation of the apolar fragment are negligible. We describe the results obtained with benzene and cyclohexane, as representative functional groups. Diphenyl ether, the fragment with the second-best binding energy in the library after compound **1**, is also analyzed to describe the behavior of slightly polar fragments docked as apolar fragments. In general, the functionality maps of nonpolar fragments are consistent with the available structural data. Moreover, they indicate that the binding affinity might be improved by supplementing the known ligands with substituents that fill the pockets 3 and/or 4.

Benzene.   A large energy gap (about 4 kcal/mol) is observed between the representative of the first cluster and the representatives of the other clusters. Moreover, the nine remaining members of the first cluster also have a more favorable energy than the positions in the following clusters. This difference is mainly due to the large favorable van der Waals term (Table 2). The first cluster of benzene is embedded in hydrophobic pocket 1, where the phenyl group of the known inhibitors is placed. The orientation of its members is also similar to that observed in the crystallographic structure (Fig. 7). The representatives of the remaining clusters display close energy values. The other apolar fragments containing a phenyl ring (e.g., naphthalene, tetraline, N-methylindole, dibenzocyclohexane) show the same behavior: there is a large energy gap (from 2.5 kcal/mol to 4.0 kcal/mol) between the position of the first cluster and the other cluster represen-

Table 2 Cluster Representatives of Fragments Docked Using the Apolar Vectors

	Intermolecular interaction energy (kcal/mol)		Electrostatic desolvation energy (kcal/mol)		ΔG_{bind}	
Rank	van der Waals	Electrostatic	Receptor	Fragment	(kcal/mol)[a]	Site[b]
Benzene						
1	−14.7	−0.6	3.4	0.3	−11.6	1
2	−10.1	0.0	2.0	0.3	−7.8	2
3	−11.3	0.3	3.1	0.3	−7.7	3
4	−10.7	0.5	2.8	0.3	−7.2	3
5	−11.0	0.1	3.6	0.3	−7.0	1
Cyclohexane						
1	−11.3	0.2	2.4	0.0	−8.6	4
2	−13.1	0.1	4.6	0.0	−8.5	3
3	−11.1	0.1	2.7	0.0	−8.4	2
4	−13.0	−0.0	4.7	0.0	−8.3	1
5	−9.7	−0.2	3.6	0.0	−7.8	2,5
Diphenyl ether						
1	−21.3	−0.9	5.5	0.5	−16.2	1,2
2	−19.9	−0.6	6.6	0.5	−13.4	1,2
3	−18.5	0.1	6.5	0.5	−11.4	3
4	−14.4	−0.5	3.8	0.4	−10.7	2,4
5	−17.0	−0.6	6.4	0.5	−10.7	3
4-(4-Fluorophenyl)-1-methyl-5-(4-pyridyl)imidazole						
1	−26.0	−1.9	9.7	1.1	−17.1	1,2
2	−20.1	0.5	4.6	0.9	−14.1	2,4
3	−18.8	−1.0	5.9	1.0	−12.9	2,4
4	−18.7	0.2	5.6	1.0	−11.8	2,4
5	−21.1	−1.9	10.9	1.0	−11.2	1,2

[a] Sum of the values in the four preceding columns (i.e., intermolecular interaction and electrostatics desolvation energies).
[b] Numbering of the hydrophobic pockets as defined in Fig. 3.

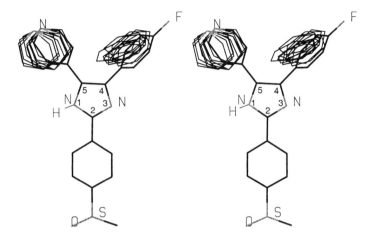

Figure 7   Relaxed-eyes stereoview of the 10 best benzenes and pyridines docked by SEED into the p38 MAP kinase. The bound conformation of the SB203580 inhibitor (94) is displayed to show that the 10 best benzenes and pyridines match the corresponding groups of SB203580.

tatives. The phenyl ring of the best position is located in the hydrophobic pocket 1, except for *N*-methylindole, whose substituted five-membered ring is placed in hydrophobic pocket 1.

Cyclohexane.   Very close binding energies are obtained for the first four cluster representatives (Table 2), although they are in four different regions of the binding site. However repartitions between the different energy terms are not similar: the fragments found in hydrophobic pockets 1 and 3 display a more favorable van der Waals term than those in pockets 2 and 4. Conversely, the electrostatic desolvation term of the receptor is more favorable in the latter regions than in the former ones. The same trend is observed for benzene and for all apolar fragments, which are small enough to fill only one pocket. This means that better van der Waals contacts can be achieved in pockets 1 and 3, to the detriment of electrostatic interactions of the receptor polar groups found in these regions. Except for adamantane and propane, no energy gap is observed between the representative positions of the first clusters of the aliphatic fragments. For the other alkyl groups the position ranking is different for each fragment type, but the total binding energies are close to one another. The large, almost spherical, adamantane fragment has a large energy gap (2.5 kcal/mol) between the second and the third cluster representatives. The first two clusters are located in pockets 2 and 4, which are large enough to contain this fragment while cluster 3 is in between. A 2.0 kcal/mol energy gap is observed between the representative of the first cluster of propane located in hydrophobic pocket 1 and the remaining ones.

Diphenyl Ether. This fragment contains two phenyl rings linked by an oxygen atom. Here, it was docked using both the apolar and polar vectors, but the best binding modes were obtained with the former. Owing to the size dependence of the van der Waals interaction, the intermolecular energy is more favorable than for benzene and cyclohexane, and diphenylether ranks second among the 70 compounds of the library. Energy gaps larger than 2.0 kcal/mol exist between the representatives of the first and the second clusters and between the second and the remaining ones. In the best binding mode, diphenyl ether matches the pyridine and phenyl ring of the SB203580 inhibitor. Some positions allow a weak hydrogen bond between the diphenyl ether oxygen and the side chains of Tyr35 and Lys53.

Fragments containing a phenyl or pyridyl ring usually have the aromatic ring in pocket 1. The position of the first cluster of 5-phenyl-1,4-benzodiazepine is very interesting: the 5-phenyl substituent is located in pocket 1, while the benzene ring is in pocket 2. For fragments with mainly apolar character, the binding energies obtained by apolar docking are more favorable than those obtained by polar seeding (data not shown), mainly because of a more favorable van der Waals term. The regions occupied by the clusters are also different. For example, large fragments, such as diphenyl ether or 5-phenyl-1,4-benzodiazepine, do not penetrate deeply into the binding site when they are docked using polar vectors. They are placed at the entry of the binding site, being hydrogen-bonded to the side chain of Tyr35 or to the backbone of the neighboring residues.

4-(4-Fluorophenyl)-1-methyl-5-(4-pyridyl)imidazole. Another interesting example of apolar docking concerns compound **1**, a close analog of a series of potent p38 inhibitors (83,88,89). Compound **1** has the best binding energy of the 70 fragments of the library and is very close to the positions of the inhibitors in the crystallized complexes with p38 (94) (the distances between corresponding heavy atoms range between 0.8 and 1.0 Å). Interestingly, although it was placed using the apolar vectors, the best binding mode has a good electrostatic interaction (−1.9 kcal/mol) and makes two hydrogen bonds, with the backbone NH of Met109 and the Lys53 side chain (Fig. 6), which are identical to those observed for the inhibitors in the crystal structures. Similarly to what was observed for other fragments docked using the apolar vectors, the representative of the first cluster displays a large gap with the other cluster representatives (3 kcal/mol).

2. Polar Groups

Pyridine and pyrrole are analyzed because they represent hydrogen bond acceptors and donors, respectively. Their electrostatic interaction energy is somewhat more favorable than that of the fragments without donors and acceptors. However, the dominant term is still the van der Waals energy, because the receptor desolvation penalty often compensates the favorable electrostatic interaction.

Pyridine. Pyridine, as well as other fragments containing a similar ring, are preferentially located in hydrophobic pocket 2, accepting a hydrogen bond from the backbone NH of Met109, as the pyridyl group of the known inhibitors (94). The main chain NH of Met109 is indeed the privileged partner of fragments with a hydrogen bond acceptor. The orientation of the members of the first cluster is very close to that of the corresponding ring in the crystallographic structure (Fig. 7). The energy gap between the first cluster representative and the others is almost 2 kcal/mol, although other cluster representatives of pyridine interact with the Met109 NH. The difference is distributed among the two intermolecular terms and the electrostatic desolvation, but the latter seems to play a significant role comparing the energy profiles of the first position and of the fourth and the fifth, given in Table 3. The other protein partners for hydrogen bond acceptors are mainly Lys53 and Tyr35 side chains. For steric reasons, the latter is preferred by large fragments, like diphenyl ether or 5-phenyl-1,4-benzodiazepine, because it is at the entry of the ATP binding site.

Pyrrole. Pyrrole and fragments with hydrogen bond donors also interact preferentially with the region around Met109. The main hydrogen bond acceptors are the main chain CO groups of Met109, Gly110, and Val30. More seldom do the side chain atoms of Tyr35, Glu71, and Asp168 play this role. Energy differences between cluster representatives are rather small. Many fragments in the

Table 3 Cluster Representatives of Fragments Docked Using Polar Vectors

Rank	Intermolecular interaction energy (kcal/mol)		Electrostatic desolvation energy (kcal/mol)		ΔG_{bind} (kcal/mol)[a]	HB partners
	van der Waals	Electrostatic	Receptor	Fragment		
Pyridine						
1	−9.1	−0.6	1.4	0.8	−7.4	Met109-NH
2	−8.5	−2.0	4.1	0.8	−5.6	Lys53
3	−7.6	0.1	1.3	0.8	−5.4	Met109-NH
4	−9.2	−1.6	4.9	0.8	−5.1	Lys53
5	−9.1	−2.2	6.1	0.8	−4.4	Lys53
Pyrrole						
1	−6.6	−0.3	1.1	0.3	−5.5	Met109-CO
2	−5.0	−0.5	0.8	0.3	−4.4	Val30-CO
3	−5.0	−0.4	1.0	0.3	−4.0	Val30-CO
4	−4.8	−0.5	1.3	0.3	−3.7	Met109-CO
5	−4.7	−0.3	1.1	0.3	−3.6	Gly110-CO

[a] Sum of the values in the four preceding columns (i.e., intermolecular interaction and electrostatics desolvation energies).

Table 4 Cluster Representatives of Charged Fragments

Rank	Intermolecular interaction energy (kcal/mol)		Electrostatic desolvation energy (kcal/mol)		ΔG_{bind} (kcal/mol)	HB partners
	van der Waals	Electrostatic	Receptor	Fragment		
Acetate						
1	−6.1	−7.5	3.6	7.6	−2.4	Arg67-H$^{\eta 21}$ Lys53-H$^{\varepsilon 3}$
2	−5.9	−7.7	3.5	7.7	−2.4	Arg67-H$^{\eta 21}$ Lys53-H$^{\varepsilon 3}$
3	−5.4	−8.1	3.4	7.9	−2.2	Arg67-H$^{\eta 21}$ Lys53-H$^{\varepsilon 3}$
4	−8.5	−4.2	3.0	7.6	−2.1	Lys53-H$^{\zeta 1}$
5	−4.9	−8.2	3.3	8.2	−1.6	Lys53-H$^{\varepsilon 3}$ Arg67-H$^{\eta 21}$
Methyl ammonium						
1	−1.9	−6.1	1.2	3.9	−2.9	Gly110-CO
2	−3.5	−3.7	1.2	3.2	−2.8	Asp112-O$^{\delta 2}$
3	−3.8	−3.6	0.9	4.3	−2.3	Gly110-CO
4	−5.4	−5.0	1.0	7.3	−2.0	Gly110-CO
5	−1.3	−6.9	1.6	4.7	−1.9	Met109-CO

[a] Sum of the values in the four preceding columns (i.e., intermolecular interaction and electrostatics desolvation energies).

library contain both donors and acceptors and occupy most of the pockets that are favorable for either pyridine or pyrrole. For example, the representative position of the first cluster of α-carboline is involved in two hydrogen bonds with the protein, with the backbone donor and acceptor of Met109.

3. Charged Groups

The binding energy of functional groups with formal charge(s) is much less favorable than the one of neutral fragments (Table 4). This is consistent with the lack of formal charges in the known inhibitors and indicates that the electrostatic desolvation penalty cannot be neglected. Small or negligible energy differences are observed between cluster representatives of charged groups, which suggests that there is no really favorable region for the charged fragments of the SEED library inside the ATP binding site of the p38 MAP kinase. The results obtained for acetate and methylammonium are described as exemplar negatively and positively charged fragments, respectively (Table 4).

Acetate. The preferred protein partners for acetate are the Lys53 and Arg67 side chains, which are located above hydrophobic pocket 3. This is also the case for most of the negatively charged fragments. The methyl group of acetate of the fourth cluster members is positioned in pocket 1. Arg173 and Tyr35 are the other donors interacting with these groups. Interestingly, the best methylphosphate (charge of -2) is at the entry of the binding cleft and interacts with the loop containing the consensus Gly-X-Gly-X-X-Gly sequence, which is known to bind the phosphate group of ATP. However it has a binding energy of about 0 kcal/mol, which is much less favorable than the one of the neutral fragments discussed earlier. This suggests that the triphosphate moiety might contribute less than the base to the binding affinity of ATP.

Methylammonium. Conversely to the negatively charged fragments, no positively charged compound is deeply buried into the binding site. Such compounds are mainly found in region 4, interacting with the backbone oxygen atom of Gly110 and Met109, as well as with the Asp112 side chain. Other partners are the Val30 carbonyl group and Asp168 side chain.

V. FUTURE DIRECTIONS

The Internet is providing an excellent opportunity for computer-aided drug design; user-friendly, interactive, and platform-independent WWW-based tools for molecular modelling have existed for several years (107–109) and more will emerge. In the near future, it is expected that multicenter ligand design will become a reality, with several researchers working on the same target by library docking and calculation of binding affinities distributed on several computers connected by very fast networks.

It is clear from this and other chapters that drug design is a really multidisciplinary research field; we hope that this book will spur interest and enthusiasm for computer-aided drug design among chemists, physicists, biologists, and computer scientists. Although the field is no longer in its infancy, new ideas and multidisciplinary approaches are required to meet the two main challenges: the accurate estimation of binding affinity (4,110), and the large amount of data emerging from the genomics and proteomics endeavors (111–113).

ACKNOWLEDGMENTS

We thank Drs. Chantal Carrez and Marco Scarsi for helpful discussions. This work was supported by Aventis Pharma (Vitry-sur-Seine, France) and the Swiss National Science Foundation (grant no. 31-53604.98 to AC).

REFERENCES

1. Caflisch A., Wälchli R., Ehrhardt C. Computer-aided design of thrombin inhibitors. News Physiol. Sci. 1998; 13:182–189.
2. Kubinyi H. Structure-based design of enzyme inhibitors and receptor ligands. Curr. Opin. Drug Design Discovery 1998; 1:4–15.

3.   Weber L., Wallbaum S., Broger C., Gubernator K. Optimization of the biological activity of combinatorial libraries by a genetic algorithm. Angew. Chem. Int. Ed. 1995; 34:2280–2282.

4.   Apostolakis J., Caflisch A. Computational ligand design. Comb. Chem. High Throughput Screening 1999; 2:91–104.

5.   Pearlman D., Murcko M. CONCEPTS: New dynamic algorithm for *de novo* drug suggestion. J. Comput. Chem. 1993; 14:1184–1193.

6.   Gehlhaar D. K., Moerder K. E., Zichi D., Sherman C. J., Ogden R. C., Freer S. T. *De novo* design of enzymes inhibitors by Monte Carlo ligand generation. J. Med. Chem. 1995; 38:466–472.

7.   Todorov N. P., Dean P. M. A branch-and-bound method for optimal atom-type assignment in *de novo* ligand design. J. Comput.-Aided Mol. Design 1998; 12: 335–349.

8.   Majeux N., Scarsi M., Apostolakis J., Ehrhard C., Caflisch A. Exhaustive docking of molecular fragments on protein binding sites with electrostatic solvation. Proteins Struct. Funct. Gent. 1999; 37:88–105.

9.   Caflisch A. Computational combinatorial ligand design: Application to human α-thrombin. J. Comput.-Aided Mol. Design 1996; 10:372–396.

10. Böhm H.-J. The computer program LUDI: A new method for the *de novo* design of enzyme inhibitors. J. Comput.-Aided Mol. Design 1992; 6:61–78.

11. Goodford P. J. A computational procedure for determining energetically favorable binding sites on biologically important macromolecules. J. Med. Chem. 1985; 28:849–857.

12. Lawrence M. C., Davis P. C. CLIX: A search algorithm for finding novel ligands capable of binding proteins of known three-dimensional structure. Proteins 1992; 12:31–41.

13. Kuntz I. D., Blaney J. M., Oatley S. J., Langridge R., Ferrin T. E. A geometric approach to macromolecule–ligand interactions. J. Mol. Biol. 1982; 161:269–288.

14. Oshiro C. M., Kuntz I. D., Dixon J. S. Flexible ligand docking using a genetic algorithm. J. Comput.-Aided Mol. Design 1995; 9:113–130.

15. Meng E. C., Shoichet B. K., Kuntz I. D. Automated docking with grid-based energy evaluation. J. Comput. Chem. 1992; 13:505–524.

16. Desjarlais R. L., Seibel G. L., Furth P. S., Kuntz I. D., Alvarez J. C., de Montellano P. R. Ortiz, DeCamp D. L., Babe L. M., Craik C. S. Structure-based design of nonpeptide inhibitors specific for the human immunodeficiency virus 1 protease. Proc. Natl. Acad. Sci. USA 1990; 87:6644–6648.

17. Shoichet B. K., Stroud R. M., Santi D. V., Kuntz I. D., Perry K. M. Structure-based discovery of inhibitors of thymidylate synthase. Science 1993; 259:1445–1450.

18. Rarey M., Kramer B., Lengauer T., Klebe G. A fast flexible docking method using an incremental construction algorithm. J. Mol. Biol. 1996; 261:470–489.

19. Böhm H.-J. The development of a simple empirical scoring function to estimate the binding constant for a protein–ligand complex of known three-dimensional structure. J. Comput.-Aided Mol. Design 1994; 8:243–256.

20. Rarey M., Kramer B., Lengauer T. Docking of hydrophobic ligands with interaction-based matching algorithms. Bioinformatics 1999; 15:243–250.

21. Rarey M., Kramer B., Lengauer T. The particle concept: Placing discrete water molecules during protein–ligand docking predictions. Proteins 1999; 34:17–28.

22. Jones G., Willet P., Willet R. C. A genetic algorithm for flexible molecular overlay and pharmacophore elucidation. J. Comput.-Aided Mol. Design 1995; 9: 532–549.

23. Jones G., Willet P., Willet R. C. Molecular recognition of receptor sites using a genetic algorithm with a description of desolvation. J. Mol. Biol. 1995; 245:43–53.

24. Clark K. P., Ajay. Flexible ligand docking without parameter adjustment across four ligand–receptor complexes. J. Comput. Chem. 1995; 16:1210–1226.

25. Jones G., Willet P., Glen R. C., Leach A., Taylor R. Development and validation of a genetic algorithm for flexible docking. J. Mol. Biol. 1997; 267:727–748.

26. Apostolakis J., Plückthun A., Caflisch A. Docking small ligands in flexible binding sites. J. Comput. Chem. 1998; 19:21–37.

27. Elber R., Karplus M. Enhanced sampling in molecular dynamics: use of the time-dependent Hartree approximation for a simulation of carbon monoxide diffusion through myoglobin. J. Am. Chem. Soc. 1990; 112:9161–9175.

28. Miranker A., Karplus M. Functionality maps of binding sites: A multiple copy simultaneous search method. Proteins 1991; 11:29–34.

29. Rosenfeld R., Zheng Q., Vajda S., DeLisi C. Computing the structure of bound peptides: Application to antigen recognition by class I MHCs. J. Mol. Biol. 1993; 234:515–521.

30. Leclerc F., Karplus M. MCSS-based predictions of RNA binding sites. Theor. Chem. Acc. 1999; 101:131–137.

31. Caflisch A., Schramm H. J., Karplus M. Design of dimerization inhibitors of HIV-1 aspartic proteinase: A computer-based combinatorial approach. J. Comput.-Aided Mol. Design 2000; 14:161–179.

32. Stultz C., Karplus M. MCSS functionality maps for a flexible protein. Proteins Struct. Funct. Genet. 1999; 37:512–529.

33. Joseph-McCarthy D., Hogle J. M., Karplus M. Use of multiple copy simultaneous search (MCSS) method to design a new class of Picornavirus capsid binding drugs. Proteins 1997; 29:32–58.

34. Bartlett P. A., Shea G. T., Telfer S. J., Waterman S. In S. M. Roberts, ed. Molecular Recognition: Chemical and Biological Problems. London: Royal Society of Chemistry, 1989, pp. 182–193.

35. Lauri G., Bartlett P. A. CAVEAT: A program to facilitate the design of organic molecules. J. Comput.-Aided Mol. Design 1994; 8:51–66.

36. Eisen M. B., Wiley D. C., Karplus M., Hubbard R. E. HOOK: A program for finding novel molecular architectures that satisfy the chemical and steric requirements of a macromolecule binding site. Proteins 1994; 19:199–221.

37. Böhm H.-J. LUDI: Rule-based automatic design of new substituents for enzyme inhibitor leads. J. Comput.-Aided Mol. Design 1992; 6:593–606.

38. Kabsch W. A solution for the best rotation to relate two sets of vectors. Acta Crystallogr. 1976; A32:922–923.

39. Klebe G., Mietzner T. A fast and efficient method to generate biologically relevant conformations. J. Comput.-Aided Mol. Design 1994; 8:583–606.

40. Klebe G., Mietzner T., Weber F. Different approaches toward an automatic structural alignment of drug molecules: Applications to sterol mimics, thrombin and thermolysin inhibitors. J. Comput.-Aided Mol. Design 1994; 8:751–778.

41. Böhm H.-J. Towards the automatic design of synthetically accessible protein ligands: Peptides, amides and peptidomimetics. J. Comput.-Aided Mol. Design 1996; 10:265–272.

42. Gillet V., Johnson A. P., Mata P., Sike S., Williams P. SPROUT: A program for structure generation. J. Comput.-Aided Mol. Design 1993; 7:127–153.

43. Sun Y., Ewing T. J. A., Skillman A. G., Kuntz I. D. CombiDOCK: Structure-based combinatorial docking and library design. J. Comput.-Aided Mol. Design 1998; 12:597–604.

44. Böhm H.-J., Banner D. W., Weber L. Combinatorial docking and combinatorial chemistry: Design of potent non-peptide thrombin inhibitors. J. Comput.-Aided Mol. Design 1999; 13:51–56.

45. Rotstein S. H., Murcko M. A. GenStar: A method for de novo drug design. J. Comput.-Aided Mol. Design 1993; 7:23–43.

46. Nishibata Y., Itai A. Automatic creation of drug candidate structures based on receptor structure. Starting point for artificial lead generation. Tetrahedron 1991; 47:8985–8990.

47. Nishibata Y., Itai A. Confirmation of usefulness of a structure construction program based on three-dimensional receptor structure for rational lead generation. J. Med. Chem. 1993; 36:2921–8.

48. Allinger N. L. MM2. A hydrocarbon force field utilizing V_1 and V_2 torsional terms. J. Am. Chem. Soc. 1977; 99:8127–8134.

49. Rotstein S. H., Murcko M. A. GroupBuild: A fragment-based method for de novo drug design. J. Med. Chem. 1993; 36:1700–10.

50. DeWitte R. S., Shakhnovich E. I. SMoG: De novo design method based on simple, fast, and accurate free energy estimates. 1. Methodology and supporting evidence. J. Am. Chem. Soc. 1996; 118:11733–11744.

51. DeWitte R. S., Shakhnovich E. I. SMoG: De novo design method based on simple, fast, and accurate free energy estimates. 2. Case studies in molecular design. J. Am. Chem. Soc. 1997; 119:4608–4617.

52. Moon J. B., Howe W. J. Computer design of bioactive molecules: A method for receptor-based de novo ligand design. Proteins 1991; 11:314–28.

53. Frenkel D., Clark D. E., Li J., Murray C. W., Robson B., Waszkowycz B., Westhead D. R. PRO_LIGAND: An approach to de novo molecular design. 4. Application to the design of peptides. J. Comput.-Aided Mol. Design 1995; 9:213–25.

54. Weiner S. J., Kollman P. A., Case D. A., Singh U. C., Ghio C., Alagona G., Profeta S. Jr., Weiner P. A. A new force field for molecular mechanical simulation of nucleic acids and proteins. J. Am. Chem. Soc. 1984; 106:765–784.

55. Ooi T., Oobatake M., Némethy M., Scheraga H. A. Accessible surface areas as a measure of the thermodynamic parameters of hydration of peptides. Proc. Natl. Acad. Sci. USA 1987; 84:3086–3090.

56. Clark D. E., Frenkel D., Levy S. A., Li J., Murray C. W., Robson B., Waszkowycz B., Westhead D. R. PRO_LIGAND: An approach to de novo molecular design. 1.

Application to the design of organic molecules. J. Comput.-Aided Mol. Design 1995; 9:13–32.

57. Shoichet B. K., Leach A. R., Kuntz I. D. Ligand solvation in molecular docking. Proteins Struct. Funct. Genet. 1999; 34:4–16.

58. Mark A. E., van Gunsteren W. F. Decomposition of the free energy of a system in terms of specific interactions. Implications for theoretical and experimental studies. J. Mol. Biol. 1994; 240:167–176.

59. So S., Karplus M. A comparative study of ligand–receptor complex binding affinity prediction methods based on glycogen phosphorylase inhibitors. J. Comput.-Aided Mol. Design 1999; 13:243–258.

60. Charifson P. S., Corkery J. J., Murcko M. A., Walters W. P. Consensus scoring: A method for obtaining improved hit rates from docking databases of three-dimensional structures into proteins. J. Med. Chem. 1999; 42:5100–5109.

61. Scarsi M., Majeux N., Caflisch A. Hydrophobicity at the surface of proteins. Proteins Struct. Funct. Genet. 1999; 37:565–575.

62. Scarsi M., Apostolakis J., Caflisch A. Continuum electrostatic energies of macromolecules in aqueous solutions. J. Phys. Chem. A 1997; 101:8098–8106.

63. Scarsi M., Apostolakis J., Caflisch A. Comparison of a GB solvation model with explicit solvent simulations: Potentials of mean force and conformational preferences of alanine dipeptide and 1,2-dichloroethane. J. Phys. Chem. B 1998; 102:3637–3641.

64. Warwicker J., Watson H. C. Calculation of the electric potential in the active site cleft due to α-helix dipoles. J. Mol. Biol. 1982; 157:671–679.

65. Gilson M. K., Honig B. H. Calculation of the total electrostatic energy of a macromolecular system: solvation energies, binding energies, and conformational analysis. Proteins Struct. Funct. Genet. 1988; 4:7–18.

66. Bashford D., Karplus M. pK_a's of ionizable groups in proteins: Atomic detail from a continuum electrostatic model. Biochemistry 1990; 29:10219–10225.

67. Davis M. E., Madura J. D., Luty B. A., McCammon J. A. Electrostatics and diffusion of molecules in solution: Simulations with the University of Houston Brownian Dynamics program. Comput. Phys. Commun. 1991; 62:187–197.

68. Still W. C., Tempczyk A., Hawley R. C., Hendrickson T. Semianalytical treatment of solvation for molecular mechanics and dynamics. J. Am. Chem. Soc. 1990; 112:6127–6129.

69. Hawkins G. D., Cramer C. J., Trulhar D. G. Pairwise solute descreening of solute charges from a dielectric medium. Chem. Phys. Lett. 1995; 246:122–129.

70. Hawkins G. D., Cramer C. J., Trulhar D. G. Parametrized models of aqueous free energies of solvation based on pairwise descreening of solute atomic charges from a dielectric medium. J. Phys. Chem. 1996; 100:19824–19839.

71. Schaefer M., Karplus M. A comprehensive analytical treatment of continuum electrostatics. J. Phys. Chem. 1996; 100:1578–1599.

72. Qiu Di, Shenkin P. S., Hollinger F. P., Still W. C. The GB/SA continuum model for solvation. A fast analytical method for the calculation of approximate Born radii. J. Phys. Chem. A 1997; 101:3005–3014.

73. Marrone T. J., Gilson M. K., McCammon J. A. Comparison of continuum and explicit models of solvation: Potential of mean force for alanine dipeptide. J. Phys. Chem. 1996; 100:1439–1441.

74. Richards F. M. Areas, volumes, packing, and protein structure. Annu. Rev. Biophys. Bioeng. 1977; 6:151–176.

75. Jackson J. D. Classical Electrodynamics. 3rd ed. New York: Wiley, 1999.

76. Luo R., Moult J., Gilson M. K. Dielectric screening treatment of electrostatic solvation. J. Phys. Chem. B 1997; 101:11226–11236.

77. Davis M. E., McCammon J. A. Solving the finite difference linearized Poisson–Boltzmann equation: A comparison of relaxation and conjugate gradient methods. J. Comput. Chem. 1989; 10:386–391.

78. Boulton T. G., Yancopoulos G. D., Gregory J. S., Slaughter C., Moomaw C., Hsu J., Cobb M. H. An insulin-stimulated protein kinase similar to yeast kinases involved in cell cycle control. Science 1990; 249:64–67.

79. Boulton T. G., Gregory J. S., Cobb M. H. Purification and properties of extracellular signal-regulated kinase 1, an insulin-stimulated microtubule-associated protein 2 kinase. Biochemistry 1991; 30:278–286.

80. Dérijard B., Hibi M., Wu I. H., Barrett T., Su B., Deng T., Karin M., Davis R. J. JNK1: A protein kinase stimulated by UV light and Ha-Ras that binds and phosphorylates the c-Jun activation domain. Cell 1994; 76:1025–1037.

81. Kyriakis J. M., Banerjee P., Nikolakaki E., Dai T., Rubie E. A., Ahmad M. F., Avruch J., Woodgett J. R. The stress-activated protein kinase subfamily of c-Jun kinases. Nature 1994; 369:156–160.

82. Kallunki T., Su B., Tsigelny I., Sluss H. K., Dérijard B., Moore G., Davis R., Karin M. JNK2 contains a specificity-determining region responsible for efficient c-Jun binding and phosphorylation. Genes Dev. 1994; 8:2996–3007.

83. Lee J. C., Laydon J. T., McDonnell P. C., Gallagher T. F., Kumar S., Green D., McNulty D., Blumenthal M. J., Heys J. R., Landvatter S. W., Young P. R. A protein kinase involved in the regulation of inflammatory cytokine biosynthesis. Nature 1994; 372:739–746.

84. Han J., Lee J. D., Bibbs L., Ulevitch R. J. A MAP kinase targeted by endotoxin and hyperosmolarity in mammalian cells. Science 1994; 265:808–811.

85. Rouse J., Cohen P., Trigon S., Morange M., Alonso-Llamazares A., Zamanillo D., Hunt T., Nebreda A. R. A novel kinase cascade triggered by stress and heat shock that stimulates MAPKAP kinase-2 and phosphorylation of the small heat shock proteins. Cell 1994; 78:1027–1037.

86. Freshney N. W., Rawlinson L., Guesdon F., Jones E., Cowley S., Hsuan J., Saklatvala J. Interleukin-1 activates a novel protein kinase cascade that results in the phosphorylation of Hsp27. Cell 1994; 78:1039–1049.

87. Zhang F., Strand A., Robbins D., Cobb M. H., Goldsmith E. J. Atomic structure of the MAP kinase ERK2 at 2.3 Å resolution. Nature 1994; 367:704–711.

88. Cuenda A., Rouse J., Doza Y. N., Meier R., Cohen P., Gallagher T. F., Young P. R., Lee J. C. SB 203580 is a specific inhibitor of a MAP kinase homologue which is stimulated by cellular stresses and interleukin-1. FEBS Lett. 1995; 364:229–233.

89. Lee J. C., Adams J. L. Inhibitors of serine/threonine kinases. Curr. Opin. Biotechnol. 1995; 6:657–661.

90. Lisnock J.-M., Tebben A., Franz B., O'Neill E. A., Croft G., O'Keefe S. J., Li B., abd S. de Laszlo C. Hacker, Smith A., abd N. Liverton B. Libby, Hermes J., LoGrasso P. Molecular basis for p38 protein kinase inhibitor specificity. Biochemistry 1998; 37:16573–16581.

91. Wang Z., Harkins P. C., Ulevitch R. J., Han J., Cobb M. H., Goldsmith E. J. The structure of the mitogen-activated protein kinase p38 at 2.1 Å resolution. Proc. Natl. Acad. Sci. USA 1997; 94:2327–2332.

92. Wilson K. P., Fitzgibbon M. J., Caron P. R., Griffith J. P., Chen W., McCaffrey P. G., Chambers S. P., Su M. S. S. Crystal structure of p38 mitogen-activated protein kinase. J. Biol. Chem. 1996; 271:27696–27700.

93. Tong L., Pav S., White D. M., Rogers S., Cywin K. M. Crane C. L., Brown M. L., Pargellis C. A. A highly specific inhibitor of human p38 MAP kinase binds in the ATP pocket. Nature Struct. Biol. 1997; 4:311–316.

94. Wang Z., Canagarajah B. J., Boehm J. C., Kassisa S., Cobb M. H., Young P. R., Abdel-Meguid S., Adams J. L., Goldsmith E. J. Structural basis of inhibitor selectivity in MAP kinases. Structure 1998; 6:1117–1128.

95. Schulze-Gahmen U., Brandsen J., Jones H. D., Morgan D. O., Meijer L., Vesely J., Kim S. H. Multiple modes of ligand recognition: Crystal structures of cyclin-dependent protein kinase 2 in complex with ATP and two inhibitors, olomoucine and isopentenyladenine. Proteins 1995; 22:378–391.

96. Mohammadi M., McMahon G., Sun L., Tang C., Hirth P., Yeh B. K., Hubbard S. R., Schlessinger J. Structures of the tyrosine kinase domain of fibroblast growth factor receptor in complex with inhibitors. Science 1997; 276:955–960.

97. Singh J., Dobrusin E. M., Fry D. W., Haske T., Whitty A., McNamara D. J. Structure-based design of a potent, selective, and irreversible inhibitor of the catalytic domain of the *erbB* receptor subfamily of protein tyrosine kinases. J. Med. Chem. 1997; 40:1130–1135.

98. Avezado W. F. De, Mueller-Dieckman H. J., Schulze-Gahmen U., Worland P. J., Sausville E., Kim S. H. Structural basis for specificity and potency of a flavonoid inhibitor of human CDK2, a cell cycle kinase. Proc. Natl. Acad. Sci. USA 1996; 93:2735–2740.

99. Bemis G. W., Murcko M. A. The properties of known drugs. 1. Molecular frameworks. J. Med. Chem. 1996; 39:2887–2893.

100. Andres C. J., Denhart D. J., Deshpande M. S., Gillman K. W. Recent advances in the solid phase synthesis of drug-like heterocyclic small molecules. Comb. Chem. High Throughput Screening 1999; 2:191–210.

101. Furka A., Bennett W. D. Combinatorial libraries by portioning and mixing. Comf. Chem. High Throughput Screening 1999; 2:105–122.

102. Gasteiger J., Marsili M. Iterative partial equalization of orbital electronegativity. A rapid access to atomic charges. Tetrahedron 1980; 36:3219–3288.

103. No K. T., Grant J. A., Scheraga H. A. Determination of net atomic charges using a modified partial equalization of orbital electronegativity method. 1. Application to neutral molecules as models for polypeptides. J. Phys. Chem. 1990; 94:4732–4739.

104. Brooks B. R., Bruccoleri R. E., Olafson B. D., States D. J., Swaminathan S., Karplus M. CHARMM: A program for macromolecular energy, minimization, and dynamics calculations. J. Comput. Chem. 1983; 4:187–217.

105. Gilson M. K., Honig B. H. Energetics of charge–charge interactions in proteins. Proteins Struct. Funct. Genet. 1988; 3:32–52.

106. Bernstein F. C., Koetzle T. F., Williams G. J., Meyer E. E. Jr., Brice M. D., Rodgers J. R., Kennard O., Shimanouchi T., Tasumi M. The Protein Data Bank: A computer-based archival file for macromolecular structures. J. Mol. Biol. 1977; 112:535–542.

107. Leach A. R. Structure-based selection of building blocks for array synthesis via the World-Wide Web. J. Mol. Graphics Modelling 1997; 15:158–160.

108. Ertl P. World Wide Web–based system for the calculation of substituent parameters and substituent similarity searches. J. Mol. Graphics Modelling 1998; 16:11–13.

109. Wild D. J., Blankley C. J. VisualiSAR: A Web-based application for clustering, structure browsing, and structure–activity relationship study. J. Mol. Graphics Modelling 1999; 17:85–89.

110. Ajay, Murcko M. A. Computational methods to predict binding free energy in ligand-receptor complexes. J. Med. Chem. 1995; 38(26):4953–4967.

111. Sánchez R., Sali A. Large-scale protein structure modeling of the *Saccharomyces cerevisiae* genome. Proc. Natl. Acad. Sci. USA 1998; 95:13597–13602.

112. Marcotte E. M., Pellegrini M., Thompson M. J., Yeates T. O., Eisenberg D. A combined algorithm for genome-wide prediction of protein function. Nature 1999; 402:83–86.

113. Enright A. J., Iliopoulos I., Kyrpides N. C., Ouzounis C. A. Protein interaction maps for complete genomes based on gene fusion events. Nature 1999; 402:86–89.

114. Press W. H., Teukolsky S. A., Vetterling W. T., Flannery B. P. Numerical Recipes in Fortran. Cambridge: Cambridge University Press, 1992.

115. Hagler A. T., Huler E., Lifson S. Energy functions for peptides and proteins. I. Derivation of a consistent force field including the hydrogen bond from amide crystals. J. Am. Chem. Soc. 1974; 96:5319–5327.

116. Pattabiraman N., Levitt M., Ferrin T. E., Langridge R. Computer graphics in real-time docking with energy calculation and minimization. J. Comput. Chem. 1985; 6:432–436.

117. Meng E. C., Shoichet B. K., Kuntz I. D. Automated docking with grid-based energy evaluation. J. Comput. Chem. 1992; 13:505–524.

118. Luty B. A., Wasserman Z. R., Stouten P. F. W., Hodge C. N., Zacharias M., McCammon J. A. A molecular mechanics/grid method for evaluation of ligand–receptor interactions. J. Comput. Chem. 1995; 16:454–464.

119. Nicholls A., Sharp K. A., Honig B. Protein folding and association: Insights from the interfacial and thermodynamic properties of hydrocarbons. Proteins 1991; 11:281–296.

120. Connolly M. L. Solvent-accessible surfaces of proteins and nucleic acids: Description of the method to calculate the "Connolly" surfaces. Science 1983; 221:709–713.

121. Kraulis P. Molscript, a program to produce both detailed and schematic plots of protein structures. J. Appl. Crystallogr. 1991; 24:946–950.

7

Quo Vadis, Scoring Functions? Toward an Integrated Pharmacokinetic and Binding Affinity Prediction Framework

Tudor I. Oprea, Ismael Zamora, and Peder Svensson*

AstraZeneca R&D Molndal
Molndal, Sweden

I. INTRODUCTION

To reach the status of marketed drug, a chemical compound must exhibit, besides high affinity and selectivity for the targeted binding site, good pharmacokinetic properties, minimal side effects, and low toxicity. A further requirement is patentability; that is, the compound must have nonobvious chemical alterations compared to previously disclosed compounds. Historically, medicinal chemistry efforts, as illustrated in Fig. 1, started from a lead structure that had (typically) poor DMPK (**d**istribution, **m**etabolism, and **p**harmacokinetic) properties and micromolar affinity. Initial synthetic efforts were aimed at improving the binding affinity (steps 1–10 in Fig. 1). Once medium- to low-nanomolar affinity was achieved, DMPK properties were optimized (steps 11–23 in Fig. 1). This process was far from trivial, since one had to preserve the molecular determinants responsible for affinity while modifying the chemical structure to achieve good DMPK properties. This often resulted in reduced binding affinity

* *Current affiliation:* Carlsson Research, Göteborg, Sweden.

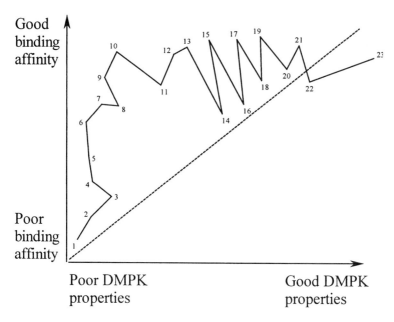

Figure 1 Schematic illustration of the traditional medicinal chemistry efforts in drug discovery. The dashed line indicates the simultaneous optimization (ideal) approach. See text for details.

(steps 14, 16), and the process before optimal structures (step 23) were found was often time-consuming.

These problems were, to a large extent, avoided by early in vivo screening for good DMPK properties—a strategy that is currently used in many of the pharmaceutical companies. While beneficial to the drug discovery process, the in vivo DMPK screening strategy cannot cope with the increasingly large numbers of compounds produced by current technologies. The advent of medium and high throughput synthesis methods such as multiple parallel synthesis and combinatorial chemistry, combined with high throughput screening (HTS), has created a increasing pressure to optimize both receptor binding affinity and DMPK properties in the very early stages of drug discovery. Both types of property have been amenable to empirical modeling within the quantitative structure–activity relationships (QSAR) paradigm (1–4). QSAR methods aim to explain and predict the target property, which may include different forms of binding affinity measurements (5), such as IC_{50} and/or K_i, but also different forms of DMPK property profile (6,7), such as passive oral absorption, passive blood–brain barrier (BBB) permeability, or active components for certain carrier system for which experimental data are available. One, however, must be careful to distinguish the active from the passive component of the permeability mechanisms, since the former are unsuitable to model in a single QSAR.

This chapter outlines key concepts related to binding affinity and pharmacokinetic property prediction. Receptor-based methods that model and predict binding affinity are briefly discussed, together with methods that estimate oral absorption and BBB permeability. We then focus on the use of QSAR descriptors relevant to DMPK models to estimate the binding affinity of small organic compounds complexed with proteins.

II. BRIEF OVERVIEW OF CLASSICAL QSAR METHODS

Typically, QSAR data are stored in a table in which rows represent compounds and columns represent physicochemical properties (descriptors). The first column after the compound identifier is usually the measured target property Y. Then a statistical procedure, such as multiple linear regression (MLR), (1), projection to latent structures (PLS), (8), or computational neural networks (CNN), (9), is used to find a mathematical model relating the observed measurement(s) with some combination of the properties represented in the other columns. There are many available QSAR descriptors, and the most appropriate choice is not always obvious (1–4). However, about 80% of the over 6000 biological QSAR equations stored in the Pomona College database (10) show significant correlation with hydrophobicity-related descriptors. Cross-correlation, or the collinearity between various descriptors, a problem that may lead to chance correlations when using MLR techniques, can be appropriately dealt with by using PLS (11).

"Classical" QSAR models, also known as Hansch analysis (1) or 2D-QSAR, are regression (MLR) equations that relate the biological activity to the effects of modifying chemical substituents on a parent structure. A typical Hansch analysis has the following form:

$$-\log [\text{Conc}] = a \left(\substack{\text{steric}\\\text{descriptor}}\right) + b \left(\substack{\text{electronic}\\\text{descriptor}}\right) + c \left(\substack{\text{hydrophobic}\\\text{descriptor}}\right) + d \qquad (1)$$

where $-\log [\text{Conc}]$ is the negative logarithm of the biological activity. The coefficients a, b, and c, and the intercept d, are usually determined by regression analysis. The quality of the resulting QSAR models can be judged by statistical means such as r^2 (the fraction of explained variance) and q^2 (the cross-validated, or predictive r^2).

The fraction of explained variance, r^2, measures of the QSAR model's ability to explain the variance in the data; in other words, it estimates the goodness of fit of the regression model (see Eq. 2). The predictive r^2, or q^2, measures the internal robustness of the QSAR model; in other words, it estimates the predictive ability of the model (see Eq. 3). Predictive estimates are obtained either by use of the cross-validation procedure (discussed shortly) or by predicting external compounds (previously not used in the model):

$$r^2 = 1 - \frac{\sum_{i=1}^{N}(Y_c - Y_a)^2}{\sum_{i=1}^{N}(Y_m - Y_a)^2} \tag{2}$$

$$q^2 = 1 - \frac{\text{PRESS}}{\text{SD}} = 1 - \frac{\sum_{i=1}^{N}(Y_a - Y_p)^2}{\sum_{i=1}^{N}(Y_a - Y_m)^2} \tag{3}$$

where Y stands for target property values [Y_a are measured (actual) values; Y_c are calculated values; Y_m is the mean binding affinity of the N given ligands; Y_p are predicted values.] PRESS is the sum of squared deviations between predicted (Y_p) and measured (Y_a) binding affinity values over N ligands, and SD is the sum of the squared deviations between the measured binding affinity values of the molecules in the test set and the average Y value (Y_m) for all N molecules in the training set. Other statistical indices used in QSAR analyses have been detailed elsewhere (12).

Cross-validation (13) is a procedure in which one or more compounds are excluded from the model, whereupon another model is derived using the reduced training set. The excluded compounds are then predicted with this second model. Cross-validation is typically performed in successive runs for randomly or systematically divided groups to test model significance. Cross-validation estimates model robustness and predictivity to avoid overfitted QSAR models (13). In particular, PLS model complexity is established by testing the significance of adding a new dimension (PLS component) to the current model. The optimal number of PLS components is then chosen from the analysis with the highest cross-validated r^2 (q^2) value (Eq. 3). In the most popular cross-validation technique, leave-one-out (LOO), each compound is left out of the model once and only once, which gives reproducible results. An extremely fast LOO method, SAMPLS (14), looks at the covariance matrix only and allows the end user to rapidly estimate the robustness of QSAR models. Stone and Jonathan, who covered statistical methods and the use of cross-validation (15), also offered a critical, comparative description for specific QSAR methods (16).

III. THERMODYNAMIC ASPECTS IN BINDING AFFINITY PREDICTION

A. General Remarks. The Bioactive Conformation

QSAR methods are based on the linear free energy relationship (LFER) formalism that relates $\Delta G^{\circ}_{\text{bind}}$, the standard free energy of binding, to the logarithm of the dissociation constant, $-\log K_D$, at thermodynamic equilibrium concentrations of the ligand, $[L]$, receptor, $[R]$, and the corresponding complex $[L-R]$, for the reaction $L + R \to L\text{-}R$:

$$\Delta G^{\circ}_{\text{bind}} = -RT \ln \frac{[L][R]}{[L-R]} = -RT \ln K_D = 2.303 RT p K_D \qquad (4)$$

where $R = 8.314$ J/mol $\cdot$ K and T is the temperature in degrees Kelvin. At 37°C, $\Delta G^{\circ}_{\text{bind}}$ is approximately 6.0 pK_D (kJ/mol), or 1.42 pK_D (kcal/mol). We often replace $\Delta G^{\circ}_{\text{bind}}$ with $\Delta G^{\text{L-R}}_{\text{bind}}$ by assuming the same reference state. The binding of a ligand to a receptor, a multistep process, evaluated by the total $\Delta G^{\circ}_{\text{bind}}$ (Eq. 4), includes sequential steps going from independent ligands and receptors in the surrounding physiological environment to the ligand–receptor complex.

Conformationally flexible compounds are mixtures of multiple conformational substates L_i, $i = 1, 2, \ldots, n$, of which only one (of very few) fit in the receptor binding site. Any solvent-accessible conformer can, in principle, transition into the active one during this process. Some ligands may already be in their lowest energy conformation when binding to the receptor, while other ligands may have to transition to the active conformation (lower in energy); hence these ligands exhibit—on the average—energies higher than the minimum energy (17). All ligands having an energy higher than the minimum energy must have a higher affinity than molecules occurring in the minimum energy conformation, independent of the energy level of the receptor-bound conformation (17,18). It follows that the concentration C_F corresponding to the bioactive conformer, L_F, represents only a *fraction* α_F of the total ligand concentration, C_T:

$$C_F = \alpha_F C_T \qquad (5)$$

As a consequence, one can adjust (18) the biological activity Y_{exp} of a ligand by taking into consideration the concentration of its bioactive conformer, L_F:

$$Y_{F,\text{adj}} = Y_{\text{exp}} - \log \alpha_F \qquad (6)$$

The relative amount, α_F, of the bioactive conformer L_F can be estimated according to Boltzmann statistics (19), using Eq. (7):

$$\alpha_F = \frac{g_F \exp(-U_F/RT)}{\sum_{i=1}^{n} g_i \exp(-U_i/RT)} \qquad (7)$$

where U_i are relative conformational energies, calculated by standard conformational analyses, and g_i are the degeneration degrees of the conformational energy levels due to the appearance of certain symmetric conformers (e.g., *gauche* forms); U_F and g_F represent the corresponding properties for the bioactive conformation. If there is more than a single bioactive conformation, the corresponding sum is performed in the denominator. Adjusted biological activities were used to select bioactive conformers for a series of 25 acetylcholinesterase (AChE) substrates in the absence of the binding site and were shown to fit well (18) into the active site of a crystallographic structure of AChE (20), thereby indicating the potential utility of this approach.

One must ascertain that no rate-limiting steps occur during intermediate stages, and that nonspecific binding does not obscure the experimental binding affinity. All the parameters that cannot be directly measured remain *hidden* (e.g., receptor-induced conformational changes of the ligand, geometric variations of the binding site). QSAR methods use a time-sliced (frozen) model; that is, the system is at equilibrium and time independent (18). Kinetic bottlenecks in the intermediate steps may occur, and $\Delta G_{\text{bind}}^{\text{L-R}}$ remains thermodynamic in nature (not kinetic).

B. Protein-Bound Waters and Hydrophobicity

Entropy is lost during reversible or irreversible ligand–receptor binding and gained during the desolvation process (21) because of the waters freed from the binding site and the ligand's hydration shell. The entropy of releasing tightly bound structural water molecules from a protein binding site into aqueous solution, which has been studied using high precision calorimetry (22), has been used to determine S_{wr}, the molar entropy of water release. The upper limit for S_{wr} is 30 J K^{-1} mol^{-1} (22). The upper limit for the molar enthalpy of water release from the protein, ΔH_{wr}, is 16 kJ/mol (22). For both S_{wr} and ΔH_{wr}, the lower limit is set at zero. The free energy of water release ΔG_{wr} has been estimated to vary between -9 and 16 kJ/mol (22). Therefore, ligand-induced displacement of water molecules from the protein binding site is not always beneficial.

A practical approach to address this issue has been proposed by Ladbury and coworkers (21). In the first step, computer models can be used to identify which water molecules are tightly bound to the protein in the binding site. A number of methods are available for this purpose (23,24). In a second step, one estimates the effect of removing those water molecules on the free energy of binding of the ligands of interest, using molecular modeling software (25). Displacing tightly bound water molecules from protein binding sites has resulted in a number of interesting ligands designed for HIV-1 protease (26) and for scyalone dehydratase (27). However, the user must be aware that deprotonated carboxylate and/or phosphate groups always have one or two water molecules in close proximity. These waters are important in maintaining the protonation state of the anions, and it is probably more difficult to replace them (28). Ligand–receptor ion pairs are likely to be hydrated; otherwise they become hydrogen-bonded, neutral interactions (28). Therefore, in the third step, one corroborates the computer calculations with experimental evidence, to define the ligand design strategy (whether to replace or not the tightly bound water). Water molecules can be ex-

plicitly incorporated in three-dimensional QSAR (3D-QSAR) analyses to augment the predictivity and interpretability of the model system (29).

Hydrophobic forces play an important role in drug–receptor interactions.
While more difficult to visualize in contrast to hydrogen bonds, hydrophobic interactions have been experimentally shown to yield higher affinity ligands than
their polar analogs, even at the expense of possible hydrogen bonds (30). The importance of hydrophobic forces during ligand–receptor interactions is further sub-
stantiated by the fractional PLS contributions (31) of the VALIDATE (32) descriptor set (described in Sect. IV.B). VALIDATE is a PLS-based model trained
on 51 diverse ligand–receptor pairs obtained from the Protein Data Bank (33). By
examining the VALIDATE contribution plot in Fig. 2, one can notice that hydrophobicity-related descriptors are dominant: The black bars represent the
lipophilic contact surface area, the intermolecular van der Waals (Lennard-Jones)
energy, and the receptor-based hydrophobicity parameter Hint LogP (34).

The importance of hydrophobic interactions is indirectly supported by two
observations (35) based on the property analysis of the Physician Desk Reference
(PDR) set of oral drugs: (1) the number of drugs in PDR that have between 0 and
2 hydrogen bond donors is four times higher than the number of drugs that have
between 3 and 5 hydrogen bond donors; and (2) over 50% of the PDR drugs have

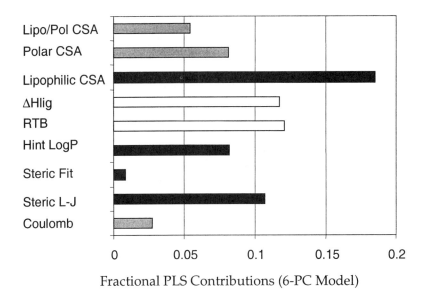

Fractional PLS Contributions (6-PC Model)

Figure 2   Fractional contributions of the VALIDATE descriptors to the final PLS model.
Black bars represent hydrophobic interactions; gray bars represent electrostatic interactions
(including hydrogen bonds); white bars represent entropy-associated interactions.

CLOGP (the logarithm of the calculated octanol/water partition coefficient: see Ref. 36) value of 1 or higher—meaning that their nature is more hydrophobic than polar.

C. Assumptions in Binding Affinity Prediction

One can use "classical" 2D-QSAR methods to investigate properties that relate to ΔG_{bind}^{L-R} at a macroscopic level. For details at the atomic level, and in particular for those related to the ligand–receptor interaction, one must resort to 3D-QSAR methods (12,37,38), or to scoring functions (39–42).

Both 3D-QSAR methods and scoring functions rely on a set of assumptions, as follows (12,41):

The modeled ligand, not its metabolite(s) or any of its derivatives, produces the observed effect.

The ligand is modeled in a single (bioactive) conformation that exerts the binding effects; the dynamic nature of this process, as shown for lactate dehydrogenase, that is likely to assume different conformational states at the binding site (43), is typically ignored.

The geometry of the receptor binding site is, with few exceptions, considered to be rigid.

The loss of translational and rotational entropy upon binding (44) is assumed to follow a similar pattern for all compounds, even though thermodynamic data suggest otherwise (45).

The entropic cost for freezing nonterminal single-bond rotors (46) is frequently estimated only by counting the number of rotatable bonds (see also Sect. IV.B.2).

The protein binding site is the same for all modeled ligands.

The on–off rate is similar for modeled compounds (i.e., the system is considered to be at equilibrium, and kinetic aspects are usually not considered).

Solvent effects, temperature, diffusion, transport, pH, salt concentrations, and other factors that contribute to the overall ΔG_{bind}^{L-R} are not considered.

Furthermore, for 3D-QSAR methods and scoring functions that are based on molecular mechanics force fields, the binding free energy is largely explained by the enthalpic component (the internal energy derived from force field calculations), which is prone to inherent force field errors.

Quite frequently, the binding free energy is expressed as the sum of the free energy components, conceptually shown (5) in the following "master equation":

$$\Delta G_{bind}^{L-R} = \Delta G_{sol} + \Delta G_{conf} + \Delta G_{int} + \Delta G_{motion} \qquad (8)$$

which accounts for contributions due to the solvent (ΔG_{sol}), to conformational changes in both ligand and protein (ΔG_{conf}), to the ligand–protein intermolecular interactions (ΔG_{int}), and to the motion in the ligand and protein once they are at close range (ΔG_{motion}). This can been rewritten as follows:

$$\Delta G_{bind}^{L-R} = \Delta G_{sol} + \Delta U_{vac} - T\Delta S_{vac} \tag{9}$$

where ΔG_{bind}^{L-R} is separated, at equilibrium, into solvation effects (ΔG_{sol}) and two components for the process in vacuum: the internal energy (ΔU_{vac}) and entropy ($T\Delta S_{vac}$). One can calculate ΔG_{sol} with a variety of methods (5), while $T\Delta S_{vac}$ is often related to the number of nonterminal single bonds (32,47). Both ΔG_{sol} and $T\Delta S_{vac}$ are assumed to have similar values for congeneric series, hence Eq. (9) is widely used in QSAR studies by expanding only the internal energy term:

$$\Delta U_{vac} = \Delta U_{vdW}^{L-R} + \Delta U_{Coul}^{L-R} + \Delta U_{distort}^{L} + \Delta U_{distort}^{R} + \Delta U_{conf}^{R} \tag{10}$$

which includes the steric (vdW) and electrostatic (Coul) aspects of the ligand–receptor interaction (ΔU_{vdW}^{L-R} and ΔU_{Coul}^{L-R}), the distortions (distort) induced by this interaction in both ligand and receptor ($\Delta U_{distort}^{L}$ and $\Delta U_{distort}^{R}$), and the ligand-induced conformational changes of the receptor (ΔU_{conf}^{R}). The agonist-induced conformational rearrangements of the receptor ΔU_{conf}^{R} may be an important component of signal transduction and are not considered to occur upon antagonist binding to the same receptor (44).

3D-QSAR methods and scoring functions (see Ref. 48 and Chap. 5, this volume) use—with different approximations—the terms of the master equation, Eq. (8), to predict ΔG_{bind}^{L-R}, once the pharmacological data have been converted to their free energy equivalent via Eq. (4).

IV. PREDICTION OF BINDING AFFINITY

A. Binding Affinity Estimation Using 3D-QSAR Methods

Comparative molecular field analysis (CoMFA), (49) computes the steric (ΔU_{vdW}^{L-R}) and electrostatic (ΔU_{Coul}^{L-R}) interactions on an uniform grid around each ligand, using hypothetical probe atoms that mimic receptor atoms (see Fig. 3). These grid point calculated interaction energies are tabulated for each molecule (row) in the series. The resulting matrix is analyzed with multivariate statistics, yielding Eq. (11), which relates CoMFA fields to ΔG_{bind}^{L-R}:

$$\Delta G_{bind}^{L-R} = \sum_{x,m}^{X,M} A_x U_{vdW}(x, L_m) + \sum_{x,m}^{X,M} B_x U_{Coul}(x, L_m) \tag{11}$$

where a number X of grid-based probes (x) interact with a number M of ligand atoms (L_m) (see Fig. 3), and A_x and B_x are regression coefficients.

In other words, CoMFA compares the molecular potential (steric and electrostatic) energy fields of a series of ligands and searches for differences and simi-

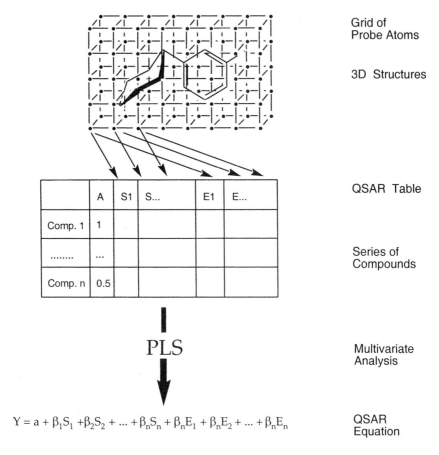

Grid of
Probe Atoms

3D Structures

QSAR Table

Series of
Compounds

Multivariate
Analysis

$$Y = a + \beta_1 S_1 + \beta_2 S_2 + \dots + \beta_n S_n + \beta_n E_1 + \beta_n E_2 + \dots + \beta_n E_n$$

QSAR
Equation

Figure 3   Flowchart of the CoMFA process: previously aligned ligands are submitted to grid-probe (field) evaluation. The tabulated field values are then submitted to multivariate analysis to derive QSAR models.

larities that can be correlated with differences and similarities in biological activity values. In CoMFA, the receptor is approximated by a rigid grid, therefore $\Delta U_{distort}^{R}$ and ΔU_{conf}^{R} are not computed. However, $\Delta U_{distort}^{L}$ is indirectly included when a particular conformer is chosen for the alignment. Other 3D-QSAR methods that have proved useful in estimating ΔG_{bind}^{L-R} are also available (50,51). While different 3D-QSAR methods appear to yield models of equivalent statistical significance, the information extracted from these models can be quite different (52,53).

Ten years after the original CoMFA report (49), over 100 CoMFA models had been reported on enzyme binding affinity, and over 100 CoMFA models on receptor binding affinity (54). An alphabetical list of the 383 CoMFA references published between 1993 and 1997 is available (55). Over 200 papers illustrate the use of CoMFA (and other 3D-QSAR methods) in estimating the binding affinity, with an average prediction error of 0.6–0.7 log unit (0.85–1 kcal) for external sets of compounds (54). This estimate, however, may not be accurate for entirely novel classes of compounds. External predictivity with CoMFA models requires adequate information to reproduce the alignment rule, but care needs to be exercised that the prediction step uses interpolation, rather than extrapolation (12). For illustrative purposes, this chapter focuses on binding affinity predictions for a diverse set of HIV-1 protease (56) inhibitors.

The training set consisted of 59 compounds from five structurally diverse transition state isostere (TSI) classes: hydroxyethylamine, statine, norstatine, ketoamide, and dihydroxyethylene (57). The availability of X-ray crystallographic data for at least one representative from each class bound to the protease provided information regarding not only the active conformation of each ligand but also, via superimposition of protease backbones, the relative positions of each ligand with respect to one another in the active site of the enzyme. Relevant statistical models (with $r^2 > 0.8$ and $q^2 > 0.6$) were obtained. The test set included 6 compounds with known or inferred binding structure, as well as 30 other compounds with unknown binding mode (58): 18 hydroxyethylureas (ligands having a different TSI) and 12 dihydroxyethylenes. The graphical output resulted from this CoMFA study (the CoMFA fields) was shown to be in good agreement with the HIV-1 protease binding site (59).

For the external set, binding affinities were adjusted according to Cheng–Prusoff equation (60):

$$IC_{50} = K_i \left(1 + \frac{S}{K_M} \right) \tag{12}$$

where K_i is the inhibition constant (the equivalent of K_D for enzyme kinetics in the presence of a reversible inhibitor), S is the concentration of substrate, K_M is the Michaelis–Menten constant, and IC_{50} is the concentration of inhibitor that reduces the reaction rate by 50%. External predictivity results for the 36 compounds, based on the CoMFA model, were very high: $q^2 = 0.814$ (see Fig. 4a). Upon closer examination, two different K_M values for the HIV-1 protease had been published by the same laboratory for the dihydroxyethylene series: 0.064 mM (61) and 1.03 mM (62), respectively.

HIV-1 protease is sensitive to Na^+ concentrations in the assays, and the discrepancy between the two reported K_M values was due to the difference in Na^+ concentrations used in the two assays (Tomasselli AG, personal communication, 1994). In fact, 2.0 mM was the correct K_M value. Equation (12), was used to correct biological activities for the dihydroxyethylenes in both the training and the test sets, and the CoMFA model was rederived. While the statistical relevance of the PLS model did not change significantly, the external predictivity dropped signif-

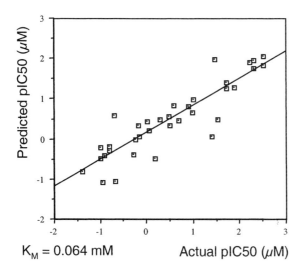

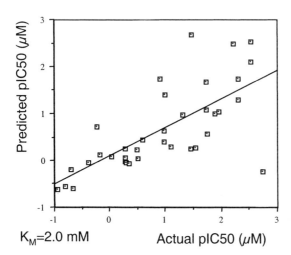

Figure 4 Experimental versus predicted pIC50 values for an external set of 36 HIV-1 protease inhibitors, using a CoMFA model. The initial K_M value for HIV-1 protease was 0.064 mM, yielding a q^2 of 0.814 (a), that was later adjusted to 2.0 mM, yielding a q^2 of 0.487 (b). See text for details.

icantly: $q^2 = 0.487$ (see Fig. 4b). This highlights the paramount importance of accurate biological data in QSAR model derivation. Upon building a training set, whether for 3D-QSAR or for scoring functions, one is rather limited by the availability of high quality experimental data.

B. Binding Affinity Estimation Using Scoring Functions

Receptor-based scoring functions that rapidly estimate binding affinity are empirical methods, which should be differentiated from free energy perturbation (FEP) methods (63,64), which are computed from first principles. It is the very fact that simplified assumptions are introduced in the empirical models that allows these methods to gain in speed what they may lack in scientific rigor. One of the fastest FEP techniques, the linear interaction energy approach (65), still requires about 1 CPU-day per compound. This method, proposed by Åqvist et al. (66), averages the interactions between the ligand and its surroundings using molecular dynamics, separately in solution and in the receptor-bound state. Intermolecular interaction energies derived from molecular mechanics have also been successful in estimating the free energy of binding for a set of HIV-1 protease inhibitors (67).

There are three major categories of scoring functions.

1. Knowledge-based methods (68–70) use Boltzmann-weighted potentials of mean force (PMF) derived from statistical analyses of ligand–receptor interatomic contacts, based on available complexes in the PDB: Wallqvist (71), Verkhivker (40,72), SMoG (73,74), Muegge (75,76), and DrugScore (77) are various implementations of this approach.

2. "Master equation" approaches (5), use semiquantitative estimates of the energetic contributions of various interaction types, as proposed by Williams (46,78) and Rose (79,80). The Williams approach (46) was targeted at peptide–peptide interactions in particular (81), whereas Rose derived a scoring function intended for high throughput virtual screening of protein targets.

3. Regression-based methods are based on available biological activity for training sets of ligand–receptor complexes extracted from the PDB: SCORE2 (82), VALIDATE (32), VALIDATEII (83,84), Pro_Score (85,86), Jain (42), Horvath (87), and SCORE (88) have been published since 1996. These methods have the disadvantage that they require biological input for parameterization, whereas the other two approaches do not.

The reader is referred to the respective references for appropriate descriptions of these approaches. In 1998 we reviewed the approaches of Wallqvist, Verkhivker, Jain, and Rose (41). For illustrative purposes, we focus on VALIDATE, an empirical paradigm that estimates $\Delta G_{\text{bind}}^{\text{L}-\text{R}}$ in the molecular mechanics approximation, based on the following equation:

$$\Delta G_{\text{bind}}^{\text{L}-\text{R}} = \beta_1 E_{\text{vdW}}^{\text{L}-\text{R}} + \beta_2 E_{\text{Coul}}^{\text{L}-\text{R}} + \beta_3 SF + \beta_4 RC$$
$$+ \beta_5 RTB + \beta_6 \Delta H_{\text{bind}}^{\text{L}} + \beta_7 CSA_{\text{lipophilic}}^{\text{L}-\text{R}} + \beta_8 CSA_{\text{hydrophilic}}^{\text{L}-\text{R}} \tag{13}$$

where $\beta_1 - \beta_8$ are PLS-fitted regression coefficients for the master equation terms briefly discussed in the remainder of this section: Eqs. (14)–(20).

1. Intermolecular Interaction Energies. The Steric Fit

Good structural complementarity is essential for the high affinity ligand–receptor interactions. Nonbonded van der Waals (vdW) interaction energies, E_{vdW}^{L-R}, are based on the Lennard-Jones potential as follows:

$$E_{vdW}^{L-R} = \Sigma_i^L \Sigma_j^R \epsilon_{ij} \left[\left(\frac{r_{ij}}{R_i + R_j} \right)^{-12} - 2 \left(\frac{r_{ij}}{R_i + R_j} \right)^{-6} \right] \tag{14}$$

where $\epsilon_{ij} = (\epsilon_i \epsilon_j)^{-1/2}$ and r_{ij} is the distance between atom center i and atom center j; R_i, ϵ_i are the vdW radius, and epsilon value of atom i, and R_j, ϵ_j are the vdW radius, and epsilon value of atom j. The electrostatic interaction energy is the explicit sum of the Coulombic potentials:

$$\Phi = \frac{1}{4\pi\varepsilon_0} \Sigma_i^L \Sigma_j^R \frac{q_i q_j}{r_{ij}} \tag{15}$$

using partial atomic charges on the ligand (q_i) and the receptor (q_j).

Steric fit *(SF)*, which describes the complementarity packing of the ligand in the binding site, computes the number of ligand–atom/receptor–atom contacts for each ligand atom (see Eq. 16). For example, the steroid ligands complexed with DB3 antibodies (89) contain a considerable percentage of their atoms outside of the active site, while HIV-1 protease inhibitors are largely surrounded by the receptor. *SF* is expressed as follows:

$$SF = \frac{\Sigma_{i,j} C_{ij}}{N} \tag{16}$$

where C_{ij} is 1 for distances smaller than or equal to the sum ($r_i + r_{ij} + \epsilon$) and 0 for distances larger than the sum ($r_i + r_{ij} + \epsilon$), N is the number of ligand atoms contained in the active site, r_i, r_j are the vdW radii of atoms i and j, and $\epsilon = 0.3$ Å.

2. Hydrophobic Interactions. Entropic and Enthalpic Aspects of Binding

For the partition coefficient, a negative value indicates a preference for a polar (e.g., aqueous) environment, and a positive value indicates a preference for a non-polar environment (e.g., octanol). VALIDATE computes the amount of hydrophilic and lipophilic surface areas as ratios to the total surface area of the binding site, attributing a value of $RC = -1$ for predominantly lipophilic binding sites (e.g., HIV-1 protease) and a value of $RC = +1$ for predominantly hydrophilic binding sites (e.g., L-arabinose binding protein). VALIDATE further computes the ligand's partition coefficient using the fragment-based $H \log P$ method in HINT (34). The final value of the partition coefficient, *PC*, is:

$$PC = RC \, H \log P \tag{17}$$

Any carbon that is covalently bound to no more than one noncarbon was considered to be lipophilic, and any hydrogen connected to such a carbon is also lipophilic. All other atoms were considered to be hydrophilic.

Changes in conformational entropy occur when freely rotating fragments of the ligand and receptor are forced to adopt rigid conformations upon binding (46). VALIDATE estimates the change in conformational entropy by counting the number of nonterminal single bonds, RTB (32,35,47). For flexible ring systems, the number of degrees of freedom is of the order $n - 4$, where n is the number of bonds in the ring. The rotatable bond count is:

$$RTB = Nr_{\text{ntsb}} + \sum_i (n_i - 4) \tag{18}$$

where Nr_{ntsb} is the number of nonterminal single bonds and n_i is the number of single bonds in ring i. A more complete version of this descriptor can be found in Ref. 35.

The change in conformational enthalpy, $\Delta H_{\text{bind}}^{\text{L}}$, was approximated by the amount of energy required for the ligand to adopt the receptor-bound conformation, defined by

$$\Delta H_{\text{bind}}^{\text{L}} = |E_{\text{bound}}^{\text{L}} - E_{\text{sol}}^{\text{L}}| \tag{19}$$

where $E_{\text{bound}}^{\text{L}}$ is the energy of the ligand's receptor-bound conformation and $E_{\text{sol}}^{\text{L}}$ is the energy of the ligand in solvent at its nearest local minimum.

3. Surface Complementarity: The Ligand–Receptor Interface

The surface complementarity between the ligand and its binding site is known to be very important (90–92) for binding affinity. However, one problem associated with the development and validation of scoring functions is that the training sets made up by known ligand–protein complexes possess almost perfect surface complementarity. As a consequence, no regression-based scoring function can deal with the fact that many complementarity mismatches in a computer-docked binding mode would be severely unfavorable for affinity. VALIDATE is no exception: of the four surface components computed at the ligand–receptor interface [i.e., lipophilic complementarity (nonpolar/nonpolar); hydrophilic complementarity (polar/polar, opposite charge); lipophilic/hydrophilic noncomplementarity (polar/nonpolar); and hydrophilic (polar/polar, like charge) noncomplementarity, respectively] only the lipophilic and hydrophilic complementarity were selected by PLS as important for the final scoring function (see also Fig. 2).

In VALIDATE two types of contact surface area (CSA) calculation are used: the absolute surface area between a ligand and a receptor and a pairwise summation of receptor–ligand contact points. Both CSA calculations start from 256 evenly distributed points placed on the vdW surface of each receptor atom whose vdW surface is within 5 Å of the atom center of any ligand atom. If a point on this surface is within a mean solvent radius (1.4 Å for water) of the vdW surface of a ligand atom, it is considered to be a contact point. The type of each point (i.e., polar or nonpolar), is then determined using the criteria discussed earlier. The contact surface on each atom for each type of contact is then computed by dividing the number of contact points of that type by 256 (the total number of possible points) and then multiplying by the total surface area of the atom:

$$CSA = \Sigma_i^R \frac{(4\pi r_i^2 CP_i)}{256} \tag{20}$$

where CP_i is the number of contact points on atom i and r_i is the vdW radius of atom i.

In an extension of VALIDATE, 15 additional descriptors have been included: 3 related to the number of hydrogen bonds of the ligand with the receptor and in water, 3 AMSOL-estimated terms related to the ligand free energy of solvation (93), the ligand's dipole moment, and the highest occupied molecular orbital (HOMO) energy, among others (83). This scoring function, termed VALIDATE II, was specifically trained on a set of 39 HIV-1 protease inhibitors extracted from the PDB and was later used to predict about 400 other inhibitors, manually docked into the binding site. Most errors of prediction were for the weak-binding ligands (affinity in the micromolar range), indicating that the goal of accurately predicting such compounds has yet to be accomplished.

Mismatched ligand–receptor surfaces, as just discussed, are not penalized in VALIDATE. One initiative to overcome this problem in training scoring functions is provided in the framework of the Prometheus software from Protherics (94). Their Property Viewer graphical interface allows the use of several types of calculated property as filters on lists of scored binding modes, as provided by the docking tools. Among these properties are a set of surface complementarity mismatch descriptors (Waszkowycz B, personal communication, 2000). Refining docking results using descriptors of this type may represent a useful way to overcome the above-mentioned deficiency, common to most scoring functions. Docking tools are discussed next.

C. Tolerance for Protein Flexibility

Receptor-based binding affinity methods depend on the accuracy of the docking method (95,96). Different approaches for docking small ligands into protein binding sites have been described in the literature. The performance of these methods varies from a few CPU-seconds per molecule, for database docking tools, up to CPU-days for methods based on molecular dynamics. The first docking tool was the program DOCK (95) by Kuntz and coworkers, which treated both ligand and protein as rigid bodies. Several methods that allow for full conformational flexibility of the ligand have recently been developed: FlexX (97), a program that makes use of an incremental docking strategy of rigid fragments of the ligand, GOLD (96), which applies an genetic algorithm (GA) to sample over all possible combinations of intermolecular hydrogen bonds and ligand conformations, and a GA-based version of DOCK (98) that treats ligands in a flexible manner. While GOLD is not as fast as FlexX, it allows rotational flexibility for the protein side chain hydrogens.

For many proteins, structural changes occur in the binding region upon ligand complexation. This problem was considered in the CASP2 contest of protein–ligand docking tools (99). To have any chance of finding the binding mode, docking methods that entered the contest had to start from the complexed protein structures (without the ligand), rather then using the uncomplexed proteins for three of the CASP2 targets. These docking tools did not allow for protein flexibility. Furthermore, examination of experimentally determined structures of protein–ligand complexes involving different ligands bound to the same target often reveals modest but significant conformational changes in the target. Therefore, to get truly predictive binding affinity scoring of combinatorial libraries, the treatment of protein flexibility around the binding site by sufficiently fast docking tools must be improved.

One approach to treat protein flexibility in docking was suggested by Kuntz and coworkers (100), who perform docking into binding site models that incorporate conformational variability from ensembles of experimental protein structures determined by either NMR or X-ray crystallography. The Internal Coordinate Mechanics (ICM) docking method allows for explicit protein flexibility by applying a Monte Carlo–based global energy optimization algorithm in internal coordinates (101). The ICM method was applied for both protein–ligand and protein–protein docking exercises.

The program SLIDE (screening ligand by induced-fit docking) provides one significant step forward in combinatorial library evaluation, since it was specifically designed for high throughput database docking (102). SLIDE first identifies a template of favorable hydrogen bond and hydrophobic interaction points in the protein binding site. For each potential ligand, all triplets of corresponding interaction points are then mapped onto triangles of template interaction points. Matches are identified via multidimensional hashing of triangles. The matched ligand interaction centers define a rigid anchor fragment. All feasible anchor fragments are then evaluated by rigid docking to the corresponding template points. SLIDE then tries to resolve any collisions of the anchor fragments with protein main chains by rigid-body translations. Flexibility is then introduced to all fragments of the ligand (except the anchor fragment) and to the protein side chains, to resolve any remaining collisions. The set of optimal rotations needed to resolve these collisions is identified by using the mean-field optimization approach. In this way, the method can deal with "induced-fit" effects in ligand binding (103,104) but does not provide a fully flexible treatment of the ligand conformations, as provided by DOCK, FlexX, or GOLD. Whatever it may lack in accuracy, SLIDE compensates in speed: it docks over 100,000 compounds a day on an average workstation; for comparison, fast docking programs can handle 1000–2000 compounds a day (105).

D. Electrons: The Unresolved Issue

There are at least two major interactions that remain poorly addressed by scoring functions, hence are likely to influence predictions of the binding affinity. These are cation–π interactions and CH–π interactions.

1.   Cation–π interactions have been shown to play a major role in the ion selectivity of potassium channels (106) and in the stabilization of the (charged) quaternary ammonium group in acetylcholine to a π-electron-rich region comprising aromatic rings (107). Over one-fourth of all tryptophan residues in the PDB were shown (108) to be involved in energetically significant cation–π interactions (mostly with arginines). In fact, whenever a cationic side chain (Lys or Arg) is near an aromatic side chain (Phe, Tyr, or Trp), favorable cation–π interaction can be observed (108).

2.   CH–π interactions have been shown to be present in a number of ligand–receptor complexes. The following complexes are covered in Ref. 109 (their PDB codes are in parentheses): heme/hemoglobin (2DHB), tri-N-acetyl-chiotriose/lysozyme (1HEW), D-sorbitol/xylose isomerase (4XIA), FK506/FK506-binding protein (1FKF), and phosphorylcholine/McPC603 immunoglobulin (2MCP). While it is true that the attractive interactions between C—H groups and π-electron systems represent an extreme case of hydrogen bonding, we are far from understanding the role of these weak, but multiple interactions, in biological systems. In fact, CH–π interactions have been suggested (109) to play a major role in guiding acetylcholine into the aromatic gorge of AChE (20), instead of (or perhaps in convergence with) the "aromatic guidance" mechanism proposed by Sussman et al. (110).

One approach that is suitable to model both cation–π and CH–π interactions is the XED (extended electron distribution) approach, developed by Andy Vinter (111,112). This approach is specifically designed to reproduce geometrical and energetic aspects of these interactions and has been also been implemented in a semiempirical quantum mechanical package (113). However, XED it is not yet suitable for predicting binding affinity in a high throughput mode and is perhaps better used in later stages of the drug discovery cycle, when lead compounds need to be refined, rather than identified or optimized (Vinter JG, personal communication, 2000).

V. PREDICTION OF DMPK PROPERTIES

A. General Comments

Since oral availability is an important desideratum for marketed drugs, the importance of pharmacokinetic property prediction during the early stage of drug discovery has been recognized. Solubility, metabolism, and permeability have been identified as important factors that influence oral availability. Although QSAR models for the oral availability of certain classes of compounds have been derived, a general model has yet to become available, probably owing to the diversity of mechanisms involved (cf. Fig. 5).

Drug permeability, a major determinant of oral drug availability, is defined as the ratio between the transport rate per area unit and the initial concentration. Several mechanisms influence drug permeability: for example, a compound can be absorbed through passive (transcellullar and paracelullar) or active mechanisms. Passive drug absorption depends on the surface characteristics of the intestinal membrane (e.g., the size of the tight junctions, the amount of surface exposed, the chemical composition of the membrane). There are two different kinds of active absorption mechanism, namely, uptake or efflux of the compound. In both cases, the active mechanism depends on the affinity and the kinetics of the drug for a certain carrier system that transports the compound through the membrane. While the driving force in the passive mechanism is the gradient concentration of the drug between the two sides of the membrane, in the case of active transport the mechanism is driven by pH gradient, ion gradient, or energy-dependent processes (114).

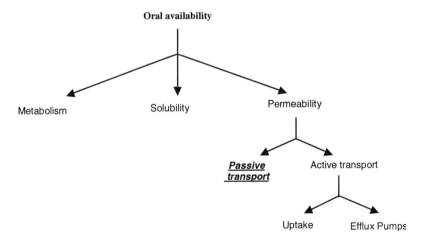

Figure 5   Some factors that influence the oral availability of drugs (136).

Several types of QSAR model predicting permeability properties from molecular descriptors are available (115–122). One limitation of these models that is not always recognized is that they can predict only the passive component of oral drug permeability. Recently, models based on the program VolSurf have successfully been used to predict and describe the passive component of the permeability in the intestine and in the blood–brain barrier (123).

B. VolSurf

VolSurf, written by G. Cruciani and coworkers (119), is a program developed to extract information from three-dimensional molecular interaction fields and convert it into a few quantitative descriptors that can be easily understood and interpreted (124). Similar to the CoMFA fields described in Section III.B, molecular interaction fields are tabulated energy values that account for the interaction between a given probe and the ligand of interest. Different molecular interaction fields have been described, including the molecular electrostatic potential (125), the molecular lipophilic potential (34,126), the steric, electrostatic, and H-bond fields in CoMFA (49), and the GRID fields (127). VolSurf has been shown to be particularly effective when used in conjunction with GRID fields.

GRID, written by P. Goodford (127), is a molecular mechanics–based program that estimates the interaction energy between different ligand atom types and specifically designed chemical probes placed at regular lattice points (in the same way as CoMFA). These interactions are parameterized on the basis of detailed information derived from crystal structures. GRID energies are the sum of the Lennard-Jones, electrostatic, and H-bond interactions between the target and the probe. In the case of the DRY probe, which represents hydrophobic interactions, the energy is computed as follows:

$$E_{DRY} = E_{entropy} + E_{LJ} - E_{HB} \tag{21}$$

where $E_{entropy}$ is the ideal entropic contribution toward the hydrophobic effect in an aqueous environment, E_{LJ} is the Lennard-Jones term that accounts for the induction and dispersion interactions, and E_{HB} is the H-bond term that estimates hydrogen bond interactions between the ligand and the GRID water probe. The program VolSurf transfers relevant information from the GRID field to specifically tailored descriptors, as shown in Fig. 6. These descriptors, listed in Table 1, have a direct chemical interpretability and are not dependent on the alignment used in the GRID lattice.

VI. VOLSURF ESTIMATES OF BINDING AFFINITY

To avoid the historical problems in medicinal chemistry research that were illustrated in Fig. 1, namely, the sequential optimization of binding properties and

Molecular interaction Field   **Surface descriptors**

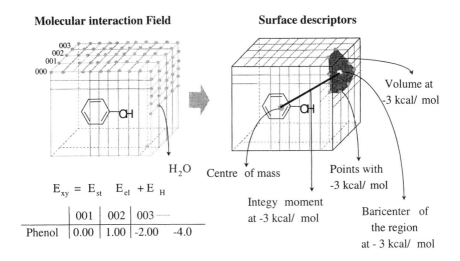

$$E_{xy} = E_{st} \quad E_{el} + E_H$$

	001	002	003 ——	
Phenol	0.00	1.00	-2.00	-4.0

Figure 6 The two-step process of VolSurf descriptor calculation: 1, Molecular interaction fields are obtained via the GRID program. 2, Surface and volume descriptors are obtained at different interaction energy levels.

DMPK properties in drug discovery, it is desirable to apply a unique set of molecular descriptors that is amenable to QSAR models for both properties. Early integration of DMPK and toxicity models in combinatorial library design has been proposed (128). Finding a unique set of descriptors to monitor these property changes is far from trivial, inasmuch as both processes are influenced by (non)specific intermolecular interactions of the ligand with different physiological microenvironments, that is, the target binding site and cellular (membrane) barriers, respectively. On the other hand, these interactions can be described by descriptors of similar types: the lipophilic CSA in VALIDATE overlaps with the low energy levels of the DRY probe in VolSurf, the hydrophilic CSA in VALIDATE matches the water probe descriptors in VolSurf, while the steric complementarity fit in VALIDATE corresponds to the critical packing descriptor in VolSurf (cf. Table 1). These properties are expected, however, to contribute differently to the binding affinity (where hydrogen bonds and hydrophobic contacts are expected to be more directional), as opposed to membrane diffusion (where directionality is less restrictive).

Since the VolSurf descriptors are well established in the area of DMPK modeling (123), we have attempted to use the same descriptors to estimate receptor binding affinity. VolSurf descriptors were calculated for the ligand, for the receptor, and for the ligand–receptor complex of 30 ligand–receptor pairs included in the VALIDATE training set (129). Data are shown in Table 2 and Fig. 7. Two sets of VolSurf descriptors (for the ligand and for the receptor) account for the sol-

Table 1 VolSurf Descriptor Definition

1. Descriptors obtained from the hydrophilic (H_2O) interaction

   V: Volume of the water molecular interaction and field at 0.2 kcal/mol energy level

   S: Surface of the water interaction field at the same 0.2 kcal/mol level

   R: Rugosity: the ratio between the volume and the surface

   G: Globularity: the ratio between the surface (S) and the surface of a sphere with the same volume (V)

   W_1–W_8: The volume of the hydrophilic interactions at 8 different energy levels: -0.2, -0.5, -1.0, -2.0, -3.0, -4.0, -5.0. and -6.0 kcal/mol

   IW_1–IW_8: The integy moments at the same energy levels than W1–W8

   CW_1–CW_8: The capacity factors are the ratio between the hydrophilic regions (W_1–W_8) and the molecular surface (S)

   E_{min1}, E_{min2}, E_{min3}: The energy values for the three lowest energy minima

   D_{12}, D_{13}, D_{23}: The distances between the three minima

2. Descriptors obtained from the hydrophobic (DRY) interaction

   D_1–D_8: The volume of the hydrophobic interactions at 8 energy levels: -0.2, -0.4, -0.6, -0.8, -1.0, -1.2, -1.4. and -1.6 kcal/mol

   ID_1–ID_8: The integy moment at the previous energy levels

   HL_1–HL_2: Hydrophilic-lipophilic balance: the ratio between the volume of the hydrophilic regions at -3 and -4 kcal/mol and the hydrophobic region at -0.6 and -0.8 kcal/mol

   A: The strength of the amphiphilic moment

   CP: Critical packing parameter

   POL: Polarizability

3. Descriptors obtained from the polar (O) interaction

   W_{p1}–W_{p2}: The volume of the interaction with the O probe at 8 different energy levels

   HB_1–HB_8: The difference between the volume of the hydrophilic interaction (W_1–W_8) and the O probe interactions (W_{p1}–W_{p8}) and represent the hydrogen bond donor capability of the target

---

vation–desolvation process that occurs during binding, while the complex-derived VolSurf descriptors further estimate ligand–receptor complementarity, as well as potential interactions in the binding site that may have not been addressed by the ligand.

The data set consists of the following classes of protein–ligand pairs: five HIV-1 protease, five thermolysin, four subtilisin, five endothiapepsin, seven L-arabinose binding protein, and four antibody–steroid complexes, respectively. The pK_i value ranged between 5.4 and 11. Ligands and ligand–receptor complexes were used in the neutral form (no functional groups were formally ionized), and water molecules were removed from each binding site. Further modeling using this data set is in progress (130).

Table 2 The PDB Access Code and p*Ki* Values for the VolSurf Data Set

PDB code	p*Ki*	PDB code	p*Ki*
1aaq	7.93	2er9	7.4
1abe	7.01	2tmn	5.89
1abf	5.42	3sic	10.20
1apb	5.82	3tmn	5.9
1dbb	9	4er1	6.62
1dbj	7.6	4er4	6.8
1dbm	9.44	4phv	9.15
1eed	4.79	4tmn	10.17
1hiv	9.15	5sic	10.20
1hvi	10.5	5tmn	8.04
1tlp	8.55	6abp	6.36
1tmn	7.3	7abp	6.46
2dbl	8.7	7hvp	9.62
2er0	6.38	8abp	8
2er6	7.22	9abp	8

The PLS coefficients for the water probe have a positive sign for the ligand (a positive contribution to the binding) and a negative sign for the receptor (a negative contribution to the binding; data not shown). This indicates that far less water probe–ligand interactions are available in the receptor-bound state than with the free ligand, illustrating the high versatility of water molecules as ligand inter-

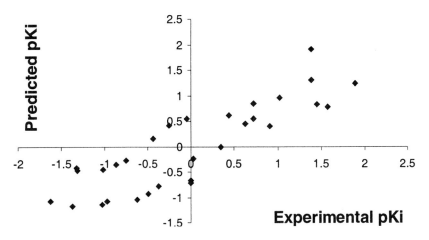

Figure 7 Experimental versus predicted pK_i values using the LOO cross-validation procedure for 30 ligand–protein complexes (see Table 2): $q^2 = 0.86$ and $r^2 = 0.73$ for a two-component model.

action partners. As discussed in 1997 by Ladbury et al. (21), water has a diminutive size (thus, few steric constraints), allowing it to adopt suitable orientations for both donating and accepting multiple hydrogen bonds. This further indicates that both the ligand and the receptor need to be evaluated to obtain accurate VolSurf models for binding affinity.

The PLS coefficients for the DRY (hydrophobic) probe have similar weights in both ligands and receptors, displaying positive contributions at low energies and negative contributions at high energies of interaction (data not shown). At low energy, the DRY probe estimates the nonpolar complementarity, whereas at high energy, its contours are related to π-electron interactions. The role of hydrophobic interactions during ligand–receptor binding was discussed in Section III.B. However, π-electron interactions are not adequately described by the current scoring functions (see Sect. IV.D).

The fact that VolSurf, a program for DMPK modeling *par excellence,* yields a statistically significant QSAR model for binding affinity suggests that it is possible to develop computer methods that will assist chemists in the simultaneous optimization of both types of properties. The VolSurf model for binding affinity is amenable to direct chemical interpretation, having the further benefit that the VolSurf output for the ligand can be directly used to estimate DMPK properties (if such QSAR models are available). This may become a drawback, since the choice of conformation needs to be addressed prior to any VolSurf calculation. It does not, however, constitute a problem in the presence of experimental evidence such as X-ray crystallography or NMR findings. The choice of conformation must be addressed by using a docking program (95,96), to place the ligand into the binding site.

The three-dimensional structures of the receptor and the receptor–ligand complex, from NMR or X-ray crystallography, or perhaps from high quality homology models and docking, are required for this combined approach—which in itself is a disadvantage. The approach further depends on the quality of the pharmacokinetic data used to derive the QSAR models, as well as on the various types of mechanism that could be involved in this process. In other words, if a compound that permeates via passive mechanisms is desired, the QSAR model must be fitted to experimental data that reflect only that mechanism, not to data that are ambiguous with respect to the absorption mechanism.

VII. FINAL REMARK

What follows are some guidelines for combinatorial library evaluation with respect to binding affinity and DMPK predictions:

> In the absence of a receptor, use available QSAR models to predict binding affinity; if no QSAR model is available, it is better to rely on molecular

diversity software (131,132), using computational filters (133) such as CLOGP (36), molecular weight, and polar surface area (134), perhaps in connection with leadlike profiling (135).

In the presence of a receptor, but without solid experimental evidence to indicate a unique binding mode, use a high throughput docking/scoring tool such as SLIDE (102), or a PMF-based scoring function such as DrugScore (77), in combination with a flexible docking tool such as GOLD (96). Use any related QSAR models (when available). DMPK property prediction tools are appropriate if micromolar leads are already available.

When unambiguous binding modes are known, and a diverse set of ligands is available in its receptor–bond conformation, use accurate scoring functions such as SCORE2 (82) or SCORE (88); an alternative is the use of VolSurf, as described in Section VI. With VolSurf, one can predict DMPK properties for each ligand simultaneously.

Water may play an important role. This needs additional investigation, since most of the docking tools do not make use of binding site waters. As discussed in Section III.B, there are three steps: (1) identify tightly bound waters using the program GRID (127), (2) deliberately replace those waters by ligands designed for that purpose, and (3) corroborate computer calculations with experimental evidence (are the designed compounds more active?) to define the combinatorial ligand design strategy (i.e., whether to replace or not replace the water). Bear in mind that the free energy of water release from a protein binding site can be anywhere from -9 to 16 kJ/mol (22), so replacing water molecules from the binding site will not always result in increased binding affinity.

In this chapter, we have advocated the simultaneous prediction of both DMPK properties (e.g., intestinal permeability) and receptor binding affinity by using a unique set of descriptors for any given ligand. The effect of using the combined DMPK/binding affinity prediction approach is illustrated in Fig. 8. The advantages of using such an approach have already been illustrated by Pickett and colleagues, who used DMPK filters to enhance the Caco-2 cell permeability for a combinatorial library (134). Such filters can be used to speed up high throughout screening campaigns by zooming in on compounds that have appropriate physicochemical characteristics, thus reducing the number of tested compounds, as well as the number of potentially interesting hits to be analyzed. In later phases of the drug discovery process, reliable chemical and biological information is available if the combined DMPK/binding affinity prediction approach can be used to select hits with good solubility, permeability, and potency. Finally, this combined approach can be applied with greater confidence in the lead optimization phase—when three-dimensional structural and biological information is available—thus further reducing the number of iterations in the drug discovery cycle.

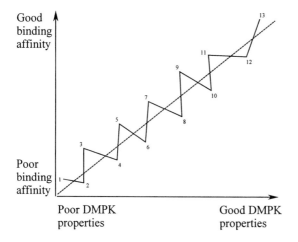

Figure 8   A possible outcome of using the combined DMPK/binding affinity prediction approach is a reduced number of iterations in the drug discovery cycle. See text for details.

ACKNOWLEDGMENTS

The authors acknowledge Dr. Anna-Lena Ungell and Dr. Thomas C. Kühler for helpful discussions.

REFERENCES

1. Hansch C, Leo A. Exploring QSAR. Fundamentals and Applications in Chemistry and Biology. Washington, DC: ACS Publishers, 1995.
2. Kubinyi H, ed. 3D QSAR in Drug Design: Theory, Methods and Applications. Leiden: ESCOM, 1993.
3. Kubinyi H, Folkers G, Martin YC, eds. 3D-QSAR in Drug Design. Vol. 2. Ligand Protein Interactions and Molecular Similarity. Dordrecht: Kluwer/ESCOM, 1998.
4. Kubinyi H, Folkers G, Martin YC, eds. 3D-QSAR in Drug Design. Vol. 3. Recent Advances. Dordrecht: Kluwer/ESCOM, 1998.
5. Ajay, Murcko M. Computational methods to predict binding free energy in ligand–receptor complexes. J. Med. Chem. 1995; 38:4953–4967.
6. Ungell AL. In vitro absorption studies and their relevance to absorption from the GI tract. Drug Dev Ind Pharm 1997; 23:879–892.
7. Ungell AL, Abrahamsson B. Biopharmaceutical support in candidate drug selection. In: Gibson M, ed. Pharmaceutical Preformulation and Formulation: A Practical Guide from Candidate Drug Selection to Commercial Formulation. Buffalo Grove, IL: Interpharm Press, 2000.

8. Wold S, Johansson E, Cocchi M. PLS—Partial least-squares projections to latent structures. In: Kubinyi H, ed. 3D QSAR in Drug Design: Theory, Methods and Applications. Leiden: ESCOM, 1993, pp. 523–550.

9. Anzali S, Gasteiger J, Holzgrabe U, Polanski J, Sadowski J, Teckentrup A, Wagener M. The use of self-organizing neural networks in drug design. In: Kubinyi H, Folkers G, Martin YC, eds. 3D-QSAR in Drug Design. Vol. 2. Ligand Protein Interactions and Molecular Similarity. Dordrecht: Kluwer/ESCOM, 1998, pp. 273–299.

10. Hansch C, Hoekman D, Leo AJ. BioByte QSAR database. Available from the BioByte Corp., 201 West 4th St. Suite 204, Claremont, CA 91711.
    http://clogp.pomona.edu/medchem/

11. Wold S, Ruhe A, Wold H, Dunn WJ. The collinearity problem in linear regression. The partial least squares approach to generalised inverses. J. Sci. Stat. Comput. 1984; 5:735–743.

12. Oprea TI, Waller CL. Theoretical and practical aspects of three dimensional quantitative structure–activity relationships. In: Lipkowitz KB, Boyd DB, eds. Reviews in Computational Chemistry. Vol. 11. New York: Wiley, 1997, pp. 127–182.

13. Wold S. Cross-validatory estimation of the number of components in factor and principal components models. Technometrics 1978; 20:397–405.

14. Bush B, Nachbar RB. Sample–distance partial least squares: PLS optimized for many variables, with application to CoMFA. J. Comput.-Aided Mol. Design 1993; 7:587–619.

15. Stone M, Jonathan P. Statistical thinking and technique for QSAR and related studies. I. General theory. J. Chemometrics 1993; 7:455–475.

16. Stone M, Jonathan P. Statistical thinking and technique for QSAR and related studies. II. Specific methods. J. Chemometrics 1994; 8:1–20.

17. Janssen LHM. Conformational flexibility and receptor interaction. Bioorg. Med. Chem. 1998; 6:785–788.

18. Sulea T, Kurunczi L, Oprea TI, Simon Z. MTD-ADJ: A multiconformational minimal topologic difference for determining bioactive conformers using adjusted biological activities. J. Comput.-Aided Mol. Design 1998; 12:133–146.

19. McClelland BJ. Statistical Thermodynamics. London.: Chapman & Hall, 1973.

20. Sussman JL, Harel M, Frolow F, Oefner C, Goldman A, Toker L, Silman I. Atomic structure of acetylcholinesterase from *Torpedo california:* A prototypic acetylcholine-binding protein. Science 1991; 253:872–879.

21. Renzoni DA, Zvelebil MJJM, Lundbäck T, Ladbury JE. Exploring uncharted waters: Water molecules in drug design strategies. In: Ladbury JE, Connelly PR, eds. Structure-Based Drug Design. Thermodynamics, Modeling and Strategy. Berlin: Springer-Verlag, 1997, pp. 161–180.

22. Connelly PR. The cost of releasing site-specific, bound water molecules from proteins: Toward a quantitative guide for structure-based drug design. In: Ladbury JE, Connelly PR, eds. Structure-Based Drug Design. Thermodynamics, Modeling and Strategy. Berlin: Springer-Verlag, 1997, pp. 143–159.

23. Helms V, Wade RC. Hydration energy landscape of the active site cavity in cytochrome P450cam. Proteins, 1998; 32:381–396.

24. García AE, Stiller L. Computation of the mean residence time of water in the hydration shells of biomolecules. J. Comput. Chem. 1993; 14:1396–1406.

25. Boyd, DB. Compendium of software and Internet tools for computational chemistry. In: Lipkowitz KB, Boyd DB, eds. Reviews in Computational Chemistry. Vol. 11. New York: Wiley, 1997, pp. 373–399.

26. Lam PYS, Jadhav PK, Eyermann CJ, Hodge CN, Ru Y, Bacheler LT, Meek JL, Otto MJ, Rayner MM, Wong YN, Chang CH, Weber PC, Jackson DA, Sharpe TR, Erickson-Viitanen S. Rational design of potent, bioavailable, nonpeptide cyclic ureas as HIV protease inhibitors. Science 1994; 263:380–384.

27. Chen JM, Xu SL, Wawrzak Z, Basarab GS, Jordan DB. Structure-based design of potent inhibitors of scytalone dehydratase: Displacement of a water molecule from the active site. Biochemistry 1998; 37:17735–17744.

28. Liljefors T, Norrby PO. An ab initio study of the trimethylamine–formic acid and the trimethylammonium ion–formate anion complexes, their monohydrates, and continuum solvation. J. Am. Chem. Soc. 1997; 119:1052–1058.

29. Pastor M, Cruciani G, Watson KA. A strategy for the incorporation of water molecules present in a ligand binding site into a three-dimensional quantitative structure–activity relationship analysis. J. Med. Chem. 1997; 40:4089–4102.

30. Davis AM, Teague SJ. Hydrogen bonding, hydrophobic interactions, and failure of the rigid receptor hypothesis. Angew. Chem. Int. Ed. Engl. 1999; 38:736–749.

31. See, for example, the Simca 8.0 manual, available from Umetrics, Umeå, Sweden. http://www.umetrics.com/

32. Head RD, Smythe ML, Oprea TI, Waller CL, Greene SM, Marshall GR. VALI-DATE: A new method for the receptor-based prediction of binding affinities of novel ligands. J. Am. Chem. Soc. 1996; 118:3959–3969. VALIDATE is available from the Washington University Center for Molecular Design. http://cmdfs1.wustl.edu/

33. Bernstein FC, Koetzle TF, Williams GJB, Meyer EF, Brice MD, Rodgers JR, Kennard O, Shimanouchi T, Tasumi M. The Protein Data Bank: A computer-based archival file for macromolecular structures. J. Mol. Biol. 1977; 112:535–542. The Protein Data Bank Web site can be accessed at: http://www.rcsb.org/pdb/

34. Abraham DJ, Kellogg GE. Hydrophobic fields. In: Kubinyi H, ed. 3D QSAR in Drug Design. Leiden: ESCOM, 1993, pp. 506–522. Hint is available from EduSoft LC: http://www.eslc.vabiotech.com/hint/

35. Oprea TI. Property distribution of drug-related chemical databases. J. Comput.-Aided Mol. Design, 2000; 14:251–264.

36. Leo A. Estimating log P_{oct} from structures. Chem. Rev. 1993; 5:1281–1306.

37. Green SM, Marshall GR. 3D-QSAR: A current perspective. Trends Pharm. Sci. 1995; 16:285–291.

38. Greco G, Novellino E, Martin YC. Approaches to three-dimensional quantitative structure–activity relationships. In: Lipkowitz KB, Boyd DB, eds. Reviews in Computational Chemistry. Vol. 11. New York: Wiley, 1997, pp. 183–240.

39. Böhm H-J. The development of a simple empirical scoring function to estimate the binding constant for a protein–ligand complex of known three-dimensional structure. J. Comput.-Aided Mol. Design 1994; 8:243–256.

40. Verkhivker G, Appelt K, Freer ST, Villafranca JE. Empirical free energy calculations of ligand–protein crystallographic complexes. I. Knowledge-based ligand–protein interaction potentials applied to the prediction of HIV-1 protease binding affinity. Protein Eng. 1995; 8:677–691.

41. Oprea TI, Marshall GR. Receptor-based prediction of binding affinity. In: Kubinyi H, Folkers G, Martin YC, eds. 3D-QSAR in Drug Design. Vol. 2. Ligand Protein Interactions and Molecular Similarity. Dordrecht: Kluwer/ESCOM, 1998, pp. 35–61.

42. Jain A. Scoring non-covalent protein–ligand interactions: A continuous differentiable function tuned to compute binding affinities. J. Comput.-Aided Mol. Design 1996; 10:427–440.

43. Xue Q, Yeung ES. Differences in the chemical reactivity of individual molecules of an enzyme. Nature 1995; 373:681–683.

44. Searle MS, Williams DH. The cost of conformational order: Entropy changes in molecular associations. J. Am. Chem. Soc. 1992; 114:10690–10697.

45. Davies TG, Hubbard RE, Tame JR. Relating structure to thermodynamics: The crystal structures and binding affinity of eight OppA-peptide complexes. Protein Sci. 1999; 8:1432–1444.

46. Williams DH, Cox JPL, Doig AJ, Gardner M, Gerhard U, Kaye PT, Lal AR, Nicholls IA, Salter CJ, Mitchell RC. Toward the semiquantitative estimation of binding constants. Guides for peptide–peptide binding in aqueous solution. J. Am. Chem. Soc. 1991; 113:7020–7030.

47. Krystek S, Stouch T, Novotny J. Affinity and specificity of serine endopeptidase–protein inhibitor interactions. Empirical free energy calculations based on X-ray crystallographic structures. J. Mol. Biol. 1993; 234:661–679.

48. Böhm H-J, Stahl M. Rapid empirical scoring functions in virtual screening applications. Med. Chem. Res. 1999; 9:445–462.

49. Cramer III RD, Patterson DE, Bunce JD. Comparative molecular field analysis (CoMFA). 1. Effect of shape on binding of steroids to carrier proteins. J. Am. Chem. Soc. 1988; 110:5959–5967.

50. Baroni M, Costantino G, Cruciani G, Riganelli D, Valigi R, Clementi S. Generating optimal linear PLS estimations (GOLPE): An advanced chemometric tool for handling 3D-QSAR problems. Quant. Struct.–Act. Relat. 1993; 12:9–20. GOLPE is available from Tripos Inc., 1699 S. Hanley Rd, St. Louis, MO 63144.

51. Doweyko A. The hypothetical active site lattice: An approach to modelling active sites from data on inhibitor molecules. J. Med. Chem. 1988; 31:1396–1406.

52. Oprea TI, Ciubotariu D, Sulea TI, Simon Z. Comparison of the minimal steric difference (MTD) and comparative molecular field analysis (CoMFA) methods for analysis of binding of steroids to carrier proteins. Quant. Struct.–Act. Relat. 1993; 12:21–26.

53. Woolfrey JR, Avery MA, Doweyko AM. Comparison of 3D quantitative structure–activity relationship methods: Analysis of the in vitro antimalarial activity of 154 artemisinin analogs by hypothetical active-site lattice and comparative molecular field analysis. J. Comput.-Aided Mol. Design 1998; 12:165–181.

54. Kim KH, Greco G, Novellino E. A critical review of recent CoMFA applications. In: Kubinyi H, Folkers G, Martin YC, eds. 3D-QSAR in Drug Design. Vol. 3. Recent Advances. Dordrecht: Kluwer/ESCOM, 1998, pp. 257–315.

55. Kim KH. List of CoMFA references, 1993–1997. In: Kubinyi H, Folkers G, Martin YC, eds. 3D-QSAR in Drug Design. Vol. 3. Recent Advances. Dordrecht: Kluwer/ESCOM, 1998, pp. 317–338.

56. Oprea TI, Waller CL, Marshall GR. Viral proteases: Structure and function. In: Ciechanover A, Schwartz A, eds. Cellular Proteolytic Systems. New York: Wiley-Liss, 1994, pp. 183–221.

57. Waller CL, Oprea TI, Giolitti A, Marshall GR. Three-dimensional QSAR of human immunodeficiency virus (I) protease inhibitors. 1. A CoMFA study employing experimentally-determined alignment rules. J. Med. Chem. 1993; 36:4152–4160.

58. Oprea TI, Waller CL, Marshall GR. Three-dimensional quantitative structure–activity relationship of human immunodeficiency virus (I) protease inhibitors. 2. Predictive power using limited exploration of alternate binding modes. J. Med. Chem. 1994; 37:2206–2215.

59. Oprea TI, Waller CL, Marshall GR. 3D-QSAR of human immunodeficiency virus (I) protease inhibitors. III. Interpretation of CoMFA results. Drug Design Discovery 1994; 12:29–51.

60. Cheng YC, Prusoff W. Relationship between the inhibition constant (K_i) and the concentration of inhibitor which causes 50 per cent inhibition (I_{50}) of an enzymatic reaction. Biochem. Pharmacol. 1973; 22:3099–3108.

61. Thaisrivongs S, Tomasselli AG, Moon JB, Hui J, McQuade TJ, Turner SR, Strohbach JW, Howe WJ, Tarpley WG, Heinrikson RL. Inhibitors of the protease from human immunodeficiency virus: Design and modeling of a compound containing a dihydroxyethylene isostere insert with high binding affinity and effective antiviral activity. J. Med. Chem. 1991; 34(8):2344–2356.

62. Tomasselli AG, Olsen MK, Hui JO, Staples DJ, Sawyer TK, Heinrikson RL, Tomich CS. Substrate analogue inhibition and active site titration of purified recombinant HIV-1 protease. Biochemistry 1990; 29:264–269.

63. Mark AE, Xu Y, Liu H, Van Gunsteren WF. Rapid non-empirical approaches for estimating relative binding free energies. Acta Biochim. Pol. 1995; 42:525–535.

64. Marrone TJ, Briggs JM, McCammon JA. Structure-based drug design: Computational advances. Annu. Rev. Pharmacol. Toxicol. 1997; 37:71–90.

65. Hansson T, Marelius J, Åqvist J. Ligand binding affinity prediction by linear interaction energy methods. J. Comput.-Aided Mol. Design 1998; 12:27–35.

66. Marelius J, Kolmodin K, Feierberg I, Åqvist J. Q: A molecular dynamics program for free energy calculations and empirical valence bond simulations in biomolecular systems. J. Mol. Graphics Modelling 1998; 16:213–225.

67. Holloway MK. A priori prediction of ligand affinity by energy minimization. In: Kubinyi H, Folkers G, Martin YC, eds. 3D-QSAR in Drug Design. Vol. 2. Ligand–Protein Interactions and Molecular Similarity. Dordrecht: Kluwer/ESCOM, 1998, pp. 63–84.

68. Sippl MJ. Boltzmann's principle, knowledge-based mean fields and protein folding. An approach to the computational determination of protein structures. J. Comput.-Aided Mol. Design 1993; 7:473–501.

69. Sippl MJ. Knowledge-based potentials for proteins. Curr. Opin. Struct. Biol. 1995; 5:229–235.

70. Domingues FS, Koppensteiner WA, Jaritz M, Prlic A, Weichenberger C, Wieder-stein M, Floeckner H, Lackner P, Sippl MJ. Sustained performance of knowledge-based potentials in fold recognition. Proteins Struct. Funct. Genet. 1999; suppl. 3:112–120.

71. Wallqvist A, Jernigan RL, Covell DG. A preference-based free-energy parameteri-zation of enzyme–inhibitor binding. Applications to HIV-1-protease inhibitor de-sign. Protein Sci. 1995; 4:1881–1903.

72. Verkhivker GM. Empirical free energy calculations of HIV-1 protease complexes. Mean field ligand–protein interaction potentials applied to prediction of binding affinity of diol-containing inhibitors. In: Geisow MJ, Epton R, eds. Perspectives in Protein Engineering and Complementary Technologies. Kingswinford (U.K.): Mayflower Worldwide, 1995, pp. 261–264.

73. DeWitte RS, Shakhnovich EI. SMoG: De novo design method based on simple, fast, and accurate free energy estimates. 1. Methodology and supporting evidence. J. Am. Chem. Soc. 1996; 118:11733–11744.

74. DeWitte RS, Ishchenko AV, Shakhnovich EI. SMoG: De novo design method based on simple, fast, and accurate free energy estimates. 2. Case studies in molecular de-sign. J. Am. Chem. Soc. 1997; 119:4608–4617.

75. Muegge I, Martin YC. A general and fast scoring function for protein–ligand inter-actions: A simplified potential approach. J. Med. Chem. 1999; 42:791–804.

76. Muegge I, Martin YC, Hajduk PJ, Fesik SW. Evaluation of PMF scoring in docking weak ligands to the FK506 binding protein. J. Med. Chem. 1999; 42:2498–2503.

77. Gohlke H, Hendlich M, Klebe G. Knowledge-based scoring function to predict pro-tein–ligand interactions. J. Mol. Biol. 2000; 295:337–356.

78. Williams DH, Bardsley B. Estimating binding constants—The hydrophobic effect and cooperativity. Perspect. Drug Discovery Design 1999; 17:43–59.

79. Rose PW. Scoring methods in ligand design. Second UCSF Course in Computer-Aided Molecular Design. San Francisco, 1997.

80. Marrone TJ, Luty BA, Rose PW. Discovering high-affinity ligands from the com-putationally predicted structures and affinities of small molecules bound to a target: A virtual screening approach. Perspect. Drug Discovery Design 2000; 20:209–230.

81. Mackay JP, Gerhard U, Beauregard DA, Maplestone RA, Williams DH. Dissection of the contributions toward dimerization of glycopeptide antibiotics. J. Am. Chem. Soc. 1994; 116:4573–4580.

82. Bohm H-J. Prediction of binding constants of protein ligands: A fast method for the prioritization of hits obtained from de novo design or 3D database search programs. J. Comput.-Aided Mol. Design 1998; 12:309–323.

83. Ragno R, Head RD, Marshall GR. A predictive model for HIV protease inhibitors. 216th National Meeting of the American Chemical Society, Boston, Aug. 23–27, 1998. Book of Abstracts, COMP-032.

84. Marshall GR, Head RD, Ragno R. Affinity prediction: The sine qua non. In: Di Cera E, ed. Thermodynamics in Biology. Oxford: Oxford University Press, 2000, pp. 87–111.

85. Eldridge MD, Murray CW, Auton TR, Paolini GV, Mee RP. Empirical scoring func-tions. I. The development of a fast empirical scoring function to estimate the bind-ing affinity of ligands in receptor complexes. J. Comput.-Aided Mol. Design 1997; 11:425–445.

86. Murray CW, Auton TR, Eldridge MD. Empirical scoring functions. II. The testing of an empirical scoring function for the prediction of ligand–receptor binding affinities and the use of Bayesian regression to improve the quality of the model. J. Comput.-Aided Mol. Design 1998; 12:503–519.

87. Horvath D. A virtual screening approach applied to the search for trypanothione reductase inhibitors. J. Med. Chem. 1997; 40:2412–2423.

88. Wang R, Liu L, Lai L, Tang Y. SCORE: A new empirical method for estimating the binding affinity of a protein–ligand complex. J. Mol. Modelling 1998; 4:379–394.

89. Oprea TI, Head RD, Marshall GR. The basis of cross-reactivity for a series of steroids binding to a monoclonal antibody against progesterone (DB3). A molecular modeling and QSAR study. In: Sanz F, Giraldo J, Manaut F, eds. QSAR and Molecular Modelling: Concepts, Computational Tools and Biological Applications. Barcelona: JR Prous Publishers, 1995, pp. 451–455.

90. Katchalski-Katzir E, Shariv I, Eisenstein M, Friesem AA, Aflalo C, Vakser, IA. The role of geometric fit between protein molecules and their ligands in determining biological specificity. Adv. Mol. Cell Biol. 1996; 15B:623–637.

91. Kubinyi, H. Computer-aided drug design: Facts and fictions. In: Ford MG, ed. Bioactive Compound Design. Oxford: Bios Scientific Publishers, 1996, pp. 1–13.

92. Sobolev V, Wade RC, Vriend G, Edelman M. Molecular docking using surface complementarity. Proteins Struct. Funct. Genet. 1996; 25:120–129.

93. Cramer CJ, Truhlar DG. AM1-SM2 and PM3-SM3 parameterized SCF solvation models for free energies in aqueous solution. J. Comput.-Aided Mol. Design 1992; 6:629–666.

94. More information about Protherics software is available at the company's Web site: http://www.protherics.com/crunch/section8.html

95. Kuntz ID, Blaney JM, Oatley SJ, Langridge R, Ferrin TE. A geometric approach to macromolecule–ligand interactions. J. Mol. Biol. 1982; 161:269–288.

96. Jones G, Willett P, Glen RC, Leach AR, Taylor R. Development and validation of a genetic algorithm for flexible docking. J. Mol. Biol. 1997; 267:727–748.

97. Rarey M, Kramer B, Lengauer T, Klebe G. A fast flexible docking method using an incremental construction algorithm. J. Mol. Biol. 1996; 261:470–489.

98. Oshiro CM, Kuntz ID, Dixon JS. Flexible ligand docking using a genetic algorithm. J. Comput.-Aided Mol. Design 1995; 9:113–130.

99. Dixon JS. Evaluation of the CASP2 docking section. Proteins Struct. Funct. Genet. 1997; suppl. 1:198–204.

100. Knegtel RMA, Kuntz ID, Oshiro CM. Molecular docking to ensembles of protein structures. J. Mol. Biol. 1997; 266:424–440.

101. Totrov M, Abagyan R. Flexible protein–ligand docking by global energy optimization in internal coordinates. Proteins Struct. Funct. Genet. 1997; suppl. 1:215–220.

102. Schnecke V, Kuhn LA. Database screening for HIV protease ligands: The influence of binding-site conformation and representation on ligand selectivity. In: Lengauer T, Schneider R, Bork P, Brutlag D, Glasgow J, Mewes HW, Zimmer R, eds. Proceedings of the Seventh International Conference on Intelligent Systems for Molecular Biology. Menlo Park, CA: AAAI Press, 1999, pp. 242–251.

103. Koshland DE. Application of theory of enzyme specificity to protein synthesis. Proc. Natl. Acad. Sci. USA 1958; 44:98–104.

104. Jorgensen WL. Rusting of the lock and key model for protein–ligand binding. Science 1991; 254:43–53.

105. Schnecke V, Kuhn LA. Virtual screening with solvation and ligand-induced complementarity. Perspect. Drug Discovery Design 2000; 20:171–190.

106. Kumpf RA, Dougherty DA. A mechanism for ion selectivity in potassium channels: Computational studies of cation–pi interactions. Science 1993; 261:1708–1710.

107. Dougherty DA, Stauffer DA. Acetylcholine binding by a synthetic receptor: Implications for biological recognition. Science 1990; 250:1558–1560.

108. Gallivan JP, Dougherty DA. Cation–pi interactions in structural biology. Proc. Natl. Acad. Sci. USA 1999; 96:9459–9464.

109. Nishio M, Umezawa Y, Hirota M, Takeuchi Y. The CH/pi interaction: Significance in molecular recognition. Tetrahedron 1995; 51:8665–8701.

110. Ripoll DR, Faerman CH, Axelsen PH, Silman I, Sussman JL. An electrostatic mechanism for substrate guidance down the aromatic gorge of acetylcholinesterase. Proc. Natl. Acad. Sci. USA 1993; 90:5128–5132.

111. Vinter JG. Extended electron distributions applied to the molecular mechanics of some intermolecular interactions. J. Comput.-Aided Mol. Design 1994; 8:653–668.

112. Vinter JG. Extended electron distributions applied to the molecular mechanics of some intermolecular interactions. II. Organic complexes. J. Comput.-Aided Mol. Design 1996; 10:417–426.

113. Rauhut G, Clark T. Multicenter point charge model for high-quality molecular electrostatic potentials from AM1 calculations. J. Comput. Chem. 1993; 14:503–509.

114. Tsuji A, Tamai I. Carrier-mediated intestinal transport of drugs. Pharm. Res. 1996; 13:963–977.

115. Lipinski CA, Lombardo F, Dominy BW, Feeney PJ. Experimental and computational approaches to estimate solubility and permeability in drug discovery and development settings. Adv. Drug Delivery Rev. 1997; 23:3–25.

116. Van de Waterbeemd H, Camenisch G, Folkers G, and Raevsky OA. Estimation of Caco-2 cell permeability using calculated molecular descriptors. Quant. Struct.–Act. Relat. 1996; 15:480–490.

117. Palm K, Luthman K, Ungell AL, Strandlund G, Artursson P. Correlation of drug absorption with molecular surface properties. J. Pharm. Sci. 1996; 85:32–39.

118. Norinder U, Österberg T, Artursson P. Theoretical calculations and prediction of Caco-2 cell permeability using MolSurf parametrization and PLS statistics. Pharm. Res. 1997; 14:1786–1791.

119. Cruciani G, Crivori P, Carrupt PA, Testa B. Molecular fields in quantitative structure–permeation relationships: The VolSurf approach. J. Mol. Struct. (THEOCHEM), 503:17–30.

120. Oprea TI, Gottfries J. Toward minimalistic modeling of oral drug absorption. J. Mol. Graphics Modelling 1999; 17:261–274.

121. Wessel MD, Jurs PC, Tolan JW, Muskal SM. Prediction of human intestinal absorption of drug compounds from molecular structure. J. Chem. Inf. Comput. Sci. 1998; 38:726–735.

122. Winiwarter S, Bonham NM, Ax F, Hallberg A, Lennernäs H, Karlén A. Correlation of human jejunal permeability (in vivo) of drugs with experimentally and theoreti-

cally derived parameters. A multivariate data analysis approach. J. Med. Chem. 1998; 41:4939–4949.

123. Guba W, Cruciani G. Molecular field-derived descriptors for the multivariate modeling of pharmacokinetic data. In: Gundertofte K, Jørgensen FS, eds. Molecular Modeling and Prediction of Bioactivity. New York: Kluwer Academic/Plenum Publishers, 2000, pp. 89–94.

124. VolSurf is available from MIA srl, Perugia, Italy.
     http://www.miasrl.com/volsurf.htm

125. Chirlian LE, Francl MM. Atomic charge derived from electrostatic potentials: A detailed study. J. Comput. Chem. 1987; 8:894–905.

126. Carrupt PA, Gaillard P, Billois F, Weber P, Testa B, Meyer C, Perez S. The molecular lipophilicity potential (MLP): A new tool for log P calculations. In: Pliska V, Testa B, Van de Waterbeemd H, eds. Drug Action and Toxicology. Weinheim: VCH, 1995, pp. 195–215.

127. Goodford PJ. Computational procedure for determining energetically favourable binding sites on biologically important macromolecules. J. Med. Chem. 1985; 28:849–857. GRID is available from Molecular Discovery Ltd, West Way House, Elms Parade, Oxford OX2 9LL, UK.

128. Darvas F, Dorman G. Early integration of ADME/Tox parameters into the design process of combinatorial libraries. Chim. Oggi 1999; 17:10–13.

129. The VALIDATE training set is available form the Washington University Center for Molecular Design: http://cmdfs1.wustl.edu/

130. Zamora I, Cruciani G, Oprea TI, Pastor M, Ungell AL. Estimating ligand–protein binding affinity with VolSurf. J. Med. Chem. (submitted)

131. Warr WA. Combinatorial chemistry and molecular diversity. An overview. J. Chem. Inf. Comput. Sci. 1997; 37:134–140.

132. Van Drie JH, Lajiness S. Approaches to virtual library design. Drug Discovery Today 1998; 3:274–283.

133. Oprea TI, Gottfries J, Sherbukhin V, Svensson P, Kühler T. Chemical information management in drug discovery: Optimizing the computational and combinatorial chemistry interfaces. J. Mol. Graphics Modelling 2000; 18:512–524.

134. Pickett SD, McLay IM, Clark DE. Enhancing the hit-to-lead properties of lead optimization libraries. J. Chem. Inf. Comput. Sci. 2000; 40:263–272.

135. Teague SJ, Davis AM, Leeson PD, Oprea TI. The design of leadlike combinatorial libraries. Angew. Chem. Int. Ed. Engl. 1999; 38:3743–3748.

136. Zamora I, Oprea TI, Ungell AL. Prediction of oral drug permeability. In: Höltje HD, Sippl W, eds. Rational Approaches to Drug Design. Barcelona: Prous Science, 2001. In press.

8

Knowledge-Based Approaches for the Design of Small-Molecule Libraries for Drug Discovery

Vellarkad N. Viswanadhan, Arup K. Ghose, and Alex Kiselyov

Amgen Inc.
Thousand Oaks, California

John J. Wendoloski

AstraZeneca
Boston, Massachusetts

John N. Weinstein

National Cancer Institute
National Institutes of Health
Bethesda, Maryland

I. INTRODUCTION

The design of small-molecule libraries has emerged in recent years as one of the important tasks aiding the multidisciplinary effort (computational, combinatorial, and medicinal chemistry and biological testing) in pharmaceutical drug design and development (1–6). In spite of the deployment of high throughout screening (HTS) strategies for testing, it proves to be practically impossible or very expensive to test large (corporate or publicly/commercially available) chemical libraries because of their sheer magnitude. In several cases, HTS strategies would not even be feasible. Thus, as a matter of economics and logistics, compound prioritization for biological testing must occur at an early stage (reaction planning) and even at subsequent stages of the drug development "funnel." The process of prioritization often involves obtaining the best compromise optimizing several factors: chemical diversity (5), druglikeness (7–10), and potential for biological/pharmacological activity (11,12).

The initial steps of the process, which would be common to all libraries are (1) identifying biologically and pharmacologically important physicochemcal properties and (2) developing a consensus definition (7,8) and quantitative score of "druglikeness" based on an analysis of known drug databases (9,10). Then, given a reaction scheme or a library (vendor or corporate), compound/reagent selections involve the following steps: (1) select initial sets of reagents compatible with the given reaction from vendor databases, excluding cross-reactive groups; (2) characterize the virtual libraries of compounds (which could include all the usable reagents) in terms of important physicochemical properties; and (3) make selections based on diversity, drugability (13) and preferably potential for biological activity. The chapter by Schnur and Venkatarangan (14) describes how descriptors can be chosen so that known "actives" are clustered. Well-known properties considered to be relevant include molecular size (represented by molecular weight, number of atoms, molar refractivity), shape (topological shape indices), hydrogen-bonding character (number of hydrogen bond donors, acceptors), conformational flexibility (number of rotatable bonds), and lipophilicity (log P). All these properties are represented in the default set of 50 descriptors available in the Cerius2 (15) package. Principal component analysis of several of these properties can then be performed for manageability and visualization (using the first three components). Consideration of "receptor relevance" led to a unique representation known as BCUT(5), wherein each molecule is represented as the eigenvalue of a symmetric matrix, generated from an "atomic detail" representation of a receptor-relevant property such as charge distribution (5).

In this chapter the Comprehensive Medicinal Chemistry (CMC) database, available as an ISIS database (16), is used as the reference for profiling physiologically and pharmacologically relevant physicochemical properties and for comparing the profiles of drugable libraries. A working definition of "druglikeness" or "drugability" is described, extending an approach described earlier (8). Furthermore, a score for "drugability" is defined based on molecular size and lipophilicity. This score differs from earlier, more elaborate scores (9,10) based on neural nets in that it uses only two properties (size and lipophilicity) and is not "trained" to discriminate between drug and nondrug databases. Nevertheless, this score offers a quantification of a Lipinski-like rule as a simple, single-number measure, as part of an algorithm (DURGA) to select reagents/compounds (17). The operation of this algorithm is demonstrated as an example of designing a drugable amide library and for selecting diverse, drugable compounds from a vendor database.

II. DIVERSIFICATION USING REAGENT/DRUGABILITY-BIASED GENERALIZED ALGORITHM (DURGA): DEVELOPMENT AND APPLICATION

The goal here is to develop a general algorithm that in conjunction with a diversity selection procedure, selects "drugable" diverse molecules. First, a brief overview of earlier work is presented, along with some recent extensions in profiling known drug databases, followed by a "consensus" definition of "drugability." Next a score is developed quantifying the drugability of a molecule relative to these drug database profiles. A procedure is then outlined for using this score as a penalty function, to bias diversity selections in a given "chemistry" space, in favor of drugable molecules. Applications of this procedure to make selections of diverse compounds from a vendor database and diverse reagents conforming to a given well-plate format to form "drugable" combinatorial reaction products are demonstrated.

A. Physicochemical and Functional Group Profiling of Known Drug Databases

The Comprehensive Medicinal Chemistry (CMC) database, version 97.1 (16), contains 7183 molecules and includes all known drugs and compounds in late-stage clinical testing. Among the several drug database considered, which include the MDL Drug Data Report (16) and the C & H Dictionary of Pharmaceutical Agents (18), the CMC database appears to be the closest to an ideal drug database with a reasonable size. The other databases contain a large fraction of compounds in the early stage of drug discovery, hence not necessarily "druglike." For this reason, the CMC database was chosen to obtain physicochemical and functional group profiles that could serve as a reference. From this database, several obvious "nondrugs" (e.g., surgical aids, surfactants, aerosol) were eliminated at the outset. This left us with 6304 compounds. Physicochemical profiling was then carried out for seven pharmacologically and biologically relevant properties: log P, molar refractivity, molecular weight, number of atoms, number of rotatable bonds, number of hydrogen bond donors, and number of hydrogen bond acceptors. Among these properties, it is easily seen that uncertainty in predictions is higher for log P than for others. Hence, it is important to choose and apply a method that is predictive for all classes of organic molecules of interest in medicinal chemistry, particularly for known drug molecules.

Lipinski's (7) characterization of the drug databases used the Moriguchi method (19), which was shown to be predictive for a small set of known drugs with accurately measured values of log P. For this property, our earlier characterization of the CMC database (8) used the ALOGP method (20,21) which showed a stable performance for molecules with diverse functionalities. We recently developed a hologram-based method called (HLOGP (22) and showed it to be more predictive than ALOGP [and an earlier version of CLOGP (23)] for a test set of 931 molecules. For a small test set of 19 drug molecules, the HLOGP method we developed gave a better performance ($r = 0.98$; SD $= 0.31$) than all other methods tested [ALOGP, CLOGP, the Moriguchi method (19), and XLOGP (24)]. Hence, HLOGP is adopted here for profiling and comparison with the log P profiles obtained using the ALOGP method. Figure 1 shows that no major differences exist between the ALOGP and HLOGP profiles, though there is a net increase in the percentage occupancy in the range of -0.5 to 5.5 with the HLOGP profile relative to the ALOGP profile.

From these profiles, we then calculated two ranges that might be useful to assess drugability: the "qualifying range," which is the minimal range accommodating 80% of the database, and the "preferred range," which is the minimal range accommodating 50% of the database. These calculated ranges are almost the same regardless of which method is used for log P estimation (qualifying range, -0.2 to 5.2 for HLOGP; -0.4 to 5.5 for ALOGP; preferred range, 1.3–4.1 for both ALOGP and HLOGP). While there are some compounds outside the qualifying range among known drugs, for most medicinal/combinatorial chemistry applica-

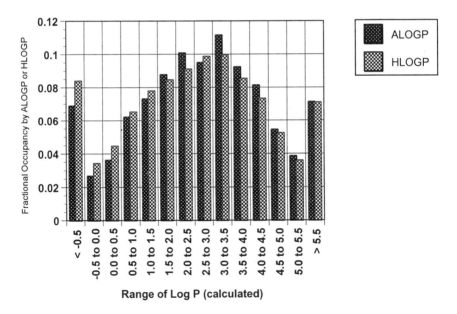

Figure 1 Histograms calculated log P's values using ALOGP and HLOGP methods in the "qualifying" range of CMC database.

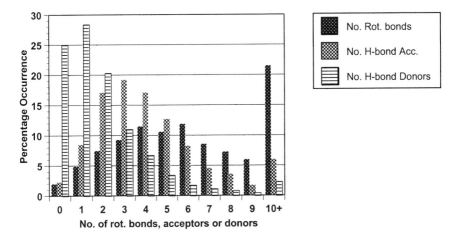

Figure 2 Histograms of the numbers of (*solid*) H-bond donors, (*dotted*) H-bond acceptors, and (*striped*) rotatable bonds in the CMC database.

tions, the log *P* ranges derived here should be appropriate. Though the HLOGP method appears more accurate, we adopt the ALOGP profiles and estimates for further analysis because of the close agreement of these two profiles and also because the ALOGP values are easily calculated using commercial modeling packages such as Cerius2 (15).

A quantitative examination of the profiles for three other properties led to the determination of the qualifying and preferred ranges for molar refractivity (AMR) 40–130 and 70–110; for molecular weight 160–480 and 230–390; and for total atom count 20–70 and 30–55. Furthermore, profiles of H-bonding character and internal flexibility (as measured by number of rotatable bonds) were also studied (Fig. 2). It is seen that each of these profiles exhibits distinct peaks that can be utilized in a quantitative definition of drugability. It is also clear from the profiles of Fig. 2 that the Lipinski criteria are too broad with respect to H-bonding characteristics. Over 80% of the compounds in the database have one to six H-bond acceptors and four or fewer donors. A more continuous distribution is seen in the case of rotatable bond counts. For the present, however, the H-bonding and flexibility properties are not monitored with respect to druglikeness, though these properties will be used (in the following sections) in the assessment of diversity as well as in library comparisons. Table 1 shows compounds identified from the CMC database that violate the ranges defined here for the four properties: ALOGP, molar refractivity, molecular weight, and number of atoms. Pharmacological/biological activities listed in the CMC database for these compounds are also given in Table 1. It can seen that most of the compounds either work by

Table 1 Compounds Identified from the CMC Database Whose Properties Are Outside the
Property Ranges Defined for Drugability

Compound name	Mol Wt	ALOGP	AMR	Atom count	Activity type
4-Amino-3-Hydroxybutyric acid	119.1	−1.4	27.4	17.0	
5-Piperidin-2-yl)-1, 4-Benzodiazepines	475.6	5.6	140.7	72.0	Cholecystokinin receptor antagonists
Acarbose	645.6	−6.6	140.0	87.0	α-Glucosidase inhibitor
Acebrochol	588.5	11.8	144.1	81.0	Hypnotic
Acetohydroxamic acid	75.1	−1.2	15.8	10.0	Urease inhibitor
Acetylcysteine	163.2	−0.7	37.6	19.0	Mucolytic
Acipimox	154.1	−0.3	34.4	17.0	Antihyperlipoproteinemic
Acivicin	178.6	−1.2	38.1	18.0	Antineoplastic
Adenine	135.1	−1.2	36.3	15.0	Vitamin B_4
Alanine	89.1	−0.6	21.5	13.0	Nutrient (amino acid)
Alanosine	149.1	−2.3	31.2	17.0	Antineoplastic
Alexidine	508.8	5.5	152.0	92.0	Antibacterial
Alinastine	433.6	6.6	133.3	71.0	Antihistaminic
Allantoin	158.1	−1.7	32.2	17.0	Vulnerary
Amikacin	585.6	−9.1	132.5	83.0	Antibacterial
Aminoethyl nitrate	106.1	−0.5	23.7	13.0	Vasodilator
Aminothiadiazole	101.1	−0.3	27.9	9.0	Antineoplastic
Arbekacin	552.6	−8.1	131.5	82.0	Antibacterial
Aspartic acid	133.1	−1.4	27.0	16.0	Nutrient (amino acid)
Azaguanidine	152.1	−1.4	36.6	15.0	Antineoplastic
Belfosdil	548.6	7.5	148.4	86.0	Antihypertensive (calcium channel blocker)
Bialamicol	436.6	7.4	140.6	72.0	Antiamoebic
Bibn-99	568.2	5.6	158.8	82.0	Muscarinic M2 antagonist
Bromperidol decanoate	574.6	7.8	152.2	78.0	Antipsychotic
Bronopol	200.0	−1.0	32.5	15.0	Antiseptic
Bucromarone	463.6	7.8	142.7	71.0	Antiarrhythmic (cardiac depressant), antidepressant
Buterizine	466.7	6.8	146.6	73.0	Vasodilator (peripheral)
Butikacin	571.6	−8.4	132.4	84.0	Antibacterial
Candocuronium iodide	642.5	−2.3	151.7	76.0	Neuromuscular blocking agent
Carbantel lauryl sulfate	520.1	7.6	141.7	76.0	Anthelminthic

Table 1 *Continued*

Compound name	Mol Wt	ALOGP	AMR	Atom count	Activity type
Carebastine	499.7	5.7	148.2	74.0	Antihistaminic
Carzolamide	111.1	−0.7	27.6	13.0	Antineoplastic
Cefbuperazone	627.6	−1.1	152.9	71.0	Antibacterial
Chloramphenicol palmitate	561.5	8.2	147.5	79.0	Antibacterial, antirickettsial
Citiolone	159.2	−0.5	39.1	19.00	Hepatic disorders (therapeutic)
Clocapramine	481.1	5.6	141.5	71.0	Neuroleptic
Clodronic acid	244.9	−2.8	35.6	15.0	Calcium regulator, osteitis deformans (therapeutic)
CP-0127	2805.3	−6.0	731.5	392.0	Bulk agent for freeze-drying
Creatinine	113.1	−0.7	28.3	15.0	Antitubercular, anthelmintic
Cyacetacide	99.1	−2.0	23.8	12.0	Antibacterial (tuberculostatic)
Cycloserine	102.1	−2.0	22.9	13.0	Detoxicant
Cysteine	121.2	−0.9	29.4	14.0	Chelating agent
Deferoxamine	560.7	−1.6	143.6	87.0	
Deloxolone	556.8	6.6	151.4	92.0	Antimigraine
Dexfosfoserine	185.1	−2.7	32.4	19.0	Antineoplastic
Dezaguanine	150.1	−1.3	39.5	17.0	Antineoplastic
Diazouracil	139.1	−1.0	32.5	13.0	Alkalizing agent
Diethanolamine	105.1	−1.2	27.5	18.0	Antibacterial
Dihydrostreptomycin	583.6	−6.8	132.3	81.0	Antiinflammatory (topical)
Dimethyl sulfoxide	78.1	−0.8	19.0	10.0	Nootropic
Dimiracetam	140.1	−0.5	33.9	18.0	Bioflavonoid
Diosmin	608.5	−1.0	145.6	75.0	
DL-Lactic acid	90.1	−0.6	18.4	12.0	Sleep enhancer
Draflazine	604.5	5.6	159.1	74.0	Antihistaminic
Ebastine	469.7	6.9	145.8	74.0	Antineoplastic
Edelfosine	524.7	6.3	147.4	94.0	Antineoplastic
Enocitabine	565.8	7.2	156.6	95.0	Aninflammatory (topical)
Enoxolone	470.7	5.3	132.7	80.0	Estrogen
Estradiol Undecylate	440.7	7.9	130.6	76.0	
Etamestrol	534.8	7.6	152.8	74.0	Calcium regulator
Etidronic acid	206.0	−2.5	32.5	19.0	Vasodilator
Fenoxedil	486.7	6.4	142.9	77.0	
Flotrenizine	492.7	7.1	144.8	74.0	Antifungal
Flucytosine	129.1	−1.4	28.1	13.0	Antineoplastic
Fluorouracil	130.1	−0.5	25.8	12.0	Antipsychotic

Table 1 *Continued*

Compound name	Mol Wt	ALOGP	AMR	Atom count	Activity type
Fluphenazine Enanthate	549.7	7.7	150.0	76.0	
Fodipir	638.5	−3.1	142.2	74.0	Antiviral
Foscarnet	126.0	−1.4	16.6	10.0	Antiviral against wild-type and mutant HIV strains
Fosfomycin	138.1	−1.0	25.0	15.0	Antibacterial
Fosfonet	140.0	−1.0	22.2	13.0	Antiviral
γ-Aminobutyric Acid	103.1	−0.4	26.0	16.0	Antihypertensive, neurotransmitter
Gefarnate	400.6	7.0	132.4	73.0	Anticholinergic (antispasmodic)
Glutamic acid	147.1	−1.2	31.6	19.0	Acidifier (gastric)
Glycerin	92.1	−1.6	20.4	14.0	Reduces intraocular and intracranial pressure
Glycine	75.1	−0.9	17.1	10.0	Nutrient
Guanazole	99.1	−0.8	28.0	12.0	Antineoplastic
Haloperidol decanoate	530.1	7.8	149.1	78.0	Antipsychotic
Hesperidin	610.6	−1.1	143.8	77.0	Bioflavonoid
Histamine	111.1	−0.5	31.9	17.0	Stimulant (gastric secretory)
Hydroxyurea	76.1	−1.7	14.5	9.0	Antineoplastic
Ilmofosine	526.8	6.6	149.8	91.0	
Imexon	111.1	−0.6	26.1	13.0	Antineoplastic
Impacarzine	493.8	7.8	148.6	90.0	Antiviral
Iodocholesterol I 131	512.6	8.9	133.3	74.0	Radioactive agent
Isepamicin	569.6	−7.9	131.5	82.0	Antibacterial
Isoniazid	137.1	−0.8	37.3	17.0	Antibacterial (tuberculostatic)
Keracyanin	632.0	−1.8	141.7	74.0	Visual adaptation to dark (agent)
L-749,805	569.6	5.4	149.1	71.0	
Lapinone	456.7	8.0	137.8	77.0	Antimalarial
Lauralkonium chloride	475.1	7.2	144.3	77.0	
Levcycloserine	102.1	−2.0	22.9	13.0	Enzyme Gaucher disease inhibitor, anticonvulsant
Lexacalcitol	460.7	6.0	137.0	81.0	Calcium regulator
Lidoflazine	491.6	5.7	143.6	71.0	Vasodilator (coronary), calcium blocker

Table 1 *Continued*

Compound name	Mol Wt	ALOGP	AMR	Atom count	Activity type
Lymecycline	602.6	−2.5	157.1	81.0	Antimicrobial
Meclorisone Dibutyrate	567.6	7.3	148.4	78.0	Antiinflammatory (topical)
Mecysteine	135.2	−0.3	33.5	17.0	Mucolytic
Meglucycline	635.6	−4.9	156.6	82.0	
Menatetrenone	444.7	6.9	146.6	73.0	Anticoagulant
Mercaptamine	77.1	−0.4	22.5	11.0	
Methimazole	114.2	−0.3	33.8	13.0	Thyroid inhibitor
Methyl methanesulfonate	110.1	−0.5	22.0	12.0	Antineoplastic
Methylformamide	59.1	−0.8	14.9	9.0	Antineoplastic
Mibefradil	495.6	7.2	140.1	74.0	Antihypertensive, vasodilator
Moctamide	473.7	9.8	154.8	82.0	Antihyperlipoproteinemic
Mofarotene	433.6	7.3	136.2	71.0	Antineoplastic
Molfarnate	454.7	7.8	151.6	83.0	
Monoethanolamine	61.1	−1.2	16.2	11.0	Sclerosing agent
Monoxerutin	654.6	−1.6	153.5	80.0	Antidote (specific)
Naboctate	511.8	11.2	157.3	90.0	Antiglaucoma agent, antiemetic
Nadide	664.4	−1.9	139.4	72.0	Narcotic antagonist, alcohol antagonist
Nafiverine	538.7	7.7	159.7	78.0	Anticholinergic (antispasmodic)
Neomycin B	614.7	−10.6	139.1	88.0	Antibacterial
Niacinamide	122.1	−0.3	32.8	15.0	Vitamin (provitamin, cofactor), Enzyme cofactor
Nicoxamat	138.1	−0.7	34.3	16.0	
Norcholestenol Iodomethyl (131I)	512.6	9.6	133.2	74.0	
Octafonium chloride	433.1	7.5	134.0	72.0	Antiseptic
Oftasceine	622.5	−1.8	152.7	71.0	
Orlipastat	495.7	7.6	141.2	88.0	Hypolipidemic
Otilonium bromide	564.6	6.1	152.6	79.0	Anticholinergic (antispasmodic)
Ouabain	584.7	−2.1	141.2	85.0	Cardiotonic
Oxidronic acid	192.0	−2.6	27.4	16.0	Calcium regulator, osteitis deformans (therapeutic)
Oxiglutathione	612.6	−4.4	137.5	72.0	Antineoplastic
Pantethine	554.7	−3.2	139.9	78.0	Antilipemic

Table 1 *Continued*

Compound name	Mol Wt	ALOGP	AMR	Atom count	Activity type
Paromomycin	615.6	−10.0	137.3	87.0	Antiamoebic
Pecocycline	585.6	−2.1	152.2	77.0	Antibacterial
Penoctonium Bromide	489.6	6.3	134.4	80.0	Anticholinergic (antispasmodic)
Pentigetide	588.6	−5.3	135.6	77.0	Antiallergic
Phytonadione	450.7	9.7	144.2	79.0	Vitamin (provitamin, cofactor), prothrombogenic
Pidolic acid	129.1	−1.3	27.8	16.0	Humectant (as Na salt)
Pinaverium bromide	592.4	6.1	143.3	74.0	Anticholinergic (antispasmodic)
Pipacycline	586.6	−3.0	156.0	80.0	Antibacterial
Piperazine	86.1	−0.7	26.3	16.0	Anthelminthic (as citrate)
Praxadine	110.1	−0.7	29.8	14.0	
Probucol	516.8	9.8	159.7	83.0	Antihyperlipoproteinemic
Pyrazinamide	123.1	−0.7	30.5	14.0	Antibacterial (tuberculostatic)
Quindecamine	454.7	8.0	142.0	72.0	Antibacterial
Ramoplanin A2	2554.1	−4.3	640.8	335.0	Antibacterial
Ristocetin	2067.9	−8.1	493.1	257.0	Antibacterial
RO-40-8757	433.6	7.3	136.6	71.0	Antineoplastic
Rolitetracycline	527.6	−1.4	140.7	71.0	Antibacterial
Ronipamil	460.8	8.4	147.7	82.0	
Rutin	610.5	−1.3	141.8	73.0	Decreases capillary fragility
Serine	105.1	−1.8	23.1	14.0	Nutrient (amino acid)
Sevopramide	481.7	5.3	145.0	78.0	
Streptomycin	581.6	−9.4	131.5	79.0	Antibacterial (tuberculostatic)
Succimer	182.2	−0.5	38.2	16.0	Metal poisoning (therapeutic)
Sucrose octaacetate	678.6	−1.1	143.8	85.0	Alcohol denaturant
Sucrosofate	982.8	−16.8	150.8	77.0	Antiulcerative (K salt)
Symetine	468.7	7.0	148.0	82.0	Antiamoebic
Taurine	125.1	−0.8	25.3	14.0	Hepatoprotectant, antiulcerative
Taurultam	136.2	−1.5	30.6	16.0	Antibacterial, antifungal
Tefenperate	534.5	7.2	144.5	73.0	
Terfenadine	471.7	5.3	146.0	76.0	Antihistaminic
Teroxalene	457.1	9.1	140.0	73.0	Antischistosomal
Teruficin	468.7	8.6	140.6	78.0	

Table 1 *Continued*

Compound name	Mol Wt	ALOGP	AMR	Atom count	Activity type
Testosterone Ketolaurate	484.7	6.3	140.2	83.0	Androgen
Tetrazolyl glycine	143.1	−1.7	35.4	15.0	NMDA agonist
Thiocolchicoside	563.6	−0.6	148.4	72.0	Muscle relaxant (general), Gynecologic disorders
Thiouracil	128.1	−0.2	34.0	12.0	Antianginal, thyroid inhibitor
Threonine	119.1	−1.6	27.5	17.0	Nutrient (amino acid)
Tidiacic	177.2	−1.0	38.2	18.0	Hepatotherapeutic
Timonacic	133.2	−0.4	32.7	15.0	Choleretic
Tiopronin	163.2	−0.5	38.0	19.0	Cystinuria (therapeutic), lead poisoning lead poisoning (therapeutic)
Toripristone	457.7	6.0	142.0	73.0	Antiglucocorticoid
Trenizine	456.7	6.8	144.1	74.0	
Trimethylcetyl ammonium	550.9	11.0	150.3	75.0	
Pentachlorophenate tromethamine	121.1	−2.0	28.6	19.0	Alkalizer
UK-81352	584.7	−0.9	145.8	80.0	Antineoplastic
Uracil	112.1	−0.8	25.7	12.0	
Urea	60.1	−1.3	13.1	8.0	Diuretic
Vapiprost	477.6	6.0	141.9	74.0	Antithrombotic
Vitamin E	430.7	11.2	137.6	81.0	Vitamin (provitamin), vitamin (cofactor)
Zafirlukast	575.7	6.0	157.1	74.0	Antiasthmatic (leukotriene antagonist)

"chemical brute force" (e.g., antineoplastics, topical anti-infectives) or work differently from most drugs (e.g., vitamins/cofactors and cholesterol).*

Profiling the variety and frequency of organic functional groups in the CMC database can be useful in designing libraries with diverse functionalities of pharmacological relevance. While the number of such groups is enormous, we identified 30 fairly common functionalities for profiling, including some reactive functionalities (25) that should be avoided in potential drug candidates. Table 2 shows normalized frequencies of these functional groups for the entire CMC database as

---

* While this is true of most medicinal chemistry applications, "drugs" such as anticancer agents may work by reactivity ("chemical brute force").

Table 2   Percentage Composition of Various Functional Groups Among Drugs Classified by Disease State: (a) Anti-inflammatory, (b) Anti-depressant, (c) Antipsychotic, (d) Antihypertensive, (e) Anticancer, and (f) the CMC Database

Number	Functional group	(a)	(b)	(c)	(d)	(e)	(f)
I	Carboxyl	38.91	2.70	1.82	23.64	9.05	15.06
II	Alcohol	23.55	11.26	18.18	22.28	37.35	25.84
III	Aldehyde	0.34	0.00	0.00	0.00	0.46	0.53
IV	Aliphatic primary amine	1.02	6.76	0.00	4.62	9.98	5.69
V	Aliphatic secondary amine	0.34	20.72	0.91	17.12	3.71	9.10
VI	Aliphatic tertiary amine	7.17	52.25	92.73	17.93	13.69	29.59
VII	Amino acid	0.68	0.00	0.00	1.36	1.16	1.49
VIII	Aromatic primary amine	2.05	1.35	1.82	4.08	8.12	5.42
IX	Aromatic secondary amine	12.97	4.50	11.82	8.15	12.30	7.16
X	Aromatic tertiary amine	4.78	18.47	36.36	10.60	10.21	10.27
XI	Carboxamide	20.48	23.87	20.91	28.80	25.29	27.15
XII	Keto	31.06	3.60	23.64	5.71	25.29	15.71
XIII	N-Oxide	0.00	0.00	0.00	0.82	0.23	0.19
XIV	Nitro	0.68	0.45	0.00	3.53	2.55	2.63
XV	Phenolic OH	6.48	1.80	2.73	6.79	13.46	10.23
XVI	Epoxy	0.00	0.00	0.00	0.00	2.32	0.76
XVOO	C—O—O—C	0.00	0.00	0.00	0.00	0.00	0.06
XVIII	C—N—O—C	0.68	0.00	0.00	0.27	0.23	0.25
XIX	C—N—N—C	5.80	2.25	0.00	3.80	0.93	1.55
XX	C—N—S—Cf	0.00	0.00	0.00	0.00	0.00	0.06
XXI	C—S—S—C	0.00	0.00	0.00	0.27	1.16	0.65
XXII	C—S—O—Cg	0.00	0.00	0.00	0.00	0.00	0.00
XXIII	Nucleoside	0.00	0.00	0.00	0.00	7.66	1.02
XXIV	Pyridine	9.90	9.01	4.55	8.15	6.26	8.07
XXV	Pyrimidine	0.68	0.90	0.00	4.08	6.26	2.45
XXVI	Pyrrole	8.19	7.21	10.91	5.98	6.96	4.43
XXVII	Benzene	76.45	92.34	97.27	79.35	43.85	70.28
XXVIII	Furan	1.71	1.35	0.91	1.36	0.46	1.98
XXIX	Thiophene	4.44	1.80	2.73	2.45	1.39	1.92
XXX	Imidazole	3.07	0.45	1.82	5.98	8.82	6.01
XXXI	Ester	25.26	2.70	7.27	25.00	19.95	18.19
XXXII	Sulfonamide	5.46	1.35	2.73	8.70	0.93	4.51
XXXIII	Sulfonic acid	0.00	0.00	0.00	0.27	0.93	0.65
XXXIV	Aliphatic ether	2.73	0.00	0.00	0.27	0.00	0.71
XXXV	Aromatic ether	0.68	1.80	0.91	7.34	0.23	1.26
XXXVI	Any heterocycle	57.68	67.12	95.45	73.37	66.59	66.84
XXXVII	Aromatic heterocycle	35.15	23.42	24.55	32.07	29.47	28.39

well as for five subclasses of drugs. Of all the various functional groups, the benzene ring appears to be the most abundant in the CMC database or any of the subclasses studied here. This observation is not surprising in view of the easy chemistry involving the benzene rings and the absence of stereochemical isomers. The phenyl ring has a marginal desolvation cost (-0.9 kcal/mol), and it can serve as a scaffold for polar or hydrophobic functionalities. Aromatic–aromatic interactions involving benzene are fairly common in protein–ligand complexes. Heterocycles are also a major structural component of drugs, though no single heterocycle dominates the list. The pyridine ring is the most common of the heterocycles studied, since the pyridine nitrogen can be a strong H-bond acceptor. Among other functional groups, the alcoholic hydroxyl and carboxamide occur with a high frequency in the CMC database as well as in the five classes shown. They are hydrophilic and chemically stable, neutral functional groups and are frequently H-bond with protein/receptor targets. Because of its basicity and biochemical stability, the aliphatic tertiary amine is also a commonly occurring organic functional group. Very few compounds were found with a single bond between heteroatoms in the database; in such relatively rare cases, they are stabilized by a conjugated $C=X$ bond.

B. Consensus Definition of Drugability

Defining a druglike molecule is a debatable, contentious issue, in as much as the drug databases are still evolving and future drugs will surely possess some characteristics not found in existing drugs. Here, we regard druglikeness or drugability as an initial constraint to the elimination of compounds that may not be useful for screening, as a first filter in the drug development funnel. Lipinski's (7) characterization, though a logical first step in this regard, offers too broad a view of drugability. As applied to our ACD sampler database containing over 36,457 compounds, less than 12.5% of the compounds failed to satisfy the Lipinski criteria. At the other extreme, the approaches of Ajay et al. (9) and Sadowski and Kubinyi (10) seek to strongly discriminate between non drug (such as ACD) databases and drug databases (such as CMC). Such a distinction may be useful in the later stage of the drug development funnel, to prioritize the medical chemistry efforts the picking potential drug candidates. However, for combinatorial chemistry efforts, diversification would be seriously compromised if only a small percentage of vendor databases were considered druglike. Hence, we attempted to arrive at a "consensus" based on the foregoing physicochemical and functional group analysis, leading to the following "consensus" definition of drugability of a molecule or chemical library: (1) calculated log P (HLOGP/ALOGP) between -0.4 and 5.5; (2) calculated molar refractivity (AMR) between 40 and 130; (3) molecular weight between 20 and 70 and (4) atom count in the range of 20–70; (5) structurally a combination of one or more of the following groups: a benzene ring, a

heterocyclic ring, an aliphatic ring (preferably tertiary), a carboxamide, an alcoholic hydroxyl and/or a keto group; and (6) chemically stable in physiological
buffer, as indicated by the absence of a reactive functional group or moiety.

C. A Quantitative Score of Drugability

To apply the consensus definition of drugability for chemical library, it is necessary to quantitate the definition in the form of a "drugability" score. This score can
then be used in conjunction with a diversity selection algorithm to select members
of a virtual library. Though a good diversity algorithm can achieve functional
group diversification, selections based on the quantitative "drugability" score will
help narrow down the compound choices to within the qualifying ranges of the relevant physicochemical properties. A simple score is devised here using the qualifying and preferred ranges of lipophilicity, molar refractivity, molecular weight,
and number of atoms.

After the "qualifying ranges" have been rescaled from 0 to 1, the final score
is given by:

$$\text{Score} = 0.4*D1*D1 + 0.2*D2*D2 + 0.2*D3*D3 + 0.2*D4*D4 \qquad (1)$$

where the D's are the distance from the center of the "preferred range." $D1$ is calculated using the following equations:

$$D1 = \frac{[\text{ALOGP}_{cal} + 0.4]}{6.0} - C1 \qquad (2)$$

where ALOGP_{cal} is the calculated log P for the compound and $C1$ is the rescaled
coordinate of the center of the preferred range. Similarly,

$$D2 = \frac{[\text{AMR}_{cal} - 40]}{90} - C2 \qquad (3)$$

$$D3 = \frac{[\text{Mol\_W}t - 160]}{320} - C3 \qquad (4)$$

$$D4 = \frac{[\text{No\_atoms} - 20]}{50} - C4 \qquad (5)$$

where AMR_{cal}, Mol_Wt, and No_atoms refer to the calculated molar refractivity,
molecular weight, and total atom count of the molecule, respectively. The rescaled
coordinates of the center of the preferred range are 0.5165 ($C1$), 0.5555 ($C2$),
0.4690 ($C3$), and 0.4500 ($C4$) for the four corresponding properties.

Additionally, compounds in the preferred range are scored "0"; for compounds in the qualifying range, the score is halved. The numerical values of the
weights assigned to the properties are somewhat arbitrary. Greater weight is given

to lipophilicity, since the other three properties are strongly interrelated and in essence represent molecular size.

D. Steps in Compound Selection

Let us begin by defining the "chemistry space" along with the drugable subspace that should be covered by selected molecules following a diversity selection procedure. The default "combichem" set of descriptors, (excluding the total charge for reasons described shortly) available in Cerius2 (version 4.0_ccH) (15) is used to define the chemistry space (see Chap. 10, this volume). This set includes topological (Information Content, Balaban, and kappa indices; PHI; subgraph counts; chi indices; Weiner and Zagreb indices) topographic (surface area), thermodynamic (ALOGP, molar refractivity), and two- or three-dimensional structural (principal moments of inertia, molecular volume, molecular weight, rotatable bonds, H-bonding) properties, all of which are shown to be of biological and pharmacological relevance. Hence this set is a good starting point to define the "chemistry" space.

For the purpose of defining the dimensions of the chemistry space, it is imperative that the reference database be sufficiently big, since the size of the virtual libraries usually considered for selections would be rather large $(10^4–10^6)$. These numbers *far* exceed the size of the CMC database. Hence, we considered the Derwent World Drug Index (WDI) database (26) as the reference database to define the dimensions of the chemistry space. A clean version (excluding high molecular weight compounds, etc.) of the database with approximately 35,000 compounds was created and used as the reference. We excluded the total charge as a descriptor in the chemistry space, in as much as we felt that the calculated values do not represent the biologically active form, and hence total charge is not a good diversity descriptor. The other 49 descriptors were computed for this reference, and principal component analysis was performed on this data set to reduce the dimensionality from 49 to 6, by selecting the 6 principal components corresponding to the largest eigenvalues explaining over 90% of the variance in the original data, and these PC dimensions then define the axes of our chemistry space. This space can be populated either by obtaining compounds from vendors or by in-house syntheses achieved by using combinatorial chemistry methods. It must be emphasized that this space is used only for the diversification of compounds using a combination of properties considered important for chemical diversity. Compounds that fall within the same general area of WDI within this space may well be "undruglike" because the earlier definition of druglikeness. On the other hand, compounds that do not overlap with WDI in this space may well be druglike.

The PC loadings of the reference database were used to obtain the distributions of compounds in the chemistry space for virtual libraries examined. Sets of reagents identified from a vendor database search are initially screened for reactive groups (other than the group of interest) and fingerprints are used to create a short list of reasonable size by means of a general diversity algorithm (e.g., Sybyl Selector). Starting from the initial sets of reagents, the first step of the algorithm is to build a virtual library with all the reagents considered initially, also recording the combination of reagents used for each molecule, in the case of a combinatorial library. Then, the corresponding three-dimensional structures are generated and important physicochemical properties are calculated. Then, the drugability score (according to Eq. 1) is calculated from these properties for each molecule. This score is used as a "penalty function" to constrain the compound selections, such that both diversity and "drugability" of the selected compounds is optimal. In the case of combinatorial libraries, the required well-plate format would be an additional restraint. For the present purposes, the Cerius2 diversity algorithm is used. It offers a choice of diversity scores and uses a Monte Carlo procedure to obtain the selections conforming to a given well-plate format and a graded weighting scheme (on a scale of 1–10) for biasing selections conform to a particular range of a property such as lipophilicity.

III. APPLICATIONS: ILLUSTRATIVE EXAMPLES

A. Comparison of ACD and CMC Databases and Selection of Druglike Compounds from the ACD Database

The version of the ACD database (16) we considered contains nearly 200,000 compounds. For the purpose of the present illustration, a randomly selected part of the ACD database (ACDS–ACD sampler) containing about 36,457 compounds is used. The loadings obtained from the principal component analysis of the 49 descriptors for the WDI database were used to compute PCs for the ACD subset. The first three components are plotted for the ACDS and CMC databases in Fig. 3; it is seen that these databases overlap considerably in the chemistry space, though significant differences are discernible.

157 compounds are selected from the ACDS database under two conditions: first, the compounds were chosen only on the basis of diversity; second, the drugability score of Eq. (1) is applied as a "mild" penalty (1 on a scale of 1–10) for obtaining diverse compound selections. In both cases, the diversity algorithm used a cell-based method (selecting one compound per cell). Figure 4 shows a histogram plot of percentage occupancies in different ranges of drugability scores for the ACDS and CMC databases and also for the compounds selected under the two different criteria. It can be seen that 70% of CMC and 66% of the ACDS databases have a drugability score less than 1, indicating that the ACDS database is not much different from the CMC database in terms of this score and the corresponding property ranges. However, the value of using the

Figure 3 Comparison of the ACDS and CMC databases in the chemistry space defined by the first three principal components from a PCA of 49 descriptors calculated for the World Drug Index (light stars represent the ACD database and dark stars represent the CMC database).

drugability score as a restraint becomes obvious when we compare the percentage occupancies of the selected compounds with and without the drugability constraint. When no constraint is applied, less than 20% of the selected compounds have a high drugability score (< 1), whereas when a mild penalty is applied, 42% of the selected compounds have a score less than 1. The reason why the selection process results in many compounds outside the "drugability" range becomes obvious upon examining the loadings (weights) for various properties for the first three principal components (PC1, PC2, and PC3). PC1 is dominated by size and shape descriptors such as principal moments of inertia, molecular volume, molecular weight, and molar refractivity, in addition to some information and connectivity measures representing molecular topology. PC2 and PC3 are dominated by various topological indices (information and graph theoretic) in addition to other structural descriptors. While the percentage of molecules in

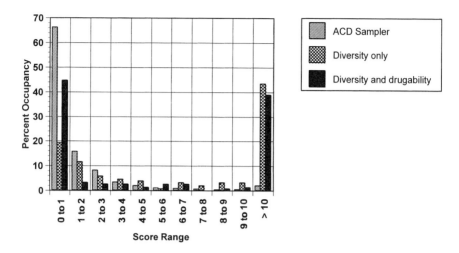

Figure 4 Histograms of drugability scores for the ACDS database and compounds selected under two different criteria with no drugability penalty and with a mild drugability penalty.

the ACDS database that depart from the property ranges used in Eq. (1) is rather small, these compounds significantly contribute to the diversity of the chemistry space represented by PC1, PC2, and PC3. This explains the selection of such non druglike compounds when the drugability score is not used in the selection process. Application of a drugability based constraint should be made an integral component of a compound acquisition strategy. The weight applied to the penalty function determines the degree of loss in diversity of the resulting compound selections, and one must balance the requirement of drugability against that of diversity.

B. Selection of Reagents for a Combinatorial Reaction Based on Analysis of Combinatorial Products

The design of combinatorial libraries involves both the selection/design of a combinatorial reaction and the selection of reagents (inputs) to the reaction. While the former aspect is important, here we are concerned only with the second aspect of reagent selections that ensure diversity and drugability of resulting libraries. Generally speaking, if the reaction is based on larger scaffolds or if the substituents tend to be large, the resulting products will also be large and must be screened for drugability. On the other hand, smaller or absent scaffolds tend to generate drugable libraries where further screening may be unnecessary. Figure 5 shows a hypothetical reaction scheme of the second category, for an amide library requiring

Figure 5 Hypothetical reaction scheme for an amide library.

an amine and an acid chloride. An ISIS search of the ACD database was conducted for amines and acid chlorides, and this was followed by reactivity screens to eliminate reactive groups undesirable in the final library. Then, 100 diverse amines and 100 diverse acid chlorides were selected, using Sybyl/Selector(27) with two-dimensional fingerprints as the variable. A virtual library of amides with 10,000 members is then constructed with these reagents using Cerius2/CombiChem/Library Builder. Using the loadings from the PCA analysis of WDI, PCs were calculated for the amide virtual library. Figure 6 shows a plot of the first

Figure 6 A comparison of the amide library selections with the CMC database in the chemistry space black dots represent compounds in the CMC database; light stars represent compounds selected with no penalty for drugability; dark stars represent compounds selected with a mild drugability penalty.

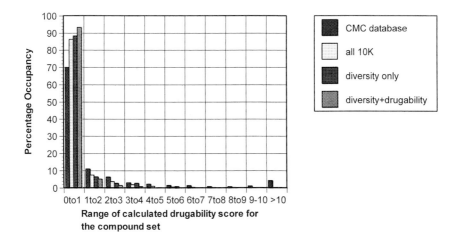

Figure 7   Histograms of drugability scores for the amide library, the CMC database, and compounds selected under two different criteria with no drugability penalty and with a mild drugability penalty.

three components of WDI and the amide library. It can be seen from this figure that while the amide library overlaps the WDI in the "chemistry space," a good number of compounds in the amide library lie outside the region of occupied by WDI.

Using Cerius2 diversity selection tools, 160 compounds were selected for 20×8 well-plate format (20 amines and 8 acid chlorides), using two different criteria: diversity of the final products and product diversity with a "mild" penalty for compounds (products) deviating from a specified range of drugability. A Monte Carlo algorithm is used for the selection of reagent subsets. Histogram plots of the drugability scores of the initial set and the resulting subsets are shown in Fig. 7, along with those of the CMC database. It is seen that the original library can be considered "drugable" because it mostly conforms to the property ranges used in Eq. (1). Still, the use of a mild drugability penalty is seen to significantly increase the percentage of products having the drugability penalty score less than 1.

IV.   CONCLUDING REMARKS

This chapter reviews recent developments in the analysis of drug databases for arriving at a consensus definition of "drugability." Quantitative criteria were developed for penalizing diversity selections for departing from this consensus definition, enhancing the conformity of the selected compounds with known drug databases in terms of physiologically relevant physicochemical properties. These criteria also reflect Lipinski's (7) "rule of five" as we found that compounds violating Lipinski's rules have high molecular weight and log P values. However, in the context of the current definition, Lipinski's rules are too broad: only 12.5% of the ACDS are excluded by the Lipinski rules, whereas approximately a third of the ACD compounds score poorly under the definition of Eq. (1). The current quantitative definition can be easily incorporated in a diversity selection procedure, offering a practical way to design combinatorial libraries or to select compounds.

Another important consideration for designing libraries is the potential for biological activity, which should be assessed within the context of a design target or mechanism of action (11,12). Though this is often considered to be a separate design methodology (directed library design), the same general principles of library design discussed here are applicable, once the right scores have been identified. When the design target is known, a structure-based scoring function may be used. This is often difficult because it depends on the quality of ligand docking, which can be hard to assess. When the target (or mechanism of drug action) is unknown, a model that predicts the mechanism or "activity" should be useful to obtain the scores for directed library design.

More elaborate, neural net–based considerations of "drugability" use ALOGP atom types, physicochemical properties, and so on (9,10; see also Chap. 9, this volume) to discriminate between a "nondrug" database such as ACD and a known drug database such as the CMC database. However, in the initial stage of finding suitable hits or even lead compounds, it is essential that the libraries be suitably diverse. Otherwise, the selected compounds/reagents will not be diverse enough. The current definition makes no assumptions on the "drugability" of vendor databases and considers only overall ranges of properties to guide selections. A function designed to discriminate between drug databases and general chemical libraries such as ACD will be useful to "funnel" hits or leads into developmental candidates. Such an effort is under way (28).

REFERENCES

1. Armstrong, RW; Combs, AP; Tempest, PA; Brown, SD; Keating, TA. Multicomponent condensation strategies for combinatorial library synthesis. Acc. Chem. Res. 1996; 29:123–131.

2. Ghose, AK; Viswanadhan, VN; Wendoloski, J. Adapting structure-based drug design in the paradigm of combinatorial chemistry and high-throughput screening: An overview and new examples with important caveats for newcomers to combinatorial library design using pharmacophore models or multiple copy simultaneous search fragments. In: Parrill, AL; Reddy, MR, eds. Rational Drug Design: Novel Methodology and Practical Applications. Washington DC: American Chemical Society, 1999, pp. 226–238.

3. Martin, EJ; Critchlow, RE. Beyond mere diversity: Tailoring combinatorial libraries for drug discovery. J. Comb. Chem. 1999; 1:32–45.

4. Caflisch, A; Karplus, M. Computational combinatorial chemistry for *de novo* ligand design: Review and assessment. Perspect. Drug Discovery Design 1995; 3:51–84.

5. Pearlman, RS; Smith, KM. Novel software tools for chemical diversity. In: Kubinyi, H; Folkers; G, Martin, YC, eds. 3D QSAR in Drug Design: Ligand–Protein Interactions and Molecular Diversity. Dordrecht: Kluwer/ESCOM, 1998, pp. 339–353.

6. Pickett, SD; Mason, JS; Mclay, IM. Diversity profiling and design using 3D pharmacophores—Pharmacophore-derived queries. J. Chem. Inf. Comput. Sci. 1996; 36:1214–1223.

7. Lipinski, CA; Lombardo, F; Dominy, BW; Feeney, PJ. Experimental and computational approaches to estimate solubility and permeability in drug discovery and development settings. Adv. Drug Delivery Rev. 1997; 23:3–25.

8. Ghose, AK; Viswanadhan, VN; Wendoloski, JJ. A knowledge-based approach in designing combinatorial or medicinal chemistry libraries for drug discovery. 1. A qualitative and quantitative characterization of known drug databases. J. Comb. Chem. 1999; 1:55–68.

9. Ajay; Walters, WP; Murcko, MA. Neural net model of druglikeness. J. Med. Chem. 1998; 41:3314.

10. Sadowski, J; Kubinyi, H. Atom-type based neural net model of druglikeness. J. Med. Chem. 1998; 41:3325.

11. Weinstein, JN; Kohn, KW; Grever, MR; Viswanadhan, VN; Rubinstein, LV; Monks, AP; Scudiero, DA; Welch, L; Koutsoukos, AD; Chiausa, AJ; Paull, KD. Neural computing in cancer drug development: Predicting mechanism of action. Science 1992; 258:476–451.

12. Weinstein, JN; Myers, TG; O'Connor, PMO; Friend, SH; Fornace Jr., AJ; Kohn, KW; Fojo, T; Bates, SE; Rubinstein, LV; Anderson, NL; Buolamini, JK; van Osdol, WW; Monks, AP; Scuderio, DA; Sausville, EA; Zaharevitz, DW; Bunow, B; Viswanadhan, VN; Johnson, GS; Wittes, RE; Paull, KD. An information-intensive approach to the molecular pharmacology of cancer. Science. 1997; 275:343–349.

13. Brown, RD; Hassan, M; Waldman, M. Tools for designing diverse, drug like, cost-effective combinatorial libraries. Chapter 10, this volume.

14. Schnur, D; Venkatarangan, P. Applications of cell-based diversity methods to combinatorial library design. Chapter 16, this volume.

15. Cerius2; 4.0_ccH version, Molecular Simulations, Inc., San Diego, CA, 1999.

16. ISIS 2.0. Integrated Scientific Information System software available from MDL Information Systems, San Leandro, CA, 1997.

17. Viswanadhan, VN; Ghose, AK; Wendoloski, JJ. A knowledge-based approach for designing combinatorial or medicinal chemistry libraries for drug discovery 2. Design of drugable libraries. (under preparation)

18. C & H Dictionary of Pharmaceutical Agents. Available from Tripos Associates. St. Louis, MD, 1998.

19. Moriguchi, I; Hirono, S; Nakagome, I; Hirano, H. Comparison of reliability of log *P* values for drugs calculated by different methods. Chem. Pharm. Bull. 1994; 42:976–978.

20. Ghose, AK; Viswanadhan, VN; Wendoloski, J. Prediction of hydrophobic properties of small organic molecules using fragmental methods: An analysis of ALOGP and CLOGP methods. J. Phys. Chem. 1998; 102, 3762–3772.
21. Viswanadhan, VN; Ghose, AK; Revankar, GR; Robins, RK. Atomic physicochemical parameters for three dimensional structure directed quantitative structure–activity relationships. 4. Additional parameters for hydrophobic and dispersive interactions and their application for an automated superposition of certain naturally occurring nucleoside antibiotics. J. Chem. Info. Comput. Sci. 1989; 29:163–172.
22. Viswanadhan, VN; Ghose, AK; Wendoloski, JJ. Prediction of solvation free energy and lipophilicity of small organic molecules: An assessment of additive constitutive methods. Perspect. Drug Discovery Design 2000; 19:85–98.
23. Leo, AJ. CLOGP. Available from Biobyte Corp., Pomona, CA, 1999.
24. Wang, R; Fu, Y; Lai, L A new atom-additive method for calculating partition coefficients. J. Chem. Inf. Comput. Sci. 1997; 37:615–621.
25. Rishton, GM. Reactive compounds and in vitro false positives in HTS. Drug Discovery Trends 1997; 2:384–386.
26. *World Drug Index.* Derwent Information, London, 1997.
27. Sybyl 6.3. Tripos Associates, St. Louis, MO, 1998.
28. Ghose, AK; Viswanadhan, VN; Wendoloski, JJ. A quantitative structure and physicochemical property based scoring scheme to evaluate druglikeness of small organic compounds. (To be published.)

9

Druglikeness Profiles of Chemical Libraries

Jens Sadowski

AstraZeneca R&D Molndal,
Molndal, Sweden

I. DRUGLIKENESS AS A GENERAL COMPOUND PROPERTY

With the avalanche of compounds that can be synthesized by automated parallel methods and can be tested by high-throughput screening, there is an increasing need for fast and reliable computational filters that can reduce these astronomical numbers to reasonable subsets. This need is manifest in areas like compound purchase, compound selection for screening, and combinatorial library design. Originally, computational chemistry came up with answers to questions like the following:

> Which subset of compounds spans the most diverse chemical space?
> Which subset fills most effectively "holes" in an existing in-house inventory?
> Which subset is most similar to a known lead compound?

These questions are focused on the diversity/similarity problem in library design. There are already many methods and computer programs available for this area (for a recent review see Ref. 1). There exist validated techniques for describing compounds by suited descriptors and for reasonable statistical assessment (2). There is evidence that diversity-driven selections can increase the chances for finding new lead compounds (3).

However, during the process of implementing approaches for compound selection and applying them routinely, it became obvious that additional decision criteria are needed which are quite different from mere diversity. It is obvious, that "druglikeness" is related to biological, chemical, and physical properties like activity, reactivity, synthesizability, bioavailability, and toxicity. Lipinski in a recent study (4) demonstrated the importance of solubility and showed approaches for assessing this problem by computational filters. Ghose et al. (5) investigated the ranges of certain properties that discriminate between drugs and nondrugs, such as log P, molar refractivity, molecular weight, and the occurrence of certain substructures.

But there seem to be a number of additional compound properties that cannot be assessed easily by computational procedures. Experienced medicinal chemists have a feeling for this, often without even being able to name substructures or rules that discriminate between drugs and nondrugs. This knowledge should nowadays implicitly be contained in chemical databases of drugs and basic chemicals. Very recently, the idea of using such databases for the construction of computational filters that recognize and rank the druglikeness of chemical compounds was almost simultaneously realized by several groups (6–8).

This chapter explains the principles of such methods and illustrates their merits with several examples.

II. KNOWLEDGE SOURCES AND COMPUTATIONAL TOOLS

A. Data Sources

As sketched in Section I, knowledge about drugs and nondrugs is implicitly contained in publicly available databases. It is obvious that databases like World Drug Index (WDI) (9), Comprehensive Medicinal Chemistry (CMC) (10), and MACCS Drug Data Report (MDDR) (11) are relatively clean collections of drugs and druglike molecules. Of course, many compounds never come into the market and some drug classes (e.g., cytostatics) are not typical drugs. But in principle, the vast majority of compounds in these databases were at least designed by medicinal chemists with the intention of making drugs. The definition and representation of nondrugs is more challenging. The only practical approach is to use databases of basic chemicals like the Available Chemicals Directory (ACD) (12) or organic reaction databases like SPRESI (13) and to assume that the rate of drugs in these databases is much smaller and can in fact be ignored. By using such databases, by removing reactive or otherwise unsuitable compounds, and by removing the exact matches of compounds of the drug database from the nondrug database, it is possible to create reasonable representations of drugs and nondrugs with several thousand compounds. Section III, Applications, provides an example of database preprocessing.

B. Descriptors

Several sets of descriptors exist for the representation of chemical structures. There are many ready-to-use systems in program packages from such vendors as Daylight, Tripos, and Molecular Design. Several of these systems have been evaluated (2). When choosing a suitable set of descriptors for the modeling of such a general property as druglikeness, a good balance between generality and specificity must be found. Sometimes, a simple atom-type descriptor with about 100 entries seems to work better than a 2000-bit fingerprint (8). Section III gives an example of such a descriptor that works.

C. Statistical Assessment

To derive knowledge from the above-mentioned databases, which are encoded by a suited set of descriptors, a statistical classification method is needed. In principle, a clustering approach or any other unsupervised learning algorithm could be used to derive clusters of similar compounds. The properties of the druglike clusters could then be used for the classification of unknown compounds. A much more straightforward approach is to use a supervised learning method like linear regression, decision trees (14), or neural networks. Neural networks seem to be superior because of their nonlinear behavior (see, e.g., Refs. 15 and 16). Professional neural network software can be found in abundance (see, e.g., Ref. 17). In principle, a neural network can be treated as a black box that resembles a very simple brain with neurons and axons and can be trained by confronting it with suitable training data. In a more elaborate view, a neural network is a complex, nonlinear equation that transforms the input data—in this case, the molecular descriptors—into output values—in this case, an estimate of druglikeness. The aim during training is to minimize the error between the given output values and the ones calculated by the neural network. The properly trained network should then be able to predict data it never saw during training.

D. Complete Classification Algorithm

The overall procedure is based on the data and methods mentioned in the preceding sections. The first step is the proper selection of databases, of descriptors, and of the classification approach. The second step is the learning phase. Here, the approach tries to extract knowledge from the databases and to translate it into a classification algorithm. In step three, this classification scheme is applied to arbitrary compounds. The resulting classification scheme is illustrated in Fig. 1. A given chemical compound is translated into a descriptor, which in turn is forwarded to the classification tool (here a trained neural network), which ends up deciding whether the compound is or is not druglike.

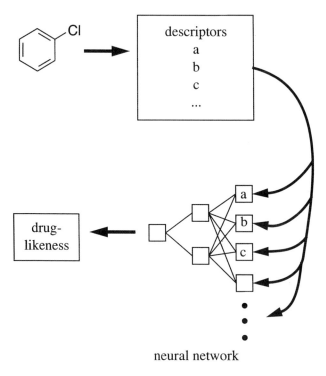

Figure 1   Classification scheme for druglike compounds.

III. APPLICATIONS

A. Published Approaches

Almost simultaneously, three different groups came up with different solutions for the problem of druglikeness. Gillet and Bradshaw (6) used global structural features like the number of hydrogen bond donors and acceptors, the numbers of rotatable bonds and aromatic rings, the molecular weights, and a shape descriptor as descriptors for the compounds in the World Drug Index and the SPRESI database as collections of typical drugs and nondrugs. A genetic algorithm was used to derive from the databases a weighting scheme that calculates the druglikeness of a compound from these features. A significant discrimination between drugs and nondrugs was achieved by this scheme. For example, when compounds from a test set consisting of 10,000 drugs from the WDI plus 168,071 nondrugs from SPRESZ were selected, the approach returned 2814 drugs within the top 10,000 compounds. This is enhancement by a factor of 5 over random selection (562 drugs expected).

Ajay et al. at Vertex (7) investigated the use of two types of descriptor (seven global molecular descriptors as in the Gillet paper and 166 MDL keys) and of two classification methods (decision trees and feedforward neural networks). The best approach was the combination of 78 descriptors with a neural network that classified 80% of the drugs in the MDDR database correctly as drugs while classifying about 90% of the ACD correctly as nondrugs. In addition, the calculated druglikeness of sets of drugs and nondrugs was compared with their diversity. The authors showed that there is no simple relationship between these two criteria.

Sadowski and Kubinyi at BASF (8) used the WDI and the ACD as drug/nondrug representations, an atom type based descriptor, and a feedforward neural network for the classification. Since this is the approach the author of this chapter is most familiar with, it is used as a detailed example of how to create such a computational filter.

B. A Recipe for a Computational Filter for Druglike Molecules

The following steps were performed.

1. The ACD and WDI databases were preprocessed by removing reactive compounds and duplicates and by removing the exact matches of WDI compounds from the ACD. This left 169,331 nondrugs and 38,416 drugs.

2. All ACD and WDI compounds were assigned druglikeness scores of 0 and 1, respectively.

3. An atom type based descriptor was calculated for each molecule. This descriptor is simply the counts of the 120 Ghose–Crippen atom types (18) in a given molecule. These atom types represent very simple functional groups (e.g., "aliphatic CH_2," "aromatic carbon," "amide nitrogen"). This descriptor—a very simple type of a fingerprint—was found to have the ideal information content. The originally 120 Ghose–Crippen types were reduced to those 92, which were populated at least 20 times in the training dataset.

4. From each of the two databases, 5000 randomly chosen compounds were forwarded to the training of a feedforward neural network based on the public domain program SNNS (17). SNNS was used to construct and train a neural network with 92 input neurons (the atom type counts), 5 hidden neurons, and 1 output neuron (the druglikeness score). The training following the "back-propagation with momentum" algorithm was performed over 2000 cycles with a learning rate of 0.2 and a momentum term of 0.1. The training pursued the aim of minimizing the classification error for the learning set of 5000 drugs and 5000 nondrugs.

The trained network was then used to reproduce the druglikeness scores of the 10,000 training compounds and as a test case to predict the druglikeness of the remaining 165,331 nondrugs from the ACD and the remaining 33,416 drugs from the WDI. Figure 2 shows the score distributions separately for the ACD and WDI compounds in the training set (solid lines) and for the remaining whole databases (dashed lines). Clearly, the network succeeded in correctly predicting the druglikeness for about 80% of the nondrugs and drugs in the training sets as well as for the much larger test sets. That means that the success rate was the same for the overwhelming majority of molecules in the whole databases the network never saw during the training phase. To increase the rate of correctly classified drugs, we currently use a score threshold of 0.3, which leads to 90% correctly classified drugs and still about 70% correctly classified nondrugs.

A simple additional verification is shown in Table 1: the druglikeness scores for a number of best-selling drugs ordered with respect to decreasing market volume (19). Again, with one exception (diclofenac), the compounds were clearly classified as drugs with a score greater than 0.5.

A number of additional statistical tests in all three papers (6–8) hardened the finding that the results of such an approach are absolutely valid. It is indeed possible to construct a filter that estimates the druglikeness of molecules just as a large collective of medicinal chemists did over many years.

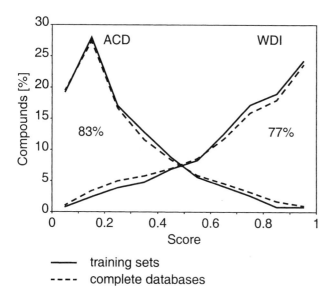

Figure 2 Distribution of calculated druglikeness over training and test sets from ACD and WDI.

Table 1 Calculated Druglikeness Score for a Number of Best-Selling Drugs

Name	Score
Ranitidine	0.78
Enalapril	0.82
Fluoxetin	0.53
Aciclovir	0.64
Simvastatin	0.80
Co-amoxiclav	
Amoxicillin	0.80
Clavulanic acid	0.68
Diclofenac	0.40
Ciprofloxacin	0.93
Nifedipin	0.76
Captopril	0.82
Diltiazem	0.80
Lovastatin	0.89
Cimetidine	0.72
Omeprazol	0.85
Cefaclor	0.67
Ceftriaxon	0.97
Estrogen	
Estrone	0.62
Equilin	0.73
Cyclosporin	0.84
Beclometason	0.82
Famotidin	0.65
Salbutamol	0.93
Sertralin	0.65

C. Retrospective HTS

To demonstrate the similarity in behavior between the trained neural network and experienced medicinal chemists, we analyzed retrospectively five high-throughput screening (HTS) runs at BASF (exact data not shown). Figure 3 shows the percentage of predicted druglike molecules (score > 0.3) over four stages in the HTS cycle: the totally screened compounds (some 100,000 compounds), the compounds above a certain level of percentage inhibition (several thousand compounds), the compounds above a certain IC_{50} level (several hundred), and finally the compounds chosen from this last list by medicinal chemists as leads for further development (about 10).

score > 0.3 [%]

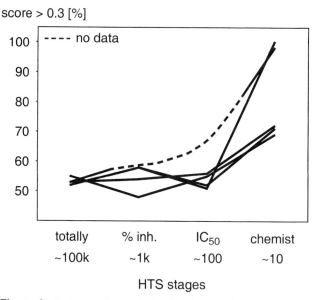

Figure 3   Retrospective analysis of high throughput screening data.

In the first stage, simply all available compounds are selected for screening without any selection criterion. The only selection criterion in the second and third stages is activity at the biological target—percentage (% inh) inhibition and IC_{50}, respectively. Only in the last stage are additional criteria applied by the medicinal chemists when selecting candidates from the lists of active compounds: synthetic accessibility and variability, bioavailability, toxicity, and so on. Obviously, for the five HTS runs the level of predicted druglikeness remains more or less the same for the first three stages—about 50–60%. It jumps significantly up to 70–100% druglikeness when the chemists select promising compounds by hand. This means that the trained network makes on average the same choices as the chemists when it comes to decision criteria beyond mere biological activity. This and not more must be expected from such an approach.

IV.   CONCLUSIONS AND OUTLOOK

A recipe was presented for a classification approach that can discriminate between drugs and nondrugs. It was shown that such approaches decide on average like experienced medicinal chemists and can classify about 80% of drugs and nondrugs correctly.

The approach is general and fast enough (0.01 s per molecule on an average desktop computer) to process large compound libraries as produced by combinatorial chemistry. It should be used simultaneously with other criteria (e.g., the diversity of the products, the price of the starting materials) for the optimization of combinatorial libraries and for the selection and prioritization of compounds for purchase, synthesis, and biological testing. This approach can significantly enrich compound collections with molecules having a better chance to become drugs. It is an additional selection and design criterion that can be combined with more specific approaches like pharmacophore models, quantitative structure–activity relationships, or protein structure–based design.

The same principles of course can be applied to combinatorial library design. It should be kept in mind that the druglikeness estimates were obtained on the basis of drug databases like the WDI on the one hand and simple chemical collections like the ACD on the other. Therefore, druglikeness is always a property of the complete synthesis product (which might be a drug) instead of a property of individual building blocks (which probably can be found among the nondrugs in the ACD). A reasonable design scheme might be to enrich combinatorial libraries with druglike compounds and to combine this criterion with other criteria like diversity or the prices of the starting materials. Indirectly, also preferable building blocks occurring most often in druglike products might be selected from the completely enumerated library. But in general, druglikeness must be assessed on the product side.

It is stressed again that druglikeness in this context is the sum of properties like protein binding groups, bioavailability, toxicity, synthesizability, and stability. A high predicted druglikeness for a given molecule does not automatically imply that this compound is biologically active. It simply means that medicinal chemists would have voted similarly.

This classification approach can in principle be applied to more specific properties like bioavailability, toxicity, or activity in certain biological assays. Another application currently pursue at BASF is the proper discrimination of crop protection compounds from basic chemicals. The only prerequisite is the availability of at least 1000 valid examples for each class (bioavailable/not available, toxic/nontoxic, etc.).

REFERENCES

1. Warr, W. A. J., Chem. Inf. Comput. Sci. 1997; 37:134.
2. Brown, R. D., and Martin, Y. C., J. Chem. Inf. Comput. Sci. 1996; 36:572.
3. Matter, H., J. Med. Chem. 1997; 40:1219.
4. Lipinski, C. A., Lombardo, F., Dominy, B. W., and Feeney, P. J., Adv. Drug Delivery Rev. 1997; 23:3.
5. Ghose, A. K., Viswanadhan, V. N., and Wendeloski, J. J., J. Comb. Chem. 1999; 1:55.

6.   Gillet, V. J. and Bradshaw, J., J. Chem. Inf. Comput. Sci. 1998; 38:165.

7.   Ajay, Walters, W. P., and Murcko, M. A., J. Med. Chem. 1998; 41:3314.

8.   Sadowski, J., and Kubinyi, H., J. Med. Chem. 1998; 41:3325.

9.   WDI: World Drug Index, version 2/96, Derwent Information, 1996.

10. CMC: Comprehensive Medicinal Chemistry; MDL Information Systems.

11. MDDR: MACCS Drug Data Report; MDL Information Systems.

12. ACD: Available Chemicals Directory; version 2/96, MDL Information Systems, 1996.

13. SPRESI: Daylight Information Systems.

14. Quinlan, J. R., C4.5: Programs for Machine Learning. San Francisco: Morgan Kaufmann, 1993.

15. Zupan, J. and Gasteiger, J., Neural Networks for Chemists: An Introduction. New York: VCH, 1993.

16. Zell, A., Simulation neuronaler Netze. Reading, MA: Addison-Wesley, 1994.

17. SNNS: Stuttgart Neural Network Simulator, version 4.0, University of Stuttgart, 1995.

18. Ghose, A. K., Viswanadhan, V. N., and Wendoloski, J. J., J. Phys. Chem. A 1998; 102:3762.

19. SCRIP No. 2040, July 7, 1995, p. 23.

10

Tools for Designing Diverse, Druglike, Cost-Effective Combinatorial Libraries

Robert D. Brown, Moises Hassan, and Marvin Waldman

Molecular Simulations, Inc.
San Diego, California

I. INTRODUCTION

High throughput technologies are rapidly becoming an established part of the drug discovery process. Combinatorial chemistry libraries are used in both the lead generation and lead optimization phases of this process, and in either case a computational library design process is typically used to aid the chemists in determining the library that should be synthesized and screened. The computational process often starts with the specification of a *virtual library,* which contains all the possible compounds that could be made within the scope of the chemistry that will be used and the reagents available, commercially or otherwise, that are compatible with that chemistry. From this virtual library one or more subsets are then selected computationally.

One can distinguish the aims of this selection process for libraries intended for lead generation and lead optimization. In the former case, one would ideally wish to sample all the biological variation in the virtual library with as few compounds as possible, thus seeking to maximize the chances of identifying all different types of active molecule in the virtual library while minimizing the synthesis and screening effort. As will be discussed later (see Sect. II.B), in practice one attempts to achieve this by sampling all chemical variation in a space that is thought to be biologically relevant. For lead generation, the concepts of diversity, coverage, and representativeness are used in the selection process. Diversity selection aims to ensure that the chosen molecules are maximally different from each other, making the implicit assumption that will enhance the probability of finding active compounds of differing types. Coverage extends this concept to suggest that one should also seek to ensure that all compounds in the virtual library are represented in the design library, in an attempt to ensure that all different types of biologically active molecule in the virtual library will be identified. Representativeness aims to sample the virtual library in such a way that the distribution of molecules in chemical space (defined in Sect. II.B) in the selected set reflects that in the virtual library.

For lead optimization, one wishes to expand the structure–activity relationship (SAR) around a series of existing hits, identifying as many other hits of the same "type" within the virtual library. The definition of type here depends on the descriptors chosen to measure molecular similarity. For example, compounds may be structurally similar in two dimensions, or have similar pharmacophore patterns in three dimensions, or have similarities in binding to the three-dimensional structure of the target.

Whether one is making a diversity or a similarity selection, it is important to take account of the space that has already been explored through screening of a historical compound collection and/or libraries that have already been synthesized. To do this one needs to be able to compare two or more sets of compounds (libraries or compound collections) in a common or reference chemical space. When making subset selections, one must consider the potential of the subset to maximally augment the information that has already been obtained from the screening effort to date. For diversity selection, one will wish to ensure that the subset samples areas of space that have not previously been covered. Note that such methods are equally applicable to compound acquisition in which one augments a compound collection through purchase of existing compounds from third-party vendors, rather than selecting virtual compounds to be synthesized.

A number of methods have been proposed to address the diversity and similarity problem, both in isolation and in the context of existing libraries, and we will discuss our own methodologies in this chapter. Historically, library design has tended to focus solely on this problem. However, many libraries resulting from design by diversity analysis or resulting from design by the intuition of a chemist have proven to be relatively poor at generating hits that were subsequently successful in medicinal chemistry. Often, the screening hits from a library either were uninteresting to medicinal chemists or suffered from absorption, distribution, metabolism, excretion, and toxicity (ADME/tox) problems that proved difficult and/or costly to resolve (1).

Recently, there has been a realization that designing a library purely for diversity or similarity is a somewhat naïve approach. Consideration must be given to the properties of the molecules forming the set selected for synthesis such that, should they prove to be hits, they are less likely to cause ADME/Tox or synthesis problems during lead optimization (2–4).

With the increasing throughput of the drug discovery process and the need for cost efficiency, a design for a library must also strive to be efficient to synthesize and cost-effective. For this, the properties of the reagents that are required to produce the products in the selected subset of the virtual library must be considered. They ought to be inexpensive, reliably and quickly obtained, and optimally combined in conjunction with the synthetic protocol, automated or manual, used to produce the library.

In summary, library design should incorporate the following:

> A diversity or similarity selection method that can sample a virtual library with the required behavior with respect to diversity, coverage, representativeness, or neighborhood sampling
>
> A bias toward the selection of a set of molecules that have the characteristics of a good leads, or "druglike" molecules
>
> A bias toward the selection of a set of molecules that can be made with reagents that are cost-effective and can be obtained within an appropriate time frame
>
> The necessary constraints to ensure that the library will be efficient to synthesize, especially when robotic synthesis is to be used

This chapter will discuss software (5) developed to address each of these factors and will show how they may be combined to produce a set of criteria for the library design that can be applied simultaneously to selection of a subset from a virtual library.

II. THEORY AND METHODOLOGY

A. Search Space and Optimization

1. The Combinatorial Constraint

The efficiency of synthesis of a library is dependent on the methodology (either automated or manual) used in its production. In some situations, particularly when one is using a manual or semiautomated parallel synthesis approach, there are no constraints on the way in which individual reagents will be combined. In this case, *cherry-picking* methodologies are appropriate for the computational design. Cherry-picking aims to select the most appropriate (diverse or similar) set of products, irrespective of the combinations of reagents that will be needed to make them. Since it is without other constraints, this approach typically leads to the most diverse (or similar) subset that can be selected. However, a cherry-picked library can be costly and inefficient in terms of the reagents it uses. In the extreme, every reagent will be used to produce only one product, and the total number of reagents will be equal to the number of products multiplied by the number of positions of diversity within the library.

In effort to address this problem and to ensure synthetic efficiency, a *combinatorial constraint* is often applied to the library design. This constraint requires that every reagent used at each diversity position be used in all combinations with all reagents at every other diversity position, and that every reagent is used in the design of full arrays and mixtures. The combinatorial constraint ensures that the maximum number of products is obtained from the minimum number of reagents and that the library will be synthetically efficient on array synthesis automation. It does, however, place a restriction on the products that can be made, since individual products can no longer be cherry-picked to be included in the library. Instead, the selection of a reagent implies that all products of that reagent with the selected reagents at all other diversity sites must be included in the design. A combinatorial subsetting problem can be expressed as the selection of, for example, a $10 \times 8 \times 12$ library with the implication that all 960 products will be made. The equivalent cherry-picking problem would be expressed as the selection of a 960-member library. The combinatorially constrained library will require 30 reagents, whereas the cherry-picked library with three positions of diversity could require up to 2880 (960×3) reagents, assuming that there are 960 or more suitable reagents available for use at each position of diversity.

A third type of design that has recently emerged is that of sparse matrix libraries (6). In this case, a strict combinatorial constraint is not applied. Instead the number of reagents is minimized (but does not have to be the absolute minimum), and the resulting library mapped as efficiently as possible onto the synthesis equipment. By allowing some relaxation of the absolute combinatorial constraint, it has been found that libraries can be designed that better meet the design criteria (diversity, properties etc.) while still being relatively efficient to synthesize. Such a design could be achieved with the methodology discussed in Section II.D.

2. Search Space

The first problem that must be faced in library subsetting is that the search spaces are extremely large, since the number of appropriate reagents that are available for many typical chemistries is high and ever increasing. As an example, consider a library with three positions of diversity in which there are 100 reagents available for use at each position, giving a virtual library of 1 million members. Consider the problem of selecting a library of 1000 molecules from this virtual library. If the library is not combinatorially constrained, then there are $C_{1000}^{1,000,000} \approx 10^{3400}$ possible libraries. If the combinatorial constraint applies (see Sect. II.A.1), then the subsetting problem is actually to select a $10 \times 10 \times 10$ library and there are $C_{10}^{100} \times C_{10}^{100} \times C_{10}^{100} \cong 5 \times 10^{39}$ possible libraries.

3. Monte Carlo Optimization

The size of the search spaces (discussed in Sect. II.A.2) and the number of constraints that must be applied (discussed in Sect. I) suggest that the design problem cannot be solved deterministically. Evolutionary algorithms have been successfully applied to many problems in which there is not only a vast search space but also a number of factors that must be simultaneously optimized (7,8). In this work we have made use of a Monte Carlo optimization procedure. Two versions of the algorithm are implemented: one for simple product selection and another that incorporates the additional problem of mapping between reagents and products in the case of combinatorial constrained selection.

Figure 1 shows a schematic view of the algorithm. For cherry-picking, an initial subset of molecules is chosen at random from the pool of all available

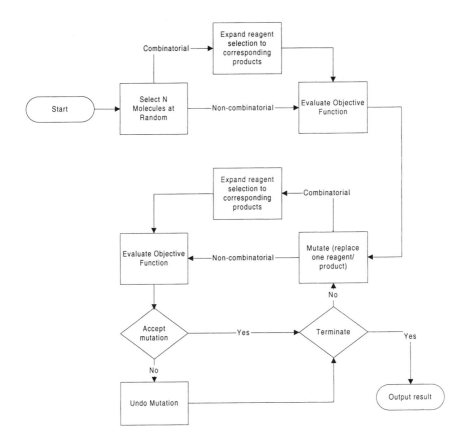

Figure 1   Monte Carlo optimization procedure for cherry-picked and combinatorial libraries.

molecules and the initial value of the objective function is calculated. The objective function (discussed in Sect. II.A.4) evaluates the quality of this subset, in terms of the design criteria (diversity, similarity, druglikeness, cost-effectiveness, etc.) that have been specified. Next, a new trial set is constructed by a single-point mutation procedure in which one of the currently selected molecules is replaced by a molecule not currently selected. Whereas the new molecule is chosen at random among those not currently selected, the selection of the old molecule is biased toward replacing a "bad" one and, therefore, improving the value of the overall function. The value of the objective function is evaluated for the trial set. If it is better than the current value, the trial set is accepted. If the new value is worse, the Metropolis criterion is used to decide whether to accept or reject the trial set. The exponential function for the Metropolis criterion is set so that a decrease of 10% in the value of the diversity function is accepted 10% of the time at a "temperature" of 300 K. If the trial set is accepted, it is compared against the optimum set found so far and, if it is better, the optimum set is replaced. The process is terminated after a maximum number of steps is reached or if the optimum value of the function does not improve after a specified number of steps.

For design including the combinatorial constraint, we have used a similar Monte Carlo strategy with the aim of optimizing the choice of R groups to perform true R-group, array-based optimization of the score of the resultant product library subset. Our goal in this case is to choose an optimal subset corresponding to a combinatorial library with a prescribed number of n_1 reagent choices for the R1 substituent, n_2 for R2, and so on. We begin the optimization by randomly selecting n_1 R1 reagents, n_2 R2 reagents, and so on, from a larger virtual pool. The corresponding set of products is identified, and the diversity of this subset is evaluated based on this resulting enumerated subset library. We then take a Monte Carlo step, but this time instead of randomly "mutating" one molecule, we "mutate" one R group by randomly replacing one of the currently selected R-group choices in our subset with an unselected R-group reagent. We then evaluate the objective function of the corresponding enumerated library subset, accept or reject the trial R-group mutation with a Monte Carlo Metropolis criterion as before. Again the process is iterated for many Monte Carlo steps until convergence (i.e., no further improvement in the value of the diversity function) has been achieved. A similar procedure has been described using a genetic algorithm (9).

4.   The Objective Function

The procedure depends on the specification of an objective function to determine the suitability of the solution proposed at each iteration. The objective function consists of a weighted combination of terms, each of which accounts for one of the constraints on the products or reagents previously described in Section I, that is,

$$F = w_{ds} (DS) + \sum_{i=1}^{n} w_i ProductPenalty_i + \sum_{j=1}^{n} w_j ReagentPenalty_j \quad (1)$$

where DS is the diversity or similarity score, w_{ds}, w_i, and w_j are weighting factors; and n and m are the number of product and reagent constraints, respectively

The diversity selection algorithms that can be used in this objective function are described in Section II.B, the penalty functions used to assess druglikeness are described in Section II.C, and the functions used to assess cost-effectiveness in Section II.D. Each of these functions returns a score that is normalized to be within a specific range and then the total objective function is calculated as shown in Eq. (1).

B. Diversity Selection Algorithms

Diversity and similarity methods are founded on the similar property principle, which notes that structurally similar molecules have similar activities (10). Whatever the selection method, the correct selection of a chemical space into which the compounds are projected is essential, so that this *neighborhood* behavior holds (11). Assessing diversity and similarity is a multistep process in which:

1. A series of descriptors is computed for each compound under consideration.
2. The compounds are projected into a common chemical space of either the original descriptors or a (usually smaller) set of descriptors derived from the originals using a technique such as principal component analysis or factor analysis.
3. A subset of compounds is selected based on the members' proximity within the chemical space. This may be either based on a distance computation or on the division of space into cells and the comparison of cell occupancies of compounds.

A variety of descriptor types have been proposed ranging from 2D and 3D fingerprints to topological and graph theory indices to BCUT descriptors and molecular properties. An excellent review of these is given in a special issue of *Perspectives in Drug Discovery and Design* (12) and in other chapters of this book, and it is not the intention of this chapter to discuss this aspect of the methodology in detail.

Two classes of diversity metrics have been widely used. The first are partitioning methods. These may employ clustering approaches, which divide the compound pool into sets of similar compounds, or cell-based approaches, which subdivide each dimension in the chemical space into bins, thereby forming hyperboxes in the space. For diversity, compounds are selected to sample different boxes/clusters, while for similarity, compounds are drawn from the box(es)/cluster(s) closest to the target. Methods in the second set are based on measuring the distance between pairs of compounds and then maximizing the distances between the selected set for diversity or minimizing the distances for similarity. In either case, metrics are available that vary in the way in which they aim to achieve diversity, coverage, and representativeness.

When selecting between the methodologies, one must consider the size of the problem, the available time for its resolution, and the accuracy required. Cell-based methods (discussed in detail in Sect. II.B.3) are the most approximate but are the fastest. The speed results from the simplicity of simply dividing space into cells. However this also introduces an approximation, since molecules in separate cells are considered to be different, irrespective of whether they are far apart, or adjacent but separated by an arbitrary cell boundary. Distance-based metrics (discussed in Sect. II.B.4) tend to be slower, since they require the computation of pairwise distances or similarities, but are less approximate, since they do not suffer from the edge effect. Clustering methods can be the most time-consuming, depending on the algorithm used.

1. Reagent vs Product Diversity

Cherry-picking clearly works in the space of the library products. That is, it is the library products whose descriptors are calculated and embedded in the chemical space from which selections are made. For combinatorially constrained design, it was initially assumed that optimizing product-based diversity could be reasonably approximated by the optimization of the diversity of the reagent R groups independently. That is, it was the reagents whose descriptors were calculated and embedded in the chemical space. Merely combining these diverse sets of reagents produced fully combinatorial libraries. Gillett and coworkers refer to this assumption as the "diversity hypothesis" (3). If true, the diversity of each substituent R group could be optimized by considering only the diversity of the possible reagents at each substitution point in a library scaffold without examining the diversity of the resulting products and without the need to enumerate a full virtual library and evaluate its descriptors. However, this assumption has been called into question (3,13). More recently, attention has been given to the problem of attempting to design combinatorial libraries by optimizing the diversity of the products (product-based diversity) under the constraint that the optimal library result from the full combinatorial explosion of a set of selected reagents (3). It is the latter problem to which our methods are addressed.

2. Requirements for a Diversity Selection Algorithm

To devise a procedure to optimize the diversity of a library in product (rather than reagent) space, we need a protocol for assessing the diversity of any possible library subset from the full virtual library being considered. We will assume that the choice of descriptors to use has already been decided, and we focus here on the problem of deciding how well the descriptor space is covered by a particular subset library. We refer to any protocol that quantitatively assesses the coverage of descriptor space as a *diversity function*. We have proposed (14) a set of requirements that a "perfect" diversity and coverage function should satisfy. It should be noted that some of these critically affect combinatorial selection but may be relaxed for cherry-picking selection. The rules mainly arise from considerations based on adding redundant or nonredundant molecules into our descriptor space and address the diversity and coverage of the resulting sets. The proposed requirements for a diversity function are as follows:

> *Requirement 1.* Adding redundant molecules to a system does not change its diversity.
> *Requirement 2.* Adding nonredundant molecules to a system always increases its diversity.
> *Requirement 3.* Space-filling behavior should be preferred.
> *Requirement 4.* Perfect (i.e., infinite) filling of a finite descriptor space should result in a finite value for the diversity function.
> *Requirement 5.* If the dissimilarity or distance of one molecule to all others is increased, the diversity of the system should increase. However, as this distance increases to infinity, the diversity should asymptotically approach a constant value.

A brief explanation for each requirement is given below; a full discussion is given by Waldman et al. (14).

Requirement 1 follows from the simple consideration that adding redundant molecules to a system does not increase coverage of the descriptor space. The diversity of the system (as opposed to the diversity per molecule) should not change, since exactly the same space is sampled before and after the addition of a redundant molecule.

Requirement 2 is almost a corollary to requirement 1. Since adding nonredundant molecules (i.e., points in the descriptor space not already present) sample or cover regions of the space not already sampled, it follows that the diversity of the system as a whole should increase. Because requirements 1 and 2 refer to the incremental change in diversity as molecules are added to the system, we have termed them *incremental diversity requirements.*

Requirement 3 encapsulates the notion that one would rather fill large voids than add new molecules that are close to already selected molecules.

Requirement 4 is particularly important in the comparison of libraries of unequal sizes. It encapsulates the notion that there is a finite limit on the diversity that can be achieved by covering a finite space and that after a point, nothing is to be gained by continuing to flood that space.

The foregoing rules state that diversity should change under certain actions (i.e. adding nonredundant molecules), but we have not stated by how much it should change. To do this we must define the "sampling radius" that provides some indication of how much of descriptor space is sampled by a particular molecule. This concept of a sampling radius in descriptor space is intimately related to requirement 5. Recall that the ultimate goal of designing diverse libraries is to maximize the probability of finding one or more biologically active compounds. It is assumed that we have already chosen a set of descriptors that correlate with biological activity in the following sense. By the similar property principle introduced in Section II.B, similar molecules (based on the values of their descriptors) are assumed to be (probabilistically) more similar in their biological activities than would be observed if molecules were selected at random. In other words, molecules similar to an active molecule should tend to be more active than a molecule chosen at random, and molecules similar to an inactive molecule should be less active (on average) than a molecule chosen at random. This means that molecules with similar diversity descriptors should have correlated biological activities, while molecules with very different values for their diversity descriptors have uncorrelated biological activities.

The transition from correlated to uncorrelated biological activities occurs when molecules no longer sample a common region of descriptor space: that is, when they are outside each other's "sampling radius." Once this has occurred, there is no further gain in diversity by making the molecules any more dissimilar, because once the molecules do not overlap in their coverage of descriptor space, there is no further gain in coverage by any additional increase in their separation from each other. Of course, it is not truly correct to assign a "hard" value to the sampling radius parameter, since the transition from correlated to uncorrelated biological activity behavior occurs gradually and continually. Consequently, rather than state that the diversity remains constant once molecules are outside each other's sampling radius, we propose as a requirement that the diversity should monotonically increase and asymptotically approach a constant as a molecule becomes very distant from all others in a diversity descriptor space. One desirable feature of the asymptotic behavior of diversity with distance is that it tends to reduce the possibility that one or a few outliers in a system can dominate the diversity score.

3. Partition-Based Selection

In clustering or cell-based methods, a library subset is selected usually with the goal of maximizing the number of occupied cells or clusters. The simplest diversity function is one that simply counts the number of occupied cells or clusters and compares that to the total number of cells occupied by the virtual library.

$$D_{\text{fraction}} = \frac{N_{\text{occupied cells}}}{N_{\text{cells}}} \tag{2}$$

This tends to reasonably obey requirements 1–5. Since redundant molecules will occupy the same cell or cluster as an existing molecule in the system, the cell count remains unchanged when redundant molecules are added, thereby satisfying requirement 1. A nonredundant molecule may belong to a previously unoccupied cell or cluster, in which case the occupied cell count will increase. If it belongs to an already occupied cell, then the cell count is unchanged. Thus, we see that requirement 2 is partially satisfied (diversity increases sometimes but not always). At least, the diversity does not decrease when nonredundant molecules are added. Requirement 3, preferring space-filling behavior, is partly adhered to in the sense that the function increases whenever an unoccupied cell becomes occupied. Perfect filling of a finite space leaves the diversity function finite, assuming that the number of cells partitioning the finite space is finite, thereby satisfying requirement 4. As two molecules become very far apart, they will occupy separate cells or clusters, but once this has occurred, the diversity function will not increase any further, partly satisfying (since the approach is not asymptotic) requirement 5. Furthermore, the cell size (actually the edge size) effectively serves as the sampling radius parameter. Thus, most of the requirements tend to be satisfied or partly satisfied. The drawback to this approach is that it is limited by the level of resolution provided by the cell divisions. Thus, a set of moderately similar molecules all falling within the same cell or cluster will appear to be less diverse than a set of molecules that are very similar but happen to lie across one or more cell boundaries thereby spanning several cells.

Clustering methods can also experience this problem, since similar molecules can sometimes be assigned to different clusters. There is no guarantee that molecules within a cluster are all more similar to each other than to any molecules outside the cluster. Thus, sets containing very similar molecules can happen to occupy more cells or clusters than a set containing only moderately similar compounds that all happen to lie within the same cell or cluster. This drawback is related to the failure to satisfy requirement 2. In these cases, adding nonredundant molecules to an already occupied cell or cluster may result in an incorrect assessment of the diversity of the system.

Another diversity function that can be used with cells or clustering methods is the chi-squared statistic (χ^2). The definition of the diversity function in this case is:

$$D_{\chi 2} = -\Sigma \left(N_i - \frac{N_{\text{sel}}}{N_{\text{cells}}} \right)^2 \tag{3}$$

where N_i is the number of molecules in cell i, N_{sel} is the total number of molecules in the sublibrary, and N_{cells} is the total number of cell partitions. The sum runs over all cells. This function attempts to produce a uniform distribution of cell occupancies, rather than simply maximize the number of occupied cells. The minus sign in front of the sum ensures that more uniform distributions are scored as being more diverse. For example, suppose diversity space was divided into two cells (or clusters) and suppose there were two libraries, each containing 10 molecules. Let the cell occupancies for the first library be (5,5). Let the occupancies for the second library be (7,3). If we simply count cells, the libraries are scored as equally diverse, since there are two occupied cells in each case. However, the chi-squared diversity function considers the (5,5) library to be more diverse, which is a desirable result. Nevertheless, the chi-squared function does not satisfy incremental diversity (requirements 1 and 2), and this may result in some problems in its behavior. Our investigations of the chi-squared function (14) have led us to conclude that it may be too strongly biased toward producing uniform distributions while not placing enough emphasis on the goal of simply maximizing cell occupancies. For example, consider a three-cell system containing six molecules. The ideal distribution is (2,2,2). The chi-squared function ranks the two distributions (3,3,0) and (4,1,1) as equivalent, whereas cell counting would prefer (4,1,1), since all three cells are occupied.

Another function that tends to show behavior intermediate between the chi-squared function and the cell occupancy count is one that measures the entropy (or information content) of the system (15,16)

$$D_{\text{entropy}} = -\Sigma \left(\frac{N_i}{N_{\text{sel}}} \right) \ln \left(\frac{N_i}{N_{\text{sel}}} \right) \tag{4}$$

This also favors more uniform distributions, but it would rank the (4,1,1) distribution ahead of (3,3,0) in the example above. This function also does not strictly obey requirements 1 and 2. However, it can be shown that requirement 2 is partially satisfied in that adding a molecule to an empty cell will increase the cell entropy diversity value. This behavior is not generally satisfied by the chi-squared function, suggesting an additional theoretical reason for preferring the cell entropy function.

The three functions are all attempting to address the diversity/coverage question and, as we have attempted to show, do so to varying extents. A final metric is aimed to address the representativeness question, this being the cell-based density

$$R_{\text{Density}} = -\Sigma N_i \ln \left(\frac{N_i}{M_i} \right) \tag{5}$$

where the sum is over all filled cells, N_i is the number of molecules in cell i for the subset, and M_i is the number of molecules in cell i for the full virtual library. This metric aims to select a subset in which the distribution of molecules in space in the subset mimics as closely as possible that in the virtual library. Dense areas of space (highly populated cells) in the latter will be more heavily sampled in the subset than sparse areas of space (lowly populated cells) that will be lightly sampled (and some may not be sampled at all). Since this is a representativeness metric and not a diversity metric, it is not instructive to consider its properties in relation to the list of requirements in Section II.B.2.

Whichever metric is used, one also needs protocols for choosing how to partition each property for the cell-based approaches. Of course, ideally one would like to use information on how the descriptors correlate with biological activity data to guide these choices, as discussed earlier, but this information is typically not available.

Consequently, after some investigation, we have settled on the following default choice for the number of cells. The value is chosen such that the number of occupied cells in the full virtual library is equal to (or just slightly greater than) the number of molecules we are selecting for our subset library. This choice means that if a combinatorial constraint was not imposed on the solution, each selected molecule could occupy a different cell. We find in practice that the combinatorial constraint rarely permits this ideal solution, and thus the fraction of occupied cells in the sublibrary (relative to the ideal value of each molecule occupying a different cell) serves as a reasonable measure of how much diversity is lost by the imposition of the combinatorial constraint.

By default the descriptor space is partitioned as follows. Each descriptor is uniformly partitioned (ranging from its minimum to maximum values) such that the cell edge lengths for the descriptor are as uniform as possible while requiring that the resulting number of occupied cells (in the virtual library) equal the target value (number of molecules to select) or just exceed it. This partitioning is achieved by means of an iterative algorithm that increases the number of bins by one for the descriptor having the currently largest cell edge until the resulting binning scheme results in the number of occupied molecules in the virtual library being equal to or just exceeding the number of molecules to select in the sublibrary.

In addition to this binning scheme, others are provided as alternatives:

Uniform, in which the number of bins for each descriptor is independently set by the user and each axis is divided into that number of bins of equal width.
Standard deviation based, in which the bin boundaries on each axis are placed at whole numbers of standard deviations of the property on that axis, centered at the mean.
Manual, in which the user defines each bin boundary for each axis.
Population weighted, in which each axis is independently divided into a user-specified number of bin in such a way an equal number of molecules falls into each bin on any given axis. (This concept was described by Bayley and Willett in Ref. 17.)

4. Distance-Based Selection

One of the first applications of distance-based selection was the use of D-Optimal design by Martin et al. (18). However this can be computationally intensive for data sets comprising large numbers of molecules and descriptors. Furthermore, it suffers from the weakness that it tends to select outliers, while leaving regions near the center or mean values of the data set unsampled, particularly when the number of molecules far exceeds the dimensionality of property space.

   We have developed a number of diversity metrics specifically designed to avoid such problems. The first is MaxMin,

$$\text{MaxMin} = \max\{\min[D_{ij}]\} \text{ where } D_{ij} = \sum (x_{ik} - x_{jk})^2 \tag{6}$$

which functions as follows. Given a set of descriptors associated with each molecule, calculate the distance between each pair of molecules in descriptor space as D_{ij}^2, the distance between molecule i and molecule j, x_{ik} is the value of the kth descriptor for molecule i, and the summation runs over all descriptors k.

   Now, for a given subset of molecules, find the minimum value of D_{ij} for all i,j pairs. Maximizing this function by using the Monte Carlo procedure described in Section II.A.3 produces the optimally diverse set.

   The PowerSum function.   In this function,

$$\text{PowerSum} = \max\left\{\left[\frac{\sum(D_{ij}^2)^{npower}}{N_d}\right]^{1/npower}\right\} \tag{7}$$

the sum runs over all i,j pairs of the subset, N_d is the number of intermolecular distances $M(M - 1)/2$, where M is the number of selected molecules
and npower is the exponent of the PowerSum function (npower $= -1$ corresponds to the reciprocal of the average of reciprocals, also known as the
harmonic mean, and is the value used by default). This function maximizes the harmonic mean of the squares of all intermolecular distances
between selected points.
   The Product Function.   In this function,

$$\text{Product} = \max\{(\Pi D_{ij}^2)^{1/N_d}\} \tag{8}$$

the product of the squares of the intermolecular distances is maximized.

While the value of the MaxMin function is not explicitly dependent on the number of models in the subset, the PowerSum and Product functions are normalized
by the number of distances in the subset. These three functions show some differences in regard to the subsets of molecules they select, but they all share the characteristic that higher values of the function correspond to subsets of molecules
with more different values of the molecular descriptors (i.e., more diverse subsets).

All these distance-based methods have been found to be very useful when applied to cherry-picking. In particular, we have found that the MaxMin approach is highly effective in selecting sets of molecules that provide optimal coverage of a descriptor space (19). However, they are not useful for combinatorially constrained libraries. In particular, they do not satisfy the incremental diversity requirements. For example, if we add a redundant molecule to the system, the MaxMin diversity value goes to zero (rather than staying constant). If we add a nonredundant molecule to the system, the MaxMin value either stays constant (the minimum distance between all possible pairs may be unchanged) or decreases (the new molecule may have a smaller distance to an existing molecule than the previous minimum value), rather than increasing as required.

The MaxMin function finds the closest pair of points in the system and assigns a diversity value based on the distance between the closest pair. This is fine when we are free to take one of the molecules involved in the closest pair and replace it with another, more diverse molecule. However, in combinatorial optimization, one does not replace a single molecule at a time, but rather attempts to find a subset of R-group choices that optimize the diversity of the enumerated sublibrary. In this case, one needs to compare the diversity of different combinatorial subsets. For example, it may be that one subset has a good overall coverage of descriptor space but has a single pair of molecules quite close to one another, which causes the MaxMin function to score the diversity as low. A second subset may have a poorer overall coverage of the space but no pair of molecules is very close to any another. MaxMin would score the second subset as more diverse than the first, and this is clearly an undesirable result. It stems from the failure of MaxMin to satisfy (even approximately) the incremental diversity requirements.

It has therefore been necessary to construct an additional diversity function specifically for combinatorial optimization (14). In attempting to construct a new function, we have found that a diversity function that exactly satisfies all the requirements 1–5 can be constructed for a one-dimensional system (i.e., where there is only a single descriptor used to characterize each molecule). In this case, each molecule can be represented as a point on a one-dimensional line where the position of the point corresponds to the value of its descriptor. A diversity function for this system satisfying requirements 1–5 is illustrated in Fig. 2. Above each point (molecule) on the line is drawn a normalized Gaussian curve. The Gaussians of all the points are allowed to overlap, and the area under the *envelope* of the Gaussians is taken as the diversity of the system. In this case, the width of the Gaussians (taken to be equal for all the points) corresponds to the sampling radius parameter discussed earlier. We can calculate the area under the curve exactly by partitioning the system into a set of Gaussian segments. This is shown by the vertical lines going through each point and through each midpoint of neighboring pairs. The areas under the far-right and far-left Gaussians are each $\frac{1}{2}$, as indicated on the figure. The area under each of the remaining Gaussian segments is $\frac{1}{2}$, the error function, erf() (20), evaluated over the length of the segment. As shown in the figure,

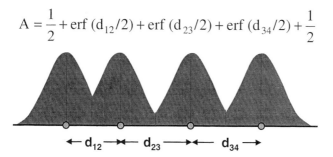

$$A = \frac{1}{2} + \text{erf } (d_{12}/2) + \text{erf } (d_{23}/2) + \text{erf } (d_{34}/2) + \frac{1}{2}$$

Figure 2   Gaussian function of straight line.

these segments occur in pairs of equal area, so the net result is a sum of error functions plus one.

The area under the Gaussian envelope satisfies all the diversity requirements 1–5. Adding redundant points makes no change to the area under the Gaussian *envelope,* since two coincident Gaussians will overlap into the envelope of a single Gaussian. Conversely, adding nonredundant points always increases the area of the Gaussian envelope. The area under the envelope is maximally increased by adding points where the largest gaps between points exist, satisfying requirement 3. The envelope of an infinite number of Gaussians over a finite line segment is a rectangle over the line segment plus two half-Gaussians extending at the right and left of the segment. The area under this region is 1 (for the two Gaussian halves at the far left and far right) plus the height of the rectangle times the length of the filled line segment. This value remains finite, thereby satisfying requirement 4. Finally, as a point is moved to the extreme left or right of the line, the area monotonically increases but asymptotically levels off (as the Gaussian of this point no longer overlaps with any of the other points). Finally, the width of the Gaussians acts as a sampling radius. Thus, this representation of diversity satisfies all the requirements stated in Section II.B.2.

We can represent the area in the general case of an arbitrary number of points on a one-dimensional line with the following formula:

$$A = 1 + \sum \text{erf} \left(\frac{\alpha d_{i,i+1}}{2} \right) \tag{9}$$

Here, the sum is over pairs of adjacent points, $(i, i + 1)$. This function obeys requirements 1–5. For example, in the case of adding a redundant point, the distance d involving the redundant point pair is zero, and $\text{erf}(0) = 0$, so there is no change in diversity. This is the mathematical equivalent of the statement that the envelope of two Gaussians centered at the same point is a single Gaussian. We have also introduced explicitly the width (or radius) parameter of the Gaussian into the argument of the error function via the parameter α. The use of the sampling radius parameter allows us to tune the behavior of the diversity function. Small values of α correspond to very wide Gaussians and lead to a nearly linear dependence of diversity on distance. Large values of α correspond to very narrow Gaussians and lead to a dependence of diversity on distance that rises rapidly and then quickly levels off to an asymptotic value.

Thus, we see that Eq. (9) provides a mathematical representation of diversity that perfectly satisfies all the recommended requirements 1–5. However, it is valid only for a one-dimensional system. The problem arises of how to generalize the result to a multidimensional system (i.e., representing molecules in diversity space using multiple properties or descriptors), which is the situation encountered in practical applications. In principle, this could be done by attempting to calculate the hyperdimensional volume of the envelope of multidimensional Gaussians centered at each of the molecules in the space. This volume can be represented as a multidimensional integral, and it does, in fact, satisfy all the five diversity function requirements. However, unlike the one-dimensional case, this multidimensional Gaussian envelope volume does not have an analytical representation. One possibility would be to evaluate the integral by approximate numerical quadrature, but such an approach may lead to problems in high dimensional spaces where numerical quadrature is difficult to perform accurately.

Instead, we have sought a different approach that results in a closed-form (but approximate) expression for the diversity of a multidimensional system. This alternate approach stems from the idea of treating the multidimensional case as a pseudo-one-dimensional system and to then make use of the one-dimensional diversity result of Eq. (9). One possible way to do this would be to connect all the points via a path and treat the line segments of the path as the quantities to sum over in Eq. (9). Since we wish redundant points to make no contribution to the diversity, the path should directly connect points that are very close (or redundant) to each other. This implies that we should seek the shortest path connecting all the points (usually referred to as the traveling salesperson problem) (21) and use the segments of this path in the diversity function of Eq. (9). However, problems arise with this approach because the computation of the shortest path through the points is a difficult problem known to be NP-complete (21). However, an alternative treatment turns out to be available which still allows for the use of the one-dimensional formula, Eq. (9), while being far more computationally tractable. Instead of requiring that the points be connected via a *path* (which implies a set of edges with no branching), we can relax this restriction and consider a set of connections in which the edges are allowed to branch. A set of edges that connect a set of points and are allowed to branch (and do not form any cycles) is known as a spanning tree (21). Instead of trying to find the shortest path through the points, we seek instead to connect the points with a minimum spanning tree, which is the spanning tree that has the smallest value for the sum of its edge lengths (21). Figure 3 illustrates the minimum spanning tree for a two-dimensional set of points.

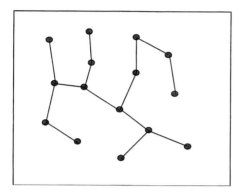

Figure 3 Minimum spanning tree.

We can use the minimum spanning tree to formulate a diversity function for the multidimensional case as follows. First, connect the molecule points with a minimum spanning tree. Assign a value for the sampling radius parameter, α. For each edge of the minimum spanning tree, calculate its error function (based on the edge length times the sampling radius parameter) and sum over all the edges. The diversity function can be represented by the following equation:

$$D = \sum \text{erf}(\alpha d_i) \qquad (10)$$

Here, the sum is over all the edges of the minimum spanning tree. The distance d_i is the length of each edge. Here, we have absorbed the factor of 2 from Eq. (9) into the radius parameter α, and we have omitted the additive constant, 1, in front of the sum, since it does not affect the relative diversity between systems. A similar function for diversity using the minimum spanning tree was proposed in 1999 by Mount and coworkers (22). The main difference between their function and ours is the use of the error function in Eq. (10) (which stems from considerations of overlapping Gaussians in one dimension). Their function involves summing up the edge lengths of the spanning tree. As such, it relates diversity linearly with distance and satisfies only requirement 1.

This diversity function of Eq. (10) also satisfies requirement 1, since adding redundant points adds edges of zero length to the minimum spanning tree, which consequently adds zero to the diversity value [since $\text{erf}(0) = 0$]. This function tends to satisfy requirement 2, but it is possible to devise cases in which adding points can lower the score. However, requirement 2 is strictly satisfied for the following cases. Adding a nonredundant point in the vicinity of an existing point will always increase the score. Adding a nonredundant point along one of the edges of the spanning tree (or in the vicinity of such an edge) will also always increase the score. The violation of requirement 2 occurs when the distances between points are very small (or α is very small), so that the error function reduces to a linear function of distance, and we return to the diversity function of Mount et al. (22). In this case, one can find examples in which adding a point into the center of several points can cause the diversity value to decrease. Fortunately, examples of this unusual behavior tend not to arise in practice. Requirement 3 is also partly satisfied in essentially the same sense as requirement 2. It can be shown that requirement 4 is not generally satisfied in the multidimensional case, but this is not important for the applications in which the combinatorial library optimizations all involve subset libraries of a fixed size. Finally, requirement 5 is fully satisfied.

In choosing the α parameter for the minimum spanning tree method, we have used a type of reasoning similar to that for choosing the cell size, discussed in Section II.B.3. Ideally, we would like each selected molecule in our subset to occupy a different region of descriptor space. In the case of the cell methods, this (ideally) corresponds to a different cell for each molecule. In the case of the spanning tree method, we need to choose a value of α that produces a reasonable result for the diversity calculated by Eq. (10). Since the distances in typical descriptor spaces are not bounded, a fixed choice for α may behave differently for different libraries and/or descriptor sets. We deal with this issue by renormalizing our descriptor space as follows. We determine the hypervolume of the descriptor space from the minimum and maximum values of each descriptor (for a given virtual library). A single scale factor is then applied to all the descriptors such that the hypervolume of the space is set to the total number of molecules to be selected. In an ideal case, each molecule would now occupy its own region of descriptor space with a hypervolume of 1 and would be separated from its nearest neighbors with a distance of about 1. With this choice for rescaling the descriptors, we then simply set $\alpha = 1$. This is equivalent to leaving the descriptors unscaled and setting α as follows:

$$\alpha = \left[\frac{N_{sel}}{\prod_{i=1}^{n} R_i} \right]^{1/n} \tag{11}$$

where the R_i's represent the range spanned by each of the n descriptors in the virtual library.

C. Biasing Selection Toward Desirable Molecular Properties

1. Druglike Libraries

Whether libraries are designed for lead generation or optimization, it is important to consider the molecular properties of the molecules selected from the virtual library. Considerable emphasis has recently been placed on identifying

the characteristics of compounds that make them "druglike" or good development candidates. The intent is to preferentially select for screening molecules
that, if found to be active and selected for further development, are not likely
to exhibit ADME/tox problems. In this way the drug discovery and development cycle should be shortened and the failure rate of compounds should be
reduced.

2. Other Molecular Properties

Other constraints on whole-molecule properties may arise from the way in which
the library will be deconvoluted or decoded to identify the active molecules following high throughput screening. Again, this will be dependent on the methodology by which the library will be synthesized and screened and whether a tagging
strategy is being used. As an example, Brown and Martin (23) describe a library-
decoding strategy for mixtures in which the molecular weights of the active
molecule(s) are identified during the screening process. To identify the active
molecules, every molecule in the screening sample with a molecular weight equal
to that of one of the active samples must be resynthesized and rescreened individually. Constraining the design library to have only a few occurrences of molecules
with any one molecular weight minimizes the effort necessary for the deconvolution process.

3. Lead Optimization Using Activity Prediction Models

While this type of analysis is typically applied to calculated physical properties, it
can equally well be applied to any computed or experimentally measured property
of each member of the library. For example, a series of virtual high throughput
screening models, derived from techniques such as recursive partitioning or hypothesis modeling, could be used to predict activities for each molecule in a virtual library. A library design could then be focused toward molecules predicted to
be active while simultaneously maintaining the combinatorial constraint and satisfying druglike or cost requirements.

4. Rule-Based Biasing

Two methods have been developed that allow for the specification of desirable or
undesirable features in the selected molecules, thereby allowing the selection to
be directed toward the former. Restraints may be applied as property ranges (discussed in this section) in which a penalty is assigned to any selected molecule
whose properties lie outside the desired range. Alternatively, libraries may be designed to mimic one or more prespecified distributions of various properties (discussed in Sect. II.C.5).

In the case of selecting druglike molecules, the best known set of formal rules are those proposed by Lipinski et al. (2), who recommend that molecules be selected according to these rules:

1. Molecular weight should not exceed 500.
2. Log P should not exceed 5.
3. The number of hydrogen bond acceptors should not exceed 10.
4. The number of hydrogen bond donors should not exceed 5.

To incorporate these and other range-based rules, product penalties are imposed by specifying a range on any calculated or measured properties of the molecules. A penalty function (shown graphically in Fig. 4) is then applied to each property in term. No penalty is incurred by any molecule that falls within the range; outside the range the penalty increases with the square of the difference between the value and the nearest bound, up to a maximum at a user specified value. Formally

$$
\begin{cases}
\Delta p = 0 & \text{for } l < x < u \\
\Delta p = \text{Min}[\text{Cap}, (l - x)^2] & \text{for } x < l \\
\Delta p = \text{Min}[\text{Cap}, (x - u)^2] & \text{for } u < x
\end{cases}
\tag{12}
$$

where Δp is the assigned penalty, l and u are the lower and upper bound, respectively, x is the value of the property for the given molecule, and Cap is the maximum penalty value at the point at which the function is capped. When a number

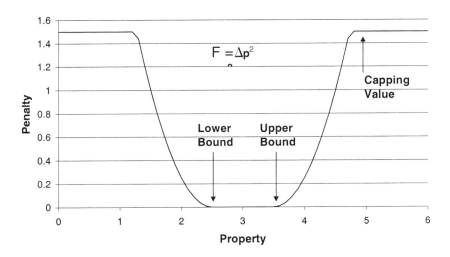

Figure 4 Rule-based penalty function of properties.

of range-based penalties are imposed, a separate weight may be set on each one to allow user control over their relative importance.

It is significant to note that Eq. (12) provides a "soft" limit on the application of rules. What this means in practical terms is that molecules that have somewhat unfavorable characteristics in some properties but not in others can still be selected. Consider the case of a diversity selection constrained by Lipinski's rules. A molecule with molecular weight 525 that is at a considerable distance from any other molecules in the diversity space may still be desirable to include in the selected set and by making a large contribution to the diversity score may do so despite the penalty score. Such an inclusion would not be possible with a hard cutoff (i.e., an elimination of all molecules >500). On the other hand, a molecule with molecular weight 525 possessing a high degree of similarity to another with molecular weight 325 is less desirable. The former compound can represented by its chemically similar neighbor, thereby avoiding the violation of the Lipinski rule for molecular weight without significantly compromising the diversity.

For a library subset, the total penalty is calculated by summing the contribution from each molecule. This value is normalized such that the total penalty equals one if each molecule violates each penalty range by one standard deviation. Formally

$$P = \frac{1}{N_{\text{prop}}} \frac{1}{N_{\text{mol}}} \sum_{i=1}^{N_{\text{prop}}} \frac{w_i}{\sigma_i^2} \sum_{i=1}^{N_{\text{mol}}} \Delta p_{ij}^2 \tag{13}$$

where N_{prop} is the number of properties being restrained, N_{mol} is the number of molecules being selected, w_i is the weight assigned for property i, σ_i is by default the standard deviation for property i evaluated over the full virtual library, and Δp_{ij} is the penalty (as defined above) for molecule j and property i. While σ can, of course, be folded into the weight w_i, its use in the formula is intended to allow the weights to be more easily determined and transferred between different virtual libraries. The capping point is also set in terms of number of standard deviations from the upper and lower bounds. The software also allows the user to specify a value for σ that may be based, for example, on another distribution (e.g., an inhouse compound collection or chemical database such as the World Drug Index: WDI).

5. Profile-Based Biasing

Other sets of rules can be inferred from an analysis (similar to Lipinski's) of a collection of molecules known to have desirable properties. These might be known drugs in the appropriate class, a list of which could be extracted from drug databases such as the WDI, or a set of known actives from screening. This type of analysis was previously suggested by Gillett et al. (3). In the Cerius (1) implementation, a set of properties is first identified as important, and these are then computed from the set of known drugs. Next a frequency histogram is computed by binning each property across those molecules. This set of histograms, one for each property, then encodes the rules that the design library should follow.

When a subset of the virtual library is to be scored during the optimization process, each histogram from the subset is compared to the equivalent histogram from the drug database. A penalty is calculated based on the difference between both the upper and lower bounds and the relative frequency for each histogram bin. An upper and lower bound is assigned on the relative frequency of each bin, and a contribution is made to the penalty any time the relative frequency of the bin in the subset library is outside this range; with the magnitude of the contribution increasing with the square of the difference. User-controlled weights can be set on the relative importance of each bin in each histogram. The penalty function is shown graphically in Fig. 5.

Again, the penalty score must be normalized so that it can be appropriately weighted against all other factors in the optimization. In this case, the score will be approximately 1 if each histogram bin of each profile violates the bin range by 10%. Formally,

$$P = \frac{1}{N_{\text{prop}}} \sum_{i=1}^{N_{\text{prop}}} \frac{1}{N_{\text{bins}}(i)} \sum_{j=1}^{N_{\text{bins}}(i)} w_{ij} \left(\frac{\Delta f_{ij}}{0.1} \right)^2 \tag{14}$$

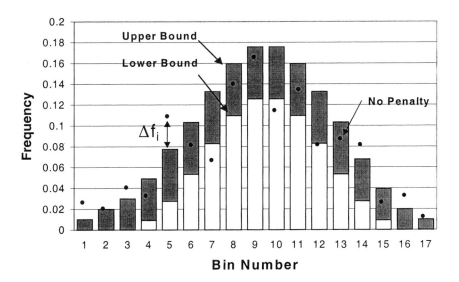

Figure 5 Penalty function for property profiles.

D. Biasing Selection Toward Desirable Reagent Properties

It is assumed in this discussion that prior to specification of the virtual library, reagent lists have been assembled to be compatible with the chemistry to be used. Furthermore, it is assumed that the lists have been filtered to remove reagents with reactive or other undesirable functionalities. In this way, every member of the reagent list is assumed, within the constraint of the current knowledge of the chemistry, to be appropriate on a chemical compatibility basis and available for use in the design library. These assumptions notwithstanding, economic and supply considerations suggest that all reagents may not be equally desirable to use.

Factors to consider in selecting reagents include the total cost of the reagents that will be used, a preference for reagents that are already available in in-house stockrooms or from preferred suppliers, and even a chemist's intuition about the desirability of using each reagent. Furthermore, it may be desirable to minimize the number of reagents used, such as in the design of sparse matrices (see Sect. II.B.1). Note that this is not the same as using the minimum possible number of reagents to make a given number of products, which is a concept captured in the combinatorial constraint also discussed in Section II.A.1. Finally, it may be desirable to minimize the number of different suppliers that are used, both to ensure timely delivery and to reduce the complexity of the ordering process.

At the reagent level, information about the source and cost of reagents is provided to allow selections to be biased toward more desirable reagents. For this, the software must be supplied with a unit cost for each reagent, the preferred supplier (including in-house if applicable), and optionally a user-defined relative penalty that might encode criteria such as ease of synthesis, toxicity, or chemist preference. Separately, a relative penalty can be assigned to the use of each vendor, allowing the user to encode preferences in the selection of vendor. Such codes might be used to indicate reliability, speed of delivery, or geographical location, for example.

In evaluating a subset library during the optimization, it is first necessary to compute the total quantity of each reagent that will be required to make the library, based on the amount of each product required. From this a number of factors can be scored.

> Total monetary cost of the reagents required
> Total number of reagents used
> Total of the user-defined reagent penalties
> Total number of suppliers used
> Total of the user-defined supplier penalties

Each or all of these can then form the basis for a total penalty based on the reagents used in that subset, and relative weights can be assigned to each factor.

Again a normalization factor is applied to the total score to allow it to be assigned a relative weighted against the based penalties

$$P_{\text{Reagents}} = N_{\text{Pen}} \left(w_1 P_{\text{cost}} + w_2 P_{\text{number\_reagents}} \right.$$
$$\left. + w_3 P_{\text{number\_suppliers}} + w_4 P_{\text{reagent\_penalties}} + w_5 P_{\text{supplier\_penalties}} \right)^{-1} \tag{15}$$

where w_1–w_5 are weight factors and N_{Pen} is the number of penalty terms used.

E. Library Comparison

The methods discussed so far in Section II have considered a single library. However, it is important to be able to consider the distribution in space of a number of libraries, or compound collections, and to make selections that complement the set of molecules already submitted to any screen of interest. Two tasks are of interest: the comparison of the overall diversity of two libraries and the selection of a subset of a library to augment an existing collection.

1. Defining a Chemical Space for Multiple Libraries

Before one can begin to analyze the overlap or diversity of a set of libraries, one must define a common chemical space in which they may be compared. For fingerprint metrics, it is sufficient to apply the same fingerprint algorithm to each set of compounds, allowing the distances between any pair of compounds to be computed irrespective of which library they are in. For continuous descriptors, one must consider how to define the chemical space when using a data reduction technique such as a principal component analysis. Two cases can be identified.

Case 1. One set of compounds defines a reference space and another is a set that is to be compared to that reference space. For example, a virtual library might be compared to a set of known drugs to understand the overlap (and thus understand how druglike the members of the virtual library might be). In this case the set of known drugs is clearly the reference space. The principal component analysis is therefore run on this set alone, and the resulting equations are then applied to the descriptors calculated from the virtual library to embed the virtual library in the space of the drugs.

Case 2. A common chemical space is to be calculated from all sets of compounds, and no one set defines a reference. For example, a number of virtual libraries can be synthesized, and the desire is to understand which ones occupy the same area of chemical space and which are fundamentally different. In this case the principal component analysis is run using the descriptors from all sets of molecules, producing a common chemical space defined by the variation of the descriptors across all the libraries.

2. Comparing Whole Libraries

Having defined a common chemical space, one may first wish to compare the diversity of a pair of libraries. One method, the diversity integral method (24), uses the following procedure to consider the amount of space spanned by each library.

1. Generate a set of random points that sample the property space covered by the two libraries and calculate the distance between each random point and the closest molecule in library 1 and library 2 (i.e., the minimum distance between the random point and each library).

2. Calculate the sum over all the minimum distances between random points and library 1 and random points and library 2, normalizing by the number of random points.

3. The library with the lowest value of this sum is considered to be more diverse; that is, it better samples the property space occupied by the two libraries. This is based on the finding that on average, one can start from any point in the property space and find a closer point in the library with the lower sum than in the other.

A second method, the similarity method, considers the distances between members of one library and another. For each member of one library, the nearest neighbor in the other library is identified and the distance recorded. A frequency distribution of these distances indicates the overlap of the libraries. A distribution skewed toward small distances indicates a high degree of overlap, since most molecules in one library have a relatively close neighbor in the other. Distributions in which the frequencies are skewed toward high distances indicate that the libraries are reasonably well separated. The mean of these minimum distances gives a single measure of the separation of the libraries.

3. Augmenting Existing Libraries

The methods described in Section II.E.2 allow the comparison of whole libraries or compound collections. Another task is to select a subset of a library to augment the space that has already been covered by existing compounds. One method to do this is to use the similarity selection methodology just described (Sect. II.E.2). Having computed the minimum distances between members of a candidate and target library, one may augment the target with all compounds in the candidate library that are greater than a threshold distance from any member of the target library, to explore only new space. For use in library optimization, the selection would be inverted to select all molecules in the candidate library within a distance threshold of members of the target library; the latter being formed from the set of currently identified hits.

An alternative to this is to use the same distance metrics and optimization process discussed in Sections II.A.3 and II.B.4. The difference in this case is that after selecting a subset of a candidate library during the Monte Carlo optimization, the diversity metric is calculated not only from the molecules in the selected subset but from a combination of the selected subset plus all the members of the reference library. In this way, the diversity metric optimizes both the distances between the members of the selected subset and the distances between the reference library and the subset. Since the reference library is fixed, the mutation step of the Monte Carlo is only allowed to switch one member of the candidate library for another. In this way, the selection is of the most diverse subset of the candidate library to best augment the reference library.

III. EXAMPLE APPLICATIONS

In this section we give two examples to demonstrate the application of some of the methods we have described. The first, in Section III.A, demonstrates the use of combined combinatorially constrained diversity, druglikeness and cost-effectiveness constraints. The second (Sect. III.B) compares the various distance and cell metrics for array optimization.

A. Diversity Analysis of an Ugi Library for Cost-Effectiveness and Druglikeness

A virtual library was prepared based on an Ugi reaction (25) using 10 acids, 10 aldehydes, 10 amines (one of which was bifunctional), and 10 isonitriles, giving a virtual library of 11,000 possible products. The design library was specified to be $4 \times 4 \times 4 \times 4$ (i.e., a combinatorially constrained selection) that would give either 256 products or, if the bifunctional amine was among the selected set, 320. Each reagent was extracted from the Available Chemicals Directory (ACD), with a preferred supplier and unit cost.

The descriptor set chosen was the default set in Cerius2, a set of 50 physicochemical properties including molecular weight, A log P (26), number of hydrogen bond donors and acceptors, number of rotatable bonds, surface area and volume, and the Balaban (27), PHI, Kappa, and CHI (28), Wiener (29), and Zagreb (30) topological indices. The descriptors used are chosen to be reasonably fast to calculate whilst still tending to group biologically similar molecules together.

A principal component analysis required five principal components to explain 93.8% of the total variance. Using the cell-based fraction measure (with mean/variance normalized PCs), the space was divided into a total of 576 cells, such that members of the virtual library occupied at least 256 cells. The Monte Carlo optimization was set to run for 1,000,000 steps at 100 K, terminating after 100,000 idle steps.

Druglikeness criteria for the selection were defined from an analysis similar to Lipinski's (2) of the World Drug Index database (31) as follows. Of the

54,944 unique molecules in the March 1998 version of the database, 7572 molecules were selected that had a USAN (U. S. adopted name) field, and 6489 were selected that had an INN (international nonproprietary name) field. The union of these two groups gave 8504 molecules, which are presumed to be drug-like. A series of descriptors was then calculated for this set, including structural descriptors (molecular weight, rotatable bonds, number of hydrogen bond donors and acceptors, AlogP and molar refractivity) and topological indices (3 kappa indices, a phi index, 5 chi connectivity indices, Balaban Jx, Wiener, Zagreb). An upper bound constraint for each property was set at the value exceeded by 10% of the molecules in the data set. These ranges were then used as penalty ranges for the design of the Ugi library subsets. Table 1 shows the range of these properties in the Ugi virtual library and the bound established by the above procedure.

The design aimed to select a combinatorially constrained $4 \times 4 \times 4 \times 4$ library to be both diverse (measured by the cell fraction), druglike (measured by having molecular properties within the 90% upper bound found in the WDI) and cost-effective (established by the total cost of all the reagents required to make the library). In addition to a run to establish the best design library to satisfy these criteria, other runs were done using either diversity alone or the total penalty alone as objective functions to establish the optimal values for each criterion.

Table 1 Property Ranges from WDI Subset

Descriptor	Range in UGI library	90% cutoff in WDI_best
MW	399–669	550
Rotlbond	10–31	13
Hbond acceptor	2–13	9
Hbond donor	1–5	5
AlogP	−4.2 to 8.2	5
MolRef	102–191	120
Balaban Jx	1.3–6.6	2.8
PHI	7.3–23.2	8
Kappa-1A	22.1–40.4	25
Kappa-2A	10.1–26.3	12
Kappa-3A	5.3–19.5	8
CHI-V-O	16.3–29.3	20
CHI-V-1	9.31–17.7	12
CHI-V-2	6.13–15.4	10
CHI-V-3P	3.88–11.1	8
CHI-V-3C	0.42–3.81	2.2
Wiener	1737–9940	4000
Zagreb	118–238	175

Table 2 Results from Design of Ugi Library

Optimization	Diversity	Penalty	Cost ($/mmol)	Cost ratio
Only diversity	0.597	1.248	94749	80
Diversity - penalty	0.438	0.062	5630	5
Only penalty	0.234	0.02	1184	1

Table 2 shows the results of these runs. Examination of the diversity scores in the table shows a range of 0.6 for the subset designed only for diversity to 0.23 for the subset designed only to minimize the penalties without regard for diversity. Even in the case in which the set is optimized for diversity, 40% of theoretically possible diversity is lost, that is, not all the possible cells of the diversity space are occupied by the subset. This is due the imposition of the combinatorial constraint, which prevents the algorithm from simply picking one compound per cell. Without this constraint (and without the penalty constraints) a diversity score of 1.0 is achieved.

The imposition of additional restraints on the products and reagents results in the loss of an additional 16% of diversity, beyond the diversity-only result. At the same time the property profile with respect to druglikeness has been greatly improved, the penalty score lowering from 1.2 to 0.06. Most remarkably the total cost of the library has been reduced by over 15-fold (and, with the loss of an additional 20% of diversity, it can be reduced a further 4-fold).

B. Comparison of Spanning-Tree and Cell-Based Metrics for Array Optimization

A library was built with a benzodiazepine core together with a set of 20 R-group substituents chosen from the available default set in the Cerius2 Analog Builder (5). The library is illustrated in Fig. 6. It has three R-group positions in the benzodiazepine core with (the same) 20 substituents available at each of the three R groups. As such, the full virtual library constitutes a $20 \times 20 \times 20$ array totaling to 8000 molecules. Of the 50 descriptors described in Section III.A, 43 were computed, and these were reduced to 3 descriptors by means of principal component analysis.

First a $5 \times 5 \times 5$ library was selected by means of the spanning-tree metric, as shown in Fig. 7. The black spheres show the selected molecules embedded in the space of the full virtual library. The most obvious omission from this library is the molecule at the top of the space, which corresponds to the choice of R1 = R2 = R3 = t-butyl. It can be seen that the selected molecules do a reasonable job of covering the space with the exception of the region near the top occupied by the single all-t-butyl compound. In Fig. 8, we show the equivalent selection using the

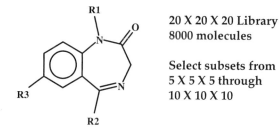

20 X 20 X 20 Library
8000 molecules

Select subsets from
5 X 5 X 5 through
10 X 10 X 10

Benzodiazepine Library (no redundant molecules)
R1,R2,R3 = (acetyl,amino,benzyl,bromo,carboxy,chloro,
        cyano,ethyl,fluoro,formyl,hydrogen,hydroxy,
        iodo,methoxy,methyl,phenoxy,
        phenyl,t-butyl,thiol,vinyl)

Figure 6   Structure of the benzodiazepine library.

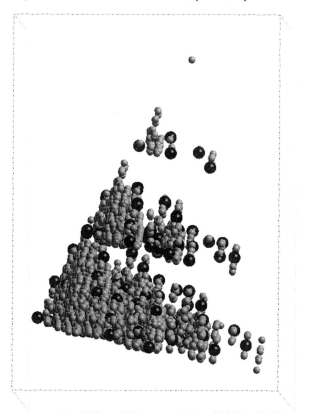

Figure 7   MaxMin spanning tree selection of a 5 × 5 × 5 library from the 20 × 20 × 20 benzodiazipines. Selected molecules are in red in the original (black here), unselected in white (gray here). Note that the uppermost molecule R1 = R2 = R3 = *t*-butyl is not selected.

330

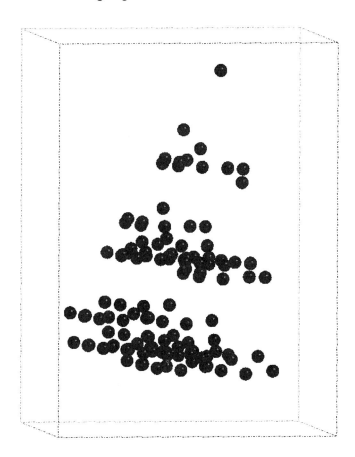

Figure 8   Cell-based selection of a 5 × 5 × 5 library from the 20 × 20 × 20 benzodi-azipines. Only selected molecules are shown. Note that R1 = R2 = R3 = *t*-butyl is selected.

cell counts function to optimize the diversity of the system. In this case, the all-*t*-butyl molecule is selected (note that *t*-butyl is present in the selection set for each of the R groups), so the question of why this molecule was not chosen with the spanning-tree method remains unanswered. It turns out that the cell counts selection gives a spanning-tree score of 52.6 while the spanning-tree score for the library optimized with the spanning tree function is 55.9, so the spanning-tree-optimized library is, in fact, more diverse based on the spanning-tree function. Close inspection of the two libraries reveals that the spanning-tree-based library actually has better coverage of the space in the lower regions (on the plot) of the diversity space, while the cell counts library has noticeable gaps in this region. Note the layering or clustering effect present in the cell counts library in Fig. 8 with the gaps between the clusters. This improved coverage of the lower part of the space turns out to be significant enough to offset the loss of diversity due to the omission of the all-t-butyl molecule for the spanning tree.

We earlier commented that in analyzing the cell-based metrics, the cell chi-squared seems to be slightly inferior to both cell counts and cell entropy in terms of producing optimally diverse libraries. To quantify this observation, libraries were selected by means of each method, and each of the libraries obtained using a cell-based method was evaluated with regard to its spanning-tree score normalized (i.e., divided) by the spanning-tree score obtained for the corresponding library optimized using the spanning-tree function itself (i.e., the optimal spanning-tree score achievable for that library). Thus, if a cell method produced a library as diverse (assessed with the spanning-tree function) as the spanning-tree library, it would achieve a (normalized) score of 1.0.

This experiment was repeated over a number of libraries sizes, using various descriptor sets (14). It is found that the average score for the cell counts (or cell fraction) method was 0.926, or equivalently, on average it achieved 92.6% of the diversity of the spanning-tree libraries. The cell entropy libraries achieved an average score of 0.920, and the cell chi-squared libraries achieved an average score of 0.910 showing very slightly improved results for the cell counts and cell entropy methods relative to cell chi-squared. However, the results are sufficiently close that it would be best to consider all the cell-based methods as yielding quite good results relative to the spanning-tree libraries. Since the spanning-tree libraries take significantly longer to calculate, these results tend to further validate the use of the cell methods for larger libraries where the spanning-tree method becomes impractical owing to its prohibitive cost in computer time.

Experiments were performed to examine the assertion that independent diversity selection at each R-group position (the "diversity hypothesis" discussed in Sect. II.B.1) is not a good protocol for selecting a diverse array-based library. For the benzodiazepine library, we have performed R-group-based diversity selection using the MaxMin method. Each R group was enumerated at a single position of the benzodiazepine core, and the MaxMin method was used to select substituents at that R-group position. The resulting R-group selections were then combined to give a combinatorial library. The diversity for the resulting libraries was then compared to the diversity for the corresponding library obtained using product-based selection with either the spanning-tree method or the cell counts method. Over a range of libraries with various descriptors and sampling methods, the diversity for the R-group-based selections ranged from 46% to 85% of the optimal diversity obtained for the product-based libraries (14). These results tend to reconfirm that product-based diversity selection can yield significantly improved results compared to R-group based selection, as has also been noted by others (9,13).

IV. SUMMARY AND CONCLUSIONS

In this chapter we have described a process for library design that incorporates consideration of the efficiency and economics of:

Chemical synthesis, by considering the economics and expediency of obtaining reagents and efficiency of the synthesis process

Biological screening, by considering the coverage of chemical space to obtain the maximum amount of information about biological activity with the minimum number of compounds

Medicinal chemistry, by building in constraints on the compounds such that they will be appealing hits for medicinal chemistry development and will have a minimized chance of causing problems (such as ADME) in development

We have described a number of options for the method by which each factor may be measured and have implemented a method by which all may be optimized simultaneously. We have described a number of diversity and similarity algorithms for both cherry-picking and array design and discussed the implementation of new algorithms for the latter that address the shortcomings of earlier approaches. We have described methods with some approximations that are fast enough to handle very large library designs typical of lead generation and others that are much more exact that can be applied to smaller lead optimization problems.

The use of this software allows the chemists and modelers involved in any drug discovery project to arrive at the library design that will satisfy the concerns of both, and will allow rapid progress to be made in such a project.

REFERENCES

1. Lipinkski, C.A., Lombardo, F., Dominy, B.W., and Feeney, P.J. Experimental and computational approaches to estimate solubility and permeability in drug discovery and development settings. Adv. Drug Delivery Rev. 1997; 23:3–25.
2. Brown, R., Hassan, M. Waldman, M. Combinatorial library design for diversity, cost efficiency and drug-like character. J. Mol. Graphics Modelling 2000, in press.
3. Gillett, V.J., Willett, P., and Bradshaw, J. Identification of biological activity profiles using substructural analysis and genetic algorithms. J. Chem. Inf. Comput. Sci. 1998; 38:165–179.
4. Martin, E., and Critchlow, R. Beyond mere diversity: Tailoring combinatorial libraries for drug discovery. J. Comb. Chem. 1999; 1:32–45.
5. MSI. All software discussed in this chapter is available in the following modules of Cerius2, version 4.5, from Molecular Simulations Inc, San Diego, CA: C2.Analog, C2.QSAR, C2.Descriptor+, C2.Diversity, C2.LibSelect, C2.LibCompare, C2.Lib Profile. 2000.

6. Smith, G. Noncombinatorial chemistry: The high-speed synthesis of drug-like molecules. In CHI Third Annual Novel Bioactive Compounds. Brussels, Belgium, 1999.

7. Clark, D.E., and Westhead, D.R. A review of evolutionary algorithms in computer-aided molecular design. J. Comput. Aided Mol. Design 1996; 10:337–358.

8. Brown, R.D., and Clark, D.E. Genetic diversity: Applications of evolutionary algorithms to combinatorial library design. Expert Opin. Ther. Patents 1998; 8:1447–1460.

9. Gillett, V.J., Willett, P., and Bradshaw, J. The effectiveness of reactant pools for generating structurally-diverse combinatorial libraries. J. Chem. Inf. Comput. Sci. 1997; 37:731–740.

10. Johnson, M., and Maggiora, G., Concepts and Applications of Molecular Diversity. Wiley, New York, 1990.

11. Patterson, D., Cramer, R., and Ferguson, A. Neighborhood behavior: A useful concept for validation of molecular descriptors. J. Med. Chem. 1996; 38:379–386.

12. Willet, P. Perspect. Drug Discovery Design 1997; 7/8 (whole issue).

13. Jamois, E.A., Hassan, M., and Waldman, M. Evaluation of reagent-based and product-based strategies in the design of combinatorial library subsets. J. Chem. Inf. Comput. Sci. 2000; 40:63–70.

14. Waldman, M., Li, H., and Hassan, M. Novel algorithms for the optimization of molecular diversity of combinatorial libraries. J. Mol. Graphics Modelling 2000, in press.

15. Roman, S. Coding and Information Theory. Springer-Verlag, New York, 1992.

16. Breiman, L., Friedman, J.H., Olshen, R.A., and Stone, C.J., Classification and Regression Trees. Chapman & Hall, New York, 1984.

17. Balley, M., and Willett, P. Binning schemes for partition-based compound selection. J. Mol. Graphics Modelling 1999; 17:10–18.

18. Martin, E.J., Blaney, J.M., Siani, M.A., Spellmeyer, D.C., Wong, A.K., and Moos, W.H. Measuring diversity: Experimental design of combinatorial libraries for drug discovery. J. Med. Chem. 1995; 38:1431–1436.

19. Hassan, M., Bielawski, J.P., Hempel, J.C., and Waldman, M. Optimization and visualization of molecular diversity of combinatorial libraries. Mol. Diversity 1996; 2:64–74.

20. Abramowitz, M., and Stegun, I.A., eds. Handbook of Mathematical Functions, Chapter 7. Dover, New York, 1972.

21. Baase, S. Computer Algorithms. Introduction to Design and Analysis. Addison-Wesley, Reading, MA, 1988.

22. Mount, J., Ruppert, J., Welch, W., and Jain, A. IcePick: A flexible surface based system for molecular diversity. J. Med. Chem. 1999; 42:60–66.

23. Brown, R.D., and Martin, Y.C. Designing combinatorial library mixtures using a genetic algorithm. J. Med. Chem. 1997; 40:2304–2313.

24. Rogers, D. Unpublished work.

25. Ugi, I., Lohberger, S., and Karl, R. The Passerini and Ugi reactions In Comprehensive Organic Synthesis: Selectivity for Synthetic Efficiency, Vol. 2, C.H.H. Herausg and B. M. Trost, eds. Pergamon, Oxford, 1991, pp. 1083–1109.

26. Ghose, A., Viswanadhan, V.N., and Wendoloski, J. Prediction of hydrophobic (lipophilic) properties of small organic molecules using fragmental methods: An analysis of ALOGP and CLOGP methods. J. Phy. Chem. 1998; 102:3762–3772.

27. Balaban, A.T. Highly discriminating distance-based topological index. Chem. Phys. Lett. 1982; 89:399–404.

28. Hall, L., and Kier, L., The molecular connectivity chi indexes and kappa shape indexes in structure–property modeling In Reviews in Computational Chemistry, Vol. 2 K.B. Lipkowitz, and D.B. Boyd, eds. VCH Publishers, New York, 1991, pp. 367–422.

29. Muller, W., Szymanski, K., Knop, J.V., and Trinajstić, N. An algorithm for construction of the molecular distance matrix. J. Comput. Chem. 1987; 8:170–173.

30. Bonchev, D. Information Theoretic Indices for Characterization of Chemical Structures. Research Studies Press, Letchworth, England, 1983.

31. World Drug Index, version 1998, Derwent Information, London.

11

Relative and Absolute Diversity Analysis of Combinatorial Libraries

Robert D. Clark

Tripos, Inc.
St. Louis, Missouri

I. SCOPE

It would be presumptuous indeed to pretend that any single chapter in one book could do full justice to the topic of molecular similarity and its distaff cousin, diversity. Several excellent books have already been devoted in whole or in part to the subject (e.g., Refs. 1–4) and several excellent general reviews are available (e.g., Refs. 5,6). The discussion here will lay out the broad range of applicable descriptors and the important (dis)similarity measures used with them in some detail, but will not delve deeply into the advantages and disadvantages of each one. Nor will the closely related question of subset selection be gone into; the interested reader is referred to Chapter 13 in this volume.

Once the general foundation has been laid, specific examples of relative and absolute diversity analysis will be presented for a specific set of descriptors and similarity measures. The concepts involved are quite general and should find application well beyond the particular application described here.

II. DESCRIPTORS

A large number of different descriptors and descriptor sets have been used or proposed for evaluating molecular diversity in a context potentially applicable to library design and comparison. These include:

Substructural keys and hashed fingerprints (7)
Atom pairs (8,9) and topological torsions (10)

Connectivity indices (11–14) and BCUT descriptors (15,16)
Estimated physical properties (17)
Pharmacophore triplets and tetrads (18–22)

Other descriptors that have been used for assessing pairwise similarities in structure are at present too slow in their calculation to be of general use in most design methods, especially when one is working with fully enumerated combinatorial libraries. Examples include molecular holograms, eigenvalues (EVA) and fields (23,24); affinity profiles derived from DOCK scores (25); and autocorrelation vectors such as WHIM descriptors (26).

Substructural keys are binary vectors for which the ith element is set to 1 if the corresponding substructure is contained within the structure in question; otherwise that element is set to 0. Such vectors can be considered to be bit sets for analytical purposes, which allows for very compact storage and easy computational manipulation. MDL (27) supports a particular, copyrighted set of such keys directly, whereas UNITY (28) software allows users to define their own. These descriptors perform well at differentiating among known classes of pharmacological agents (7).

Fingerprints are also binary vectors but include information on *all* fragments within a set of specified classes (e.g., all fragments made up of three non-hydrogen atoms). Each such fragment sets one (more or less arbitrarily assigned) bit in the fingerprint. This is a many-to-one mapping ("hashing"), so different fragments can set the same bit.

The loss of information due to hashing can be minimized by hashing only within classes for fragments made up of two to six atoms. In UNITY fingerprints, for example, all fragments composed of two heavy atoms map to the first 85 bits of the 998 in the default fingerprint definition, whereas fragments made up of three heavy atoms map to bits 86–184. Hydrogens are included only for four-atom fragments, which map into the block from bit 185 to bit 333. Heavy atom fragments of four to six atoms map to bits 334–928.

The fingerprints also encode specific atom counts for O, N, S, P, and individual halogens. Separate generic counts are included for heteroatoms and halogens, and for distinct phenyl, five-membered, and six-membered rings. Bit 959 is set for structures that contain silicon, and bit 960 is set for those that contain an element not covered elsewhere in the definition.

UNITY and SYBYL (28) fingerprint definitions can be modified by editing the corresponding standard.2DRULES file.

Atom pair vectors are binary vectors in which each bit corresponds to a particular pair of atom types separated by a particular number of bonds. Atom types for each element may be characterized in terms of the number of bonds they form with heavy atoms, the number of π-electron pairs they bear, and the number of bonded hydrogen atoms. Since these three numbers sum to the valence of the atom, only two need be specified to fully characterize each atom type; the number of bonded heavy atoms is usually used as one index. The separation is always tallied in terms of topological distance, that is, the number of intervening bonds (or atoms) in the shortest connecting path.

In the SYBYL implementation, a more complete atom typing system is used, so that each atom in a molecule is classified as belonging to one of 15 distinct types:

> C as sp^3, sp^2, aryl ($\langle$C.3$\rangle$, $\langle$C.2$\rangle$, or $\langle$C.ar$\rangle$, respectively) or *other*
> N as sp^3, sp^2, aryl ($\langle$N.3$\rangle$, $\langle$N.2$\rangle$, or $\langle$N.ar$\rangle$) or *other* ($\langle$N.am$\rangle$ etc.)
> O as sp^3, sp^2 ($\langle$O.3$\rangle$ or $\langle$O.2$\rangle$) or *other*
> S ($\langle$S$\rangle$)
> P ($\langle$P$\rangle$)
> Halogen ($\langle$F$\rangle$, $\langle$Cl$\rangle$, $\langle$Br$\rangle$, or $\langle$I$\rangle$)
> Other heavy atoms

Each pairwise combination is allocated 10 bits in the atom pair vector to indicate separations of from 1 to 10 bonds, giving a vector 1200 bits in length. These are default values; the maximum degree of separation can be changed by editing SYBYL's standard.aprules file, as can the atom types considered.

Topological torsions are similar to atom pairs except that they encode the presence of pairs of features separated by a particular topological distance; they differ in that for torsions, the features are bonds rather than atoms. In the simplest implementation, bond types are differentiated by both orbital type (e.g., C—C vs C=C vs C≡C vs C:C [aryl]) and constituent elements (C≡C vs C≡N), with bonds to hydrogen not taken into account explicitly. Atoms at branch points and in rings set more bits than do atoms in open-chain substructures, particularly if separations of 0 bonds (i.e., atom triples) are counted. In addition, heterologous bivalent bond types (e.g., C=N) generally set more bits than do homologous bond types (e.g., C—C) because they bear distinct attachment points.

Connectivity indices are obtained by treating chemical structures as graphs. The Wiener index W was devised as a way to quantitatively account for the effect of branching on the boiling points of alkanes. It is simply the sum across all heavy-atom pairs of the number of bonds in the minimum length path connecting those pairs (14). Kier split out the contributions from each contributing path length and introduced the minimum and maximum possible values as a way to normalize the series of indices $^l\kappa$ obtained:

$$^l\kappa = \frac{2P_l^{min}\,P_l^{max}}{P_l} \tag{1}$$

where P_l is the number of paths of length l connecting distinct atom pairs within the molecule of interest (11); the minimum and maximum possible values are straightforward functions of the number of heavy atoms (vertices) in the structure. The W and $^l\kappa$ indices are purely graph-theoretical shape descriptors, however, which do not differentiate between carbon and other heavy atoms, or among single, double, and triple bonds. Kier therefore proposed a subsidiary set of indices, $^l\kappa_\alpha$, which incorporate a correction term α based on deviations in bonded radii from that of aliphatic carbon (12).

Molecular connectivities χ were developed for similar purposes, with the generalized definition given by:

$$^l\chi = \sum \frac{1}{\sqrt{\Pi \delta_i}} \tag{2}$$

where the summation is taken over all paths of length l in a molecule, the product is taken over all atoms i in each path, and δ_i is the degree of the vertex corresponding to atom i - i.e., the total number of bonds to heavy atoms in which that atom participates. The index originally proposed by Randic took no account of differences in atom type (14). This was subsequently addressed by Kier and Hall (13), who proposed treating δ as an adjusted chemical valence δ^v rather than as a graph-theoretical quantity:

$$\delta_i^v = Z_i - n_i^H \tag{3}$$

where Z_i is the number of valence electrons and n_i^H is the number of hydrogens borne by the atom in question. Replacing the δ_i term in Eq. (2) with δ_i^v yields valence connectivities $^l\chi^v$, which are also widely used. Commercial packages such as MolConn-Z (29) support the rapid calculation of these and hundreds of other connectivity indices for any given molecule.

Such connectivity indices can all be obtained from a few simple matrices widely used to characterize graphs in general. For a graph $\mathbf{G}$ of n vertices connected by m edges, these are as follows:

1.   The *connectivity matrix* $\mathbf{C}$ is an $n \times n$ matrix in which the ijth element is 0 if and only if no edge connects vertices i and j. In the simplest ("uncolored") case, the $2m$ elements for which an edge does connect vertices i and j are set to 1.

Powers of the connectivity matrix $\mathbf{C}^l$ are $n \times n$ matrices obtained by multiplying $\mathbf{C}$ by itself l times. The nonzero off-diagonal elements in $\mathbf{C}^l$ indicate the number of paths of length l that connect vertices i and j, whereas each diagonal element ii indicates the number of cycles of length l that include vertex i. Note that, since each edge is itself a cycle of length 2, the diagonal elements in $\mathbf{C}^2$ are the degrees of the corresponding vertices.

2.   The *distance matrix* $\mathbf{D}$ is an $n \times n$ matrix in which the value of the ijth element is the length of the shortest connected path ("topological distance") between i and j.

3.   The *adjacency matrix* $\mathbf{A}$ is edge-oriented. It is an $m \times m$ matrix in which the ijth element is 0 if and only if edges i and j do not share a common vertex.

Connectivity indices are then formally obtained by pre- and postmultiplying these characteristic matrices by weighting vectors **w**, where the ith element in **w** is the relevant property of vertex (atom) or edge (bond) i. The second-order connectivity index $^2\chi$, for example, can be obtained from the "square" of **C** by setting the ith element of **w** to $1/\sqrt{\delta_i}$:

$$^2\chi = \mathbf{wCCw}^T \tag{4}$$

Similarly, Randic cites Lovasz and Pelikan's suggestion of using the leading eigenvalue derived from the adjacency matrix **A** as a descriptor (14).

BCUT descriptors correspond to the most positive and most negative eigenvalues derived from modified connectivity matrices. Here, a weighting vector of atomic charges, polarizabilities, or the ability to accept or donate hydrogen bonds is incorporated directly into **C** by setting each diagonal element ii equal to the ith element in the corresponding **w**. Off-diagonal elements reflect the nature (single, double, etc.) of the corresponding bond. Further "flavors" of descriptor are obtained by varying how the off-diagonal elements are scaled with respect to the property weights.

The DiverseSolutions (30) software package includes tools for choosing two to three pairs of descriptors to be able to maximize the dispersion within each descriptor and to minimize the correlation between descriptors across any particular data set.

The **physical properties** used in diversity analysis have by and large been those previously found useful in studies of quantitative structure–activity and structure–property relationship (QSAR and QSPR), including octanol/water partition coefficient (log P), molar refractivity (MR), molecular volume, dipole moment, and various partial molecular surface areas. In principle, these can be obtained by physical measurement. In practice, they are nearly always estimated by using schemes based on fragment additivity or connectivity, and are supplemented with connectivity indices of one kind or another.

Connectivity indices have been projected into three dimensions by replacing topological distances with geometrical distances, as have atom pairs (31). BCUTs can be similarly extended by replacing bond orders on the off-diagonals with pairwise Euclidean (geometrical) distances. Both extensions require the specification of reference conformations; typically those obtained from CONCORD (32) work quite well for this purpose.

Just as substructural fingerprints are binary vectors with roots in two-dimensional database searching, **pharmacophore triplets** and **tetrads** have their roots in three-dimensional database searching. When such searches are run, substructures are assigned to feature classes based on their potential modes of interaction with protein binding sites via hydrophobicity and hydrogen bonding. Individual features are not discriminating enough to be useful; neither are pairs of features. Picking up triplets and tetrads clearly has more potential in this regard (see Chapter 14, this volume).

The triangular and tetrahedral structures formed by pharmacophore triplets and tetrahedra cannot readily be characterized in terms of topological distances, however. Instead, the geometric distances between features are used, with the lengths of each edge binned, typically with a resolution comparable to that used in three-dimensional searches (i.e., about 0.5 Å).

The triplet-generating software package available from ChemX (33) utilizes four feature types by default (hydrophobic, H-bond donor atom, H-bond acceptor atom, and positive nitrogen) and divides the edge lengths up into 30 bins (19). The UNITY and Selector (28) software recognizes three atomic feature types by default (hydrophobic centers, donor atoms and acceptor atoms), as well as two types of extension point (acceptor sites and donor sites), which indicate the position a complementary donor or acceptor atom on the protein is likely to occupy. With 27 half-angstrom bins for each edge, this produces descriptors comprised of about 300,000 bits, one for each nondegenerate, geometrically allowed triangle.

Four edges plus a chirality flag are required to define a tetrad. Taken together with the combinatorial increase in the kinds of tetrahedra that can be defined, this leads to prohibitively cumbersome vectors. As a result, fewer, larger bins are generally used, along with a somewhat coarser classification scheme for feature types.

The primary motivation behind using pharmacophore triplets and tetrads is the desire to efficiently capture the many different ways a ligand can interact with a target protein, and the method used to sample the available conformational space is critical to accomplishing this. The triplet generation approach taken by ChemX is to "exhaustively" sample each molecular torsion at fixed intervals, typically in 30 or 60° increments, though researchers at Rhône-Poulenc and elsewhere have created more complex schemes (22). The UNITY implementation, in contrast, performs a random sampling of conformer space using a directed tweak algorithm (34).

III. DESCRIPTOR TRANSFORMATIONS

The large ensembles of descriptors obtained by combining connectivity indices of many different types tend to be redundant owing to correlation among the indices: χ and κ, for example, are highly correlated both within and between series. Such redundant data are inefficient to store, to manipulate, and to visualize. To minimize such problems, several groups have applied **principal components analysis** (PCA) to extract composite descriptors of greatly reduced dimensionality from such ensembles (35,36). Each of the elements in these short vectors (typically six dimensions or fewer) is a linear combination of the descriptors in the original ensemble, selected so as to be mutually orthogonal.

Others have used **nonlinear mapping** (NLM) as a less population-dependent way to reduce the effective dimensionality of descriptors (5). In this approach, a set of artificial descriptors is created for each molecule, and the ensemble is perturbed to minimize the deviation from the distances found in the original, high-dimensional space.

Binning is often used to convert continuous, real-valued descriptors into integer indices representing discrete intervals along the respective coordinate axis. There is an implicit (and sometimes overlooked) assumption made in applying this transform that proximity in terms of each descriptor implies a comparable degree of similarity between the original structures, regardless of differences in the values of other descriptors. In other words, binning assumes that the descriptors involved are mutually independent in a statistical sense.

In many cases, intervals of constant size are used. Doing so entails an implicit assumption that differences in value have essentially the same meaning irrespective of where the values being compared fall along a particular axis. Variable intervals can be used instead, but their boundaries are generally set by considerations of ease of analysis rather than by applying knowledge specific to the underlying descriptor. Choosing interval boundaries such that each bin is equally populated, for example, is a coarse form of ranking.

IV. RELATIONSHIPS AMONG DESCRIPTORS

It is tempting to simplify the discussion of relationships among the various descriptors cited above by categorizing them as "2D" or "3D" depending on whether their calculation requires specification of three-dimensional structure. This approach can be quite misleading, however. Because the analyses they are being used for are needed to make decisions about which compounds to synthesize, it must be possible to accurately estimate any descriptor used from structural formulas. Hence, the underlying information content is the same for all of them. Rather, the descriptors differ in their level of generalization and in their level of structural specificity. This is best illustrated by considering some examples, such as the three compounds shown in Fig. 1: 2-ethoxyethylmethylamine (**1**), *N*-methyl morpholine (**2**), and *p*-toluamide (**3**).

Substructural fingerprints are completely specific. Molecule **1** contains 3 distinct two-atom fragments (CC, CN, and CO), 4 three-atom fragments (CNC, COC, NCC, and CCO), 19 four-atom fragments (including hydrogen atoms), 3 five-atom fragments, and 2 six-atom fragments. Hence the total number of bits set to 1 in the corresponding UNITY fingerprint (its cardinality) is 35. There are more distinct fragments in **2** because the cyclization introduces branch points, so its fingerprint has a higher cardinality. The mix of bond types in **3** introduces yet more complexity, which is reflected in the fingerprint (Fig. 1) (29,37).

Descriptor	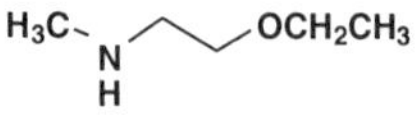 1	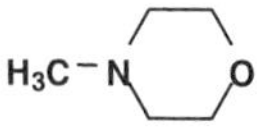 2	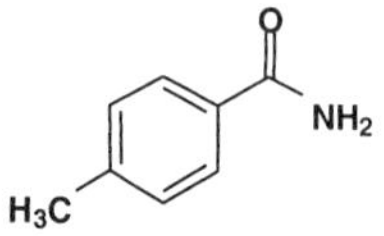 3
$H_3C-NH-CH_2CH_2-OCH_2CH_3$			
UNITY fingerprint[a]	35	42	56
Atom pairs[a]	14	9	25
Topological torsions[a]	12	18	29
ClogP[b]	0.16	0.18	1.15
CMR[b]	3.02	2.84	4.02
Molecular Volume (Å^3)	116	108	126
Connectivity indices[c]			
$^1\kappa$; $^2\kappa$; $^3\kappa$	7.00; 6.00; 6.00	5.14; 2.34; 1.50	8.10; 3.41; 2.29
$^1\kappa_\alpha$; $^2\kappa_\alpha$; $^3\kappa_\alpha$	6.92; 5.92; 5.92	5.06; 2.28; 1.45	6.96; 2.62; 1.66
$^1\chi$; $^3\chi$; $^5\chi$	3.41; 1.21; 0.35	3.39; 1.89; 0.90	4.70; 3.00; 1.45
$^1\chi^v$; $^3\chi^v$; $^5\chi^v$	2.64; 0.70; 0.14	2.66; 1.23; 0.40	3.06; 1.34; 0.48
Pharmacophore triplets[a]			
10 conformers	385	219	101
100 conformers	958	430	129

[a] Cardinality of the specified binary vector. [b] Calculated using BioByte software [37]
[c] Calculated using MolconnZ software [29]

Figure 1   Descriptor values taken on by selected compounds.

Atom pair descriptors are less literal than substructural fingerprints. Fragments containing C.ar, for example, share many physical properties but would be distinct in fingerprints. Moreover, the presence or absence of any particular atom pair is relatively independent of the atoms in the intervening structure.

There are 45 ways to pair up the 10 heavy atoms in **3**, but the corresponding atom pair vector has a cardinality of only 25 because of symmetry. The five distinct types (sp^3, sp^2, and aromatic carbons, one amide nitrogen, and one sp^2 oxygen) can pair up in 10 different ways. Some of these pairs are uniquely represented and so set only one bit, but other pairs show up more than once and so set more than one bit. The lone N.am–O.2 atom pair in **3**, for instance, is separated by two bonds and so sets the second bit in that block of the vector. Three bits are set for the C.ar–C.ar atom pair (one, two, and three bond separation), and four bits are set for each pairing of C.ar with another atom type.

Note that the carbon in the amide carbonyl is C.2 and so is distinct from the aromatic ring carbons. This increases the resolution of the atom pair vectors produced. If the structure is entered in its Dewar form with localized single and double bonds, each ring carbon becomes C.2 rather than C.ar, and fewer bits get set in the atom pair vector (22 instead of 25).

The presence or absence of rings is not encoded explicitly in atom pair vectors. Ring closure, however, often alters some atom types (e.g., from sp^2 to aromatic) and always shortens the topological distance between some atom pairs. As a result, the number of rings is encoded in part implicitly, by bits that are *not* set. The atom pair vector for **2,** for example, has a cardinality of 9, whereas its alicyclic analog **1** sets 14 bits—the same 9 bits set for **2** plus 5 bits for four-, five-, and six-bond separated pairs "short-circuited" by ring closure.

Structure **2** also has more distinct topological torsions (viz., 18) than does **1** (which has 12), but far fewer than does **3** (with 29). Aromatization has relatively little effect on the cardinality of topological torsions.

Figure 1 also presents values of c log P, CMR, molecular volume, and some connectivity indices for the three model compounds under consideration; all were calculated using ChemEnlighten (28). Note that discrimination between the compounds increases as the path length considered increases.

Pharmacophore triplets are the most generalized structural descriptors listed in Fig. 1. The entries shown reflect surveys across 10 or 100 conformers in SYBYL. Though the number of triplets identified does not increase linearly with the number of conformations sampled, it does increase steadily for flexible molecules like **1.** This is especially striking when one realizes that **1** contains no hydrophobic centers to enrich the triplets. The values obtained illustrate the point that the number of triplets found is dependent not only on the number of conformations examined but also on the rigidity of the molecule in hand.

V. SIMILARITY AND DISTANCE MEASURES

Among the many possible ways to define pairwise similarity and dissimilarity (38), the most familiar is the **Euclidean distance** d, which is also the most demanding in what is required of the underlying descriptors. Euclidean distance is usually defined in terms of its square, both because it is slightly simpler to express in that way and because working with the square wherever possible avoids the computational cost involved in calculating square roots.

$$d^2(\mathbf{x}, \mathbf{y}) = \sum (x_i - y_i)^2 \tag{5}$$

If the Euclidean distance is to be meaningful, all contributing descriptors should be set to a common scale—that is, made commensurate. This is often done by **autoscaling** each descriptor: subtracting off the sample mean and dividing the residual obtained by the sample standard deviation. Many descriptors do have a roughly normal (i.e., Gaussian) distribution that justifies this treatment, but some important ones do not.

Ranking is a more robust alternative to autoscaling, especially when the ranks produced are expressed as *percentiles* to normalize them to the size of the data set of interest. The rank-transformed descriptor is then relatively independent of sample size, and the relationship between naive and rank-transformed descriptors will vary little from sample to sample. Rank transforms are particularly well-behaved for the large data sets of interest in library design—provided, of course, that the large data sets being considered represent statistically valid samplings of the structural "universe" about which conclusions are to be drawn at the end of the day.

Principal components analysis assumes that an underlying Euclidean distance measure is appropriate.

In the case of binary vectors, d^2 reduces to the **Hamming distance,** H:

$$H(\mathbf{x}, \mathbf{y}) = \Sigma \, | \, x_i - y_i \, | = | \, \mathbf{x} \cup \mathbf{y} \, | - | \, \mathbf{x} \cap \mathbf{y} \, | \tag{6}$$

where the vertical bars within the summation denote absolute value of the differences between elements and those to the left denote the cardinality of bit sets. H is sometimes referred to as the *Manhattan* or *city-block* distance.

Note that neither d nor H distinguishes between 0 and any other value of a descriptor element. This implies that here we have interval- or ratio-scale descriptors, which is the case for physical properties and BCUTs but is not necessarily so for most of the other descriptors described above.

Other measures are more appropriate when the constituent descriptor elements are categorical in scale but are not tightly coupled, as is the case with fingerprints and other bit set descriptors. In such cases, sharing the ith attribute ($x_i = y_i = 1$) is strongly indicative of similarity, whereas a difference in the ith attribute ($x_i = 1$, $y_i = 0$, or vice versa) is strongly indicative of dissimilarity. The shared absence of a trait ($x_i = y_i = 0$), on the other hand, is very weakly discriminating.

The **cosine coefficient** Cos is a *similarity* measure better suited for applications involving such binary vectors (39,40); it, too, has a simplified form (the right-hand side of Eq. 7) specifically applicable to bit sets:

$$\mathrm{Cos}(\mathbf{x}, \mathbf{y}) = \frac{\Sigma x_i y_i}{} = \frac{| \, \mathbf{x} \cap \mathbf{y} \, |}{\sqrt{| \, \mathbf{x} \, \| \, \mathbf{y} \, |}} \tag{7}$$

The definition implies that the value of *Cos* runs from -1 to 1 for real-valued descriptors and falls between 0 and 1 for descriptors (including bit sets) that are strictly nonnegative. Negative correspondences ($x_i = y_i = 0$) contribute to neither the numerator nor the denominator of *Cos*. Note, too, that it scales the pairwise similarity by the number of substructures each contains, which parallels the size and complexity of the structures in question. Cos $(\mathbf{x}, \mathbf{y})$ is, in fact, equal to the cosine of the angle formed between the rays running from the origin out to $\mathbf{x}$ and to $\mathbf{y}$.

The Dice coefficient is a related measure employed by Carhart et al. (8) in which the similarity is normalized with respect to the algebraic rather than to the geometric mean of the size of the individual vectors:

$$\text{sim}(\mathbf{x}, \mathbf{y}) = \frac{2 \sum x_i y_i}{} = \frac{2 \mid \mathbf{x} \cap \mathbf{y} \mid}{\mid \mathbf{x} \mid + \mid \mathbf{y} \mid} \tag{8}$$

The similarity measure most widely used for comparing bit sets (especially fingerprints) is the **Tanimoto coefficient** T, whose general form is given by:

$$T(\mathbf{x}, \mathbf{y}) = \frac{\sum x_i y_i}{} \tag{9}$$

The Tanimoto coefficient is particularly appropriate for assessing sparse binary vectors such as substructural fingerprints (41). Treating such vectors as bit sets yields a simplified form of the Tanimoto, which always falls between 0 and 1:

$$T(\mathbf{x}, \mathbf{y}) = \frac{\mid \mathbf{x} \cap \mathbf{y} \mid}{\mid \mathbf{x} \cup \mathbf{y} \mid} = \frac{\mid \mathbf{x} \cap \mathbf{y} \mid}{\mid \mathbf{x} \mid + \mid \mathbf{y} \mid - \mid \mathbf{x} \cap \mathbf{y} \mid} \tag{10}$$

where the vertical bars again indicate cardinality. This calculation can be done very quickly, which has helped make it popular for carrying out diversity analyses on very large data sets.

The weakest and least demanding similarity measure derives from the simplest possible assumption about similarity, namely, that two molecular structures are similar if they generate identical descriptor values. This defines a **dichotomous similarity** measure d_0:

$$d_0(\mathbf{x}, \mathbf{y}) = \begin{cases} 1 : x_i = y_i & \text{for all } i \\ 0 : x_i \neq y_i & \text{for some } i \end{cases} \tag{11}$$

In other words, two structures are the same if all their descriptors are the same; otherwise they are different. This measure is appropriate when the descriptors involved are categorical and extremely context dependent. It has seen a great deal of direct use with pharmacophore multiplets, but it also underlies many cell-based analysis methods. In the latter case, there is an implicit intermediate transform of the descriptors into vectors of bin indices along each descriptor axis.

VI. NEIGHBORHOOD BEHAVIOR

In the end, biochemistry defines the only measure of similarity and dissimilarity relevant to library design for drug discovery and development. Medicinal chemical experience clearly shows that molecular structure is closely related to biological activity, but not all descriptors derived from structure work equally well in bridging that gap. This is less often due to any flaw in the descriptors than it is a seductive tendency to misinterpret the similarity principle, which states that similarity in structure implies similarity in biological activity (1). It does *not* follow that similarity in biological activity implies similarity in molecular structure, however much we may want that to be the case.

Of the descriptors considered here, only fingerprints and, to a lesser extent, atom pairs have been shown to display consistent neighborhood behavior (42). The converse, on the other hand, is *not* necessarily true: molecules that exhibit similar biological activity may have similar fingerprints, but they may not. Rather, there are likely to be many disjoint islands of activity distributed across fingerprint space for any particular biological target.

When a strong neighborhood property holds, only one compound need be tested from each region in the descriptor space to determine with confidence whether it is an "inactive" neighborhood. For fingerprints, this neighborhood radius (one minus the Tanimoto similarity) is about 0.15. If a compound is inactive, compounds whose fingerprints have a Tanimoto similarity of 0.85 or greater are usually inactive as well, in that they are very likely to all exhibit potencies within about two orders of magnitude of the first (26,42).

Much more generalized descriptors serve to draw such islands of activity together and thereby expedite "island hopping" (43) between literal structural classes. As this is done, however, inactive compounds generally get drawn in as well, so more compounds need to be drawn from a given neighborhood to be confident that the area is inactive.

If, for example, half of the compounds near a potent active are likely to be effectively inactive in a given descriptor space, each area in that space must be sampled three times to be reasonably confident ($> 85\%$ confidence) that the area in question does not include actives.

Pharmacophore descriptors are the most generalized of those discussed here. When used as a basis for 3D searches, they do well if more than 10% of the structures they return are active. At this "density" in activity space, each area would have to fail to produce an active compound 18 times before a researcher could conclude with 85% confidence that the area is indeed devoid of activity. Worse, those 18 samples have to be drawn as *independent* random samples to give this level of certainty, a goal not easily achieved in combinatorial libraries.

The specificity of fingerprints makes for a relatively large number of separate potential activity islands needing to be considered. In our experience, however, no descriptor put forward to date for use in comparing libraries provides a high enough level of "island" consolidation vis à vis fingerprints to justify the reduction in efficiency produced by the uncertainty described above. The methods to be described shortly focus on using fingerprints to keep interpretation as straightforward as possible. The methods presented can be equally well applied to other descriptors, though the interpretation may not be as straightforward.

Recent work suggests that a modified Tanimoto similarity function may give improved performance in compound selection (44). In particular, these authors recommend allowing bits not set in either binary vector to make a substantive contribution to similarity.

VII. RELATIVE DIVERSITY: COMPARING LIBRARIES TO ONE ANOTHER

Measures of library similarity and diversity can be expressed in purely mathematical and schematic terms. It is often useful, however, to connect the summary statistics produced to particular examples as well. Here, we consider relationships between three combinatorial libraries built in Legion (28) using similar substituents on a cyclohexane, pyrimidine, or pyridine core. The Chex and Pym cores are each asymmetrically substituted at three positions to yield libraries comprised of 2244 ($11 \times 12 \times 17$) compounds each. The Pyr library is similar in the kinds of substituent employed but is larger—four substitution sites yield 6000 ($6 \times 10 \times 10 \times 10$) compounds. The particular substituents, which include halogens, pseudohalogens, alkyls, aryls, arylalkyls, ethers, thio ethers, and acyl groups, are laid out in Fig. 2.

A. Union Bit Sets

One simple way to compare two libraries **A** and **B** is to treat each as the sum of its parts. This has proven particularly appealing when one is using pharmacophore multiplet descriptors as the basis for comparison. The binary vector for a library is then simply taken as the union of the bit sets for the compounds that comprise the library, and the library that includes more pharmacophore triplets or tetrads— that is, the library with the bigger bit set cardinality—is the better library. When comparing candidate libraries to those already in hand, the "better" candidate is taken as that which includes more pharmacophore multiplets that are "new" with respect to the reference library of compounds already in hand.

One drawback of this approach is that pharmacophore multiplets may well be present in compounds that do not set the corresponding "bit." With discrete torsional sampling, this will occur when the requisite combination of torsions does not fall close enough to the specified increments. For the fully flexible stochastic search implemented in Selector (28), it occurs because complete sampling of conformational space would take a prohibitively long time.

Even were conformational sampling truly complete, a problem would arise because even four features is a pretty minimal size for a biochemically specific

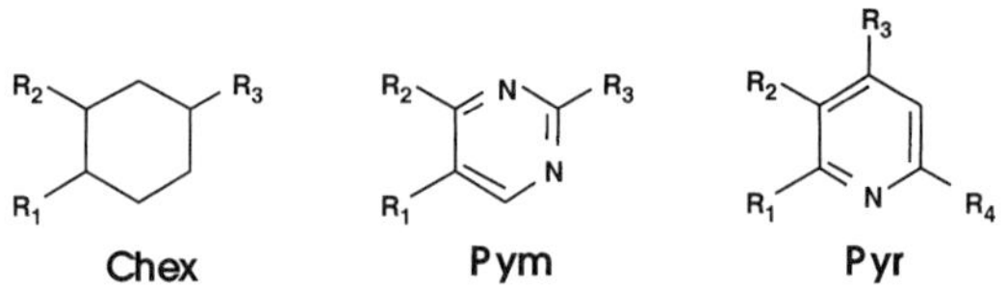

Position	All libraries	Chex & Pym	Pyr only
R1	F, Br, NO_2, Et NMe_2, Ac, $COCF_3$ SPh, OPh, CH_2Ph	H, Cl, CF_3 Me, iPr, SMe Ph	none
R2	F, Et, CF_3, $COCF_3$ OPh, CH_2Ph	Br, NO_2, NMe_2 Ac, SPh	Cl, Me, SMe, Ph
R3	CF_3, Ac, $COCF_3$	F, Br, NO_2 Et, NMe_2, Ac SPh, OPh, CH_2Ph	CN, CO_2Me, $CONH_2$
R4	none	none	F, iPr, CF_3, SMe Ac, $COCF_3$, Ph SPh, OPh, CH_2Ph

Figure 2 Structural definitions of combinatorial example libraries.

pharmacophore. Moreover, the meaning of a given triplet is quite dependent upon the molecular context in which it appears. The latter point, which is particularly pertinent with regard to steric factors, can cause the same pharmacophoric bit to be set by compounds with radically different biochemical properties.

Similar caveats apply when the descriptors in question are substructural keys, whether they are specific or generalized—for example, in terms of the presence or absence of certain ring systems (45).

B. Pairwise Identity

Another approach is to apply the dichotomous similarity measure d_0 to all possible pairs of compounds such that one member of the pair is drawn from **A** and the other is drawn from **B**. Any member of **A** that is not identical to some member of **B** scores 1 dissimilarity "point" for **A,** and any member of **B** that is not identical to some member of **A** scores 1 "point" for **B**. Such comparisons are indistinguishable from bit set comparisons, and summary statistics can then be obtained just as for single compounds. The Hamming distance between sets, for example, is then given by:

$$H(\mathbf{A}, \mathbf{B}) = |\mathbf{A} \cup \mathbf{B}| - |\mathbf{A} \cap \mathbf{B}| \tag{12}$$

whereas the corresponding Tanimoto and cosine similarity coefficients are given by:

$$T(\mathbf{A},\mathbf{B}) = \frac{|\mathbf{A} \cap \mathbf{B}|}{|\mathbf{A} \cup \mathbf{B}|} = \frac{|\mathbf{A} \cap \mathbf{B}|}{|\mathbf{A}| + |\mathbf{B}| - |\mathbf{A} \cap \mathbf{B}|} \tag{13}$$

and

$$\text{Cos}(\mathbf{A},\mathbf{B}) = \frac{|\mathbf{A} \cap \mathbf{B}|}{\sqrt{|\mathbf{A} \| \mathbf{B}|}} \tag{14}$$

This rationale is often not applied explicitly, but nonetheless it lies at the heart of many cell-based analyses. The key to seeing this is to recognize the intermediate binning transform (see above) applied implicitly when the cells are being defined. Edge effects can introduce serious distortions into the statistics produced, particularly when the bin boundaries used are a function of the population being analyzed. Of more concern is the implicit assumption that occupancy of cells that are close together but not identical is no more indicative of similarity than is occupancy of cells that are far apart. Hence this approach discards a substantial amount of potentially useful information.

No compound in any of the three example libraries is identical to any in one of the others. The UNITY fingerprints reflect this fact, so the three are completely dissimilar by this criterion.

C. Average Pairwise Similarity

More information can be obtained from library comparisons involving all pairwise similarities across a pair of data sets. One such approach exploits the fact that the average of the cosine coefficient is equal to the cosine coefficient of the averages. If the centroid of each library is defined as the vector of descriptor values wherein each element bears the (weighted) average value for that element across all compounds in that library, then the cosine coefficient between the centroids from two libraries $\mathbf{A}$ and $\mathbf{B}$ will be equal to the average cosine coefficient obtained across all pairs of compounds such that one is drawn from $\mathbf{A}$ and the other is drawn from $\mathbf{B}$ (39,40).

Centroid similarities can be calculated rapidly, but the centroids themselves are, to some degree, artificial constructs. This is so for bit set descriptors, since the "centroid" is almost always a vector of fractional values and cannot itself be thought of as a bit set. Nonetheless, this similarity measure provides useful information regarding the relative separation between two libraries (39). The average cosine coefficient calculated between the centroid of each library and its constituent compounds provides a summary measure of the size of each library.

D. Nearest-Neighbor Profiles

Pairwise identity is a measure of the "exact redundancy" between libraries, whereas average pairwise similarity considers aggregate properties only and takes no account of redundancy. An intermediate, more detailed approach is to consider the relative extent of *near* redundancy between two libraries as a measure of their similarity. The simplifying assumption made here is that the similarity of any compound to a reference library is dominated by its similarity to the most similar compound contained in that library, that is, by its **nearest-neighbor** (NN) **similarity.**

Figure 3 shows the nearest-neighbor profiles obtained by using the *dbcmpr* function in Selector (28) for the Chex, Phym, and Pyr libraries with respect to one another. Note that NN similarity is not symmetric, in that the profile of **A** with respect to **B** (written as **B ← A**) is in general not the same as the profile of **B** with respect to **A** (i.e., **A ← B**).

The asymmetry and complexities possible under nearest-neighbor similarity measures are well illustrated by the interrelationships found among the Chex, Pyr, and Pym libraries. When Chex is taken as the reference library, the average nearest-neighbor Tanimoto similarity for Pyr is 0.311 (Fig. 3A), which is somewhat greater than the corresponding average for Pym (0.271: Fig. 3B). The complementary value obtained for Chex when the Pyr library is taken as reference is 0.268 (Fig. 3C), which is significantly smaller than 0.311. The average NN similarity for Chex with respect to Pym, on the other hand, is 0.278 (Fig. 3E), which is slightly higher than the complementary value of 0.271 (Fig. 3B).

As one would expect based on their respective cores, the Pyr and Pym libraries are much more similar to each other than either is to Chex. Again, the average NN similarity is smaller when the Pyr library is taken as reference (0.550; Fig. 3D) than when Pym takes that role (0.587; Fig. 3F).

Such differences indicate that one library may be "larger" than the other in some sense. One way for this to occur is for one library to be contained within another, as is illustrated schematically in Fig. 4. In such a situation, every compound in the smaller set (hatched circles) will be similar to some compound in the larger one (open circles), but there will be compounds in the larger set that are relatively dissimilar to anything in the smaller. The net result is that the NN similarity will be less when it is calculated using the "smaller" (less extensive) library as reference. Though it is less obvious, something similar occurs whenever one library covers a significantly greater region of the descriptor space.

Note that this effect can be severely exacerbated by outliers in the larger library. The distortion produced is minimal if the cosine or Tanimoto coefficient is used as the measure of pairwise similarity and sparse bit sets such as fingerprints

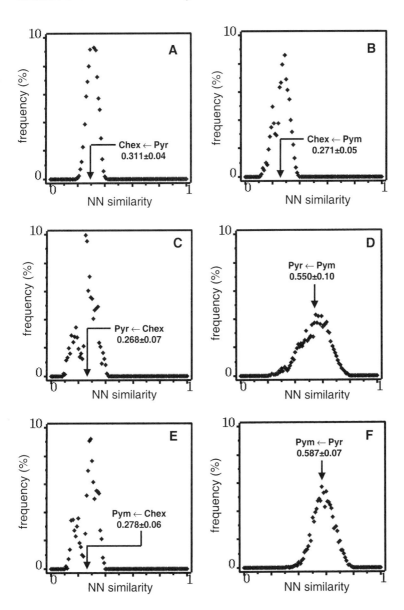

Figure 3 Tanimoto nearest-neighbor profiles for combinatorial libraries using UNITY fingerprints and Tanimoto similarity obtained using *dbcmpr* from Selector. (A, B) Chex taken as the reference database. (C, D) Pyr taken as the reference database. (E, F) Pym taken as the reference database.

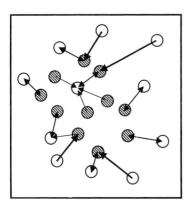

Figure 4 Effect of spread on NN similarity. Two-headed arrows indicate mutual nearest-neighbor distances (dissimilarities) that contribute equally to both sets. Light single-headed arrows indicate distances that contribute only when the open circles are used as the reference data set, whereas bold single-headed arrows indicate distances that are relevant only when the hatched circles are used as reference.

are used as descriptors. These measures expand distances that are small under Euclidean or Hamming metrics, compressing smaller similarities (larger distances) down into the lower end of the 0–1 range of similarities possible (Eqs. 7 and 10). Outlier effects can be a severe problem with open-ended measures such as the Euclidean metric.

Do the profiles shown in Fig. 3 prove, then, that the Chex and Pym libraries are "bigger" than Pyr? Not quite, unfortunately. A similar effect occurs when libraries are comparable in *extent* but differ in *intensity*—typically simply because one includes more compounds. This is most readily appreciated by considering the case of one library that is a random subset of another, as illustrated in Fig. 5 for a 500-compound subset (Pyr500) drawn at random from the 6000-member Pyr library and compared back to its complement, Pyr5500. Note that the effect is consistent with that discussed above: taking the "smaller" library (here, in terms of compound count rather than scope of chemistry) as reference yields a lower average NN similarity than the reverse (i.e., mapping the smaller library into the larger one).

Fortunately, ambiguity due to differences in membership size can be circumvented by running the NN similarity profile against a random subset of the larger library which is comparable in count to the smaller. When Chex and Pym are compared to a 2200-compound subset of Pyr (Pyr2k) the respective average NN similarities are nearly identical to those obtained with the full pyridine library.

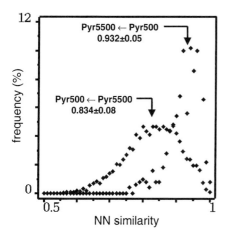

Figure 5 Similarity of a subset to its complement. Pry500 is a subset drawn at random from among the 6000 compounds in Pyr, and Pyr5500 is the complementary subset.

VIII. ABSOLUTE DIVERSITY OF SINGLE LIBRARIES

The binary vector for a library can be taken as the union of the bit sets for the compounds that comprise the library. Then the library that includes more pharmacophore triplets or tetrads—that is, the library with the bigger bit set cardinality—is the bigger (and presumably better) library. This approach does not return much detailed information about the libraries under consideration, however.

   Comparing a library to itself is an obvious extension of the approach described above to calculating the diversity of an individual library. In fact, this proves rather disappointing, as is illustrated for the Chex, Pyr, and Pym libraries in Fig. 6. Particularly distressing is the fact that the self-similarity of the subset is lower (and the diversity therefore higher) than is that of the full set (Fig. 6C and 6D); in fact, the smaller the subset is, the more diverse it will appear to be. This makes more sense, however, when the self-similarity is seen as an intensive measure diversity—that is, as a measure of density.

   The *total* NN self-*dis*similarity [obtained by subtracting the similarity from 1; for the Tanimoto coefficient operating on bit sets, this is known as the **Soergel distance** (38,46)] is a better-behaved measure of *extensive* diversity in that the expected diversity of a randomly drawn subset is less than or equal to the diversity of the full set from which it is drawn (47,48). Unfortunately, inequality does not hold for all subsets, as one would wish it to.

   The average distance from the "center" of a library is an obvious measure of **dispersion,** hence can be a useful measure of extensive diversity. For the cosine coefficient, an appropriate centroid can be calculated directly (40); with the

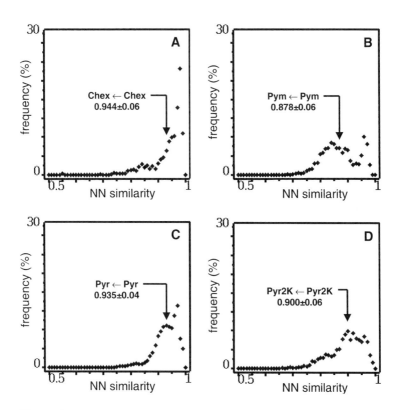

Figure 6   Self-similarity profiles for the example libraries: (A) Chex, (B) Pym, (C) Pyr, and (D) a 2200 compound subset (Pyr2κ) drawn at random from Pyr.

Tanimoto coefficient, the "most central" compound is used, that is, the compound for which the average similarity to others in the data set is greatest.

The total edge length of the **minimal spanning tree** (MST) is an elegant alternative that has also been suggested as a measure of absolute library diversity (49), where the MST is defined as the minimum possible sum of edge lengths across an entire complete acyclic subgraph. The difference between this approach and the total NN dissimilarity is illustrated schematically in Fig. 7, where lightweight edges contribute equally to both measures. The dark bold lines connect pairs of points that are each other's nearest neighbors—that is, **reciprocal nearest neighbors** (RNNs). Such edges count twice toward the total NN dissimilarity, but only once toward the MST. The lighter bold lines correspond to extensions needed to pull isolated "islands" together into a single graph; these edges contribute to the MLT but are not nearest-neighbor distances.

For the set shown in Fig. 7, the aggregate length of the "extra" connections is mostly offset by the elimination of double counting for RNN edges, so MST yields a diversity statistic only 7% greater than the total NN distance method. The discrepancy becomes large when the distance between clusters is much greater than the average NN distance within clusters.

Neither dispersion nor MST is well suited for use with fingerprints, however, because the two share an implicit assumption: that the compounds making up a data set convey substantial information about the properties of compounds falling in the descriptor space between compounds in the set. But "between" has little meaning in fingerprint space, and the relevance of exact values for relatively large dissimilarities is questionable (50,51).

A better strategy is to select a maximally diverse (minimally redundant) subset (52) of standard size from the library of interest, then determine *its* NN self-similarity (47). This is analogous to measuring areas of rooms (or any other enclosed space) by spreading a fixed number of dimes across the floor, keeping them as far apart as possible, then determining the average distance between the dimes. The bigger the floorspace available, the farther apart the dimes will be.

To illustrate this approach, maximally diverse subsets Chex*, Pym*, and Pyr* comprising 100 compounds each were drawn from Chex, Pym, and Pyr, respectively, using the *dbdiss* tool in Selector. Each was converted into a UNITY database, then analyzed by using *dbcmpr*. The profiles obtained are shown in Fig.

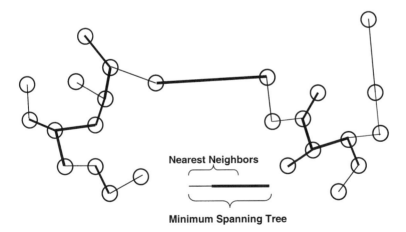

Figure 7 Schematic illustration of a minimal spanning tree (MST) for an arbitrary set. The distances marked by dark bold lines count twice when the average NN dissimilarity is calculated but only once when the size of the MST is calculated. The lighter bold lines indicate distances that count toward the MST but do not contribute to the average NN dissimilarity. The narrow lines are distances that contribute equally to both statistics.

8. The average NN self-similarity values obtained indicate that the three libraries do indeed differ appreciably in diversity, with the pyrimidines being somewhat more diverse (less self-similar) than the pyridines, and the cyclohexanes being considerably less diverse. The sharp profiles obtained here make the ordering obtained quite convincing, in contrast to the roundabout arguments involved in interpreting the complex profiles in Figs. 3 and 6.

Figure 8 also includes the series of plots for **A*** ← **A.** These can serve to show up cases of "hollow" data sets. A smooth hump centered about halfway between the self-similarity peak for the maximally diverse subset and 1 indicates a relatively even distribution across the structural space circumscribed by the maximally diverse subset. All three libraries examined fulfill this criterion nicely, from which one can conclude that they lack large "holes." Getting the NN simi-

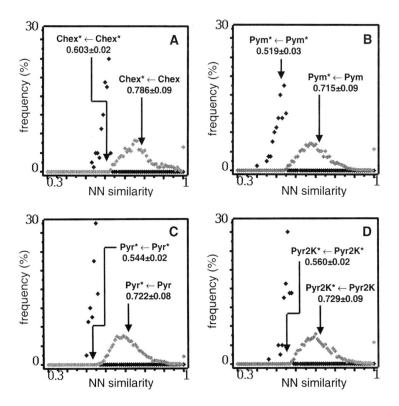

Figure 8 Self-similarity profiles (dark symbols) for 100-compound, maximally diverse subsets from each example library, along with comparisons against the parent set (lighter symbols); an asterisk indicates a diverse subset. (A) Chex* taken as reference. (B) Pym* taken as reference. (C) Pyr* taken as reference. (D) Pyr2κ taken as reference.

larity profile of **A** with respect to **A*** is analogous to spreading pennies at random around the room described above, then calculating the average distance from each penny to the nearest dime so as to determine how open the room is.

Finally, note that the profiles obtained for Pyr and for Pyr2K are very similar indeed, with the random subset looking slightly less diverse, just as one would expect. Such independence from the number of elements making up the set is key to having a robust measure of absolute diversity.

IX. SUMMARY

Many classes of descriptor are available to the computational chemist for reducing chemical structures to readily manipulatable numerical representations. Among these, substructural fingerprints exhibit good neighborhood behavior when used in conjunction with Tanimoto similarity. Nearest-neighbor similarity profiles give a good qualitative picture of the similarity between two libraries, with the average NN similarity providing a reasonable quantitative measure of relative diversity. Self-similarity profiles provide an indication of intensive diversity (nonredundancy), but are too locally focused to be useful for determining extensive diversity. Profiles for maximally diverse subsets, on the other hand, are well defined and can provide useful insight into absolute diversity and how evenly distributed a library is.

REFERENCES

1. Johnson M, Maggiora GM. Concepts and Applications of Molecular Similarity. New York: Wiley, 1990.
2. Dean PM, ed. Molecular Similarity in Drug Design. London: Chapman & Hall, 1995.
3. Kubinyi H, Folkers G, Martin YC, eds. 3D QSAR in Drug Design. Vol. 2. Ligand Protein Interactions and Molecular Similarity. Dordrecht: Kluwer/ESCOM, 1998.
4. Dean PM, Lewis RA. Molecular Diversity in Drug Design. Dordrecht: Kluwer, 1999.
5. Agrafiotis D. Diversity of chemical libraries. In: Allinger NL, Clark T, Gasteiger J, Kollman PA, Schaefer HF III, Schreiner PR, eds. Encyclopedia of Computational Chemistry. Chichester: Wiley, 1998, pp 742–761.
6. Lajiness MS. Applications of molecular similarity/dissimilarity in drug research. In: van de Waterbeemd, ed. Structure–Property Correlations in Drug Research. Austin, TX: Academic Press, 1996, pp 179–205.
7. McGregor MJ, Pallai PV. Clustering of large databases of compounds using MDL keys as structural descriptors. J Chem Inf Comput Sci 1997; 37:443–448.
8. Carhart RE, Smith DH, Venkataraghavan. Atom pairs as molecular features in structure–activity studies: Definition and applications. J Chem Inf Comput Sci 1985; 25:64–73.

9. Sheridan RP, Nachbar RB, Bush BL Extending the trend vector: The trend matrix and sample-based partial least squares. J Comput-Aided Mol Design 1994; 8:323–340.

10. Nilakantan R, Bauman N, Dixon JS, Venkataraghavan R. Topological torsions: A new molecular descriptor for SAR applications. Comparison with other descriptors. J Chem Inf Comput Sci 1987; 27:82–85.

11. Kier LB. Shape indexes of order one and three from molecular graphs. Quant Struct–Acit Relat 1986; 5:1–7.

12. Kier LB. Distinguishing atom differences in a molecular graph shape index. Quant Struct–Act Relat 1986; 5:7–12.

13. Kier LB, Hall LH. Molecular Connectivity in Chemistry and Drug Research. New York: Academic Press, 1976.

14. Randic M. Topological indices. In: Allinger NL, Clark T, Gasteiger J, Kollman PA, Schaefer HF III, Schreiner PR, eds. Encyclopedia of Computational Chemistry. Chichester: Wiley, 1998, pp 3018–3032.

15. Pearlman RS, Smith KM. Novel software tools for chemical diversity. In: Kubinyi H, Folkers G, Martin YC, eds. 3D QSAR in Drug Design. Vol. 2. Ligand–Protein Interactions and Molecular Similarity. Dordrecht: Kluwer/ESCOM, 1998, pp 339–353.

16. Pearlman RS, Smith KM. Metric validation and the receptor-relevant subspace concept. J Comput-Aided Mol Design 1999; 39:28–35.

17. Shemetulskis NE, Bunbar JB Jr, Dunbar BW, Moreland DW, Humblet C. Enhancing the diversity of a corporate database using chemical database clustering and analysis. J Comput-Aided Mol Design 1995; 9:407–416.

18. Good AC, Kuntz ID. Investigating the extension of pairwise distance pharmacophore measures to triplet-based descriptors. J Comput-Aided Mol Design 1995; 9:373–379.

19. Ashton MJ, Jaye MC, Mason JS. New perspectives in lead generation. II. Evaluating molecular diversity. Drug Discovery Today 1996; 1:71–78.

20. McGregor MJ, Muskal SM. Pharmacophore fingerprinting. 1. Application to QSAR and focused library design. J Chem Inf Comput Sci 1999; 39:569–574.

21. Pickett SD, Luttmann C, Guerin V, Laoui A, James E. DIVSEL and COMPLIB—Strategies for the design and comparison of combinatorial libraries using pharmacophoric descriptors. J Chem Inf Comput Sci 1998; 38:144–150.

22. Mason JS. Absolute vs relative similarity and diversity. In: Dean PM, Lewis RA, eds. Molecular Diversity in Drug Design. Dordrecht: Kluwer, 1999, pp 67–91.

23. Clark RD, Brusati M, Jilek R, Heritage T, Cramer RD. Validating novel QSAR descriptors for use in diversity analysis. In: Gundertofte K, Jorgensen FS, eds. Molecular Modeling and Prediction of Bioactivity. New York: Kluwer Academic/Plenum Publishers, 2000, pp 95–101.

24. Cramer RD III, Clark RD, Patterson DE, Ferguson AM. Bioisosterism as a molecular diversity descriptor: Steric fields of single topomeric conformers. J Med Chem 1996; 39:3060–3069.

25. Good AC, Ewing TJA, Gschwend DA, Kuntz ID. New molecular shape descriptors: Application in database screening. J Comput-Aided Design 1995; 9:1–12.

26. Matter H. Selecting optimally diverse compounds from structure databases: A validation study of two-dimensional and three-dimensional molecular descriptors. J Med Chem 1997; 40:1219–1229.

27. MDL Information Systems, San Leandro, CA.
http://www.mdli.com
28. UNITY®, SYBYL®, Selector™, Legion™, and ChemEnlighten™ are distributed by Tripos, Inc., St. Louis, MO 63144.
http://www.tripos.com
29. MolConnZ is a product of Hall Associates Computing, Quincy MA 02170.
30. DiverseSolutions was developed by RS Pearlman and KM Smith at the University of Texas, Austin, and is distributed by Tripos, Inc., St. Louis, MO 63144.
31. Sheridan RP, Miller MD, Underwood DJ, Kearsley, SK. Chemical similarity using geometric atom pair descriptors. J Chem Inf Comput Sci 1996; 36:128–136.
32. CONCORD® was developed by RS Pearlman, A Rusinko, JM Skell, and R Balducci at the University of Texas, Austin, and is distributed by Tripos, Inc., St. Louis, MO 63144.
http://www.tripos.com
33. Chem-X is distributed by Oxford Molecular Group PLC, Oxford, England.
http://www.oxmol.com
34. Hurst T. Flexible 3D searching: The directed tweak technique. J Chem Inf Comput Sci 1994; 34:190–196.
35. Martin EJ, Critchlow RE. Beyond mere diversity: tailoring combinatorial libraries for drug discovery. J Comb Chem 1999; 1:32–45.
36. Xue L, Godden J, Gao H, Bajorath J. Identification of a preferred set of molecular descriptors for compound classification based on principal component analysis. J Chem Inf Comput Sci 1999; 39:699–704.
37. BioByte, Inc., Pomona, CA.
38. Gower JC. Measures of similarity, dissimilarity and distance. In: Kotz S, Johnson NL, eds. Encyclopedia of Statistical Sciences. Vol. 5. New York: Wiley, 1985, pp 397–405.
39. Turner DB, Tyrrell SM, Willett P. Rapid quantification of molecular diversity for selective database acquisition. J Chem Inf Comput Sci 1997; 38:18–22.
40. Holliday JD, Ranade SS, Willett P. A fast algorithm for selecting sets of dissimilar structures from large chemical databases. Quant Struct–Act Relat 1996; 14:501–506.
41. Willett P, Winterman V. A comparison of some measures of intermolecular structural similarity. Quant Struct–Act Relat 1986; 5:18–25.
42. Patterson DE, Cramer RD, Ferguson AM, Clark RD, Weinberger LE. Neighborhood behavior: A useful concept for validation of molecular diversity descriptors. J Med Chem 1996; 39:3049–3059.
43. Andrews KA, Cramer RD. Toward general methods of targeted library design: Topomer shape similarity searching with diverse structures as queries. J Med Chem 2000; 43:1723–1740.
44. Dixon SL, Koehler RT. The hidden component of size in two-dimensional fragment descriptors: Side effects on sampling in bioactive libraries. J Med Chem 1999; 42:2887–2900.
45. Nilakantan R, Bauman N, Haraki KS. Database diversity assessment: New ideas, concepts, and tools. J Comput-Aided Mol Design 1997; 11:447–452.
46. Cheng C, Maggiora G, Lajiness M, Johnson. Four association coefficients for relating molecular similarity measures. J Chem Inf Comput Sci 1996; 36:909–915.

47. Clark RD, Langton WJ. Balancing representativeness against diversity using optimizable K-dissimilarity and hierarchical clustering. J Chem Inf Comput Sci 1998; 38:1079–1086.
48. Cramer RD, Patterson DE, Clark RD, Soltanshahi F, Lawless MS. Virtual compound libraries: A new approach to decision making in molecular discovery research. J Chem Inf Comput Sci 1998; 38:1010–1023.
49. Mount J, Ruppert J, Welch W, Jain AN. IcePick: A flexible surface-based system for molecular diversity. J Med Chem 1999; 42:60–66.
50. Sello G. Similarity measures: Is it possible to compare dissimilar structures? J Chem Inf Comput Sci 1998; 38:691–701.
51. Flower DR. On the properties of bit string-based measures of chemical similarity. J Chem Inf Comput Sci 1998; 38:379–386.
52. Lajiness M, Johnson MA, Maggiora GM. Implementing drug screening programs using molecular similarity methods. In: Fauchere JL, ed. QSAR: Quantitative Structure–Activity Relationships in Drug Design. New York: Liss, 1989, pp 173–176.

12

Rational Combinatorial Library Design and Database Mining Using Inverse QSAR Approach

Alexander Tropsha

University of North Carolina
Chapel Hill, North Carolina

I. INTRODUCTION

Rapid development of combinatorial chemistry and high throughput screening techniques in recent years has provided a powerful alternative to traditional approaches to lead generation and optimization. In traditional medicinal chemistry, these processes frequently involve the purification and identification of bioactive ingredients of natural, marine, or fermentation products or random screening of synthetic compounds. This is often followed by a series of painstaking chemical modification or total synthesis of promising lead compounds, which are tested in adequate bioassays. On the contrary, combinatorial chemistry involves systematic assembly of a set of "building blocks" to generate a large library of chemically different molecules that are screened simultaneously in various bioassays (1,2). In the case of targeted library design, the lead identification and optimization task then becomes that of generating libraries with structurally diverse compounds that are similar to a lead compound; the underlying assumption is that structurally similar compounds should exhibit similar biological activities. Conversely, structurally dissimilar compounds should exhibit very diverse biological activity profiles; thus the goal of the diverse library design is to generate libraries with maximum chemical diversity of the composing compounds (3).

In many practical cases, the exhaustive synthesis and evaluation of combinatorial libraries is prohibitively expensive, time-consuming, or redundant (4). Theoretical analysis of available experimental information about the biological target or pharmacological compounds capable of interacting with the target can significantly enhance the rational design of targeted chemical libraries. In many cases, the number of compounds with known biological activity is large enough to develop viable quantitative structure–analysis relationship (QSAR) models for such data set. These models can be used as a means of selecting virtual library compounds (or actual compounds from existing databases) with (high) predicted biological activity. Alternatively, if a variable selection method has been employed in developing a QSAR model, the use of only selected variables can improve the performance of the rational library design or database mining methods based on the similarity to a probe. This procedure of using only selected variables in similarity searches in the descriptor space is analogous to the more traditional use of conventional chemical pharmacophores in database mining.

The main objective of this chapter is to illustrate the principles of using QSAR models for rational design of targeted chemical libraries and database mining. We refer to these applications of QSAR models to predict biologically active structures in virtual or actual chemical libraries as the inverse QSAR method. We describe the development of algorithms for QSAR based on the principle of stochastic optimization of variable selection to achieve highest value of cross-validated R^2 (q^2). Using selected examples and methodologies developed in our laboratory (5–7), we show that the use of preconstructed QSAR models improves the efficiency of rational targeted library design and database mining.

II. GENERAL COMPUTATIONAL DETAILS AND MOLECULAR DESCRIPTORS

Library design was based on the analysis of whole molecular structures as opposed to fragments; thus we used molecular connectivity indices (MCI) and atom pair (AP) descriptors. MolConnX program (8) was used to generate topological indices for all data sets and virtual library compounds. The AP descriptors were generated as follows, using an approach initiated by Carhart et al. (9).

The key components of defining a set of atom pair descriptors include the definition of atom types and the classification of distance bins. An atom pair is a simple type of substructure defined in terms of the atom types and the shortest path separation (or graph distance) between two atoms. The graph distance is defined as the smallest number of atoms along the path connecting two atoms in a molecular structure. The general form of an atom pair is as follows:

atom type i—(distance)—atom type j

where (distance) is the graph distance between atom i and atom j in the case of two-dimensional atom pairs description. (The distance can also be defined as the physical distance between atoms i and j in the case of three-dimensional atom pair description.)

SYBYL atom types (mol2 format) (10) were utilized as the starting point. In principle, all SYBYL atom types can be used in the generation of atom pair descriptors. To reduce the number of atom pair descriptors, however, we used only 10 atom types: (1) C.ar, aromatic carbons; (2) C.na, nonaromatic carbons; (3) N.ar, aromatic nitrogen atoms; (4) N.na, nonaromatic nitrogen atoms; (5) O.3, oxygen atoms in sp^3 hybridization state; (6) O.2, oxygen atoms in sp^2 hybridization state; (7) S, all sulfur atoms; (8) P.3, phosphorus atoms; (9) X, halogen atoms; (10) other atoms. The total number of pairwise combinations of all 10 atom types is 55. Furthermore, 15 distance bins were defined in the interval between graph distance zero (i.e., zero atoms separating an atom pair) to 14. Thus, a total of 825 (55×15) atom pair descriptors are generated for each molecular structure.

III. RATIONAL LIBRARY DESIGN USING ACTIVITY PREDICTION FROM QSAR MODELS

To illustrate this approach, we consider the design of a pentapeptide combinatorial library with the bradykinin activity using a QSAR model derived for a small bradykinin peptide data set. Figure 1 shows the schematic diagram illustrating the targeted pentapeptide combinatorial library design using the FOCUS-2D method

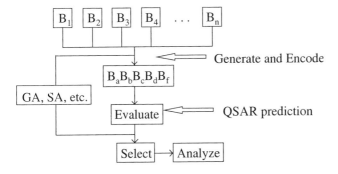

Figure 1 Schematic diagram of the library design by FOCUS-2D.

developed in our laboratory (5,6). The algorithm includes the description, evaluation, and optimization steps.

To identify potentially active compounds in the virtual library, FOCUS-2D employs stochastic optimization methods such as simulated annealing (SA) (11,12) and genetic algorithms (GA) (13–15). The latter algorithm was used for targeted pentapeptide library design as follows. Initially, a population of 100 peptides is randomly generated and encoded using topological indices or amino acid dependent physicochemical descriptors. The fitness of each peptide is evaluated by its biological activity predicted from a preconstructed QSAR equation (see below). Two parent peptides are chosen by using the roulette wheel selection method (i.e., high-fitting parents are more likely to be selected). Two offspring peptides are generated by a crossover (i.e., two randomly chosen peptides exchange their fragments) and mutations (i.e., a randomly chosen amino acid in an offspring is changed to any of 19 remaining amino acids). The fitness of the offspring peptides is then evaluated and compared against parent peptides fitness, and two lowest scoring peptides are eliminated. This process is repeated for 2000 times to evolve the population.

A. GA-PLS QSAR Method

The algorithm of the GA-PLS method (16) is implemented in six steps as follows.

1.   The MolConnX program (8) was applied to generate descriptor variables (460 topological indices) automatically for each data set represented in the SMILES notation. All descriptors that were dependent on atom identification (atom id: 150 descriptors) and descriptors with zero variance were removed.

2.   An initial population of 100 different random combinations of subsets of these descriptors (parents) was generated as follows. Each parent was described by a string of random binary numbers (i.e., one or zero), with the length (total number of digits) equal to the total number of descriptors selected for each data set. The value of one in the string implied that the corresponding descriptor was included for the parent, and the value of zero meant that the descriptor was excluded.

3.   For every combination of descriptors (i.e., every parent), a QSAR equation was generated for the training data set using the partial least-square (PLS) algorithm (17). Thus, for each parent a q^2 value was obtained, which was further used to calculate the following fitness function: $1 - (n - 1)(1 - q^2)/(n - c)$, where q^2 is the cross-validated R^2, n is the number of compounds, and c is the optimal number of components from PLS analysis. This fitness function was used to guide GA. See our earlier paper (16) for more discussion on the selection of the fitting function.

4.   Two parents were selected randomly and subjected to a crossover (i.e., the exchange of the equal-length substrings), which produced two offspring. Each

offspring was subjected to a random single-point mutation; that is, randomly selected one (or zero) was changed to zero (or one) and the fitness of each offspring was evaluated as described above (cf. step 3).

5. If the resulting offspring were characterized by a higher value of the fitness function, then they replaced parents; otherwise, the parents were kept.

6. Steps 3–5 were repeated until a predefined convergence criterion was achieved. As the convergence criterion, we used the difference between the maximum and minimum values of the fitness function. Calculations were terminated when this difference was less than 0.02.

In summary, each parent in this method represents a QSAR equation with randomly chosen variables, and the purpose of the calculation is to evolve from the initial population of the QSAR equations to the population with the highest average value of the fitness function. In the course of the GA-PLS process, the initial number of members of the population (100) is maintained, while the average value of the fitness function for the whole population converges to a high number.

B. Development of a QSAR Model

Pentapeptide analogs of BK have been described either by topological indices or by a combination of physicochemical descriptors, generated for each amino acid. The topological indices of virtual pentapeptides were calculated using the MolConn X program (8). The MOLCONN format (8), which is the standard input file format for MolConn X, was used to input the structure of each peptide: atom id, the number of hydrogens connected, atom type, and atom ids of all other heteroatoms were listed in a connection table, separated by a comma for each heteroatom of the peptide. Amino acids were predescribed in this way, and the connection tables of selected amino acids were combined as necessary to construct the input file for MolConn X (8).

We have also employed several amino acid based descriptors, including Z_1, Z_2, and Z_3 descriptors (related to hydrophilicity, bulk, and electronic properties of individual amino acids, respectively) reported by Hellberg et al. (18), as well as isotropic surface area (ISA) and electronic charge index (ECI) descriptors reported by Collantes and Dunn (19). In this case, virtual pentapeptides were encoded in the form of a string of descriptor values. Each string consisted of 15 descriptor values (five blocks of three descriptors per amino acid) when Z descriptors were used, or 10 descriptor values (five blocks of two) for ISA-ECI descriptors.

Ufkes et al (20) used 28 BK potentiating pentapeptides as a training set to develop a QSAR equation that was employed to predict the bioactivity of virtual library peptides. The log relative activity index (RAI) values of bradykinin potentiating pentapeptides were used as dependent variables. The original publication

(20) gives a detailed description of the assay as well as the calculation of relative activity index values.

The two most active compounds, VEWAK and VKWAP, were excluded from the training set. The calculated log RAI values compared favorably with the experimental data (data not shown). Although the activities of the two excluded peptides were underestimated (the experimental values of log RAI were 2.73 and 2.35 for VEWAK and VKWAP, respectively), the QSAR equations correctly predicted them to have activities higher than those of compounds in the training set. Thus, the log RAI values of 1.79, 1.48, and 1.47 were obtained for VEWAK using ISA-ECI, Z_1-Z_2-Z_3, and topological indices, respectively, and the log RAI values of 1.80, 1.74, and 1.95 were obtained for VKWAP using ISA-ECI, Z_1-Z_2-Z_3, and topological indices as descriptors, respectively.

The statistics obtained from the PLS regression analyses and the GA-PLS method applied to the training set using ISA-ECI, Z_1-Z_2-Z_3, and topological indices are shown in Table 1. To test the reliability of the prediction using preconstructed QSAR equations with these descriptors, we incorporated the modified

Table 1 Summary of Statistics

	PLS			GA-PLS: Topological indices[c]	
	ISA-ECI[a]	Z_1-Z_2-Z_3[b]			
Number of crossovers	0	0	0	2000	10000
Number of compounds	28	28	28	28	28
Number of variables	10	15	160	45	23
ONC[d]	3	2	1	2	5
q^{2e}	0.725	0.633	0.367	0.533	0.845
SDEP[f]	0.410	0.464	0.598	0.524	0.322
Fitness[g]	0.702	0.619	0.367	0.515	0.818
RSD of the **X** matrix[h]	0.886	0.818	0.381	0.134	0.195
SDEE[i]	0.313	0.315	0.544	0.466	0.260
R^2	0.840	0.831	0.476	0.630	0.899
F values	42.020	61.355	23.575	21.289	38.984

[a] ISA-ECI ($n = 28$, $k = 3$).
[b] Z_1-Z_2-Z_3 ($n = 28$, $k = 2$).
[c] Topological indices: $n = 28$, $k = 1$ for 0 crossover; $n = 28$, $k = 2$ for 2,000 crossovers; and $n = 28$, $k = 5$ for 10,000 crossovers.
[d] The optimal number of components.
[e] Cross-validated R^2.
[f] Standard error of prediction.
[g] $[1 - (n - 1)(1 - q^2)]/(n - c)$.
[h] The residual SD of the **X** matrix.
[i] Standard error of estimate.

"degree of fit" condition originally developed by Lindberg et al. (21). According to this condition, if RSD of dependent variables of a virtual peptide is less than the RSD of the X matrix of the training set, the predicted values are considered to be reliable. If this condition is not met, the log RAI of the virtual peptide is not predicted or is set to a low log RAI number to avoid selecting it. The condition does not allow the FOCUS-2D program to overextrapolate. Since the number of peptides in the training set is very small compared to theoretical number of different pentapeptides (3.2 million), the extrapolation of QSAR relationship should be done very carefully in small increments, and the "degree of fit" condition implemented here allows us to do this. The RSD values of the X matrix of the training set of 0.886, 0.818, and 0.195 were obtained for ISA-ECI, Z_1-Z_2-Z_3, and topological indices description methods, respectively, and used to test the reliability of the prediction (Table 1).

C. Design of a Targeted Library with Bradykinin (BK) Potentiating Activity

The results obtained with FOCUS-2D and a QSAR-based prediction are shown in Fig. 2 for Z_1-Z_2-Z_3 descriptors. The position-dependent frequency distributions of amino acids in highest scoring pentapepeptides are shown before (Fig. 2a) and after (Fig. 2b) FOCUS-2D. To evaluate the efficiency of stochastic sampling, the entire pentapeptide library (that includes as many as 3.2 million molecules) was also generated and subjected to evaluation using the same QSAR model, and the results are shown in Fig. 2c. Apparently, the populations after FOCUS-2D and the exhaustive search were very similar. With Z_1-Z_2-Z_3 descriptors, FOCUS-2D analysis selected amino acids E, I, K, L, M, Q, R, V, and W. Interestingly, these selected amino acids included most of those found in two most active pentapeptides, VEWAK and VKWAP (excluded from the training set for the QSAR model development). Furthermore, the actual spatial positions of these amino acids were correctly identified: the first and fourth positions for V; the second and fifth positions for E; the third position for W; and the second and fifth positions for K. More detailed analysis of these results (cf. Fig. 2b, 2c) may suggest which residues should be preferably chosen for each position in the pentapeptide to achieve a library of limited size with high predicted bradykinin activity.

IV. RATIONAL DATABASE MINING USING DESCRIPTOR VARIABLES SELECTED BY QSAR MODELS

The search for active compounds in chemical databases can be also conducted on the basis of chemical similarity to an active compound (probe) calculated in the descriptor space. The protocol for the similarity search is given in Fig. 3. First, a

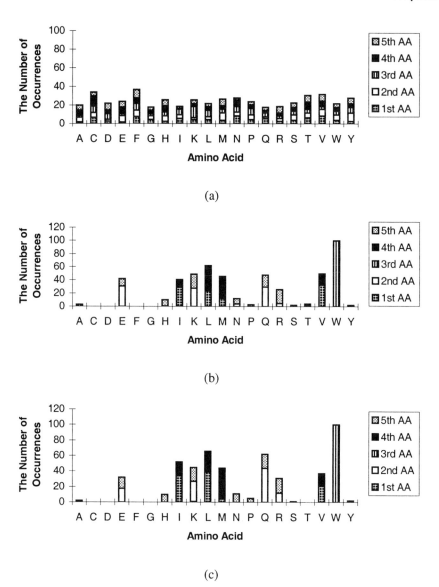

Figure 2 FOCUS-2D using Z_1-Z_2-Z_3 description method and a QSAR equation: (a) initial population, (b) final population after FOCUS-2D, and (c) final population after the exhaustive search.

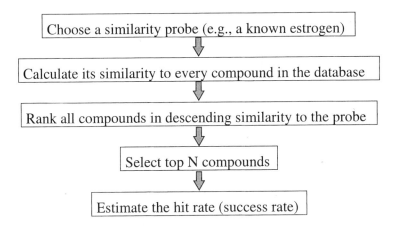

Figure 3   Flowchart for similarity-based database search.

similarity probe is chosen and its numerical descriptors are calculated. Then, using either the whole set of descriptors or descriptors chosen by a variable selection QSAR, the similarity of this probe molecule to every molecule in the database is computed as the value of the Euclidean distance between the two molecules in multidimensional descriptor space. All compounds in the database are sorted in descending similarity to the probe molecule, and a certain number of top-ranking compounds are suggested as the active compounds. The hit rate is evaluated as the number of known active molecules found in the set of selected compounds.

To demonstrate the effectiveness of the method, a QSAR analysis of a set of estrogen receptor ligands was performed, with MCI and AP serving as molecular descriptors. The usefulness of selected descriptor-based similarity searching is demonstrated in a database mining experiment. A known estrogen receptor ligand was chosen as a probe molecule, and molecular similarity was calculated by means of either the whole set of connectivity descriptors, or a subset of descriptors selected by the variable selection QSAR model.

A.   QSAR Based on the *K* Nearest-Neighbors (KNN) Principle

In principle, the KNN technique is a conceptually simple, nonlinear approach to pattern recognition problems. In this method, an unknown pattern is classified according to the majority of the class labels of its K nearest neighbors of the training set in the descriptor space.

The assumptions underlying KNN-QSAR method are as follows. First, structurally similar compounds should have similar biological activities, and the activity of a compound can be predicted (or estimated) simply as the average of the activities of similar compounds. Second, the perception of structural similarity is relative and should always be considered in the context of a particular biological target. Since the physicochemical characteristics of receptor binding sites vary from one target to another, the structural features that can best explain the observed biological similarities between compounds are different for different biological end points. These critical structural features are defined in this work as the topological pharmacophore (TP) for the underlying biological activity. Thus, one of the tasks of building a KNN-QSAR model is to identify the best TP. This is achieved by the "bioactivity-driven" variable selection, that is, by selecting a subset of molecular descriptors that afford a highly predictive KNN-QSAR model. Since the number of all possible combinations of descriptors is huge, an exhaustive search of these combinations is not possible. Thus, a stochastic optimization algorithm (i.e., simulated annealing) has been adopted for an efficient sampling of the combinatorial space. Figure 4 shows the overall flowchart of KNN-QSAR method, which involves the following steps.

1. Select a subset of n descriptors randomly (n is a number between one and the total number of available descriptors) as a hypothetical topological pharmacophore (HTP).

2. Validate this HTP by a standard cross-validation procedure, which generates the cross-validated R^2 (or q^2) value for the KNN-QSAR model built using this HTP. The standard leave-one-out procedure has been implemented as follows.

a. Eliminate a compound in the training set.

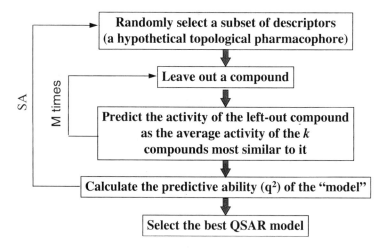

Figure 4 Flowchart for the construction of a KNN-QSAR model.

b.   Calculate the activity of the eliminated compound, which is treated as an unknown, as the average activity of the K most similar compounds found in the remaining molecules (K is set to 1 initially). The similarities between compounds are calculated using only the selected descriptors (i.e., the current trial HTP) instead of the whole set of descriptors.

c.   Repeat this procedure until every compound in the training set has been eliminated and predicted once.

d.   Calculate the cross-validated R^2 (or q^2) value using Eq. (1), where y_i and $\hat{y}_i$ are the actual activity and the the predicted activity of the ith compound, respectively, and $\bar{y}$ is the average activity of all the compounds in the training set. Both summations are over all the compounds in the training set.

$$q^2 = 1 - \frac{\Sigma(y_i - \hat{y}_i)_2}{\Sigma(y_i - \bar{y})^2} \tag{1}$$

Since the calculation of pairwise molecular similarities, hence the predictions, are based upon the current HTP, the obtained q^2 value is indicative of the predictive power of the current KNN-QSAR model.

e.   Repeat calculations for $K = 2, 3, 4, \ldots, n$. The upper limit of K is the total number of compounds in the data set; however, the best value is found empirically between 1 and 5. The K that leads to the best q^2 value is chosen for the current KNN-QSAR model.

3.   Repeat steps 1 and 2, the procedure of generating trial HTP's and calculating corresponding q^2 values. The goal is to find the best HTP that maximizes the q^2 value of the corresponding KNN-QSAR model. This process is driven by a generalized simulated annealing using q^2 as the objective function.

B.   Application of the KNN Method to Estrogen Receptor Ligands

Fifty-eight estrogen receptor ligands were chosen as a comprehensive test case for the KNN-QSAR technique. This data set was successfully analyzed earlier by Waller et al. (22), who used the comparative molecular field analysis (CoMFA) method.

In the KNN-QSAR method, n_{var} (the number of descriptors to be selected) can be set to any value that is less than the total number of descriptors generated by a molecular description method. Since the optimum value of n_{var} is not known *a priori*, several runs are usually needed to examine the relationship of the predictive power of a model (quantitated by a q^2 value) and the number of descriptors selected (n_{var}). Figure 5 shows this relationship when MCI was used to describe each of the estrogen receptor ligands.

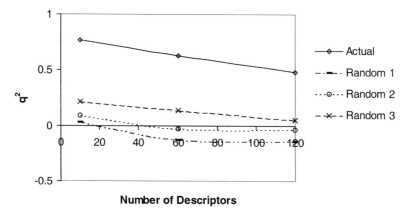

Figure 5   The relationship of q^2 to n_{var} for estrogen receptor ligands using MCI as molecular descriptors, (n_{var} is the number of descriptors selected for a KNN-QSAR model). The results for both the actual estrogen data set and three data sets with random activity values are included.

When the real activity values for estrogen receptor ligands were used in the KNN-QSAR analysis, the q^2 values was 0.77, 0.63, and 0.48 for a 10-descriptor model, a 60-descriptor model, and a 120-descriptor model, respectively. To show the robustness of KNN-QSAR analysis, one needs to demonstrate that no comparable q^2 values can be obtained when randomly shuffled activity values or randomly assigned activity values (but within the same range as the real activity) are used in KNN-QSAR analysis. Figure 5 also includes the q^2 vs n_{var} relationships when three randomly assigned activity values were used in the KNN-QSAR analysis. Overall, these q^2 values are very low in comparison to those of the actual data set. This suggests that the KNN-QSAR models obtained for the real data set are distinguishable from those for random data sets. One can also observe that the q^2 values decrease when the number of descriptors increases. On the surface, this may be counterintuitive. The intuition may come from the fact that the more descriptors are used in multiple linear regression analysis, the higher regression coefficient is normally obtained. However, it should be kept in mind that the KNN-QSAR is not based on a regression method, but rather on the similarity principle. Theoretically, there should be no apparent trend in q^2 vs n_{var} relationships, although in many practical situations, q^2 tends to decrease slightly when the number of descriptors increases. Conceivably, there should be one optimum number of descriptors, where either the q^2 is the highest or the separation between the q^2 for the real data set and those for random data sets is the largest.

Figure 6 plots the predicted vs actual activity during the cross-validation process for a 10-descriptor model. Apparently, the trend of the predicted values is similar to that of the real activity values. The results of KNN-QSAR modeling were in fact better than those reported by Waller et al. (22) using CoMFA analysis in terms of the q^2 values (0.77, vs 0.59 from Waller et al.).

C. Database Mining Using Estrogen Receptor Ligands as Probe Molecules

To demonstrate the effectiveness of molecular similarity based database mining, eight molecules were chosen arbitrarily from the known estrogen receptor ligands as the query molecules (or probes), one from each of the eight compound classes described elsewhere (22). These were 2,4,6-trichloro-4'-biphenylol, butylbenzylphthalate, 2-*tert*-butylphenol, coumestrol, DES, estradiol, HPTE, and M1. The typical result of the similarity search is given in Fig. 7, which shows the following four curves: (1) the hit rates obtained in the ideal case, where every compound found at the top 58 is actually a known estrogen receptor ligand (this is the upper limit that anyone would like to reach), (2) the hit rates obtained by a random search, (i.e., a random sampling of a number of compounds followed by an examination of how many known estrogen receptor ligands are found), (3) the results obtained from similarity searches based on the whole set of descriptors and (4) the results based the KNN-QSAR selected descriptors. Our results show that in most cases,

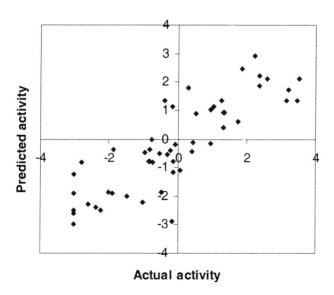

Figure 6 Predicted vs actual activity obtained from a 10-descriptor model for estrogen receptor ligands using MCI as molecular descriptors.

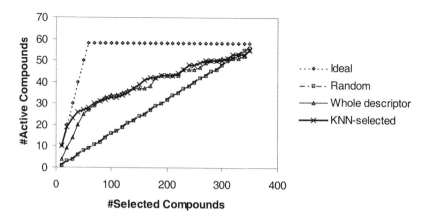

Figure 7   Comparison of hit rates for known estrogen receptor ligands for ideal, random, and similarity-based search with butylbenzylphthalate as the probe.

the hit rates obtained by similarity search for the known estrogen receptor ligands are more than two times higher than what would be expected by a random search. This demonstrates the effectiveness of the similarity search strategy. It should be noted that in five out of eight cases, the hit rates obtained using KNN-QSAR selected descriptors were better than those obtained using the whole set of descriptors, indicating that KNN-QSAR selected descriptors are better suited for similarity searches. It also implies that the KNN-QSAR method captures the critical structural features that distinguish the activities of the underlying compounds.

V.   CONCLUSIONS AND PROSPECTUS

One important aspect of any QSAR investigation is the potential application of the derived QSAR models. It is common to think that in the case of three-dimensional QSAR such as CoMFA, the results could be used to predict the modifications of known compounds that may lead to more potent ligands. Such applications are not possible using GA-PLS or KNN-QSAR methods, since the relationship between molecular descriptors such as connectivity indices or atom pairs and the underlying chemical structures are not obvious: although molecular descriptors can be calculated for any molecular structure, the opposite is not straightforward. However, as we demonstrate in this chapter, the results of QSAR analyses could be used, in a fairly straightforward manner, to search for biologically active molecules in existing databases or virtual chemical libraries. Thus, a precon-structed QSAR model can be used to directly predict biological activity of sampled chemical structures, and the selection of actual or virtual compounds can be based on the (high) value of predicted biological activity. Alternatively, the selection of active compounds can be based on their similarity to a known active probe (lead) molecule. We showed that similarity searches using the descriptor variables selected by a QSAR model (termed here topological pharmacophore), are in general more efficient than using all available descriptors. We believe that the use of QSAR modeling in rational library design (or database mining) as illustrated in this chapter should be increasingly explored to improve the design of chemical libraries with desired activity.

REFERENCES

1. Gallop MA, Barret RW, Dower WJ, Fodor SPA, Gordon EM. Applications of combinatorial technologies to drug discovery. 1. Background and peptide combinatorial libraries. J Med Chem 1994; 37:1233–1251.
2. Gordon EM, Barret RW, Dower WJ, Fodor SPA, Gallop MA. Applications of combinatorial technologies to drug discovery. 2. Combinatorial organic synthesis, library screening strategies, and future directions. J Med Chem 1994; 37:1385–1401.
3. Johnson M. Maggiora GM. Concepts and Applications of Molecular Similarity. New York: Wiley, 1990.
4. Sheridan RP, Kearsley SK. Using a genetic algorithm to suggest combinatorial libraries. J Chem Inf Comput Sci 1995; 35:310–320.
5. Zheng W, Cho SJ, Tropsha A. Rational combinatorial library design. 1. *FOCUS-2D:* A new approach to the design of targeted combinatorial chemical libraries. J Chem Inf Comput Sci 1998; 38:251–258.
6. Cho SJ, Zheng W, Tropsha A. Rational Combinatorial Library Design. 2. Application of Genetic Algorithms and the Inverse QSAR in the Design of Targeted Combinatorial Chemical Libraries. J Chem Inf Comput Sci 1998; 38:259–268.
7. Tropsha A, Cho SJ, Zheng W. "New tricks for an old dog": Development and application of novel QSAR methods for rational design of combinatorial chemical libraries and database mining. In: Parrill AL, Reddy MR, eds. Rational Drug Design: Novel Methodology and Practical Applications: ACS Symposium Series No 719. Washington, DC-American Chemical Society, 1999, pp 198–211.
8. MolConn, X 2.0. Hall Associates Consulting, Quincy, MA.
9. Carhart RE, Smith DH, Venkataraghavan R. Atom pairs as molecular features in structure–activity studies: Definition and applications. J Chem Inf Comput Sci 1985; 25:64–73.
10. The program SYBYL 6.6 is available from Tripos Associates, 1699 South Hanley Road, St Louis, MO 63144.
11. Bohachevsky IO, Johnson ME, Stein ML. Generalized simulated annealing for function optimization. Technometrics 1986; 28:209–217.
12. Kalivas JH, Sutter JM, Roberts N. Global optimization by simulated annealing with wavelength selection for ultraviolet–visible spectrophotometry. Anal Chem 1989; 61:2024–2030.

13. Goldberg DE. Genetic Algorithm in Search, Optimization, and Machine Learning. Reading, MA: Addison-Wesley, 1989.
14. Holland JH. Genetic algorithms. Sci Am 1992; 267:66–72.
15. Forrest S. Genetic algorithms: Principles of natural selection applied to computation. Science 1993; 261:872–878.
16. (a) Available from the authors at
    http://mmlin1.pha.unc.edu/~jin/QSAR/.
    (b) Hoffman BT, Cho SJ, Zheng W, Wyrick S, Nichols DE, Mailman RB, Tropsha, A. QSAR modeling of dopamine D_1 agonists using comparative molecular field analysis, genetic algorithms–partial least squares, and K nearest neighbor methods. J Med Chem 1999; 42:3217–3226.
17. Wold S, Ruhe A, Wold H, Dunn WJ III. The collinearity problem in linear regression. The partial least squares (PLS) approach to generalized inverses. SIAM J Sci Stat Comput 1984; 5:735–743.
18. Hellberg S, Sjöström M, Skagerberg B, Wold S. Peptide quantitative structure–activity relationships, a multivariate approach. J Med Chem 1987; 30:1126–1135.
19. Collantes ER, Dunn WJ III. Amino acid side chain descriptors for quantitative structure–activity relationship studies of peptide analogues. J Med Chem 1995; 38:2705–2713.
20. Ufkes JGR, Visser BJ, Heuver G, Van Der Meer C. Structure–activity relationships of bradykinin potentiating peptides. Eur J Pharm 1978; 50:119–122.
21. Lindberg W. Persson JA, Wold S. Partial least-squares method for spectrofluorimetric analysis of mixtures of humic acid and ligninsulfonate. Anal Chem 1983; 55:643–648.
22. Waller CL, Minor DL, McKinney JD. Using three-dimensional quantitative structure–activity relationships to examine estrogen receptor binding affinities of polychlorinated hydroxybiphenyls. Environ Health Perspect 1995; 103:702–707.

13

Dissimilarity-Based Compound Selection for Library Design

Valerie J. Gillet and Peter Willett

University of Sheffield
Sheffield, England

I. INTRODUCTION

A. Molecular Diversity Analysis

One of the principal objectives of molecular diversity analysis is to devise computational methods that ensure coverage of the largest possible expanse of chemical space in the search for bioactive molecules. The concept of diversity is normally quantified by using techniques derived from those developed for similarity searching in chemical databases, which involves measuring the degree of structural similarity (or dissimilarity) between two molecules by a comparison of the sets of descriptors that characterize those molecules (1). There has thus been much interest in measures of structural similarity (including both the descriptors that are employed to characterize molecules, and the coefficients that are employed to quantify the degree of resemblance between two molecules' sets of associated descriptors) and in ways in which such measures can be used in diversity analyses (2,3), in particular in methods for selecting compounds to maximize their structural diversity (4).

The similar property principle states that structurally similar molecules are likely to have similar properties and activities (5,6), with the result that unconstrained use of the combinatorial synthesis and high throughput screening processes that characterize modern pharmaceutical and agrochemical research could lead to very large amounts of redundant SAR information. Considerations of cost-effectiveness hence dictate that as few compounds as possible should be selected for synthesis and testing while still maintaining structural diversity (i.e., covering the full range of structural types that are present in a database, either real or virtual). This simple requirement presupposes a quantitative definition of structural diversity, or diversity index, and many such definitions have been described: thus, Gillet references eight different indices (2), and new ones continue to be reported in the literature. Assume that an index, I, has been defined that can be readily calculated for a set of compounds that have been chosen by some selection algorithm. Then the maximally diverse, n-compound subfile from an N-compound parent file (with, typically, $n \leq 10^4$ and $N \leq 10^6$) can be obtained by the simple procedure shown in Algorithm I.

The reader should note that this procedure will return just a single set of compounds, *Subfile,* but there may well be several, or many, that all possess the same maximal value for the chosen diversity index I; similar comments apply to several of the other algorithms discussed in this chapter.

The problem with Algorithm I is that it requires the generation and evaluation of no less than $N!/[n!(N - n)!]$ index values, and this factorial time dependency makes it computationally infeasible for values of n and N typical of those encountered in the context of chemical libraries; nonchemical studies of the maximum diversity problem are discussed by Kuo et al. (7) and Ghosh (8). The need for cost-effective approaches to the design of combinatorial syntheses has led to considerable interest in more efficient methods for selecting diverse sets of molecules, with three principal approaches having been described thus far: cluster-based, partition-based, and dissimilarity-based (DBCS) compound selection. We shall describe the first two approaches briefly before focusing upon DBCS algorithms in the remainder of this chapter. Sections II–IV review the range of algorithms that are available, and Section V then discusses one particular DBCS approach that has been developed in Sheffield.

B. Cluster-Based and Partition-Based Methods for Compound Selection

Cluster analysis, or clustering, involves subdividing a group of objects (chemical molecules in the present context) into groups, or clusters, of objects that exhibit a high degree of both intracluster similarity and intercluster dissimilarity (9,10). It is thus possible to obtain an overview of the range of structural types present within a data set by selecting one, or some small number, of the molecules from each of the clusters resulting from the use of an appropriate clustering method on that data set.

Cluster-based methods have been widely used for molecular diversity studies (see, e.g., Refs. 11–13), but they are increasingly being supplanted by dissimilarity-based and partition-based methods for compound selection. The latter approach requires the identification of a set of p characteristics, these typically being molecular properties that would be expected to affect the ability of a small molecule to bind to a protein (14). The range of values for each such characteristic is divided into a set of subranges (15). The combinatorial product of all possible subranges then defines a p-dimensional grid of bins (or cells) that is referred to as a partition, and each molecule is assigned to the bin that matches that molecule's set of characteristics. A subset is obtained by selecting one (or some small number) of the molecules from each of the bins. Partition-based selection is much faster than the other two approaches and has certain advantages: it permits the rapid identification of the sections of structural space that are underrepresented, or even unrepresented, in a database, and it is easily applied to tasks such as the identification of structural overlap in databases and the mapping of structural space (16).

C. Dissimilarity-Based Methods for Compound Selection

Cluster-based and partition-based approaches identify a set of dissimilar molecules indirectly, since the approaches involve the identification of clusters or bins, respectively, of similar molecules. DBCS, conversely, tries to identify a set of dissimilar molecules in a data set directly, using some quantitative measure of intermolecular structural dissimilarity (17). However, as will be illustrated by the algorithms discussed in this chapter, the way in which this identification is effected will be determined by several factors: whether a library (i.e., a set of compounds) or a combinatorial library (i.e., a set of compounds that can be generated by a combinatorial synthesis procedure using the minimum number of reactants) is required; and by the size and precise nature of the data set from which the compounds are to be selected. This data set may be a real data base, such as a corporate structure file, a carefully selected file of reactants, such as all the commercially available primary amines, or a virtual database, consisting of the fully enumerated set of products for a combinatorial synthesis. A structurally diverse library can be created very simply by means of what is commonly referred to as "cherry-picking," the selection of individual compounds without any consideration of the structural relationships between those selected. Cherry-picking is conceptually simple but is most unlikely to result in a combinatorial library, even if extensive subsequent processing is applied to identify commonly occurring reactants. Thus alternative approaches are required, depending upon whether the selection is carried out in reactant space or product space. The relative merits of reactant-based and product-based selection [first described by Martin et al. (18) and by Good and Lewis (19), respectively] have been much debated, particularly with regard to the greater level of diversity achievable with product-based selection (20–22).

The factorial time dependency of Algorithm I has meant that algorithms that are to be used on an operational basis for library design are, of necessity, unable to guarantee the identification of the most diverse set of compounds; instead, they achieve efficiencies of operation by means of various heuristics that provide good, but nonoptimal, sets of molecules. Many different DBCS algorithms have been reported in the literature (see, e.g., Refs. 17–20, 22, 25–46), but it is possible to identify three broad classes that encompass the various procedures that have been described: maximum dissimilarity algorithms, sphere-exclusion algorithms, and optimization algorithms (although there are, as will be seen, strong connections between these approaches).

II. MAXIMUM DISSIMILARITY ALGORITHMS

The basic maximum dissimilarity algorithm was first described by Kennard and Stone (47) and is shown in Algorithm II, where an n-molecule *Subfile* is to be chosen from an N-molecule *File*. This greedy algorithm was used in the early studies of Lajiness (25) and Bawden (26) and permits many variants to be implemented, depending upon the precise way in which steps 1 and 3 are implemented.

The starting molecule in step 1 can be obtained by, inter alia: choosing it at random, choosing that molecule that is most dissimilar to the other molecules in *File,* or choosing that molecule that is nearest to the center of of *File* [assuming a quantitative definition of the center of a set of molecules, as is required in some chemical applications of clustering (48)]. Step 3 in Algorithm II requires a quantitative definition of the dissimilarity between a single molecule in *File* and the molecules currently comprising *Subfile,* so that the most dissimilar molecule in *File* can be identified in each iteration of the algorithm. The constitution of *Subfile* will thus be determined by the way in which "most dissimilar" is defined, with different definitions leading to different final sets of molecules (29): this situation is analogous to that encountered in hierarchic agglomerative clustering, where a whole family of different, but related, classifications can be obtained by using slightly different interobject similarity coefficients (49).

Examples of dissimilarity measures that have been used for DBCS include MaxMin (31) and MaxSum (37) [these two definitions have also been studied by Ghosh in an analysis of exact solutions to the maximum diversity problem (8)]. Let $DIS(A,B)$ represent the dissimilarity between two molecules, or sets of molecules, A and B. Consider a single compound, J, taken from *File* and the m molecules that form the current membership of *Subfile* at some stage in the selection process shown in Algorithm II; then the dissimilarity between J and *Subfile,* $DIS(J, Subfile)$, is given by

$$\sum DIS\,(J,K)$$

using the MaxSum definition, and

$$minimum\{DIS(J,K)\}$$

using the MaxMin definition, with K ($1 \leq K \leq m$) ranging over all of the m molecules in *Subfile* at that point. The molecule chosen for addition to *Subfile* in step 3 (i.e., the molecule that is considered the most dissimilar) is then that which has the largest value of $DIS(J, Subfile)$. In this respect, the MaxMin and MaxSum definitions can be regarded as the DBCS equivalents of the single-linkage and group-average hierarchic agglomerative clustering methods, and it would be possible to generate other types of maximally diverse subset by using the definitions that underlie other such clustering methods, such as the median or centroid methods (9,49).

A single-linkage clustering is closely related to the minimal spanning tree (MST) for the same set of objects (50) and it is thus hardly surprising that an MST has been used for compound selection (45). Let M be an $N \times N$ dissimilarity matrix in which the element $M(I,J)$ contains the dissimilarity between the Ith and Jth molecules in a file of size N molecules. Then a spanning tree is a set of $N - 1$ matrix elements that connects all the molecules, and the MST is the spanning tree with the minimum sum of weights (i.e., minimum sum of dissimilarities in the present context); thus the diversity here is the sum of just some of the intermolecular dissimilarities, rather than all of them as in the MaxSum algorithm, and the diversity will be maximized if this minimum sum is as large as possible. This approach is appealing given that there are several fast algorithms available for generating an MST (see, e.g., Refs. 51–53). The selection algorithm of Mount et al. starts by calculating the MST for a randomly selected set of n molecules; it then tries to improve the diversity score by swapping in new molecules and calculating the resulting MST each time, and continuing until no further increase in diversity can be obtained (45). The procedure is closely related to the optimization algorithms described later in Section IV.

The basic maximum dissimilarity algorithm shown in Algorithm II has an expected time complexity of $O(n^2N)$; since n is normally some small fraction of N (such as 1 or 5%), this represents a running time that is cubic in N, which makes it extremely demanding of computational resources if *File* is at all large. Holliday et al. (28) described a MaxSum selection algorithm with a time complexity of $O(nN)$, using an equivalence that had been developed by Voorhees for the rapid implementation of group-average document clustering (54). However, analyses of the MaxSum definition by Agrafiotis and Lobanov (41) and by Mount et al. (45) showed that its use for DBCS could result in subfiles containing pairs (or larger groups) of closely related molecules. This limitation was demonstrated experimentally by Snarey et al. (46) in a comparison of different DBCS algorithms, which showed that MaxMin was more effective than MaxSum in identifying database subsets exhibiting a range of biological activities. The MaxMin dissimilarity definition is not normally as fast in practice as MaxSum, but it also can be implemented with an $O(nN)$ time complexity (31,34). In fact, since MaxMin is based on identifying nearest neighbors, it is possible to use the extensive work that has been carried out on fast algorithms for nearest-neighbor searching in low-dimensional spaces, and Agrafiotis and Lobanov have used one such approach, based on k-d trees, to obtain an algorithm with a time complexity of only $O(n \log N)$ for this purpose (41).

III. SPHERE-EXCLUSION ALGORITHMS

A significant modification of the maximum dissimilarity algorithm is obtained by specifying a threshold dissimilarity t and then rejecting the molecule selected in step 2 of Algorithm II if it has a dissimilarity less than t with any of the compounds already in *Subfile*. This is analogous to enclosing the current contents of *Subfile* within hyperspheres of radius t [in a manner reminiscent of the neighborhood criterion advocated by Patterson et al. (6) for the evaluation of diversity metrics] and yields the sphere-exclusion approach first described by Hudson et al. (30). Here, a molecule is selected, either at random or using some rational basis, for inclusion in *Subfile,* and the algorithm then excludes from further consideration all other compounds within the sphere centered on that selected compound, as shown in Algorithm III. It is not possible to derive an expected time complexity for the running time of this algorithm, since such an analysis would require information about the distribution of intermolecular dissimilarities in the file that was being processed (to be able to predict the fraction of the file excluded by each iteration of step 3). This distribution is dependent both upon the natures of the molecules involved and upon the nature of the coefficient that is used for the calculation of the dissimilarities (55,56).

Many variants are again possible, depending upon the manner in which stage 2 is implemented. Thus, one can choose the molecule that is most dissimilar to the existing *Subfile,* in which case different results will be obtained (as with the maximum dissimilarity algorithms) depending upon the dissimilarity definition that is adopted. Alternatively, a compound can be selected at random, as in the DIVPIK algorithm reported by Nilakantan et al. (35), which is very fast in operation but in which the random element results in nondeterministic solutions. Several sphere-exclusion algorithms were included in the comparative evaluation of Snarey et al. (46), who found that they were broadly comparable in effectiveness to the MaxMin algorithm.

The close relationship that exists between the maximum dissimilarity and sphere-exclusion algorithms is evidenced in work by Clark and Langton on their

OptiSim (for optimizable K-dissimilarity selection) program (36,38). Algorithm IV shows that OptiSim makes use of an intermediate pool of K selected compounds, here called *Sample*. The mode of processing and the characteristics of the final set of selected molecules are determined by the value of K that is specified, with values of K equal to 1 and to N corresponding to (versions of) the sphere-exclusion and maximum dissimilarity algorithms, respectively. Clark suggests that the latter mode of processing will yield sets of molecules that are structurally more diverse than those resulting from the sphere-exclusion approach, which are likely to be more representative of the original file from which they were selected (36); a further discussion of these complementary characteristics is presented by Clark and Langton (38).

The final algorithm to be discussed in this section can be regarded as being closely related to Algorithm III, since it again makes use of a dissimilarity threshold to exclude closely related compounds; however this threshold is used in a very different way. In addition, the algorithm provides an exact solution, rather than adopting the greedy heuristics that underlie the preceding algorithms, albeit at the cost of a large reduction in the file sizes that can be processed. We have noted that the identification of the maximally diverse subset (or subsets) by enumeration is infeasible because of the factorial nature of Algorithm I; indeed, the diverse subset problem is known to belong to the class of NP-complete algorithms for which no efficient polynomial time algorithm exists (or is even expected to exist) (7,8). However, it is possible to use the experience gained with other types of NP-complete problem (57) to devise a DBCS algorithm that is capable of identifying the maximally diverse subset when n and N are sufficiently small: for example, one might wish to select the most diverse set of 25 haloketones from all those in the Available Chemicals Directory (58) to act as one of the sets of reactants for the synthesis of a thiazoline-2-imine combinatorial library (59) (as shown in Fig. 1). This is in marked contrast to the algorithms discussed thus far, which are suitable for cherry-picking on a very large scale (given a sufficiently rapid calculation of the dissimilarities that provide the inputs to these algorithms).

Brown and Martin, who carried out a detailed verification of the similar property principle (13,60), showed that given a molecule of known activity, there is a high a priori probability that any of that molecule's near neighbors will also be active [where a near neighbor is deemed to be one that has a Tanimoto simi-

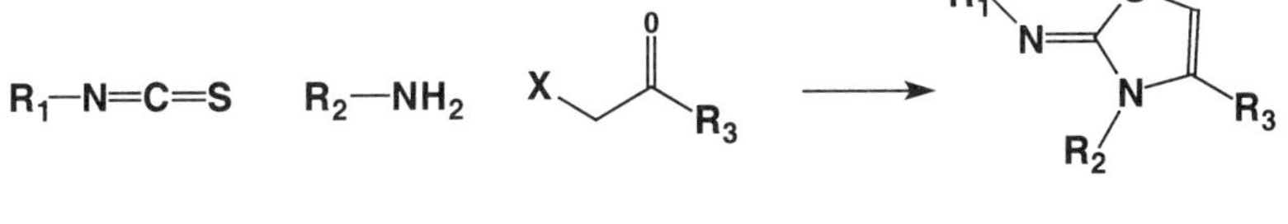

Figure 1 Combinatorial synthesis of a thiazoline-2-imine library.

larity (1) of at least 0.85 when the molecules are characterized by Tripos UNITY 2D fingerprints (61)]. This observation provides a simple basis for DBCS by ensuring that no two molecules in the final *Subfile* will be strongly similar to each other. The dissimilarity criterion is implemented by the excluded-sphere technique in Algorithm III or by applying a threshold in step 2 of Algorithm II, but it will result in the identification of only a single set of molecules meeting the criterion: the algorithm of Gardiner et al. (39) is designed to identify all such sets that satisfy it.

As with the MST algorithm described in Section II.A, let M be an $N \times N$ dissimilarity matrix in which $M(I,J)$ contains the dissimilarity between the Ith and Jth compounds in a file of N compounds; here, these molecules will typically be all the available reactants of some particular type, as described earlier. A graph G, which we refer to as a subset selection graph, is created from M by applying a threshold dissimilarity t and then setting each element $M(I,J)$ to unity (or zero) depending upon whether it is greater than (or not greater than) the threshold. The complete set of subsets satisfying the dissimilarity criterion is then the set of n-vertex cliques in G, where a clique is a subgraph in which every vertex is connected to every other vertex and is not contained in any larger subgraph with this property (62); that is, each clique in the subset selection graph thus denotes a set of n molecules for which none of the $n(n - 1)$ dissimilarities are less than t.

Gardiner et al. report a comparison of several clique detection algorithms when applied to the processing of subset selection graphs (39) and suggest that one due to Babel (63) is sufficiently fast to enable the procedure to be applied to the selection of reactants for combinatorial synthesis. Once all the subsets have been generated by the procedure, which is summarized in Algorithm V, a further filtering step, based on criteria such as cost, physicochemical parameters, or diversity index values, can be employed to identify the particular set of molecules that will be chosen for use in some application.

Clique detection is known to be NP complete, except in the case of special types of graph, and any specific NP-complete problem can be translated into any other such problem (57). It is thus to be hoped that the use of clique detection (as here) will spur researchers to consider other types of procedure that might be applied to the identification of maximally diverse sets of reactants.

IV. OPTIMIZATION ALGORITHMS

Algorithm V is able to identify all the maximally diverse subfiles only because the search space that it explores (i.e., the space of all possible n-molecule subfiles selected from a limited number of reactants) is extremely restricted, and there are no obvious algorithmic enhancements that could possibly enable it to process data sets of the size that can be encompassed by the maximum dissimilarity and sphere-exclusion algorithms discussed earlier. Similar comments would seem to apply to other exhaustive search algorithms: for example, Ghosh reports experiments to obtain exact solutions using the MaxMin and MaxSum criteria and found that they were only feasible when $n < N \leq 50$ (8).

Several researchers have thus developed methods for sampling these large chemical search spaces using techniques for combinatorial optimization. One approach has involved the theory of D-optimal and S-optimal designs (18,43,64), but this has been criticized on the grounds of selecting extreme outliers (34) and (like MaxSum) molecules that are very similar to each other (45); accordingly, we shall focus attention here on the several studies that have been carried out using genetic algorithms and simulated annealing (19,20,32,33,40,42,65).

We shall introduce the optimization approach by describing the genetic algorithm (GA) that we have developed in Sheffield for product-based library design, where the full set of products from a combinatorial synthesis is enumerated and molecules are then selected subject to the combinatorial constraint (20). Assume a two-component combinatorial synthesis in which n_1 of the N_1 possible first reactants are to be reacted with n_2 of the N_2 possible second reactants. The chromosome in the GA contains $n_1 + n_2$ elements, each specifying one possible reactant, and the cross-product of these two sets of reactants then specifies one of the size $n_1 n_2$ possible combinatorial libraries that could be synthesized, given the two complete sets of reactants. The fitness function for the GA is a diversity index quantifying the diversity of the size $n_1 n_2$ library encoded in each chromosome, and the GA thus tries to evolve chromosomes that maximize the value of this index. In principle, any index could be used for this purpose: that initially adopted (20) was the mean pairwise dissimilarity index of Turner et al. (66), which can be calculated very rapidly (a prerequisite for use in a GA-based application where very large numbers of fitness values may need to be calculated). SELECT, a development of this basic GA, is discussed in more detail later, in Section V.

Another early example of an optimization approach to product-based library design is provided by the HARPick program of Good and Lewis (19). Here, a molecule is characterized by its constituent three-point pharmacophores, these being generated from an approximate 3D structure, and the diversity of a putative combinatorial library is given by a function based on the number of distinct pharmacophores present in that particular library. Simulated annealing (SA) is used to optimize the combinatorial libraries [a GA-based version has also been described (65)], but the scoring function for the annealing not only tries to maximize the pharmacophore-based diversity but also tries to ensure an approximately even distribution of the molecules comprising the encoded library across three properties (the number of heavy atoms in a molecule, the largest triangle perimeter for any of the three-point pharmacophores in that molecule, and the largest triangle area for any of these pharmacophores) that provide crude, but rapidly computable, measures of molecular shape.

SA-based selection algorithms have also been described by Hassan et al. (32) and by Agrafiotis (33), who characterize molecules by principal components derived from calculated physical properties (topological and information content indices, and electronic, hydrophobic, and steric descriptors) or by low-dimensionality autocorrelation vectors describing the distribution of the electrostatic potential over the van der Waals surface of a molecule, respectively. The SA in both cases is driven by a scoring function that uses one of several different intermolecular distance coefficients, with Agrafiotis noting the benefits to be gained from a clear separation of selection algorithm and scoring function (33).

The principal aim of most of the selection algorithms discussed thus far is to maximize the diversity of the subfiles that result from their use. There are, however, many other criteria that may need to be considered in the design of computer bioactive molecules, such as reactant costs, synthetic feasibility, and "rule-of-five" considerations. Such criteria can be included in a lead discovery program in an initial filtering step [as reviewed by Dunbar (67) and by Walters et al. (68)] or, as advocated by Agrafiotis (33) and Good and Lewis (19), in the fitness or penalty function that drives a selection procedure. This can be achieved by drawing upon several recent "drug-ability" studies, that is, statistical analyses of chemical databases that seek to identify criteria that can discriminate between druglike and non-drug-like molecules (69–73). SELECT provides a typical example of work in this area; other such programs are HARPick, as mentioned previously, and PICCOLO developed by Zheng et al. (74).

V. DESIGNING COMBINATORIAL LIBRARIES WITH SELECT

SELECT performs product-based selection taking direct account of the combinatorial constraint. The GA that lies at its heart employs a multiobjective fitness function that allows many properties to be optimized simultaneously, facilitating the design of combinatorial libraries that are, by definition, synthetically efficient and are optimized with respect not just to diversity but also to other user-defined properties of importance in designing bioactive molecules. The program can also be used to design libraries that complement existing libraries and to explore different library configurations, as discussed in detail by Gillet et al. (42).

SELECT can be used with a range of types of descriptor: thus far, we have employed both Daylight (75) and UNITY (61) fingerprints and Molconn-Z parameters (which are real numbers that have been standardized to fall in the range $0 \ldots 1$) (76). It is also hospitable to a range of diversity indices, with those studied thus far being the sum-of-pairwise dissimilarities calculated using the cosine coefficient and implemented using the centroid algorithm of Turner et al. (66), SUM_{COS}; the sum-of-pairwise dissimilarities using the Tanimoto coefficient,

SUM$_{TAN}$; and the average nearest-neighbor distance using the Tanimoto coefficient, NN. SUM$_{COS}$ for a library of N molecules is defined as follows:

$$\frac{\sum_{J=1}^{N} \sum_{K=1}^{N} 1 - Cos(J,K)}{N^2}$$

where $Cos(J,K)$ is the similarity between molecules J and K defined using the cosine coefficient; SUM$_{TAN}$ is defined in just the same way, but with $Tan(J,K)$, the similarity between molecules J and K defined using the Tanimoto coefficient, replacing $Cos(J,K)$; and NN for a library of N molecules is defined as follows:

$$\frac{\sum_{J=1}^{N} \min(1 - Tan(J,K))}{N}$$

where $\min(1 - Tan(J,K))$ is the distance from molecule J to its closest neighbor, K.

We noted in Section I the differences between reactant-based and product-based selection, and Gillet and Nicolotti report detailed experiments demonstrating the increased level of structural diversity that can be obtained when SELECT is used to implement product-based, as against reactant-based, DBCS (22). Here, we focus upon SELECT's multicomponent fitness function which enables it to design libraries in product space where the properties of individual molecules within these libraries are optimized simultaneously with the library's structural diversity. Specifically, the physicochemical property profiles of the libraries are optimised with respect to the profile of the same property in some reference collection, for which we normally use the *World Drugs Index* (hereafter WDI) database of known drugs (77) (although any other specific collection could be used for this purpose).

The fitness function is of the form

$$w_D(D) + w_C(C) + w_{f1}\Delta f1 + w_{f2}\Delta f2 \ldots$$

where the first term, $w_D(D)$, describes the diversity of the library that is being designed, using one of the three diversity indices listed above. The second term, $w_C(C)$, is designed to force the library to be different from some existing reference collection; for example, it may be desirable to ensure that the library is maximally dissimilar from a library that has already been synthesized and tested. This weight can be set to zero if there are no such additional libraries that need to be considered. The remaining terms in the fitness function, $w_{f1}\Delta f1$, $w_{f2}\Delta f2$, and so on, relate to physical properties of molecules that are thought to affect their ability to function as a drug (such as the molecular weight, the numbers of rotatable bonds, hydrogen donors and acceptors, and the octanol/water partition coefficient) and can be calculated rapidly enough to permit the processing of libraries of realistic size. A physical property of the library is optimized by comparing the distribution of its values in the library with the distribution of values of the same property in the WDI. The various w terms act as weights that reflect the relative importance of each of the various components of the fitness function, thus allowing the designer to control the characteristics of the libraries that are produced.

The effectiveness of this procedure will be illustrated with reference to a three-component library that is based on a thiazoline-2-imine template (59), as shown in Fig. 1. Here, the R1 reactants are isothiocyanates, the R2 reactants are amines, and the R3 reactants are haloketones. Ten isothiocyanates, 40 amines, and 25 haloketones were selected at random to give a fully enumerated virtual library of 10,000 thiazoline-2-imines. The molecules in this set were represented by Daylight fingerprints and the diversity measure used was SUM_{COS}, with the number of rotatable bonds and the molecular weight as the physicochemical properties of interest; other studies of this data set are described by Gillet and Nicolotti (42). The experiments reported here compared the physicochemical property profiles of diverse libraries selected by analyzing reactant space with the profiles of the same physicochemical properties in libraries selected from product space that are optimized on property and diversity, simultaneously. In each case, the profile is recorded in a series of 20 bins, where each bin represents the percentage of compounds in the library having a given number of rotatable bonds or having molecule weight within a given range. In the case of rotatable bond profiles the bins represent the occurrence of $0, 1, 2, \ldots, > 19$ rotatable bonds, while in the case of molecular weight profiles the bins cover the following ranges: $0 \ldots 49, 50 \ldots 99 \ldots \geq 950$.

SELECT was first used to generate diverse sets of reactants (6 isothiocyanates, 10 amines, and 15 haloketones) and hence to generate a combinatorial library in reactant space containing 900 thiazoline-2-imines, for which the profiles of rotatable bonds and molecular weights were then calculated. SELECT was next run to choose an analogous 900-molecule library in product space, with the library optimized on both diversity and the rotatable bond profile, and finally in the same way but using both diversity and the molecular weight profile. In each case, the fitness function consisted of the sum of two weighted terms, the diversity term and the relevant property term. The property was included in the fitness function as the rmsd between the distribution of the property in the library represented in a chromosome and the distribution of the property in WDI, where the distributions are given as percentages. The weight assigned to diversity was 1.0 and the weight assigned to the rmsd of the property was 0.1, these weights being chosen so that the rmsd property values were approximately in the same range of values as diversity. The results of these runs are illustrated in Fig. 2, where it can be seen that that simple reactant-based selection often results in libraries with poor physicochemical property profiles. The product-based selection, conversely, has enabled the construction of libraries with profiles that are much more "WDI-like" and are thus more likely to contain bioactive compounds.

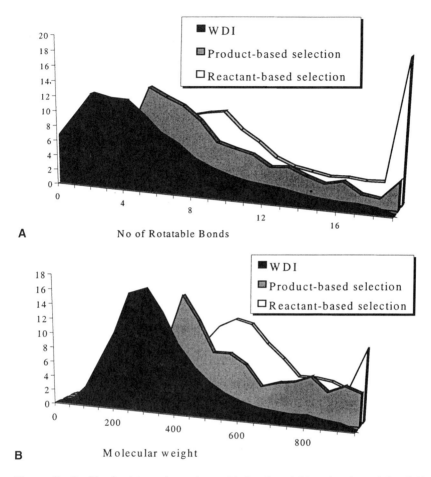

Figure 2 Profiles for (a) numbers of rotatable bonds and (b) molecular weight of thiazoline-2-imine libraries.

VI. CONCLUSIONS

A selection procedure for library design involves three major components: the descriptors that are used to characterize the molecules in a data set, a subset selection algorithm that operates on these descriptions, and a diversity index that quantifies the degree of structural diversity in the resulting subset. This chapter highlights the second of these components, summarizing some of the many DBCS algorithms that have been reported in the literature and are now being increasingly widely used for library design. This algorithmic focus means that little attention has been given here to the biological activities of the molecules selected by these procedures. There are, however, an increasing number of empirical studies now appearing in the literature that take account of bioactivity data in evaluating selection algorithms: reviews of this work up till mid-1998 are presented by Willett (4) and by Snarey et al. (46), while more recent studies involving DBCS include those by Potter and Matter (78,79) and by Bayada et al. (80).

Work in Sheffield has involved several different DBCS algorithms, but we are now focusing attention upon the GA-based approach embodied in SELECT. GAs have been widely adopted for various aspects of molecular diversity analysis in addition to those discussed here (see, e.g., Refs. 81–84): they seem particularly well suited to the present application owing to the ease with which it is possible to include a range of additional molecular features in the optimization. SELECT currently employs just a few, rapidly calculated physicochemical properties, but many other characteristics could be included, such as the reliability of a reactant supplier or the synthetic feasibility of each product molecule. We hence believe that this approach can provide the chemist with a powerful tool to design libraries that not only are diverse but also contain molecules that are typical of those encountered in realistic lead development programs.

ACKNOWLEDGMENTS

We thank Peter Artymiuk, John Bradshaw, Eleanor Gardiner, Darren Green, John Holliday, Orazio Nicolotti, Mike Snarey, Nick Terret, and David Wilton for their contributions to the work on dissimilarity-based compound selection that has been carried out in Sheffield. This work has been funded by the Engineering and Physical Sciences Research Council, GlaxoWellcome Research and Development, Pfizer Central Research, and Tripos, Inc., with software support being provided by Daylight Chemical Information Systems and Tripos, Inc. The Krebs Institute for Biomolecular Research is a designated center of the Biotechnology and Biological Sciences Research Council.

ALGORITHMS

Algorithm I. DBCS Using Enumeration

1. Initialize I to $-\infty$.
2. Generate a new n-molecule *Subfile* and calculate the value of its diversity index, i.
3. If $i > I$ then $I: = i$ and store *Subfile*.
4. Return to step 2 if not every possible distinct *Subfile* has been tested.

Algorithm II. DBCS Using a Maximum Dissimilarity Algorithm

1. Initialize *Subfile* by transferring a molecule from *File*.
2. Calculate the dissimilarity between each remaining molecule in *File* and the molecules in *Subfile*.
3. Transfer to *Subfile* that molecule from *File* that is most dissimilar to *Subfile*.
4. Return to step 2 if there are fewer than *n* molecules in *Subfile*.

Algorithm III. DBCS Using a Sphere-Exclusion Algorithm

1. Define a threshold dissimilarity *t*.
2. Transfer a molecule from *File* to *Subfile*.
3. Remove from *File* all molecules that have a dissimilarity with the transferred molecule of less than *t*.
4. Return to step 2 if there are molecules remaining in *File*.

Algorithm IV. DBCS Using the OptiSim Algorithm (36,38)

1. Define a threshold dissimilarity *t*.
2. Initialize *Subfile* by transferring a molecule from *File*.
3. Select a molecule from *File*. If it has a dissimilarity less than *t* with any molecule in *Subfile* then remove it from *File;* otherwise add it to *Sample*.
4. Repeat step 3 until *Sample* contains *K* molecules.
5. Transfer to *Subfile* that molecule from *Sample* that is most dissimilar to *Subfile*. Return the remaining members of *Sample* to *File*.
6. Return to step 3 if there are fewer than *n* molecules in *Subfile*.

Algorithm V. DBCS Using a Clique Detection Algorithm to Identify All Subsets Meeting a Dissimilarity Criterion

1. Define a threshold dissimilarity *t*.
2. Generate an $N \times N$ dissimilarity matrix in which $M(I,J)$ contains the dissimilarity between molecules *I* and *J*.
3. Generate a graph *G* from *M* by setting each element $M(I,J)$ to one (or zero) if it is greater than (or not greater than) *t*.
4. Use a clique detection algorithm to identify the set of size-*n* cliques in *G*.

REFERENCES

1. Willett P, Barnard JM, Downs GM. Chemical similarity searching. J Chem Inf Comput Sci 1998; 38:983–996.
2. Gillet VJ Background theory of molecular diversity. In: Dean PM, Lewis RA, eds. Molecular Diversity in Drug Design. Dordrecht: Kluwer, 1999, pp 43–65.
3. Clark RD. Relative and absolute diversity analysis of combinatorial libraries. This volume, Chap. 11.
4. Willett P. Subset-selection methods for chemical databases. In: Dean PM, Lewis RA, eds. Molecular Diversity in Drug Design. Dordrecht: Kluwer, 1999, pp 115–140.
5. Johnson MA, Maggiora GM, eds. Concepts and Applications of Molecular Similarity. New York: Wiley, 1990.
6. Patterson DE, Cramer RD, Ferguson AM, Clark RD, Weinberger LE. Neighbourhood behaviour: A useful concept for validation of "molecular diversity" descriptors. J Med Chem 1996; 39:3049–3059.
7. Kuo C-C, Glover F, Dhir KS. Analyzing and modelling the maximum diversity problem by zero-one programming. Decision Sci 1993; 24:1171–1185.
8. Ghosh JB. Computational aspects of the maximum diversity problem. Oper Res Lett 1996; 19:175–181.
9. Sneath PHA, Sokal RR Numerical Taxonomy. San Francisco: Freeman, 1973.
10. Everitt BS. 1993. Cluster Analysis. 3rd ed. London: Edward Arnold, 1993.
11. Shemetulskis NE, Dunbar JB, Dunbar BW, Moreland DW, Humblet C. Enhancing the diversity of a corporate database using chemical database clustering and analysis. J Comput-Aid Mol Design 1995; 9:407–416.
12. Dunbar JB. Cluster-based selection. Perspect Drug Discovery Design 1997; 7/8:51–63.
13. Brown RD, Martin YC. Use of structure–activity data to compare structure-based clustering methods and descriptors for use in compound selection. J Chem Inf Comput Sci 1996; 36:572–584.
14. Mason JS, Pickett SD. Partition-based selection. Perspect Drug Discovery Design 1997; 7/8:85–114.
15. Bayley MJ, Willett, P. Binning schemes for partition-based compound selection. J Mol Graph Modelling 1999; 17:10–18.
16. Pearlman RS, Smith KM Metric validation and the receptor-relevant subspace concept. J Chem Inf Comput Sci 1999; 39:28–35.
17. Lajiness MS. Dissimilarity-based compound selection techniques. Perspect Drug Discovery Design 1997; 7/8:65–84.
18. Martin EJ, Blaney JM, Siani MA, Spellmeyer DC, Wong AK, Moos WH. Measuring diversity: Experimental design of combinatorial libraries for drug discovery. J Med Chem 1995; 38:1431–1436.
19. Good AC, Lewis RA. New methodology for profiling combinatorial libraries and screening sets: Cleaning up the design process with HARPick. J Med Chem 1997; 40:3926–3936.
20. Gillet VJ, Willett P, Bradshaw J. The effectiveness of reactant pools for generating structurally diverse combinatorial libraries. J Chem Inf Comput Sci 1997; 37:731–740.

21. Jamois EA, Hassan M, Waldman M. Evaluation of reagent-based and product-based strategies in the design of combinatorial library subsets. J Chem Inf Comput Sci 2000; 40:63–70.
22. Gillet VJ, Nicolotti O. New algorithms for compound selection and library design. Perspect Drug Discovery Design 2000; 20:265–287.
23. Reference deleted in proofs.
24. Reference deleted in proofs.
25. Lajiness MS. Molecular similarity-based methods for selecting compounds for screening. In: Rouvray DH, ed. Computational Chemical Graph Theory. New York: Nova Science Publishers, 1990, pp 299–316.
26. Bawden D. Molecular dissimilarity in chemical information systems. In: Warr WA, ed., Chemical Structures. 2. The International Language of Chemistry. Heidelberg: Springer-Verlag, 1993, pp 383–388.
27. Marengo E, Todeschini R. A new algorithm for optimal, distance-based experimental design. Chemomet Intell Lab Syst 1992; 16:37–44.
28. Holliday JD, Ranade SS, Willett P. A fast algorithm for selecting sets of dissimilar structures from large chemical databases. Quant Struct–Act Relat 1995; 14:501–506.
29. Holliday JD, Willett P. Definitions of 'dissimilarity' for dissimilarity-based compound selection. J Biomol Screen 1996; 1:145–151.
30. Hudson BD, Hyde RM, Rahr E, Wood J. Parameter based methods for compound selection from chemical databases. Quant Struct–Act Relat 1996; 15:285–289.
31. Polinsky A, Feinstein RD, Shi S, Kuki A. LiBrain: Software for automated design of exploratory and targeted combinatorial libraries. In: Chaiken IM, Janda KD, eds. Molecular Diversity and Combinatorial Chemistry. Libraries and Drug Discovery. Washington, DC: American Chemical Society, 1996, pp 219–232.
32. Hassan M, Bielawski JP, Hempel JC, Waldman M. Optimization and visualization of molecular diversity of combinatorial libraries. J Comput-Aid Mol Design 1996; 2:64–74.
33. Agrafiotis DK Stochastic algorithms for maximising molecular diversity. J Chem Inf Comput Sci 1997; 37:841–851.
34. Higgs RE, Bemis KG, Watson IA, Wikel JH. Experimental designs for selecting molecules from large chemical databases. J Chem Inf Comput Sci 1997; 37:861–870.
35. Nilakantan R, Bauman N, Haraki KS. Database diversity assessment: New ideas, concepts and tools. J Comput-Aid Mol Design 1997; 11:447–452.
36. Clark RD. OptiSim: An extended dissimilarity selection method for finding diverse representative subsets. J Chem Inf Comput Sci 1997; 37:1181–188.
37. Pickett SD, Luttman C, Guerin V, Laoui A, James E. DIVSEL and COMPLIB—Strategies for the design and comparison of combinatorial libraries using pharmacophore descriptors. J Chem Inf Comput Sci 1998; 38:144–150.
38. Clark RD, Langton WJ. Balancing representativeness against diversity using optimizable *k*-dissimilarity and hierarchical clustering. J Chem Inf Comput Sci 1998; 38:1079–1086.
39. Gardiner EJ, Holliday JD, Willett P, Wilton DJ, Artymiuk PJ. Selection of reagents for combinatorial synthesis using clique detection. Quant Struct–Act Relat 1998; 17:232–236.

40. Zheng W, Cho SJ, Tropsha A. Rational combinatorial library design. 1. Focus-2D: A new approach to the design of targeted combinatorial libraries. J Chem Inf Comput Sci 1998; 38:251–258.

41. Agrafiotis D, Lobanov VS. An efficient implementation of distance-based diversity measures based on k-d trees. J Chem Inf Comput Sci 1999; 39:51–58.

42. Gillet VJ, Willett P, Bradshaw J, Green DVS. Selecting combinatorial libraries to optimise diversity and physical properties. J Chem Inf Comput Sci 1999; 39:169–177.

43. Martin EJ, Crichlow RW. Beyond mere diversity: Tailoring combinatorial libraries for drug discovery. J Comb Chem 1999; 1:32–45.

44. Flower DR. DISSIM: A program for the analysis of chemical diversity. J Mol Graph Modelling 1998; 16:239–253.

45. Mount J, Ruppert J, Welch W, Jain AN. IcePick: A flexible surface-based system for molecular diversity. J Med Chem 1999; 42:60–66.

46. Snarey M, Terret NK, Willett P, Wilton DJ. Comparison of algorithms for dissimilarity-based compound selection. J Mol Graph Modelling 1998; 15:372–385.

47. Kennard RW, Stone LA. Computer aided design of experiments. Technomet 1969; 11:137–148.

48. Willett P. Similarity and Clustering in Chemical Information Systems. Letchworth: Research Studies Press, 1987.

49. Lance GN, Williams WT. A general theory of classificatory sorting strategies. I. Hierarchical systems. Comput J 1967; 9:373–380.

50. Gower JC, Ross GJS. Minimum spanning trees and single linkage cluster analysis. Appl Stat 1969; 18:54–64.

51. Prim RC. Shortest connection networks and some generalizations. Bell Syst Tech J 1957; 37:1389–1401.

52. Bentley JL, Friedman JH. Fast algorithms for constructing minimal spanning trees in coordinate spaces. IEEE Trans Comput 1978; C27:97–105.

53. Whitney VKM. Algorithm 422. Minimal spanning tree. Commun ACM 1972; 15:273–274.

54. Voorhees EM. Implementing hierarchic clustering algorithms for use in document retrieval. Inf Process Manag 1986; 22:465–476.

55. Flower DR. On the properties of bit string-based measures of chemical similarity. J Chem Inf Comput Sci 1998; 38:379–386.

56. Godden JW, Xue L, Bajorath J. Combinatorial preferences affect molecular similarity/diversity calculations using binary fingerprints and Tanimoto coefficients. J Chem Inf Comput Sci 2000; 40:163–166.

57. Garey MR, Johnson DS. Computers and Intractability: A Guide to the Theory of NP-Completeness. San Francisco: Freeman, 1979.

58. The Available Chemicals Directory is available from MDL Information Systems, Inc. http://www.mdli.com/

59. Watson S. Solution phase synthesis of libraries based on thiazole templates. Presented at the Strategic Research Institute, Third Annual Random and Rational Conference, Geneva, 1996.

60. Brown RD, Martin YC. The information content of 2D and 3D structural descriptors relevant to ligand–receptor binding. J Chem Inf Comput Sci 1997; 37:1–9.

61. The UNITY system for chemical information management is available from Tripos, Inc. http://www.tripos.com/
62. Pardalos PM, Xue J. The maximum clique problem. J Global Optimiz 1994; 4:301–328.
63. Babel L. Finding maximum cliques in arbitrary and special graphs. Computing 1991; 46:321–341.
64. Andersson PM, Linusson A, Wold S, Sjostrom M, Lundstedt T, Norden B. Design of small libraries for lead exploration. In: Dean PM, Lewis RA, eds. Molecular Diversity in Drug Design. Dordrecht: Kluwer, 1999, pp 197–220.
65. Lewis RA, Good AC, Pickett SD. Quantification of molecular similarity and its application to combinatorial chemistry. In: van de Waterbeemd H, Testa B, Folkers G, eds. Computer-Assisted Lead Finding and Optimization. Weinheim: Wiley-VCH, 1997, pp 137–155.
66. Turner DB, Tyrrell SM, Willett P. Rapid quantification of molecular diversity for selective database acquisition. J Chem Inf Comput Sci 1997; 37:18–22.
67. Dunbar JB. Compound acquisition strategies. In: Altman RB, Dunker AK, Hunter L, Lauderdale K, Klein TE, eds. Pacific Symposium on Biocomputing 2000. Singapore: World Scientific, 2000, pp 555–565.
68. Walters WP, Stahl MT, Murcko MA. Virtual screening—An overview. Drug Discovery Today 1998; 3:160–178.
69. Gillet VJ, Willett P, Bradshaw J. Identification of biological activity profiles using substructural analysis and genetic algorithms. J Chem Inf Comput Sci 1998; 38:165–179.
70. Ajay, Walters WP, Murcko MA. Can we learn to distinguish between "drug-like" and "non-drug-like" molecules? J Med Chem 1998; 41:3314–3324.
71. Sadowski, J. Druglikeness profiles of chemical libraries. This volume, Chap. 9.
72. Clark DE, Pickett SD. Computational methods for the prediction of 'druglikeness'. Drug Discovery Today 2000; 5(2):49–58.
73. Wang J, Ramnarayan K. Toward designing drug-like libraries: A novel computational approach for prediction of drug feasibility of compounds. J Comb Chem 1999; 1:524–533.
74. Zheng W, Hung ST, Saunders JT, Seibel GL. PICCOLO: A tool for combinatorial library design via multicriterion optimization. In: Altman RB, Dunker AK, Hunter L, Lauderdale K, Klein TE, eds. Pacific Symposium on Biocomputing 2000. Singapore: World Scientific, 2000, pp 588–599.
75. Daylight Chemical Information Systems, Inc. http://www.daylight.com/
76. The MolConn-Z software is available from eduSoft LC. http://www.eslc.vabiotech.com/
77. The World Drugs Index is available from Derwent Information. http://www.derwent.com/
78. Potter T, Matter H. Random or rational design? Evaluation of diverse compound subsets from chemical structure databases. J Med Chem 1998; 41:478–488.
79. Matter H, Potter T. Comparing 3D pharmacophore triplets and 2D fingerprints for selecting diverse compound subsets. J Chem Inf Comput Sci 1999; 39:1211–1225.

80. Bayada DM, Hamersma H, van Geerestein VJ. Molecular diversity and representativity in chemical databases. J Chem Inf Comput Sci 1999; 39:1–10.

81. Gobbi A, Poppinger D. Genetic optimization of combinatorial libraries. Biotechnol Bioeng 1998; 61:47–54.

82. Sheridan RP, Kearsley SK. Using a genetic algorithm to suggest combinatorial libraries. J Chem Inf Comput Sci 1995; 35:310–320.

83. Liu DX, Jiang HL, Chen KX, Ji RY. A new approach to design virtual combinatorial library with genetic algorithm based on 3D grid property. J Chem Inf Comput Sci 1998; 38:233–242.

84. Weber L, Almstetter M. Diversity in very large libraries. In: Dean PM, Lewis RA, eds. Molecular Diversity in Drug Design. Dordrecht: Kluwer, 1999, pp 93–113.

14

Pharmacophore-Based Approaches to Combinatorial Library Design

Andrew C. Good

Bristol-Myers Squibb
Wallingford, Connecticut

Jonathan S. Mason

Bristol-Myers Squibb
Princeton, New Jersey

Darren V. S. Green and Andrew R. Leach

Glaxo Wellcome
Stevenage, England

I. INTRODUCTION

Ligand–protein binding interactions are primarily governed by three-dimensional (3D) spatial interactions. Consequently, in an ideal world one would wish to apply 3D descriptors when undertaking molecular selection for most computer-aided molecular design (CAMD) problems. Unfortunately, even the fastest approaches typically require molecular superimposition in coordinate space before most 3D molecular descriptors can be utilized (1). As a result, the CPU time needed for such calculations is too large to cope with the virtual data set sizes of most combinatorial library calculations or compound selection exercises. To overcome this problem, a number of descriptors with 3D content have been devised that circumvent the need for molecular superposition. In their original guises such descriptors were primarily designed as shape similarity measures based on matching interatomic distance distributions (2–5). It was soon realized, however, that elements of molecular chemistry could be incorporated to produce full pharma-cophoric fingerprints (6–10). A pharmacophore is defined here as a critical three-dimensional geometric arrangement of molecular fragments forming a necessary but not sufficient condition for biological activity (11,12). The use of such descriptors has formed a mainstay of ligand-based virtual screening for much of the last decade. These descriptors have proven to be excellent for divorcing the 3D structural requirements for biological activity from the 2D chemical makeup of a ligand. The resulting measures are thus able to exploit even limited data regarding a target to discover structurally novel active chemotypes. This proven ability, together with calculation speeds that permit their use on data sets deemed too large for most other 3D measures, make them attractive as combinatorial library descriptors. In this chapter we further highlight reasons for their utility and detail the techniques applied thus far in their exploitation.

II. WHY USE 3D PHARMACOPHORES? A SHORT HISTORY OF THEIR APPLICATION IN CAMD

Single-pharmacophore searching has been a successful method of virtual screening for many years now. As we have already mentioned, we are focusing on pharmacophores based on 3D arrangements of critical functionality. In general such functionality is atom based and is defined in terms of generic chemical properties (e.g., acid, base, hydrophobe, aromatic, etc.), although other properties (e.g., planes, normals, potential target atom positions) are sometimes used. The relative positions of these features can be described using interfeature distances and/or by specifying 3D coordinates (typically with some spherical tolerance). Such methods provide an excellent paradigm for discovering structurally novel active chemotypes based on potential ligand binding modes, with hit rates for selected data sets of 1–20% (depending on the quality of the pharmacophore and care taken in compound filtering) (13–18). The importance of these techniques is such that they have been reviewed many times (19–21) and continue to be an active area of research (22–29).

Such techniques have also been extended to encompass structure-based virtual screening. By exploiting structural information taken directly from the target active site, it is possible to discover ligands with both diverse chemotypes and binding modes. As a result, structure-based screening is potentially even more powerful than ligand-based pharmacophore searching (30–34). This advantage is also the source of a major problem for structure-based searching tools, however, since the thousands of putative ligand–receptor orientations tested make the searching process very CPU intensive. In the past, it was not uncommon to spend one or two weeks searching a database of 100,000 structures [DOCK (35) searching, single conformers only]. To combat this, methods were developed to introduce active site pharmacophore constraints into the searching model (36–40). When combined with the definition active site critical regions (e.g., a crucial salt bridge interaction)

(41), the resulting site point definition is essentially a series of *n*-point pharmacophores (for a DOCK search, *n* would be the number of matching nodes requested). Using such an approach dramatically decreases the number of nonproductive (low chemical complementarity) ligand–receptor orientations, producing increases in search speed (up to two orders of magnitude) and better hit rates (42).

The searches described thus far generally require either a postulated crucial pharmacophore for binding or a target active site. It is often the case, however, that this amount of structural information is not available, with only a single lead (e.g., competitor ligand or peptide substrate) being accessible. With technologies such as genomics lining up an ever increasing queue of potential targets with only limited biological information, this scenario is likely to become increasingly common. Further, the virtual data set sizes that must be analyzed for combinatorial chemistry library are generally beyond the 3D search techniques requiring molecular superpositions. As a result, alternative 3D technologies are required for application in these situations. This has been accomplished primarily through the creation of pharmacophore descriptors using atom pair histograms (7) and pharmacophore triplet (6,8) and quartet (9,10) binary fingerprints (see Fig. 1 for a sum-

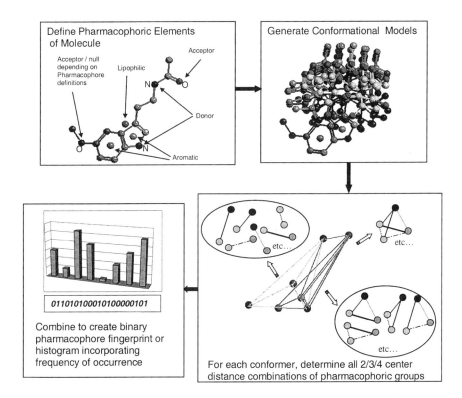

Figure 1   General paradigm for pharmacophore-based fingerprint/histogram creation.

mary of descriptor generation). These techniques have been shown to extract active compounds with diverse structures from databases using the fingerprint of a single lead [substrate (including peptide) and small-molecule inhibitor] (7,10).

The foregoing examples clearly illustrate that pharmacophore descriptors contain important information relating to potential biological activity. Together with the relatively rapid search speeds and alignment-independent nature of pharmacophore fingerprints, this makes such descriptors natural choices for combinatorial library design. Such descriptors can be used in several ways. They can be used in a simple way as a biologically relevant measure of diversity. The goal is then to select reagents that give products maximizing the coverage of possible pharmacophore shapes. Alternatively, focused libraries can be created by using pharmacophore constraint fingerprints. These can be based on an active site (43), target family (9), or something more generic—for example, biasing based on comparison with the pharmacophores of a large set of druglike molecules (9,43) [such as the MDL Drug Data Report (44) (MDDR)]. The pharmacophore fingerprint can encode multiple hypotheses, with the optimization goal of maximizing library overlap with this fingerprint. In the text that follows, we examine a number of such techniques.

III. PHARMACOPHORES IN COMBINATORIAL LIBRARY DESIGN: THE DIVISION OF DESCRIPTOR SPACE AND OTHER IMPORTANT ISSUES

The two most widely applied methods for dividing descriptor space in combinatorial library design involve the application of clustering techniques (45–48) and cell-based partitioning (49,50) (see also Chap. 16, this volume). Clustering methodology can be defined as the division of a group of objects into clusters with high intracluster similarity and intercluster dissimilarity. Partitioning involves the subdivision of property space into a number of regions (bins). Partition-based profiling is then generally defined as the selection of an object subset for maximal coverage of these property bins.

The primary reason for constructing large diverse libraries is to enhance the diversity of in-house screening databases. Understanding the descriptor space already covered is thus crucial to ensuring diverse selections relative to an existing data set. Similarly for focused libraries it is often important to understand which areas of descriptor space are already covered, since in general it is these regions that we wish to occupy with the new data set. Classic clustering techniques (51,52) do not usually encompass the concept of data set comparison. As a consequence, while they are fine for determining an internally diverse set, they are not adept at calculating interlibrary diversity. It is possible to use clustering in terms of minimizing (diverse library) and maximizing (focused library) similarity relative to an existing data set (53). While this technique is sound for focused libraries, it is not generally suitable for database enhancement. This is because the CPU bottleneck for such a calculation is the comparison of potential library compounds with the existing data set. If this reference molecule group is large (as is the case for in-house screening databases), the time required for library analysis can become prohibitively long. The calculation times of partitioning techniques tend to scale linearly with the number of molecules being processed. As a result of this advantage, the paradigm is more suitable (faster) for the large data sets that must be analyzed in combinatorial library design. Further, the resultant cells provide a consistent frame of reference in property space, making comparison and visualization between molecular data sets a simple process.

Pharmacophores provide an excellent measure for use with partitioning, since each pharmacophore naturally forms a single cell within descriptor space. With clustering, the use of pharmacophores is trickier, since the molecular fingerprints tend to be both too sparse and too sensitive:

1. *Too sparse.* For example, the average "druglike" molecule in the Standard Drug File (SDF) (59) has about 250 pharmacophores in a three-point pharmacophore descriptor regimen with 184,884 theoretically accessible pharmacophores (55), making the similarity measure very discontinuous. This problem is made serious because many implementations of pharmacophore fingerprint (56,57) use a large number of distance bins—for example, the 32 bin settings (with narrow 0.1–1.0 Å bin ranges used by default in Chem-Diverse (56,58)—to increase pharmacophore diversity. Unfortunately, to ensure rapid keying times, conformational searching tends to be coarse. As a result, the extra molecular separation produced by high bin count leads to descriptor artifacts, since a small difference in distance can easily lead to different distance bin assignments.

2. *Too sensitive.* The number of pharmacophores present in a given molecule varies as $n(n - 1)/2!$, $n(n - 1)(n - 2)/3!$ and $n(n - 1)(n - 2)(n - 3)/4!$ times the number of pharmacophore points, for two-, three-, and four-point pharmacophore descriptors, respectively. Small molecular differences can thus potentially lead to large differences in fingerprint.

Since clustering methods generally employ data from molecular similarity comparisons, these issues require mitigation before pharmacophore can be employed in this manner. This has been accomplished in three ways.

1. The assignment of coarser bins (8,9), with as few as six bins used with sizes of 2.5–5.0 Å.

2. The application of bin "bleeding." This technique increments bins on either side of the bin in which the distance is found to reside (7,59).

3. Modification of the Tanimoto index (9,52) denominator (the most commonly applied equation for similarity calculations) to reduce sensitivity to total pharmacophore count. Equations (1) and (2) highlight these potential modifications.

$$T_{AB} = \frac{\sum_{i=1}^{N} P_{A_i} P_{B_i}}{\sum_{i=1}^{n} P_{A_{i_2}} + \sum_{i=1}^{n} P_{A_{i_2}} - \sum_{i=1}^{n} P_{A_i} P_{B_i}} \rightarrow T_{AB}^{M} \tag{1}$$

$$T_{AB}^{M} = \frac{\sum_{i=1}^{n} P_{A_i} P_{B_i}}{0.25 \sum_{i=1}^{n} P_{A_i} \left(1 - \frac{\sum_{i=1}^{n} P_{A_i} P_{B_i}}{\sum_{i=1}^{n} P_{B_i}}\right) + 0.05 \sum_{i=1}^{n} P_{B_i} + \sum_{i=1}^{n} P_{A_i} P_{B_i}} \tag{2}$$

where T = Tanimoto similarity

T^{M} = modified Tanimoto similarity

P = property (in this case 0/1 from binary pharmacophore fingerprint)

A = database molecule

B = reference molecule

n = number of theoretically accessible pharmacophores

i = ith pharmacophore

With suitable preparation, pharmacophores are thus amenable to both methods for dividing descriptor space, examples of which will be given later in the chapter.

Another important issue to consider is that of descriptor validation. This is a complex subject that contains many potential pitfalls for those attempting to address it. Sections IV–VIII describe a number of methods used to apply pharmacophores to library design. The methodologies covered should allow the reader to gain a deeper understanding of the issues involved before the topic of validation is tackled in more detail (Sect. IX).

IV. INITIAL APPROACHES TO PHARMACOPHORE LIBRARY DESIGN: CHEM-DIVERSE

The program Chem-Diverse (58,59) was the first commercial offering (60) to exploit 3- (and more recently 4-) point pharmacophore information in diversity analysis. The software provides a number of useful tools for library design and is probably the most widely used software for this form of profiling. The Chem-Diverse protocol for molecular diversity uses an algorithm designed to maximize coverage of pharmacophore space by potential combinatorial chemistry products (Fig. 2). A core paradigm in Chem-Diverse is the generation and comparison of pharmacophore fingerprints on the fly. The pharmacophores of any accepted molecule are added to a binary ensemble pharmacophore key describing the entire data set selected so far. Compounds are deemed acceptable if the pharmacophores they express overlap with the ensemble key by less than a user-defined amount: that is, the molecule contains a significant number of previously unseen pharmacophores. As a consequence, the results of such searches are dependent on the order in which

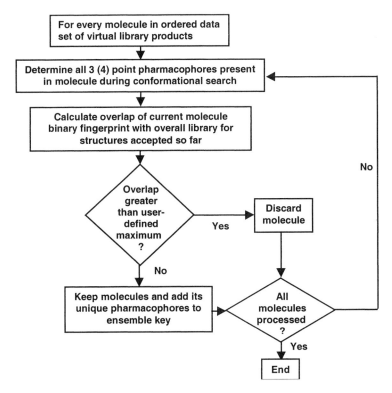

Figure 2 Outline of Chem-Diverse compound selection strategy. (Adapted from Ref. 55.)

the molecules are selected from the database of potential library products. In its default form the program relies exclusively on the binary ensemble fingerprint to guide explicit selection. Additional constraints regarding, for example, druglikeness, flexibility, and size must be treated implicitly through filtering and virtual library preordering (e.g. from smallest to largest). The binary form of the ensemble fingerprints also limits their utility, since such a key registers only whether a particular pharmacophore exists in the selected molecular ensemble, not how many times it is found. As a result, the ensemble key is prone to saturation, even when artificially small distance bins are applied, as they are in the default Chem-Diverse settings (32 bins with bins widths as small as 0.1 Å). Further, Chem-Diverse chooses molecules based purely on what it considers to be the most diverse set of products ("cherry-picking"), with no explicit reference to the constituent reagents. Consequently, the program will often make a combinatorially inefficient selection of products. By inefficient we mean that, for example, when selecting 100 products for a two-component combinatorial library, rather than choosing a 100% efficient 10×10 reagent set, the software will choose products containing a larger reagent subset, say 28×17. Using such a selection would be both costly and complicated to program up on the synthesis robot and is thus termed inefficient (and is generally unpopular with chemists). To achieve an efficient subset using Chem-Diverse, increasingly constrained successive selections must be executed and analyzed to determine the most frequently occurring reagents.

V. ADDRESSING DEFICIENCIES IN THE CHEM-DIVERSE APPROACH: THE HARPICK PROGRAM

To address many of these Chem-Diverse issues, Good et al. (55,61,62) created the HARPick program. The basic outline of HARPick is illustrated in Fig. 3. This

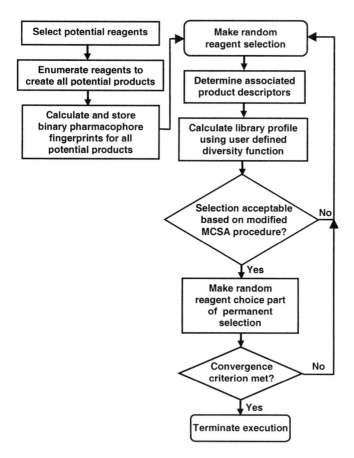

Figure 3 Outline of HARPick compound selection strategy. (Adapted from Ref. 55.)

software was designed to overcome many of the issues associated with the Chem-Diverse protocol described above.

A. Introducing Stochastic Optimization

The main structure of the HARPick program revolves around the application of stochastic optimization [in this case Monte Carlo simulated annealing (63): MCSA]. The use of such methodology offers a number of potential advantages.

1. Reagent selection and diversity evaluation can be decoupled from each other. It is then possible to make selections in reagent space, while diversity is calculated in product space. Consequently, the user still has direct control over the number of reagents selected from each component of a combinatorial synthesis. As a result it is possible to undertake product-based calculations that ensure reagent selections of maximum combinatorial efficiency.

2. Because of this independence, it is a simple task to introduce flexible scoring functions that include many diverse properties. This is important, since the nature of a diverse molecule is such that it will often tend to be an outlier in property space. For example, in pharmacophore space polyfunctional large flexible molecules, which tend to have large diverse fingerprints, are a popular choice in an unconstrained calculation. Thus, as well as including pharmacophores as our primary descriptor, it is important to add secondary descriptors. The real advantage of this is that such properties do not necessarily need to be made optimally diverse. Rather, they can be designed to act as moderating descriptors, ensuring a balance between a desired diversity profile and sensible compound selection.

B. Pharmacophore Profiling Alterations

While the standard three-point, seven-type (hydrogen bond donor, hydrogen bond acceptor, basic, aromatic, hydrophobe, acidic, variable, e.g., hydrogen bond donor and acceptor, privileged: (see Sect. VII) Chem-Diverse pharmacophores were also used by HARPick, numerous changes were made to improve performance.

1. The primary alteration was to reduce from 31 to 17 the number of distance bins used in the key creation. This increased coarseness was justified by the coarseness of the Chem-Diverse conformational search. The 17 bins have been tailored to approximate the 20% tolerance determined experimentally for 3D database searches involving rule-based conformational analyses (64) (it may be that this is still not coarse enough, since as we mentioned in Sect. III, some applications have reduced the bin count as low as 6). The resulting combination of 84 variants of three points with 17 bins produces a total of 184,884 geometrically accessible pharmacophores.

2. To use a stochastic optimization algorithm for reagent selection, molecular pharmacophore keys need to be accessed as and when required. It was thus not possible to use the on-the-fly generation procedure employed in Chem-Di-

verse. Rather, procedures were developed to store the keys in compressed form ahead of the HARPick run.

3. The majority of pharmacophores present in a molecule tend to be small relative to the largest pharmacophore in the structure, and thus they are less likely to play a key role in ligand binding. A novel self-consistent method for pharmacophore removal was thus developed to permit the user to strip these pharmacophores from the fingerprint. The technique allows the user to determine the minimum ratio required for pharmacophore perimeter, relative to the largest perimeter found for all pharmacophores in the current structure. Since all calculations are carried out with respect to the internal pharmacophore geometries of each individual molecule, no structure can lose all its pharmacophores, only the ones defined as small. The resulting fingerprints are more compact (often by $> 60\%$) and should provide a better description of the relevant molecular properties.

C. Implementation of a Customizable Scoring Function

To fully exploit the advantages of the stochastic optimization paradigm, a variety of property measures were incorporated into the HARPick profiling function:

1. The main descriptor *Unique,* is equivalent to the Chem-Diverse scoring function. That is, *Unique* keeps a count of the number of pharmacophores bins occupied in the selected set of library products.

2. Three crude measures of molecular shape were introduced, in conjunction with a partition function designed to ensure even descriptor distribution across selected products. The measures were heavy-atom count (ha), largest triangle perimeter present for all pharmacophores found (pp), and largest triangle are present for all pharmacophores found (pa).

$$\text{Partscore} = \frac{\max_o - \sum_{j=1}^{P} \sqrt{(\bar{o} - o_j)^2}}{} \tag{3}$$

where *partscore* = partition score

$\quad\quad\quad \max_o$ = maximum possible mean absolute deviation (when all molecules occupy a single partition)

$\quad\quad\quad \bar{o}$ = mean molecule occupation across all partitions p

$\quad\quad\quad o_j$ = number of molecules occupying partition j

3. To allow control over molecular flexibility, a function incorporating the number of calculable conformations for each molecule (as defined by the conformational search criterion used in fingerprint creation) was included.

$$\text{Flex} = \frac{\sum_{i=1}^{N} f_i}{n} \tag{4}$$

where   Flex = flexibility score

f_i = number of calculable conformations for molecule i

n = number of selected molecules

4. To fully exploit the pharmacophoric data available, it is was considered necessary to move beyond binary key descriptions of libraries. This was accomplished by storing the pharmacophores of the selected products as a histogram rather than a simple fingerprint. In this way the frequency with each pharmacophore was hit could be taken into account. To this end a constraining function was added to the profiling routine, weighting pharmacophore selection toward filling relative diversity voids (diverse libraries) or frequently hit regions (focused libraries):

$$\text{Conscore} = \sum_{i=1}^{a} O_i S_i \tag{5}$$

where Conscore = constraint score

O_i = number of times pharmacophore i has been hit for molecules selected from current data set

S_i = score associated with pharmacophore i for the constraining library

a = number of accessible pharmacophores

$$S_i = [\max(0,(\text{avcov} - Oc_i))]^{\nu} \tag{6}$$

where $\max(0,(\text{avcov} - Oc_i))$ = maximum of the values 0 and avcov $- Oc_i$

avcov = average pharmacophore count across all occupied pharmacophores in constraining library

Oc_i = number of molecules containing pharmacophore i in the constraining library

ν = user-defined weight

$$\text{avcov} = \frac{\sum_{i=1}^{a} \min(Oc_i,\beta)}{\text{Unique}_c} \tag{7}$$

$\min(Oc_i,\beta))$ = minimum of the values Oc_i and β

β = user-defined maximum contribution to avcov by any single pharmacophore

Unique_c = number of pharmacophore bins occupied in constraining library

5. To provide a weighting against promiscuous molecules (structures exhibiting many pharmacophores), the total number of pharmacophores present in all currently selected molecules (Totpharm) was also included in the function denominator.

6. The total number of scoring molecules (S) in the selected set was also included to minimize the number of unacceptable molecules (products that fail user-defined bounds on properties such as maximum flexibility and pharmacophore promiscuity).

These features were combined to create a single overall scoring function.

$$\text{Energy} = \frac{\text{Unique}^w \times \text{Conscore} \times \text{Partscore}_{pp}^x \times \text{Partscore}_{pa}^x \times \text{Partscore}_{ha}^x \times S}{\text{Totpharm}^y \times \text{Flex}^z \times n} \quad (8)$$

where w, x, y, z = user-defined weights.

D. HARPick Application

The utility of HARPick was tested in a number of ways. Two data sets were used in the studies: (1) 20,168 molecules taken from the SDF and (2) a simple hypothetical combinatorial library comprising two components undergoing amide bond formation (Fig. 4). The second data set comprised 67 amino acids and 505 acids, giving a total library size of 33,835 products.

In the first investigation, the importance of a nonbinary pharmacophore scoring function was studied. This was undertaken through an analysis of the SDF structures. Over 68% of all HARPick theoretically accessible pharmacophores were found to be present in the data set. Further, the distribution of pharmacophores is far from even, with ~56,000 hit 1–10 times and ~27,000 hit more than 50. The full results of this analysis are shown in Fig. 5. When one considers that the average size of a company compound collection will be far in excess of 20,000, this highlights the issue of binary fingerprint saturation.

A second study was run to test the behavior of the secondary descriptors in the HARPick scoring function. Chem-Diverse was used to select a diverse

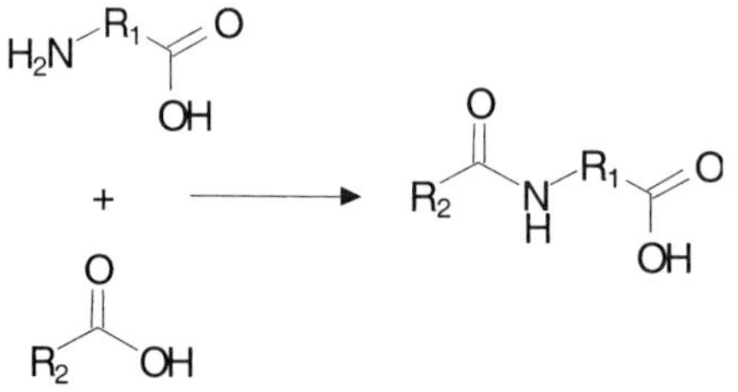

Figure 4 Combinatorial library reaction used in HARPick analyses. (Adapted from Ref. 55.)

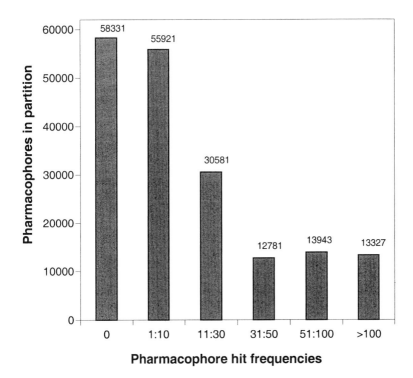

Figure 5 Results of first HARPick investigation: three-point pharmacophore frequency distribution histogram for 20,169 SDF molecules; number of theoretically accessible pharmacophores, 184,884, total pharmacophore count, 4,797,745 (using a perimeter cutoff of 0.7), number of unique pharmacophere triplets in library, 126,553 (this represents more than 68% (126,553/184,884) of the theoretically accessible pharmacophores. (Adapted from Ref. 55.)

subset of molecules from the SDF data based on pharmacophore diversity. Molecules were ordered according to heavy-atom count (smallest to largest), with the maximum permitted pharmacophore overlap percentage between each keyed molecule and the total library key set to 60%. All molecules in the set were processed, with any passing the selection criterion being added to the selected subset. This calculation resulted in the selection of 372 molecules. HARPick runs were then initiated on the same SDF data set, with the program configured to select an identical set size (372) using different diversity criteria. Three random runs were also undertaken to provide a baseline comparison. The property distributions for these calculations are shown in Fig. 6 and Table 1. In the Chem-Diverse and in all HARPick runs the number of pharmacophores determined are significantly greater than in random runs. All the HARPick runs

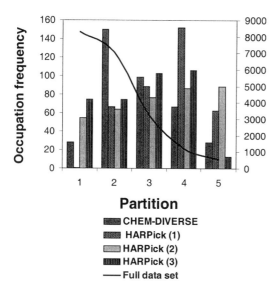

Figure 6   Data from second HARPick study. The left-hand y axis is for the library selections, while the right is for the full data set. Note the ability of HARPick to redistribute its selections more evenly throughout the partitions, In contrast, the Chem-Diverse selections have a distribution similar to that seen in the full data set. (Adapted from Ref. 55.)

were run with the shape partition functions active: Eq. (3), weight x set > 0 in Eq. (8). It can be seen that for all the runs the resulting partition scores are significantly improved versus Chem-Diverse. When a constraint on the total pharmacophore count is introduced in run 2 active (y set > 0 in Eq. 8), the total

Table 1   Data from Second HARPick Study

Calculation	Unique pharmacophores	Total pharmacophores	Property partition function scores			Calculable conformer count	HARPick scoring function weights w, x, y, z
			ha	pp	pa		
Chem-Diverse	49,829	68,987	0.33	0.66	0.43	2.2×10^8	Not applicable
HARPick 1	105,222	419,870	0.97	0.69	0.70	4.0×10^8	1, 1, 0, 0, 0
HARPick 2	61,913	93,977	1.00	0.90	0.85	6.4×10^8	1, 1, 1, 0.75, 0
HARPick 3	70,656	137,801	0.90	0.79	0.54	2.4×10^6	1, 0.5, 0.75, 0.33
Random	39,625	80,556	0.54	0.47	0.34	1.1×10^8	Not applicable

[a] Pharmacophore counts data and HARPick constraint weights are included. Incorporate these constraints into Eqs. (3)–(8) to better determine their effect. A perfect partition score is 1.0.
Source: Adapted from Ref. 55.

pharmacophore count drops by greater than a factor of 4 relative to run 1. In run 3, when the flexibility constraint is activated (Eq. 4, weight z set > 0 in Eq. 8), the total calculable conformer count is reduced by factor of ~90. These results illustrate the ability of a multicomponent scoring function to control the properties in the resulting library selection. They also highlight the need for multiple runs to tweak the scoring function weights sufficiently to give the desired result. This provides another major reason for storing pharmacophore fingerprint data ahead of time, since fingerprint calculation is the major CPU bottleneck for any pharmacophore-based calculation.

For the final investigation, the behavior of the nonbinary pharmacophore scoring function was tested (Eqs. 5–7). To accomplish this, two selections of 1000 compounds (20×50 reagent array) from data set 2 was undertaken against the SDF data set. In the first, *Conscore* was deactivated (ν weight set to 0), allowing selections based purely on internal diversity. For the second run, an avcov β value of 10 was set (which for the SDF set translates to pharmacophore bins with an occupation level of 7 or less scoring), and ν was set to 1. The resulting selections were thus constrained to find pharmacophores present seven or fewer times in the SDF data set. In addition three random 20×50 reagent selections were run to provide baseline results. The results (see Table 2) highlight the utility of the nonbinary term. For ~20% additional total pharmacophores in the library, a greater than 100% increase in pharmacophores occupying low occupancy SDF pharmacophores is achieved. By inverting the *Conscore* term, one can use the procedure to convert the method to select for high occupancy bins, a requirement for focused library designs.

The foregoing results illustrate the great flexibility of this kind of stochastic optimization methodology, variants of which are now being employed by many other researchers for library design (65–67). As with HARPick, most such implementations have been undertaken "in-house" by pharmaceutical companies. One commercial variant of this approach is soon to made available, however (67).

Table 2   Results of the Third HARPick Analysis[a]

	Pharmacophores		
HARPick Run	Unique	Total	Found in scoring bins
Unconstrained	78,567	806,539	99,696 (12%)
Conscore, constrained	79,125	1,079,998	203,061 (19%)
Random	51,791	1,033,772	51,837 (5%)

Source: Adapted from Ref. 55.

VI. LIBRARY PROFILING FUNCTIONS BASED ON STATISTICAL ANALYSIS OF PHARMACOPHORE FINGERPRINT ENSEMBLES: PHARMPRINT

McGregor and Muskal (68) recently proposed a new variant of three-point pharmacophore fingerprint analysis for library design. Of the seven pharmacophoric atom types described in Section V, B, six were applied, and the "donor and acceptor" definition was replaced by "all remaining unassigned atoms." It is of interest that the addition of this extra type resulted in improved QSAR model statistics, the reason for which was ascribed to indirect volume description. There were six distance bin (8), giving a total of 10,549 accessible pharmacophores that passed the triangle rule (the length of one side cannot exceed the length of the other two) and/or were redundant by symmetry. To validate the descriptors, they were applied as PLS (69) QSAR variables for three estrogen receptor (ER) data sets previously analyzed using other QSAR descriptors (70). The resultant models were found to be more predictive than those developed using CoMFA, CODESSA, and holographic QSAR approaches (70). In the fourth test a model was derived from 15 actives (activity set to 1.0) from data set 1 of the initial QSAR studies, plus 750 "inactives" extracted from the non-ER active structures in the MDDR (activity set to 0.0). This manipulation of the activity data was designed to mimic the kind of "noisy" results one would see from high throughput screens. Analysis of the resultant model was initiated by scoring 250 MDDR ER active molecules, 86 ER actives from a combinatorial library, and 8290 "inactives" from the remainder of the MDDR not used in model training. Using a cutoff value of 0.2 to define the boundary between active and inactive, in all three cases more than 87% of each test set was assigned correctly. It is the intention of the authors that such models be applied as the primary constraint in focused library designs. An added attraction of this technique is its ability to turn the QSAR model into a graphic based on the highly weighted pharmacophores. This is illustrated in Fig. 7, which shows the highest

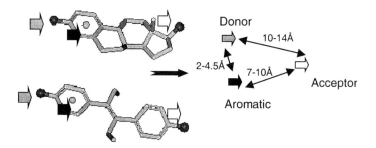

Figure 7 Most positively weighted pharmacophore in study 4 of McGregor and Muskal, mapping the natural ligand estradiol (top) and the most potent data set compound, diethylstilbestrol (bottom). (Adapted from Ref. 68.)

positive weighted pharmacophore for the test 4 QSAR equation in two diverse ER-α active chemotypes.

VII. COMBINATORIAL LIBRARY DESIGN BASED ON RELATIVE DIVERSITY/SIMILARITY: EXPLOITATION OF "PRIVILEGED" SUBSTRUCTURES

Another potentially powerful use for pharmacophore fingerprints is in the application of pharmacophoric descriptors altered to provide a "relative" measure of diversity/similarity (9,43). In this method one of the points in the pharmacophoric description is forced to be a group, substructure or site point of interest. The fingerprint thus obtained describes the possible pharmacophoric shapes from the viewpoint of that special point/substructure, creating a "relative" or "internally referenced" measure of diversity (see Fig. 8). This method has been extensively used to design combinatorial libraries focused with respect to 7-

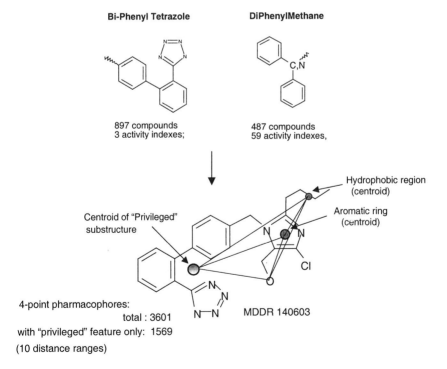

Figure 8 Privileged substructures derived from analyses of GPCR druglike ligands in the MDDR (44). A sample privileged pharmacophore is shown for one of these ligands. (Adapted from Ref. 9.)

trans membrane G-protein-coupled receptors (7TM-GCPRs) (9). Such receptors form a large family of important biological targets for which no high resolution experimental 3D structures are published. Design must therefore generally be focused around the ligands, and this can be usefully addressed through the application of 7TM-GPCRs "privileged" substructures (often spanning several targets—see Fig. 8). By focusing this measure around the privileged substructure of interest, it is possible to obtain a novel quantification of all the 3D pharmacophoric shapes containing the substructure. The goal of the published library design was to synthesize novel structures containing privileged substructure reagents/cores. Reagents were selected both to enrich the relative 3D pharmacophoric shapes of known ligands, together with the pharmacophores not in existing structures. In this way the resulting library could explore new 3D pharmacophoric diversity focused around features known to be important for biological activity. The Ugi reaction, a four-component condensation reaction was chosen for the library, and over 100,000 compounds were synthesized. Privileged substructures such as biphenyl tetrazole were used, for example, at the amine position (see Fig. 9).

The focus of the work was on GPCRs with peptidic endogenous ligands, so only ligands reported to be active for receptors in this class were used. A "privileged" pharmacophore fingerprint was calculated for this subset of structures, focused around the substructure of interest (a dummy atom was placed as a centroid of the substructure and assigned a special atom type that was forced to be one of the four points of the pharmacophore tetrahedron: Fig. 8).

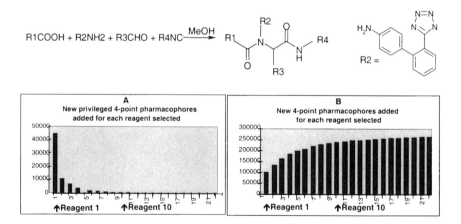

Figure 9 Library design using privileged substructures with Chem-Diverse. The Ugi reaction used is illustrated, together with graphs highlighting pharmacophore selection performance. (Adapted from Ref. 9.)

The next stage of the design process was to create a virtual combinatorial library. This was accomplished by optimizing each reagent position in turn, through the calculation of privileged pharmacophore fingerprints for each candidate reagent of all the products in the potential library. The privileged substructure was included in one or more of the reagents for the other positions, and either already selected or representative sets of reagents were used for the other three components to generate the Ugi products. The combinatorial library was then designed by using Chem-Diverse to compare each reagent fingerprint with the fingerprint for the known drug ligands (MDDR fingerprint). The first reagent selected was the one that would give library compounds with the most number of privileged pharmacophores in common with the drug set. These pharmacophores exhibited by the library compounds from the selected reagent were then removed from the fingerprint of the drug set. The next reagent was selected based on the maximum number of common privileged pharmacophores with the new smaller MDDR fingerprint. This process was repeated until no more reagents could be found that contributed a nontrivial number of new privileged pharmacophores. The total number of new pharmacophores (in this case with no restriction to contain privileged substructures) was also monitored and optimized. Figure 9 illustrates example results from one of the Ugi library optimizations.

Cell-based analysis of the pharmacophore fingerprint was used to monitor progress, particularly with regard to whether a given chemistry could yield further compounds matching the design criteria. The authors used these monitors to show that the same Ugi chemistry could indeed yield significant new diversity for multiple 14,000-compound libraries, but that diminishing returns were obtained after three libraries. By analyzing the remaining MDDR-pharmacophore fingerprint, it was shown that most of the remaining pharmacophores to be matched contained acids and/or bases, and the chemistry approach was modified to use masked acids (t-butyl esters) and bases (BOC-protected) in the Ugi reaction. This example illustrates how the ability to rapidly determine which cells/partitions are empty, and by extension which pharmacophores are missing, provides a natural basis for iterative library design of further libraries.

VIII. SPEEDING UP PHARMACOPHORE DESCRIPTOR ANALYSIS: GRIDDING AND PARTITIONING (GAP) APPLIED TO REAGENT SELECTION

While whole-molecule pharmacophore descriptors provide a viable technique for library design, the methodology is still computationally intensive relative to 2D descriptor-based technology. With a 190 MHz R10000 CPU, HARPick can process about 20,000 druglike compound phamacophore fingerprints per day (55). With multiprocessor servers it is possible to process up to a million fingerprints per day.

Nevertheless, for some larger libraries, the possible reagent combinations can be counted in the hundreds of millions. It is thus inevitable that for many libraries some reagent prefiltering is required. In the case of HARPick, 2D fingerprint clustering was used to preselect a subset of reagents before the pharmacophore analysis was undertaken (55). It was realized early on, however, that a technique more consistent with the whole-molecule descriptor would be preferable for this task.

Of equal or greater importance is the fact that as combinatorial chemistry departments have matured, a need has developed for an easily accessible set of in-house monomers available for library generation. These monomers again need to be diverse and able to probe regions of space via attachment to known leads. All the advantages inherent in 3D descriptors are equally applicable to monomer selection and to whole molecules. Further, the combinatorial issues of monomer selection conformational searching are far less daunting, since the compounds are much smaller. As a consequence, a more detailed conformational search paradigm may be employed in descriptor creation.

To exploit 3D descriptors for monomer selection, the GaP program was developed (71). The GaP procedure is outlined in Fig. 10. The methodology is similar to that employed in HARPick, with a stochastic optimization technique em-

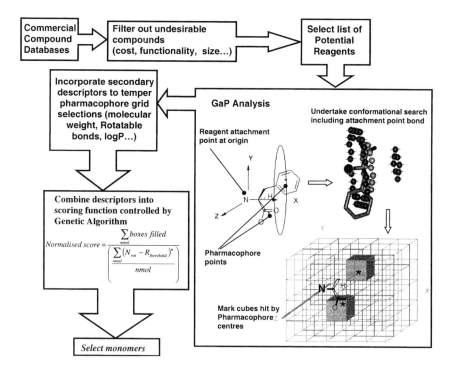

Figure 10   Outline of GaP compound selection strategy.

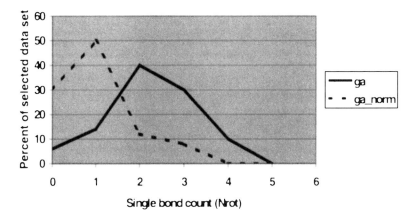

Figure 11 Graph highlighting the effect of scoring function modification on reagent selection in GaP. By normalizing the pharmacophore count with an N_{rot} denominator term (ga modified to ga_norm), the resultant compound selections are significantly more rigid.

ployed in conjunction with multiple component scoring function dominated by a pharmacophore-based descriptor. The specifics of GaP operations are as follows.

1. Orient all reagents into a common frame of reference. Here this is achieved via alignment of the reagent attachment bond along the x axis from the coordinate origin.
2. Undertake a systematic conformational analysis of the reagent, allowing free rotation (1- or 2-degree increments) about the attachment bond. This is undertaken within a 1 Å density rectilinear grid centered on the origin. A different grid is used for each pharmacophore atom type (donor, acceptor, acid, base, aromatic, combined donor/acceptor, and heavy atom).
3. For each molecule, map the position of the relevant pharmacophore types within each grid, marking the cubes within which they fall. Pairs of points are also tracked [e.g., "basic group in cube centered at (4, 4, 4) and aromatic ring centered in cube at (8, 0, 3)].
4. Combine the resultant pharmacophore descriptors with other secondary measures to ensure sensible selections [molecular weight, flexibility, numbers of acidic/basic/donor/acceptor groups, c log P and a "complexity index" (71)].
5. Use the scoring function in conjunction with a genetic algorithm to make the reagent selections.

As with HARPick, unconstrained (pharmacophore-only) searches were found to produce selections dominated by large flexible molecules rich in pharmacophoric functionality. Figure 11 illustrates the effect of secondary descriptor incorporation. The results show the average flexibility of 50 monomers selected by means of a genetic algorithm from a set of 385 primary amines. It is clear that the introduction of appropriate secondary descriptors significantly constrains the relative flexibility of the resulting selections. These results again illustrate the utility of secondary descriptor incorporation, ensuring that diverse selections are made within a more sensible property space.

IX. WHY USE 3D PHARMACOPHORES? DESCRIPTOR VALIDATION

Understanding the utility of a selected descriptor is an important aspect of any library design. There are many facets to this understanding, and it is hoped that having read the preceding sections, the reader will be better equipped to deal with those pertaining to pharmacophores. The following section highlights these issues through reference to general CAMD application experience and by means of explicit validation examples.

A. Implicit Validation

From our own perspective, much implicit descriptor validation can be drawn from knowledge regarding classical CAMD studies. Section IX.B highlights many of the useful properties of pharmacophores, together with numerous references to their successful application. Their ability to determine 3D similarities in structurally diverse molecules is a key feature of these descriptors, no matter what problem they are applied to. Further, the hit rates for many pharmacophore searches have been published or are known from in-house work. As a consequence, we have an objective criterion upon which to determine the quality of our internal screening databases. For example, given a 1% hit rate, we could objectively set 100 instances of any biologically relevant pharmacophore as the target frequency of occurrence for the database. Given that most in-house databases are heavily biased toward particular historical targets, one would actually expect the general hit rate to increase as more diverse libraries are added (at least for novel targets, e.g., those determined from genomics).

The issues of implicit descriptor properties have been discussed in the literature. We have already alluded to some of these. Clark et al. (72) also raised the issue of pharmacophore descriptors when making comparisons to the topoisomer approach. In their work they raise a number of points regarding pharmacophore measures:

1.   The authors recognized the potential for oversensitivity in the descriptors.

2. They surmised that since such descriptors comprise potentially many thousands of theoretically accessible pharmacophores, they would tend to have a lower information density than their field-based counterparts.

3. The multiconformational information content of such descriptors was said to potentially blur meaningful distinctions between molecules.

4. Finally, they did not see an obvious way in which such descriptors could be used for visual inspection of any relationships between molecules.

All these points have been addressed in various part of this chapter.

The problems of oversensitivity and information content (points 1 and 2) are addressed by careful fingerprint construction and similarity index use (Sec. III). Further, the QSAR results of McGregor and Muskal (68) (Sec. 6) would suggest that their informational content is potentially superior to that of their 2D and field-based counterparts.

While it is true that there is a blurring of molecular properties in a multi-conformational content (point 3), that information content is more complete than to that achieved from a single conformation. Further, the resulting descriptors also permit alignment independent analysis, which forms an important property, particularly for product-based designs.

With respect to point 4, again the Pharmprint (Sec. VI) and HARPick (Sec. V) techniques illustrate how it is possible to visualize pharmacophoric descriptors.

These points illustrate that, while there are issues with the exploitation of any descriptor, careful application can lead to success.

B. Explicit Validation

A number of papers have also been written that attempt to address the validation issue in an explicit manner (57,73–76). Within these publications, two primary concepts have been applied to analysis of biological data. In the first the idea of "neighborhood" behavior (73) is promoted as a measure of descriptor utility. The idea is that if a descriptor is able to cluster molecules with a particular biological activity, the descriptor encodes information regarding the requirements for that activity, and by extension is a useful measure of diversity. Analysis using this technique comparing 2D fingerprints [e.g., MACCS (77) and Unity fingerprints (78)] with pharmacophore fingerprints led to the conclusion that 2D descriptors performed better than their 1D and 3D counterparts (57,74). Unfortunately there are a number of issues with the studies undertaken. The first relates to the data sets employed. Most have been taken from databases of drugs/compounds from the clinic (e.g., the MDDR). The problem with such an approach is that the resultant biological activity class sets contain many closely related analogs, which by their nature are very similar in 2D terms. As a result, one would expect at least a slight biasing of the results toward 2D descriptors, making the study somewhat rhetorical. Further, the parameters used to set up the pharmacophore descriptors were less than ideal. For example, in one study only single conformations were used in the pharmacophore fingerprint calculation (74), and in the other the oversensitive Tanimoto index (Eq. 1) was employed with too tight a bin setting (27 bins from 2.5–15 Å in steps of 0.5 Å, with five types: acceptor atoms, acceptor sites, donor atoms, donor sites, and hydrophobes, giving a total of 307,020 bits) (57). The problems with such settings have already been extensively highlighted (see Secs. III and V). Limited studies of this kind have also been undertaken within Bristol-Myers Squibb using more optimum settings for pharmacophore fingerprint generation [four-point pharmacophores, seven distance bins, and full conformational analysis (9)], with quite different results. An example is shown in Fig. 12. This study illustrate the hit rates obtained by similarity ranking of roughly 150,000 compounds containing some 250 known melatonin antagonists. The graph shows the number of active compounds located across the first 1000 compounds in the list when the data set was ranked using melatonin as the probe molecule. It is clear that of the 2D descriptors shown, only the most generic atom pair (79) descriptor (which in many respects is essentially a two-point pharmacophore fingerprint with

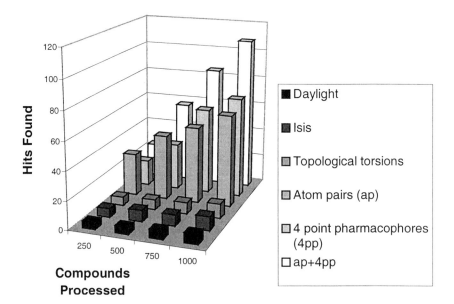

Figure 12 Active compound selection hit rates for similarity analysis of database containing around 250 known melatonin antagonists. The excellent performance of the pharmacophore descriptors is clear. The complementarity of atom pair and pharmacophore descriptors when combined is also noteworthy.

intercenter distance replaced by bond count) produces comparable results. Further, running a 2D Unity similarity search with 50% similarity cutoff produced a hit list of 1669 compounds containing 10 melatonin actives. In contrast, the pharmacophore search finds 93 actives in the first 1669 compounds, an enrichment of exceeding ninefold. We are currently undertaking preliminary studies on systems with a much wider divergence in active ligand chemotypes. Preliminary results here suggest hit rates even more favorable to pharmacophore fingerprints, with no 2D methods (including atom pairs) able to better random hit rates. (as one would expect given the wide range of structural types). It is of interest that averaging the four-point pharmacophore/atom pair rankings leads to even better results in the melatonin investigation, highlighting the potential advantages of combined descriptors [something that was also noted in one of the published validation studies (57)].

The second technique for validating descriptors may be termed "coverage analysis" (76). With this method the ability to include as many molecules with distinct biological activity classes as possible in a diverse selected set is used as a measure of descriptor quality. The technique has the advantage of not being biased by the presence of close analogs in the data sets, since it is biological class coverage rather than clustering capability that is being tested. Published studies using this technique again suggest a better performance using 2D descriptors (57) (e.g., 58% vs. 49% coverage from 55 biological classes, for a selection of 60 molecules from a 1268-compound subset), but these results are once more subject to the caveat of suboptimal pharmacophore fingerprint construction.

Similar studies have been undertaken at Glaxo Wellcome, and again quite different results were obtained. A set of known drugs was extracted from the World Drug Index (54), with each molecule having one, and only one, reported mechanism of action. The set was filtered to remove alkylating agents and molecules where the mechanism of action was felt to be too generic, such as "antifungal." The final set contained 653 compounds with 104 reported mechanisms. A random selection was made from the set of 653 drugs, with multiple subsets of 1 to 50% being constructed. The number of unique biological activities was determined for each subset, and each selection was performed 1000 times. A variety of commonly applied chemical descriptors and selection methods were applied to the data set. These include 2D MACCS public keys and Wards clustering of Tanimoto similarity, BCUT descriptors (50) with a cell-based selection and Cerius2 combinatorial chemistry consortium (67) default properties with principal component analysis. The two 3D pharmacophore key methods studied were Chem-Diverse sequential selection procedure (default three-point, seven-type pharmacophore keys), and an extension of the GaP program (71) using three-point pharmacophore keys from Catalyst (60). Pharmacophore center types similar to Chem-X (see Sects. IV–V) with bin distances as reported by Mason et al. (80) (creating a key based on six distance bins with ~6000 bits) were employed. For the GaP keys, Wards clustering and Tanimoto similarity were again used. BCUT descriptors were calculated with Diverse Solutions software, which yielded a five-dimensional property space. The cell-based selection was performed in the Cerius2 software (81), to aid comparison with other properties (not shown). The Cerius2 default properties are mainly topological indices, supplemented by 1D terms such as molecular weight and rotatable bonds, totaling 49 descriptors. PCA was used to reduce the dimensionality, yielding a set of six principal components that cover more than 90% of the variance in the data. Before application of the Chem-X selection method (discussed in Sect. IV), the molecules were first sorted by the length of their SMILES strings. Other sorting procedures (random, pharmacophore promiscuity) failed to improve the results. To select different percentages of the set, the acceptance criterion for new molecules was changed from the default "10% of pharmacophores must be new" down to the simple "displays at least one new pharmacophore".

The results of this analysis are shown in Fig. 13. The reported "gold standard" for this type of work is Wards clustering with MACCS keys (74). As can be seen, the 3D GaP descriptors are at least as good (despite the use of the oversensitive Tanimoto index), with both methods giving improvements over random of several standard deviations. The deficiencies of the Chem-Diverse selection method are well illustrated. Although initially very good, as the pharmacophore keys are saturated the selection becomes poorer, until eventually it is little better than random (highlighting the descriptor saturation problem). The widely used BCUT descriptors do not perform well in this study, and the superiority of MACCS keys over simple molecular properties is exemplified.

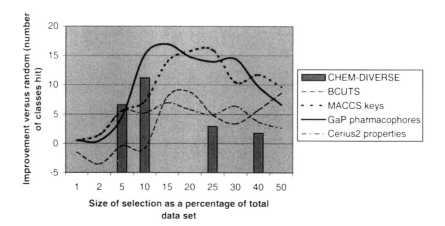

Figure 13 Results of biological activity class hit rate analysis for diverse compound subset selections using a variety of descriptors. Again the excellent performance of pharmacophore keys is worthy of note.

Work still needs to be done with such validation experiments. For example, further investigation is required into the performance of combined descriptors. Also much is left to study regarding the relative performance of different pharmacophore fingerprints (e.g. the utility of two- vs. three- vs. four-point descriptors). Nevertheless, these studies again demonstrate the biological relevance of 3D pharmacophore keys. Further, they highlight the potential pitfalls of descriptor validation, since it is clear that not all descriptors (pharmacophores and others) are created equal.

X. CONCLUSIONS

This chapter highlights the potential utility in library design of pharmacophores, which possess a proven ability to divorce the 3D structural requirements for biological activity from the 2D chemical structure. Such descriptors can be alignment independent, with calculation speeds rapid enough to permit their use on library profiling data sets, a task deemed too large for most other 3D descriptors. When carefully applied and used in combination with other properties to ensure a balanced selection, pharmacophore-driven scoring functions thus provide a powerful tool for library design.

REFERENCES

1. N.C. Perry, V.J. Van Geerestein, J. Chem. Inf. Comput. Sci., 1992; 32(6):607–616.
2. G.W. Bemis, I.D. Kuntz, J. Comput.-Aided Mol. Des. 1992; 6:607–628.
3. R. Nilakantan, N. Bauman, R. Venkataraghavan, J. Chem. Inf. Comput. Sci. 1993; 33:79–85.
4. A.C. Good, T.J.A. Ewing, D.A. Gschwend, I.D. Kuntz, J. Comput.-Aided Mol. Des. 1995; 9:1–12.
5. W. Fisanick, K.P. Cross, A. Rusinko III, J. Chem. Inf. Comput. Sci., 1992; 32:664–674.
6. A.C. Good, I.D. Kuntz, J. Comput.-Aided Mol. Design, 1995; 9:373–379.
7. R.P. Sheriden, M.D. Miller, D.J. Underwood, S.K. Kearsley, J. Chem. Inf. Comput. Sci. 1996; 36:128–136.
8. S.D. Pickett, J.S. Mason, I.M. Mclay, J. Chem. Inf. Comput. Sci., 1996; 36:1214–1233.
9. J.S. Mason, I. Morize, P.R. Menard, D.L. Cheney, C. Hulme, R.F. Labaudiniere, J. Med. Chem. 1999; 42(17):3251–3264.
10. S.D. Pickett, I.M. McLay, D.E. Clark, J. Chem. Inf. Comput. Sci. 2000; 40(2):263–272.
11. P. Gund, *Progress in Molecular and Subcellular Biology,* Vol. 5. Springer-Verlag, Berlin, 1977; pp. 117–143.

12. G.R. Marshall, 3D *QSAR in Drug Design.* ESCOM, Leiden, 1993, pp. 80–116.
13. S. Wang, G.W.A. Milne, X. Yan, I.J. Posey, M.C. Nicklaus, L. Graham, W.G. Rice, J. Med. Chem. 1996; 39(10):2047–2054.
14. M.C. Nicklaus, N. Neamati, H. Hong, A. Mazumder, S. Sunder, J. Chen, G.W.A. Milne, Y. Pommier, J. Med. Chem. 1997; 40(6):920–929, 1997.
15. H. Hong, N. Neamati, S. Wang, M.C. Nicklaus, A. Mazumder, H. Zhao, T.R. Burke Jr., Y. Pommier, G.W.A. Milne, J. Med. Chem. 1997; 40(6):930–936.
16. N. Neamati, H. Hong, A. Mazumder, S. Wang, S. Sunder, M.C. Nicklaus, G.W.A. Milne, B. Proksa, Y. Pommier, J. Med. Chem. 1997; 40(6):942–951.
17. J.J. Kaminski, D.F. Rane, M.E. Snow, L. Weber, M.L. Rothofsky, S.D. Anderson, S.L. Lin, J. Med. Chem. 1997; 40(25):4103–4112.
18. D.P. Marriott, I.G. Dougall, P. Meghani, Y.J. Liu, D.R. Flower, J. Med. Chem. 1999; 42:3210–3216.
19. A.C. Good, J.S. Mason, in Reviews in Computational Chemistry, Vol. 7, K.B. Lipkowitz and D.B. Boyd, eds. VCH, New York, 1995, pp. 67–127.
20. G.W.A. Milne, M.C. Nicklaus, S. Wang, SAR QSAR Environ. Res. 1998, 9, 23–38.
21. W.A. Warr, P. Willett, in Descriptions of Bioactive Molecules. American Chemical Society, Washington, DC, 1998, pp. 73–95.
22. P. Traxler, P. Furet, Pharmacol. Ther., 1999; 82:195–206.
23. R.C. Glen, P. Willett, G. Jones, Book of Abstracts, 213th ACS National Meeting, San Francisco, April 13–17, 1997, American Chemical Society, Washington, DC, COMP-007.
24. J.D. Holliday, P. Willett, J. Mol. Graphics Modelling, 1997; 15(4):221–232.
25. D. Barnum, J. Greene, A. Smellie, P. Sprague, J. Chem. Inf. Comput. Sci., 1996; 36(3):563–571.
26. X. Chen, A. Rusinko III, A. Tropsha, S.S. Young, J. Chem. Inf. Comput. Sci. 1999; 39(5):887–896.
27. A.K.T. Ting, P. Johnson, S. Green, R. McGuire, Book of Abstracts, 218th ACS National Meeting, New Orleans, August 22–26, 1999, American Chemical Society, Washington, DC, COMP-141.
28. I.B. Bersuker, S. Bahceci, J.E. Boggs, R.S. Pearlman, SAR QSAR Environ. Res. 1999; 10(2–3):157–173.
29. M.D. Miller, R.P. Sheridan, S.K. Kearsley, J. Med. Chem. 1999, 42(9):1505–1514.
30. I.D. Kuntz, *Science,* 1992; 257:1078–1082.
31. B.K. Shoichet, R.M. Stroud, D.V. Santi, I.D. Kuntz, K.M. Perry, Science. 1993; 259:1445–1450.
32. C.S. Ring, E. Sun, J.H. McKerrow, G.K. Lee, P.J. Rosenthal, I.D. Kuntz, F.E. Cohen, Proc. Natl. Acad. Sci. USA, 1993; 90:3853–3857.
33. I.D. Kuntz, E.C. Meng, B.K. Shoichet, Acc. Chem. Res. 1994; 27:117–123.
34. P. Burkhard, U. Hommel, M. Sanner, M.D. Walkinshaw, J. Mol. Biol. 1999; 287(5):853–858.
35. T.J.A. Ewing, I.D. Kuntz, J. Comput. Chem., 1997; 18(9):1175–1189.
36. M.D. Miller, S.K. Kearsley, D.J. Underwood, R.P. Sheridan, J. Comput.-Aided Mol. Design 1994; 8(2):153–174.
37. M.C. Lawrence, P.C. Davis, Proteins 1992; 12:31.
38. P. Burkhard, P. Taylor, M.D. Walkinshaw, J. Mol. Biol. 1998; 277(2):449–466.

39. Cerius 2/Structure-Based Focussing. Developed and distributed by MSI Inc., San Diego, CA 92121. http://www.msi.com
40. Design in Receptor (DiR). Developed and distributed by Oxford Molecular PLC, the Medewar Centre, Oxford Science Park, Oxford OX4 4GA, UK. http://www.ox-mol.com
41. R.L. Desjarlais, J.S. Dixon, J. Comput.-Aided Mol. Design. 1994; 8:231–242.
42. In-house results at GlaxoWellcome and Bristol-Myers Squibb using DOCK.
43. J.S. Mason, D.L. Cheney, Pac. Symp. Biocomput. '99, 1999:456–467.
44. MDDR. Developed and distributed MDL Information Systems, Inc., San Leandro, CA. http://www.mdl.com
45. R.D. Brown, Y.C. Martin, J. Chem. Inf. Comput. Sci., 1996; 36:572–584.
46. P. Willett, Perspect. Drug Discovery Design. 1997; 7/8:1–11.
47. R.D. Cramer, R.D. Clark, D.E. Patterson, A.M. Ferguson, J. Med. Chem. 1996; 39:3060–3069.
48. D.E. Patterson, R.D. Cramer, A.M. Ferguson, R.D. Clark, L.E. Weinberger, J. Med. Chem. 1996; 39:3049–3059.
49. J.S. Mason, S.D. Pickett, Perspect. Drug Discovery Design. 1997; 7/8:85–114.
50. D. Schnur, J. Chem. Inf. Comput. Sci., 1999; 39(1):36–45.
51. G.M. Downs, P. Willett, in Reviews in Computational Chemistry, Vol. 7, K.B. Lipkowitz and D.B. Boyd, eds. Wiley-VCH, New York, 1995, pp. 1–66.
52. P. Willett, in Molecular Similarity in Drug Design, P.M. Dean, ed., Blackie Academic and Professional, Glasgow, 1995, pp. 110–137.
53. S.D. Pickett, D.E. Clark, R.A. Lewis, Book of Abstracts, 215th ACS National Meeting, March 29–April 2, 1998, Dallas, American Chemical Society, Washington DC, COMP-009.
54. Standard Drug File (now known as the World Drug Index), Derwent Publications Ltd., London.
55. A.C. Good, R.A. Lewis, J. Med. Chem. 1997; 3926–3936.
56. Chem-Diverse, part of the Chem-X package developed and distributed by Oxford Molecular (see Ref. 40).
57. H. Matter, T. Potter, J. Chem. Inf. Comput. Sci. 1999; 39(6):1211–1225.
58. E.K. Davies, in Molecular Diversity and Combinatorial Chemistry: Libraries and Drug Discovery, American Chemical Society, Washington DC, 1996, pp. 309–316.
59. A recent addition to Chem-Diverse. See Ref. 56.
60. Cerius/Catalyst, developed and distributed by MSI (see Ref. 39) and Pharmacophore Triplets (developed and distributed by Tripos Inc. (see Ref. 78) also provide implementations of pharmacophore fingerprint descriptors.
61. R.A. Lewis, A.C. Good, S.D. Pickett, in Computer-Assisted Lead Finding and Optimization, 11th European Symposium on Quantitative Structure–Activity Relations. Verlag Helvetica Chimica Acta, Basel, 1997, pp. 135–156.
62. A.C. Good, S.D. Pickett, R.A. Lewis, Book of Abstracts, 213th ACS National Meeting, San Francisco, April 13–17, 1997, American Chemical Society, Washington DC, COMP-339.
63. S. Kirkpatrick, C.D. Gelatt, M.P. Vecchi, Science. 1983; 220:671–680.
64. J.S. Mason, in Molecular Similarity in Drug Design, P.M. Dean, ed., Blackie Academic and Professional, Glasgow, 1995, pp. 138–162.

65. V.J. Gillet, P. Willett, Peter; J. Bradshaw, D.V.S. Green, J. Chem. Inf. Comput. Sci. 1999; 39(1):169–177.
66. L. Weber, Drug Discovery Today. 1998; 3(8):379–385.
67. Cerius2, version 4.5, distributed by MSI (see Ref. 39) includes a customizable scoring function involving MCSA library analysis with pharmacophore constrains available for inclusion.
68. M.J. McGregor, S.M. Muskal, J. Chem. Inf. Comput. Sci. 1999; 39:569–574.
69. S. Wold, M. Sjöström, L. Eriksson, in Encyclopedia of Computational Chemistry, P. N. Schleyer, ed., Wiley, New York, 1998, pp. 2006–2021.
70. W. Tong, D.R. Lowis, R. Perkins, Y. Chen, W.J. Welsh, D.W. Goddette, T.W. Heritage, D.M. Sheehan, J. Chem. Inf. Comput. Sci. 1998; 38:669–677.
71. A.R. Leach, D.V.S. Green, M.M. Hann, A.C. Good, D.B. Judd, *Book of Abstracts,* 218th ACS National Meeting, August 22–26, 1999; New Orleans, CINF-002.
72. R.D. Clark, A.M. Ferguson, R.D. Cramer, Perspect. Drug Discovery Des. 1998; 9/10/11; 213–224.
73. D.E. Patterson, R.D. Cramer, A.M. Ferguson, R.D. Clark, L.E. Weinberger. J. Med. Chem. 1996; 39:3049–3059.
74. R.D. Brown, Y.C. Martin, J. Chem. Inf. Comput. Sci. 1997; 37(1):1–9.
75. Y.C. Martin, M.G. Bures, R.D. Brown, Pharm. Pharmacol. Commun 1998; 4(3): 147–152.
76. M. Snarey, N. Terrett, P. Willett, D. Wilton, J. Mol. Graphics. Modelling. 1998; 15(6):372–385.
77. MACCS keys, available in ISIS-3D, developed and distributed by MDL Information Systems Inc. See ref. 44.
78. Unity available from Tripos Inc., 1699 South Hanley Road, Suite 303, St Louis, MO 63144. http://www.tripos.com
79. S.K. Kearsley, S. Sallamack, E.M. Fluder, J.D. Andose, R.T. Mosley, R.P. Sheridan, J. Chem. Inf. Comput. Sci. 1996; 36(1):118–27.
80. M.J. Ashton, M. Jaye, J.S. Mason, Drug Discovery Today. 1996; 1(2):71–8.
81. Cerius 2/Diversity, developed and distributed by MSI Inc. See Ref. 39.

15

High Throughput Conformational Sampling and Fuzzy Similarity Metrics: A Novel Approach to Similarity Searching and Focused Combinatorial Library Design and Its Role in the Drug Discovery Laboratory

Dragos Horvath

CEREP
Rueil-Malmaison, France

I. INTRODUCTION

Combinatorial chemistry, a major drug discovery tool, heavily relies on chemoinformatics (1,2) and molecular modeling to manage the huge flux of structural information related to potentially feasible combinatorial products, and to intelligently direct synthesis efforts toward products with a maximal chance of fulfilling the stringent conditions required of a drug molecule. Until recently, even the number of combinatorial products that potentially could have been obtained on the basis of commercially available starting materials and relatively simple two- or three-step chemistries would have largely exceeded the available modeling capacities. In response to these novel constraints, molecular modeling tools dedicated to combinatorial chemistry (3–5) have been successfully developed. Software packages aimed at processing large sets of molecules are nevertheless restricted to the fast bidimensional (topological) (6–8) description of combinatorial products, thus avoiding the computational effort due to geometry buildup and conformational sampling. Conformer generation may require seconds to minutes of CPU time per molecule, depending on the effort spent to score the relative relevance of the visited phase space region (using a simple bump check criterion to reject impossible geometries vs. performing a full-blown potential energy evaluation). Therefore, 3D descriptors may be routinely used to characterize libraries containing 10^4–10^5 compounds.

Ingenious solutions featuring combinatorial product descriptors defined in terms of the descriptors of their corresponding building blocks (9), therefore not requiring explicit construction of products, are successfully exploiting the "combinatorial advantage" to significantly reduce the cost of descriptor evaluation. The generation of 3D descriptors in the latter context is especially rewarding, since the computationally intensive 2D $\rightarrow$ 3D conversion is performed at a building-block level, scaling as the sum of implied combinatorial reagents. However, such an approach accounts for only the structural features confined within every building block, but not for the global structural features of the final combinatorial product, since no information about the relative positioning of the building block moieties in the final product is provided.

The use of geometry-dependent descriptors to characterize molecules may prove an important source of artifacts, if the typical variance of such descriptors with respect to a set of conformers of a same molecule is comparable to the variance expected within a set of single conformers of different molecules. In particular, a molecule for which single conformations are generated according to different geometry builders may be the source of significantly different 3D descriptor values in function of the used 2D $\rightarrow$ 3D conversion strategy. A similarity search routine based on these 3D descriptors may actually rank two such conformers as "dissimilar" species, failing to recognize that they represent the same molecule!

There is an ongoing debate (10,11) with regard to the superiority of 3D over 2D descriptors, since the benefits of including geometric information may be counterbalanced by previously discussed artifacts. Nevertheless, 3D information is essential for the understanding of structure–property relationships, justifying the investment of research effort into developing novel 3D descriptors that are less dependent on conformational sampling. The present work focuses on this important problem, describing the following:

> An algorithm for the generation of multiconformational models of the combinatorial products by concatenation of building-block conformers obtained from an adapted conformational sampling approach. This scheme successfully opens the way to the characterization of the global

structural product features, while fully preserving the computational advantage due to relegation of the 2D → 3D conversion step at a building-block level.

The definition of fuzzy bipolar pharmacophoric autocorrelograms (FBPA), fuzzy logic-based 3D descriptors, taken as *averages* with respect to all the sampled conformers.

Based on this high throughput methodology, we will discuss drug design tools dedicated to the buildup and exploitation of a virtual library of chemically feasible compounds, such as the following:

Rule-based chemical filters designed to allow the enumeration of the chemically feasible combinatorial products in contrast to all the mathematically possible pairs, triplets, or multiplets of building blocks

The use of similarity metrics based on the previously defined FBPA, for high throughput similarity searches

The aim is to provide an overall view of the potential applications of the computational tools in drug design.

In particular, a critical analysis of the potentialities and limitations of similarity-based searches for active analogs will try to situate this methodology within the everyday context of the drug discovery activity. Real-life validation studies must go beyond the estimation of some statistical quality parameter of the algorithm according to ever so ingenious *in silico* simulations, and they must account for the manifold heuristic factors interfering with the research process.

First, an *in silico* "seeding" experiment was performed to demonstrate the ability of the FBPA metric to retrieve structurally different known farnesyl protein transferase inhibitors-"hidden" among large compound collections-on the basis of their low dissimilarity score with respect to an active reference molecule. This simulation also provided a benchmark against which to compare the discriminant powers of different "classical" dissimilarity metrics against the FBPA approach.

Second, the design and testing results of a small focused library of dopamine transporter (DAT) ligands, relying on combinatorial FBPA-selected nearest neighbors from the virtual collection of synthetizable products, were shown as a typical example of the practical results that can be achieved with this methodology.

Eventually, a more complex lead optimization program in which high throughput modeling and similarity searches were synergistically combined with substructure searching techniques and—most important—medicinal chemistry know-how, lead to the discovery of novel, nanomolar μ-opiate receptor ligands.

II. METHODS

A. From Reagents to Building Blocks to Combinatorial Products: The "Virtual Chemistry Engine"

A first task of molecular modeling in combinatorial chemistry is the automated enumeration of the expected combinatorial products of a library, on the basis of the reagents used to build it. Several approaches to this problem have already been reported and implemented in different commercially available software packages.

"Analog builders" (12) ornate a common scaffold of the library with the corresponding "groups" provided by each of the synthons for each predefined "substitution point" of the scaffold [provided different software (13) was employed earlier to automatically "extract" these substructures that actually appear in the combinatorial products out of the actual synthon molecules submitted to the given chemical transformation]. This buildup of combinatorial products is a purely topological operation, involving the updating of the connectivity tables to include the newly formed bonds between the scaffold and the added fragments. Some software packages also generate a very crude 2D/3D model of the resulting compound by adjusting the available coordinates of the group atoms to ensure a reasonable length and orientation of the newly formed bonds, but to obtain clash-free 3D geometries of the combinatorial products, no attempt is made to optimize the relative orientations of the introduced substituents.

In contrast to analog builders, general methods for the modeling of chemical transformations are based on open-ended "chemical languages" (14) with a well-defined grammar, allowing the description of virtually any chemical transformation, including those involving ring closures. This description implies the characterization of the common chemical environment of each functional group implied in the transformation, to allow the software to automatically detect the corresponding representatives in every submitted synthon molecules. Furthermore, mapping rules outlining the equivalencies between the groups in the starting and final products must be provided. Graphical interfaces allow the input of a simple reaction sketch and automatically convert it to the corresponding expression processed by the chemical language interpreter. The enumeration of combinatorial products by this algorithm is a purely topological operation.

A different combinatorial product buildup strategy has been adopted in the present work, formally representing each product as resulting from the coupling of two "building blocks" by means of a single, double, or partially double bond. In this context, the term "building block" is used to design any conveniently chosen molecular fragments allowing the representation of the combinatorial products as pairs of such fragments, linked by only one bond. A rule-based "chemical filter" performs the automated conversion of the input reagent structures into the corresponding "building blocks," according to the following steps.

> Detection and deletion of counterions if the synthon appears under the form of a salt in the input files.

Detection of the reactive center in a synthon molecule according to the input list of (electrophilic or nucleophilic) functional groups that display an appropriate reactivity in the chemical synthesis envisaged. Nucleophilic groups may be specified in choosing the type of the nucleophilic center and its substitution count, whereas the nature of the leaving group and the nature of the electrophilic center must be specified for a complete designation of electrophilic groups. For example, "Hydroxyl group@Carbonyl group" describes the carboxylic group in an acylation process (Fig. 1). Recognition of the outlined functional groups is rule-based, and the functional groups of significantly different reactivity, such as amines, ani-

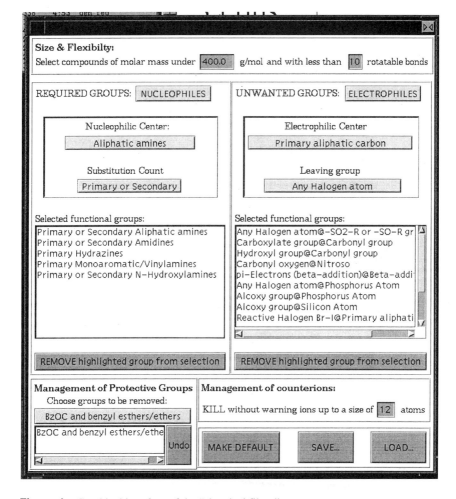

Figure 1 Graphical interface of the "chemical filter."

lines, and amides, are distinguished. After having detected a putative reactive center (RC), in its default mode, the software breaks the bond between the reactive center and the leaving group (if necessary) and labels
the reactive center by adding a "D" dummy atom to it (Fig. 2A). However, in an alternative "RC cutoff" mode, the bond between the reactive
center and the rest of the molecule will be broken, and the reactive center
deleted (Fig. 2B). If multiple potentially reactive centers exist in a
molecule, the software generates all the possible building blocks, but
flags those species and issues a warning diagnostic allowing the user to
check whether some or all must be discarded. An error flag is attached to
all the synthons in which no reactive centers have been detected.

Optional "transformation" of the building blocks, by replacing the dummy
atom, denoting the free valence of the RC by a "transformer" moiety that
can be assimilated to the common scaffold of analog builders. These
transformer moieties may represent any valid chemical substructures,
with two specifically labeled centers: the anchoring point to the building
blocks to be transformed and the new "reactive center," evidenced by its
bond to a "D" atom and denoting the point at which the coupling to the
combinatorial partners will occur.

Molecular mass and rotatable bond number evaluation, issuing a warning
message if the user-specified threshold values are exceeded.

Detection of unwanted groups, potentially harmful for the synthesis yield or
for the biological properties of the combinatorial products. If such groups
are detected, a specific warning flag will be attached to those building
blocks.

Optional automated detection and deletion of classical protective groups
such as BOC (benzyloxycarbonyl), FmOC (fluorenylmethyloxycar-
bonyl), or *t*-butyl esters.

As shown in Fig. 2, this approach can be consistently used to generate any
combinatorial library that could be constructed by analog builders, having the special advantage of the straightforward functional group–based definition of the required chemical profiles. This is unlike classical software (15) requiring multiple
substructure-based queries to perform the selection of synthons qualifying for a
given chemical synthesis. To maximize the relevance of the selected building
blocks at every step of the synthesis, the algorithm is completely interactive and
the automated decisions to select/discard a given synthon may be easily overrid-
den by the user. The approach does not however support ring closure reactions involving polydentate synthons. The major advantage of this library construction
philosophy resides in the fact that it can be generalized to handle molecular geometries and to rapidly build 3D multiconformational models of combinatorial
products starting from conformer families of the constituent "building blocks."

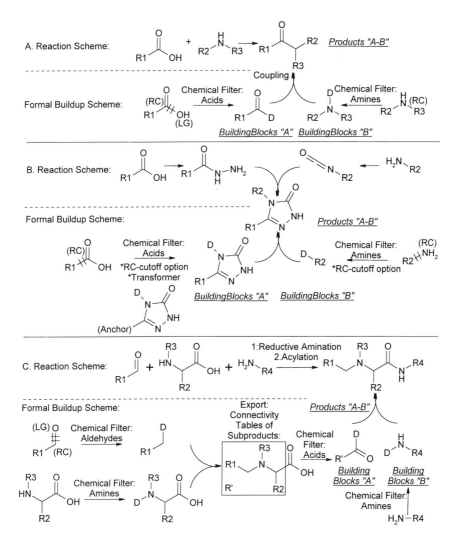

Figure 2 Various applications of the "synthon–building-block–combinatorial product" buildup strategy. (A) Building an amide library with building-block linker bond identical to the one formed in the chemical transformation. (B) Modeling of a library of heterocycles, involving a ring closure step. The formal building-block linker bond is chosen to be exocyclic by using the "RC cutoff" and "transformer" options to conveniently define building blocks already including the heterocyclic scaffold. (C) Modeling of a library involving multiple synthesis steps. Dimeric combinatorial subproducts are generated at a first stage and then submitted to the "Chemical Filter" to be converted into building blocks.

B. Exploiting the "Combinatorial Advantage" for High Throughput Generation and Analysis of 3D-Multiconformational Models of Combinatorial Products: The "Ghost Database" Mechanism

The key advantage of parallel high throughput organic synthesis (HTOS) of combinatorial libraries over classical medicinal chemistry is the ability to generate a large quantity of compounds. This quantity scales as the product of used synthons of each type, whereas the up-front, labor-intensive steps including the preparation (purchasing, weighing, dissolving) of the synthons only scales as the *sum* of synthon numbers. Synthesis robots are in charge of the "combinatorial explosion" step at a relatively low cost. By analogy, an optimal molecular modeling scheme of combinatorial products would maximally benefit from this "combinatorial advantage" if the rate-limiting step of geometry buildup and conformational sampling could be performed at building-block level (16). This would allow the combinatorial building mechanism to use *predefined* building-block geometries to rapidly construct conformer families of the combinatorial products.

Two main problems have to be solved before such an approach can be enabled:

1. Building blocks must be defined to ensure that in order to form a combinatorial product, only one new bond needs to be established between two partners.
2. The predefined building block conformers must be representative of the geometries adopted by that substructure in the combinatorial products, not of the geometries adopted by the synthon molecule itself.

Although there is no absolutely general solution satisfying the first issue, the library enumeration strategy described previously specifically focuses on the description of combinatorial products as building block pairs linked by a unique bond and may cover a wide range of arbitrarily complex chemistries. To address the second issue, a conceptually simple "maximum constraint conformational sampling" approach of building-block geometries has been developed as part of the present work, relying on the standard conformational sampling procedure used by the Catalyst (catConf) software (13). Prior to submission to the conformational sampling algorithm, the dummy "D" atom representing the free valence of the RC of every building block is replaced by a "spaceblocker" template, and the resulting "chimerical" molecule is submitted to catConf, using the "best" sampling mode and allowing for 20 kcal/mol of excess strain energy with respect to the best conformer or for a maximal number of typically 20 conformers. In the resulting conformers, the spaceblocker template is deleted and replaced again by the "D" marker atom of the free valence of the building block. The spaceblocker moiety was chosen to satisfy the following constraints:

> Being bulkier than most of the combinatorial partners a building block may encounter in a combinatorial library, so that the space it occupies in the vicinity of the RC, which remains free of atoms after deletion of the spaceblocker, is large enough to accommodate these combinatorial partners.
>
> Being apolar, to avoid the formation of fictive hydrogen bonding or charge–charge interactions between the functional groups of the building block and the spaceblocker atoms, forcing the former to fold over the latter.
>
> Being void of internal degrees of freedom, to minimize the computer time spent to sample the meaningless conformational space of the spaceblocker itself. Since the catConf software recognizes topological symmetry and avoids sampling of "degenerate" rotatable bonds, a fully symmetrical moiety in which all the σ bonds are 3-fold symmetry axes represents an elegant solution to this problem.

A species thought to best fulfill these conditions is the tris(triiodomethyl)silyl moiety —$Si(CI_3)_3$.

The restricted conformations of the building blocks obtained after the spaceblocker moiety has been severed are next oriented with their free valence along the z axis, with the RC at the origin of the coordinate system, the "D" free valence label at positive z values and one of the remaining RC substituents in the y–z plane. Whenever the RC may adopt different hybridizations in function of the nature of the coupling chemistry, distinct sets of coordinates reflecting the possible local RC geometries are created. For example, if amine building blocks are modeled, the resulting hybridization of the nitrogen atom RC in R_1R_2N—D after the conformational sampling of R_1R_2NH—$Si(CI_3)_3$ will be sp^3. An alternative set of coordinates in which the geometry of the N atom is set (by relocating the dummy atom D) to the planar sp^2 configuration is also created and stored on disk to be used for the buildup of amides. The original set of building-block geometries with an sp^3 nitrogen will be used to generate conformers of combinatorial products of N-alkylation reactions.

The set of routines in charge of building multiconformational 3D models of combinatorial products on the basis of the previously described sets of restricted conformations of building blocks will be further on referred to as the "ghost database" mechanism (GDbM). This mechanism is completely transparent to the user and automatically takes over control to generate the required structures of combinatorial products whenever the system encounters a molecule ID that can be successfully interpreted as a combinatorial compound name of the following form:

[First Building Block ID]-(special character denoting the type of the linker bond: single, double, partially double)-[Second Building Block ID]

Performing the actual buildup of the conformers of the combinatorial product is actually as fast as the input of the corresponding molecular file, and therefore the system actually behaves as if the combinatorial product structures were actually stored on disk in a fictive "ghost database." This buildup occurs in several steps.

Syntactic analysis of the combinatorial compound name, broken up into the two names of component building blocks and the type of the newly formed bond.

Retrieval of the sets of conformers stored for each of the building blocks, corresponding to the proper hybridization status of the reactive centers as defined by the order of the newly formed bond.

Pairwise coupling of the current conformers of the building blocks. Given that all building-block geometries have their "free valence" oriented along the z axis, this operation resumes to a mirroring of the coordinates of the second building block with respect to the x–y plane, followed by their translation along the z axis to ensure a natural length for the newly formed bond.

A two- or three-step torsional driving around this newly formed bond, in function of the nature of the newly formed bond.

Bump check of the rotamers resulting from the previous step. If no bad contacts are observed, the conformers of the combinatorial product are registered, otherwise they are discarded. The loop continues with the next possible combination of building block conformers, until a predefined number (typically 40) of valid product geometries have been registered.

To evaluate the quality of the geometries provided by the "ghost database" mechanism, a series of 350 randomly cherry-picked combinatorial products issued from four different libraries (amides, carbamates, ureas, and hydrazones) have been built using this approach. A single conformer per compound out of the set produced by the GDbM was subjected to minimization using the CVFF force field (17). The change in internal energy accompanying the rearrangement of the molecular geometry during minimization has been monitored and stored for further analysis. At each iteration it, the average drift of atomic coordinates (ADAC):

$$ADAC(it) = \frac{\sum_{i=1,N} \sum_{j=1,3} (x_{i,j}^{(it=0)} - x_{i,j}^{(it)})^2}{3N} \qquad (1)$$

and the maximal interatomic distance change (MICD, where $d_{i,k}$ represents the Cartesian distance between two atoms i and k):

$$MICD(it) = \max_{k>i; i=1,N} | d_{i,k}^{(it=0)} - d_{i,k}^{(it)} | \qquad (2)$$

have been monitored against the excess of strain energy $\varepsilon(it)$ of the current geometry with respect to the found energy minimum. These values were extracted at the points of the minimization trajectory intersecting the excess strain energy thresholds of $\Delta = (+20, +10, +5, +2, 0$ kcal/mol) with respect to the CVFF optimum. Their averages over the molecules in this set, as well as their maximal values observed within that set are reported in Table 1. In parallel, the same simulation was run with departure geometries taken (1) as the lowest energy conformers found by the catConf software and (2) as minimum energy conformers according to the universal force field (UFF) (18).

C. Fuzzy Bipolar Pharmacophore Autocorrelograms (FBPA)

In a similarity space based on 3D descriptors, a single molecule may be alternatively positioned at different points of this space, if different geometries were used to evaluate the 3D indices. Therefore, a key requirement of any similarity metric—the identity of any molecule to itself ([A = B] => [dissimilarity(A,B) = 0])—may be violated if no measures are taken to ensure that strictly the same 3D models were generated for every occurrence of a molecule in a study. In particular, topologically highly similar pairs of compounds that might in principle also adopt similar 3D folds may no longer be recognized as being similar species if the 2D → 3D conversion algorithm produced a completely different geometry for

Table 1 Monitoring the Average Drift of Atomic Cooridnates (ADAC) and Maximal Interatomic Distance Change (MIDC)[a]

Δ (kcal/mol)	$\langle ADAC(it)|_{\varepsilon(it) = \Delta}\rangle$			$\langle MID(it)|_{\varepsilon(it) = \Delta}\rangle$			$\max(ADAC|_{\varepsilon(it) = \Delta})$			$\max\langle MIDC|_{\varepsilon(it) = \Delta}\rangle$		
	GDbM	cat	UFF	GDbM	cat	UFF	GDbM	cat	UFF	GDbM	cat	UFF
20	0.11	0.03	0.03	0.59	0.18	0.20	1.02	0.44	0.52	5.73	2.21	2.92
10	0.25	0.07	0.11	1.18	0.37	0.53	1.46	0.70	1.81	7.71	3.45	7.67
5	0.41	0.16	0.25	1.87	0.81	1.18	1.85	1.22	2.16	8.88	6.10	8.40
2	0.66	0.35	0.45	2.83	1.64	2.06	2.27	1.73	2.28	11.92	8.20	13.13
0	0.93	0.59	0.73	3.75	2.67	3.20	2.43	2.18	2.71	13.61	8.50	13.98

[a] These represent the degree of structural change required to reach the nearest point within Δ kcal/mol from the next local optimum on the CVFF-based potential energy surface, starting from initial molecular geometries obtained from the ghost database mechanism (GDbM), the catConf sampling algorithm (lowest energy conformer obtained in "best" sampling mode, abbreviated "cat"), or energy-optimized catConf geometries, using the Universal Force Field (UFF). The table shows both average values of these descriptros taken over the 350 combinational product structures used in this study and the maximal fluctuations.

each one of them. The original 3D fingerprints introduced in this paper, the **fuzzy bipolar pharmacophore autocorrelograms (FBPAs)** and the FBPA-based dissimilarity metric, have been defined to minimize the chance of such artifacts. The main measures taken to achieve this goal were as follows.

1. Molecular fingerprint buildup as an *average* of the individual conformational fingerprints of the diverse sampled conformers. While any two conformational fingerprints may widely differ, the "consensus" descriptor built by averaging over a representative, albeit not exhaustive, set of diverse conformers is less likely to show a marked dependence on used the sampling scheme.

2. The use of fuzzy logic to build and compare these 3D fingerprints, avoiding the "all-or-nothing" bitwise match of binary 3D fingerprints (19), in which sampling artifacts triggering differences in geometry may result in reversing the on/off status of the corresponding bits. By contrast, the fuzzy similarity score defined here *continuously* decreases as the observed differences between the monitored geometrical features increase.

1. Pharmacophoric Features: Bipolar Pharmacophore Types (BPT)

Six classical pharmacophoric features—hydrophobicity (H), aromaticity (AR), hydrogen bond acceptor (HBA) and donor (HBD) character, as well as cationic (POS) and anionic (NEG) character—are considered for the construction of the FBPA. An automated feature assignment routine is used to detect all the heavy atoms representing each of these features, according to empirical rules defining the most probable ionization status of the functional groups at pH 7. Labeled as *hydrophobic* are all the halogen atoms, sulfur, and carbons, except for the carbons in aromatic systems, guanidines, and carboxylate groups. Every individual carbon or heteroatom member of an aromatic ring is taken as a representative of the *aromatic* feature (aromaticity is checked for any ring system involving only sp^2-hybridized atoms according to the hybridization flag in Cerius2, by summing the number of π electrons and applying Hückel's "$4n + 2$" rule). Any O or N atoms with a free electron pair are considered to be *hydrogen bond acceptors,* while any heteroatom explicitly bound to a hydrogen is labeled as *donor.* Aliphatic amino groups (including hydrazines) and guanidines/amidines, but not aniline groups or amides, are considered to be protonated, that is, carriers of a *positive charge,* accordingly acting as hydrogen bond donors and not acceptors. The positive charge feature in guanidines/amidines is assigned to the central C atom. Aniline and amide—N atoms are on the contrary labeled as (neutral) hydrogen bond acceptors. Carboxylate groups (and the equivalent tetrazole rings) are representatives of the *anionic* feature (associated with the C atom). Carboxylate oxygens are marked as hydrogen bond acceptors. Imidazoles (but not benzimidazoles or imidazoles substituted by electron-attractive groups such as phenyl or nitro), are reserved for special treatment, owing to ambiguity concerning their protonation status at pH 7:

two different fingerprints will be generated for such compounds, the first assuming a neutral, the second a protonated imidazole ring.

The $\frac{1}{2}N_f(N_f - 1) = 21$ pairwise combinations of the N_f pharmacophoric features we consider (H–H, H–AR, H–HBA, . . . , H–NEG, AR–AR, AR–HBA, . . ., AR–NEG, . . . , . . . , POS–NEG, NEG–NEG) will be referred to later as "bipolar pharmacophore types" (BPTs). A pair of atoms in which the first partner possesses one, and the second the other feature defining a given BPT is said to "match" or "represent" the given BPT. For example, an atom pair might consist of an alcoholic–OH group and the >NH group of an indole ring represents the (HBD–HBD), (HBD–AR), (HBA–HBD), and (HBA–AR) bipolar pharmacophore types. Formally, if BPT $(a, b) = (f_a, f_b)$ is a pair of two specified features f_a and f_b $(a, b = 1, . . ., N_f)$, for a pair of atoms (i, k) with $i \sim f_a$ and $k \sim f_b$ (e.g., i having feature f_a and k feature f_b), we can write $(i, k) \sim$ BPT (a, b).

2. FBPAs Represent the Conformer-Averaged Distance Distribution Densities of the Sets of Distances Between the Atom Pairs Representing Every BPT

For each available conformation of the current molecule, a complete interatomic distance matrix d_{ik} is first calculated. At next step, the algorithm sequentially browses through the 21 BPTs, selectively analyzing the pairs of atoms matching the current BPT to build the conformational distance distribution histogram of this BPT. It uses $N_{bin} = 12$ distance bins with a width of 1 Å, homogeneously spanning a range between 3 and 15 Å, to classify all the atom pairs matching the current BPT (a, b). The most straightforward scheme to assign the pairs $(i, k) \sim$ BPT (a, b) into these distance bins is to consider all the pairs with $\Delta \leq d_{ik} < \Delta + 1$ as members of bin Δ. However, this leads to "binning artifacts" for pairs separated by distances that are very close to an integer value, their classification in the lower or upper bin being decided by irrelevant fluctuations setting the actual distance slightly above or below the threshold Δ. The "smoothing out" of binning artifacts used to build property-weighted autocorrelograms as input for neural networks (20) led to powerful molecular descriptors. This work features a slightly different solution to this problem, introducing a probability $p(d, \Delta)$ of association of an atom pair at distance d to a bin Δ:

$$p(d^*, \Delta) = \begin{cases} [d^* -0.5] - d^* + 1.5 & \text{if } \Delta == [d^* - 0.5] - 2 \\ d^* - 0.5 - [d^* -0.5] & \text{if } \Delta == [d^* -0.5] - 1 \\ 0 & \text{otherwise} \end{cases} \quad (3)$$

where the notation $[x]$ stands for the integer (truncated) part of x. Here d^* represents a normed interatomic distance forced to take values between 3.5 and 15.5 Å, so that atom pairs at $d < 3.5$ Å or pairs at $d > 15.5$ Å will be counted, respectively, in the first and last bin rather than being ignored for falling out of the considered binning range:

$$d^* = \begin{cases} 3.5 & \text{if } d < 3.5 \\ d & \text{if } (3.5 \le d \le 15.5) \\ 15.5 & \text{if } d > 15.5 \end{cases} \qquad (3')$$

Accordingly, an atom pair separated by a distance of exactly 5.5 Å will be assigned to a single distance bin $\Delta = 3$ with a probability of $p(5.5, 3) = 1$. At 5.4 Å however, the atom pair would have been classified in bin $\Delta = 2$ with a probability of $p(5.4, 2) = 0.1$ and in bin $\Delta = 3$ with a probability of $p(5.4, 3) = 0.9$. All the short-distance pairs of bonded and geminal atoms ($d < 3.5$) are assigned to bin $\Delta = 1$, whereas all the pairs exceeding the maximal distance range are grouped in the last bin, $\Delta = N_{bin}$.

The molecular fuzzy bipolar pharmacophore autocorrelogram is defined as the vector of the fuzzy (assignment probability-weighted) number of atom pairs of the given bipolar pharmacophore type BPT(a, b) assigned to each distance bin Δ, averaged over the set of sampled conformers:

$$\Psi_M(a, b, \Delta) = \langle \Sigma_{(i,k) \sim BPT(a,b)} \, p(d^*_{ik}, \Delta) \rangle_{conformers \; of \; M} \qquad (4)$$

The FBPA of a compound M consists of $\frac{1}{2}N_f(N_f + 1) \times N_{bin} = 21 \times 12 = 252$ components $\psi_M(a, b, \Delta)$ and can be represented as a set of 21 distance distribution histograms. Each of these histograms is associated with a pharmacophoric feature pair BPT(a, b) = (f_a, f_b), with $1 \le a \le b \le N_f$, and monitors the population of matching atomic pairs within each of the N_{bin} distance bins. Figure 3 provides a comparative display of the populated histograms from the FBPA of three compounds (histograms corresponding to feature pairs represented in at least one of the three compounds).

3. FBPA-Based Similarity Metric: Definition and Calibration

A quantitative measure $\delta(m, M)$ of the dissimilarity of two molecules m and M can now be defined as the "distance" between their representative points in the space of the FBPA descriptors. However, not all the possible metrics that could be defined in the 252-dimensional space of the FBPA vectors are equally relevant measures of molecular similarity. A convenient way to define a metric accounting for the specific structure of the FBPA descriptors is the introduction of *feature pair partial dissimilarity scores* $\delta_{a,b}(m, M)$, each comparing the distance distribution histogram associated to a BPT (a, b) in m to its counterpart in M, defined as follows:

$$\delta_{a,b}(m, M) = 1 - \frac{2(\Psi_m \otimes \Psi_M)_{a,b}}{(\Psi_M \otimes \Psi_M)_{a,b} + (\Psi_m \otimes \Psi_m)_{a,b}} \qquad (5)$$

where

$$(\Psi_m \otimes \Psi_M)_{a,b} =$$
$$\sum_{\Delta_1=1}^{N_{bin}} \sum_{\Delta_2=1}^{N_{bin}} \Psi_m (a, b, \Delta_1) \Psi_M (a, b, \Delta_2) \exp[-\alpha(\Delta_1 - \Delta_2)^2] \qquad (6)$$

It can be seen that at very large values of the "fuzziness" factor α, the convolution product (Eq. 6) resumes to a scalar product between the vectors describing the distance distribution density of the given BPT (a, b) and Eq. (5) becomes the classical Dice dissimilarity metric (21). The intuitive example of four hypothetical molecules A, B, C, D, each displaying a unique peak corresponding to an atom pair of given BPT at 4, 6, 12, and 14 Å, respectively, offers a straightforward explanation of the importance of an appropriate calibration for the fuzziness factor α. The molecules are predicted to be completely dissimilar with respect to each other ($\delta = 1.0$) at very large α values and perfectly identical ($\delta = 0.0$) at $\alpha = 0$. The fuzziness factor should be set to obtain dissimilarity coefficients agreeing with the intuitive observation that A is similar to B and practically equally dissimilar with respect to C and D [e.g., $0 \approx \delta(A, B) \ll \delta(A, C) < \delta(A, D) \approx 1$]. The sensitivity of the metric to conformational sampling artifacts decreases with decreasing α (note that in the preceding example, molecule B could have actually represented the same chemical species A, in which the small offset of peaks was due to the use of different sampling procedures to build the geometries). However, too low an α value would force the metric to completely ignore the differences in distance separating the atom pairs of given type, degenerating into a dissimilarity score based on the differences of total numbers of pairs of the given BPT, disregarding their interatomic distances. Formally, increasing the tunable parameter α from 0 to large values gradually turns the FBPA dissimilarity score from a feature pair count-based metric (like the ISIS key similarity score, with the difference that the monitored "fragments" represent pairs of pharmacophoric features) into a full-blown 3D metric. The fitting of α is expected to reveal an optimal configuration at which the FBPAs best combine the robustness of substructure key fingerprints and the information richness of 3D descriptors.

The global dissimilarity score $\delta(m, M)$ can now be introduced as a weighted average of the partial dissimilarity scores with respect to all the bipolar pharmacophore types BPT (a, b) represented in at least one of the molecules m and M (otherwise, the denominator of Eq. 5 being zero, the corresponding partial dissimilarity score is not defined):

$$\delta(m, M) = \frac{\sum_{a=1}^{N_f} \sum_{b=a}^{N_f} w(a)w(b)\delta_{a,b} (m, M) \mid (\Psi_M \otimes \Psi_M)_{a,b} + (\Psi_m \otimes \Psi_m)_{a,b} > 0}{\sum_{a=1}^{N_f} \sum_{b=a}^{N_f} w(a)w(b) \mid (\Psi_M \otimes \Psi_M)_{a,b} + (\Psi_m \otimes \Psi_m)_{a,b} > 0}$$
$$(7)$$

The feature weighing factors $w(a)$; $a = 1, \ldots, N_f$ represent five supplementary fittable parameters of the dissimilarity metric, encoding the relative impor-

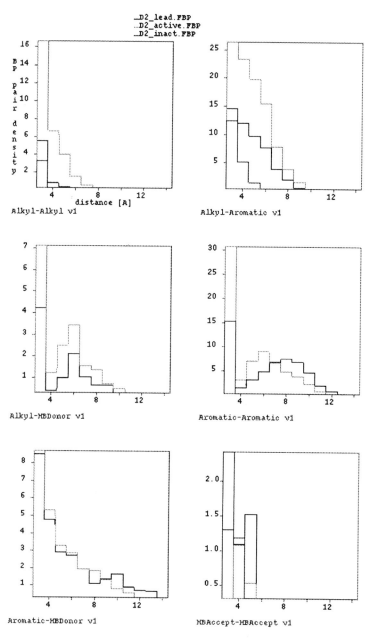

Figure 3 The FBPA of two biologically related molecules (D2 inhibitors) display an easily observable covariance, by contrast to that of a third randomly picked compound void of D2 inhibitory activity. Each histogram is associated to a feature pair (a,b), displaying on Y the "fuzzy" total number of atom pairs of corresponding type classified in each of the 12 distance bins on X.

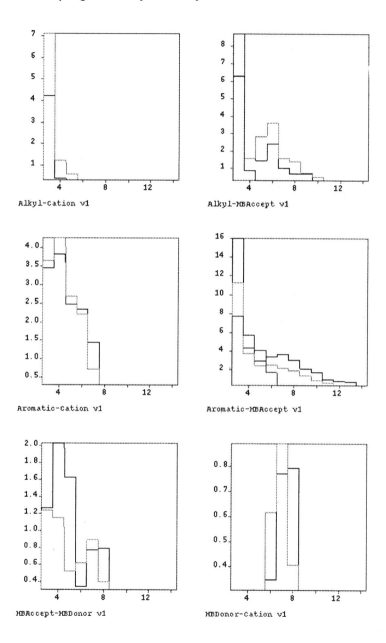

Figure 3 Continued. Each histogram is associated to a feature pair (a,b), displaying on the *y* axis the "fuzzy" total number of atom pairs of corresponding type classified in each of the 12 distance bins shown on the *x* axis.

tance of the six pharmacophoric features considered (Eq. 7 being invariant to the rescaling of the weighting parameters, one of the w coefficients can be arbitrarily set to 1.0).

In the present work, the weighing factors and the fuzziness factor α were simultaneously optimized with respect to the following maximal diversity selection problem (22,23): a set of $N_{lig} = 75$ different reference ligands (24), subdivided into $N_{fam} = 14$ activity families, each containing three to five ligands of a given receptor or enzyme, has been encoded under the form of FBPA. Pairwise dissimilarity scores $\delta(m, M) \mid_{w,\alpha}$ have been generated at all possible configurations of the fitted parameters given by a systematic grid sampling a judiciously chosen parameter space ($0 < w(a) \leq 1; 0.1 < \alpha < 3$). At each such point in parameter space, a "spread" diversity selection (25) defined the "most diverse" subsets of $N_{div} = 10, 11, \ldots, 25$ members out of N_{lig} compounds. As shown in Fig. 4, the number

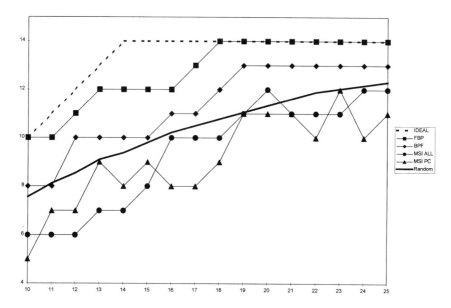

Figure 4 Number of distinct activity families $n_{fam}(N_{div})$ (y axis) for which at least one representative ligand has been included in the selection of the N_{div} (x axis) most diverse ligands out of a total set of $N_{lig} = 75$ ligands belonging to a total of $N_{fam} = 14$ activity families. Different selections of most diverse ligands were performed with various dissimilarity metrics based on pharmacophoric fingerprints (FBPA and BPF) and, respectively, topological and shape descriptors implemented in the QSAR package of Cerius2 from MSI, with (MSI-PC) or without (MSI ALL) reduction of the dimensionality of the corresponding Euclidean descriptor space to five principal components. The results obtained with FBPA correspond to the optimal setup of its fittable parameters, maximizing the average ratio $\langle n_{fam}/N_{div} \rangle$ at all $N_{div} \leq N_{fam}$.

of activity families $n_{fam}(N_{div})$ represented in each of these subsets reflects the overall quality of the dissimilarity metric: an optimal metric would pick each molecule from another family [e.g., $n_{fam}(N_{div}) = N_{div}$ until $N_{div} \leq N_{fam}$]. A unique quality criterion was chosen as the average ratio $\langle n_{fam}/N_{div} \rangle$ at all $N_{div} \leq N_{fam}$, and the parameter setup yield maximizing this criterion has been adopted as a "general diversity" setup of the FBPA-based metric.

4. FBPA-Based Query Scores: Asymmetric Similarity Scores

The previously defined similarity metric can be used to quickly retrieve the "nearest neighbors" m of a reference compound M, out of a very large collection of molecules described by their FBPA. However, the set of potentially interesting analogs of M may be expanded to include the m that are not *strictly* similar, but contain all the pharmacophoric features of M and also include supplementary pharmacophoric features not present in M. These latter would be a source of dissimilarity between m and M according to Eq. (5). The previously discussed similarity metric can be easily converted into a "pharmacophore query scoring" function $\delta^*(M,m)$ quantitatively expressing the extent to which the pharmacophoric pattern of M is represented in m, without introducing penalties if m actually possesses an even richer pharmacophore:

$$\delta^*_{a,b}(m, M) = 1 - \frac{2(\Psi^*_m \otimes \Psi_M)_{a,b}}{(\Psi_M \otimes \Psi_M)_{a,b} + (\Psi^*_m \otimes \Psi^*_m)_{a,b}} \tag{8}$$

where

$$\Psi^*_m (a, b, \Delta) = \min[\Psi_m(a, b, \Delta), \Psi_M (a, b, \Delta)] \tag{9}$$

Actually, the impact of the presence of supplementary pharmacophoric patterns in the candidate molecule on its ranking as a potentially interesting analog of M could be fine-tuned by means of a subunitary control parameter ζ, using:

$$\Psi^{(\zeta)}_m (a, b, \Delta) = (1 - \xi)\Psi^*_m (a, b, \Delta) + \xi\Psi_m (a, b, \Delta) \tag{10}$$

At $\zeta < 1$, the corresponding similarity score is asymmetric (26,27)—dependent, for example, on the choice of reference and candidate compound. The setup $\zeta = 1$ corresponds to the use of the original symmetric similarity metric.

5. FBPA Robustness with Respect to the Conformational Sampling Used: Comparison of GDbM-Derived and catConf-Derived Autocorrelograms

The extremely fast access to 3D-multiconformational models mainly serves to characterize the tens to hundreds of millions of potentially feasible combinatorial compounds in terms of FBPA (in more rigorous modeling studies, the GDbM conformer sets might be used as a diverse collection of raw starting geometries to be energy-minimized prior to their use, at a computational cost that may appear reasonable in the context of, e.g., binding energy predicting simulations). Therefore, the quality of GDbM conformational sampling was also evaluated in terms of the reproducibility of the obtained FBPA: Would the FBPA obtained from sets of conformers from different sampling algorithms be significantly different from the ones built from GDbM geometries? Is the number of considered conformations important to ensure the convergence to a reproducible average molecular FBPA? To answer these questions, a set of 200 randomly cherry-picked combinatorial products were submitted to three different conformational sampling calculations: (A) generation of up to 40 conformers/molecule with GDbM, (B) generation of up to 40 conformers/molecule with catConf (13) ("best" sampling mode), and (C) generation of a single conformer/molecule with catConf ("best" sampling mode). For each molecule, the FBPA obtained on the basis of the conformational family

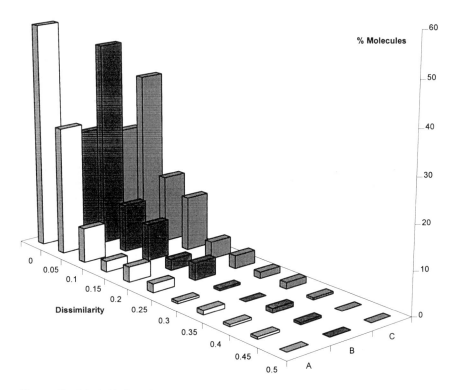

Figure 5 Dissimilarity of the FBPA obtained from GDbM conformer collections vs. those of different conformer families of the same molecule. This statistical study involved 200 combinatorial products, the z axis displaying the fraction of molecules for which the conformational sampling artifacts led to dissimilarity scores within each bin shown on the x axis. (A) Multiple catConf conformers (up to 40) with optimal fuzziness factor $\alpha = 0.32$. (B) Same as (A) but with $\alpha = 5$. (C) Single catConf geometries, with $\alpha = 0.32$.

A (GDbM) was compared to its counterparts built from conformer sets B and C, in terms of the FBPA dissimilarity metric, at "standard" fuzziness factor $\alpha = 0.32$ and at $\alpha = 5.0$, respectively (Fig. 5). The monitored fraction of molecules (%) for which the artifactual dissimilarity between the fingerprints based on different sampling approaches falls within each of 10 dissimilarity classes defined within a dissimilarity range of $[0, 0.5]$ represents a distribution of the probability of failure to generate a consensus descriptor for the molecule. Ideally, if both sets A and B exhaustively cover the conformational space of the molecule, they should lead to identical 3D fingerprints, which may significantly differ from the ones generated on the basis of a single conformer. A higher degree of "fuzziness" of the dissimilarity metric is expected to reduce the impact of the conformational sampling noise.

D. Virtual Screening of the Potential Fingerprint Library of Combinatorial Compounds (PFL) Based on FBPA Query/Similarity Scores

By using chemical filtering tools to enumerate all the viable reagents for each chemical synthesis protocol, and using the ghost database mechanism to generate conformers, the construction of a potential fingerprint library (PFL) enumerating the FBPAs of up to 10^8 combinatorial products can be achieved in about 1 month on a single-processor workstation.

By relying on the "similarity paradigm" or "neighborhood behavior" (28), a central working hypothesis in medicinal chemistry according to which "similar" molecules will display similar activities, the PFL can be mined for compounds that display pharmacophore similarity with respect to reference structures of desired physicochemical and biological properties.

Since certain aspects of molecular structure (such as chirality) are not encoded by FBPA, and given the unavoidable loss in information occurring when a structure is reduced to a molecular size–independent fingerprint, FBPA matching is best used as a high throughput filter to discard obviously dissimilar candidates. Shrinking the initial search space (the PLF, 10^8 compounds) concentrates the best matching candidates into a subset of few thousand, also including some "false similars" due to fingerprint artifacts. ComPharm, a refined similarity metric based on molecular overlay (29), is applied at a later stage to discard the latter. To retrieve the corresponding matches from the PLF, this two-stage virtual screening approach (Fig. 6) may use different pharmacophoric hypotheses as starting points.

The *structures of active compounds* can be used for strict similarity searches as well as for pharmacophore queries. This represents a default option when a shortage of examples of actives renders structure–activity data unavailable for the biological target of interest. The obvious drawback of this approach stems from the implicit assumption that *all* the pharma-

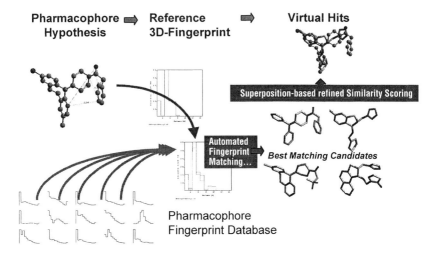

Pharmacophore ⇒ Reference ⇒ Virtual Hits
Hypothesis   3D-Fingerprint

Superposition-based refined Similarity Scoring

Automated
Fingerprint   Best Matching Candidates
Matching...

Pharmacophore
Fingerprint Database

Figure 6 Virtual screening approach using the FBPA metric as a high throughput filter aimed at reducing the scope of the initial search against 10^8 combinatorial compounds down to a few thousand best matching candidates among which the most relevant virtual hits can be selected on the basis of refined, overlay-based pharmacophore similarity calculations.

cophore elements present in the lead structure are *relevant* for the biological activity under consideration and must be matched by the virtual hits. Active compounds that fail to match *irrelevant* pharmacophoric features of the lead will be discarded by the pattern recognition algorithm.

If a set of diverse active leads is available, the subset of pharmacophoric features shared by all the actives can be retrieved using any hypothesis generation software (or plain common sense) to define the basic pharmacophore that is supposed to yield the activity of interest. Such a *consensus pharmacophore hypothesis* can be easily encoded as a FBPA and used to query the PFL, while direct confrontation of the hypothesis to 10^8 3D structures would be unfeasible—this is actually performed at the next step, when the FBPA-filtered best matching candidates are superimposed on the hypothesis.

Site- or ligand-based hypotheses are appropriate when a 3D crystal structure of the macromolecular target is available and can be used to define the coordinates and pharmacophoric nature of anchoring points at which the presence of specific ligand groups is expected. These can be defined from the actual binding models of ligands in ligand–target complexes or from the distribution of the functional groups of the assumed site walls that are potentially accessible to the ligand. The candidate structures that were successfully overlaid on the pharmacophore hypothesis can be easily docked into the active site.

E. Validation Studies of the Virtual Screening Approach

The validation of the virtual screening techniques is a nontrivial task, implying many different aspects, many of which are heuristic and cannot be unambiguously quantified. Two main categories of validation studies are reported here.

1. "Seeding" Experiments: An In Silico Validation of the FBPA Similarity Metric

Two structurally different classes farnesyl protein transferase (FPT) inhibitors (30,31), were used in a "seeding" study designed to answer the following question: Supposing that only one representative family A is known, while the "unknown" inhibitors of chemotype B are members of a compound library, would the FBPA-based selection and testing of nearest neighbors of lead A from this library be a winning strategy over the "blind" high throughout screening of the full collection, providing a more effective discovery of B inhibitors? Figure 7 shows the two classes, A and B.

In a first experiment, two large collections of combinatorial compounds—a part of our corporate collection (150,000 molecules) and a general lead-seeking library (24,000 molecules), both including several sublibraries issued from various connecting chemistries—were "seeded" with the 37 FPT inhibitors of family B, most of which displayed IC_{50} values in the micromolar range. FBPAs were generated for both these collections and for the reference member of family A. Eventually, the similarity scores with respect to the reference were calculated according to Eq. (7) for each member of the "seeded" sets (original libraries including FPT inhibitors "hidden" among the combinatorial compounds), using the weigh-

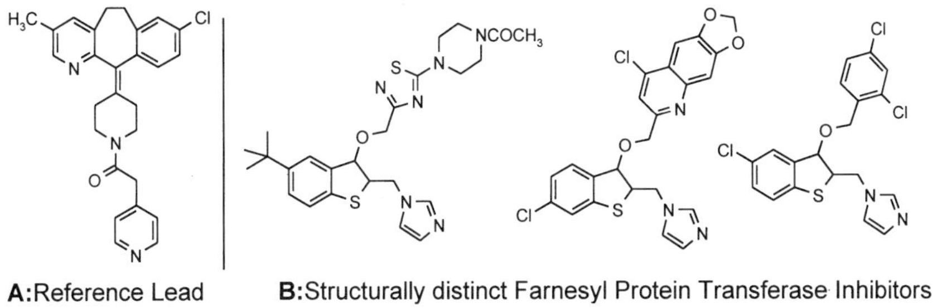

A:Reference Lead   **B:Structurally distinct Farnesyl Protein Transferase Inhibitors**

Figure 7   (A) The reference lead used as a starting point for the "discovery" of an alternative structural family "hidden" among large collections of combinatorial compounds. (B) Examples of some family members.

ing factors obtained as shown in Section II.C.3. The members were ranked in decreasing order of their similarity scores, and the fraction of family B ($y\%$) found among the best ranked $x\%$ of the library was plotted as a function of $x\%$ (Fig. 8).

A comparative study of various diversity metrics was then performed, using a smaller set of 629 drug molecules from the U.S. Pharmacopeia as the basis library to be seeded with the B family. The metrics (Fig. 9) were tested for their ability to retrieve the "B" family among the nearest neighbors of the reference lead, as described earlier.

The FBPA-based similarity metric.

Euclidean metric involving descriptors calculated with the Cerius2 software including the electrotopological state keys (32), charge and dipole, structural and shape terms, A log P98, and topological indices (12).

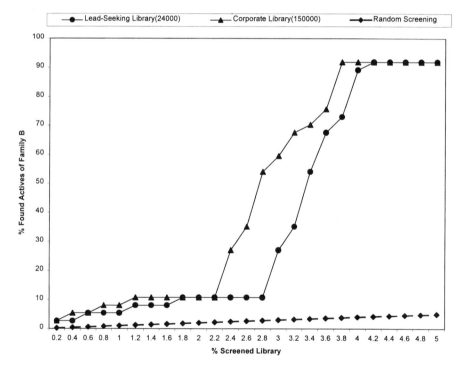

Figure 8   Results of the seeding experiment show (y axis) the percentage of the active FPT inhibitor family B that would have been discovered by screening the corresponding percentage of the total library given on the x axis if high throughput screening of this library had been prioritized to test compounds with high FBPA similarity scores with respect to the reference lead first. By contrast, the retrieval rate of the family B ligands would increase linearly with the fraction of the screened collection if compounds were screened in random order.

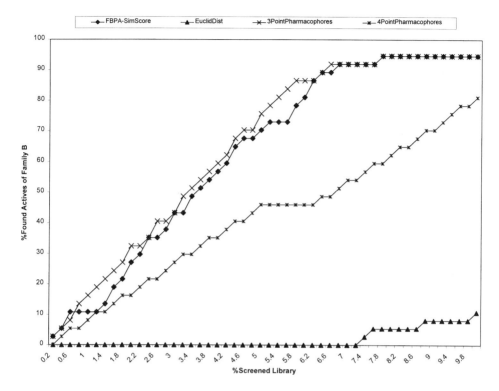

Figure 9 The ability of various similarity metrics to "discover" family B of FPT inhibitors, "hidden" among 629 drugs of the U.S. Pharmacopeia, by selection of nearest neighbors of the reference compound (Fig. 7A).

Three- and four-point (33) pharmacophore fingerprint-based similarity metrics, where the fingerprints were built on the basis of multiconformational models involving up to 20 conformers, using the catDB and Cerius2 software tools. A Dice (21) metric has been employed to evaluate the corresponding similarity scores.

2. Similarity-Search-Directed Synthesis of Active Lead Analogs

In a first example reported here, the structure of a dopamine transporter (DAT) ligand reported in literature (34) was used as the source of an FBPA-based "strict" similarity query returning structures of the nearest neighbors from a potential fingerprint library that, at the time of this experiment, featured only 5 million combinatorial candidate compounds. Typically, several practical design and synthesis constraints eventually define the exact set of molecules that are actually made on the basis of the list of the virtual hits. In this situation, the best virtual hits were seen to belong to a quite homogeneous family of amides and a list of the most often recurring building blocks found in the majority of virtual hits resumed to a pair of carboxylic acids and 21 aromatic amines, all of them previously obtained by reductive amination of aromatic aldehydes with BOC-piperazine. The resulting small combinatorial library of 21 × 2 amides was integrally synthesized, purified, and tested, with the results shown in Table 2. The most active analog structures are given in Table 3. To evaluate the relative potency of the lead vs. the virtual hit with the largest percent of inhibition at 10 μM, the IC_{50} value of the former was reevaluated under the experimental conditions used for the latter.

3. Applications of the Virtual Screening Approach in Lead Optimization

As part of a discovery program of potent μ-opiate receptor ligands, a lead optimization experiment described in more detail elsewhere (35) started from a mi-

Table 2 Reference DAT Ligand Structure and Screening Results of a Small Combinational Library Designed to Encompass Most of the Nearest Neighbors of the Reference Structure Selected by Using the FBPA Metric from the Potential Fingerprint Library

Reference inhibitor	Screening results of a focused combinatorial library (2 acids x 21 amines) issued from reductive amination of aromatic aldehydes

$(IC_{50} = 0.1 \ \mu M)$

Table 3 Structures and Potencies of the Most Active
Members of the Focused Library from Table 2

Formula	Inhibition at 10 µM (%)
	93% $IC_{50} = 1\ \mu M$
	76%
	76%
	70%

cromolar hit (Fig. 10) obtained from the primary screening of a lead-seeking carbamate library. Interestingly, this compound does not include any basic amino
group thought to be an important pharmacophore element for µ-opiate affinity.
Therefore, the research for more potent analogs of this lead has been conducted
along two conceptually "orthogonal" directions.

　　1. *Searching for novel molecular topologies compatible with the given
pharmacophore pattern.* Using the virtual screening methodology as shown in
Fig. 6, the best matching candidates of the reference hit were selected out of the
PFL of 80 million combinatorial candidates by an FBPA similarity search and

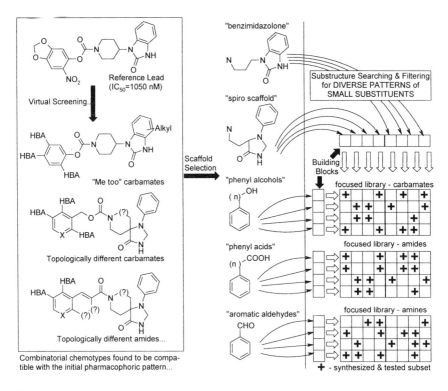

Figure 10 A lead optimization program starting from a micromolar μ-opiate primary hit and using both pharmacophore-based and substructure-based searches to discover a structurally diverse set of potent leads.

then ranked in terms of their ComPharm overlap scores with respect to the reference. These "virtual hits" may show widely different molecular topologies and/or originate from different connecting chemistries. However, since the reference hit itself is a combinatorial member of the PFL, this compound as well as many other "me too" analog carbamates (issued from alcohols and amines with *slightly* different substitution patterns not impacting the overall pharmacophore) are also expected to figure among the virtual hits.

2. *Searching for optimized pharmacophore patterns by insertions, substitutions, and deletions of pharmacophoric substituents of the main building block scaffolds.* Rather than using *only* the building blocks appearing in the virtual hits (see the example of DAT ligands), the main scaffolds represented in these building blocks were used to conduct *substructure* queries in the corporate reagent database and in commercial databases such as ACD, employing the ISIS/Base software. The resulting *extended* building block families around every scaffold

yield combinatorial products featuring pharmacophoric patterns that are either *enriched, conserved,* or *improverished* with respect to the reference. Changes affecting the nature of the substituent acting as *reactive* group require, if possible, an appeal to other connecting chemistries. In particular, the retrieval of *aldehydes* R_i—CH=O with the same scaffolds R_i found in the alcohols R_i—CH_2—OH gives way to the design of *reductive amination products.* These feature a supposedly important *basic amino group,* in contrast to the amide-type nitrogen in the corresponding carbamates.

Eventually, all the building blocks selected were merged and used to design combinatorial focused sublibraries, representing all the implied retrosynthetic routes. However, these sublibraries were *not* synthesized and tested as such. A total of 344 representative compounds were "cherry-picked" to minimize the synthesis effort, including (1) all the analogs explicitly retrieved by the virtual screening approach and (2) other combinatorial compounds chosen such that each of the building blocks selected is represented several times.

III. RESULTS AND DISCUSSION

A. Validation of the Molecular Geometries Sampled by the Ghost Database Mechanism

It is difficult to define an absolute "quality" criterion of a molecular geometry. Neither the potential energy differences, nor the rms deviations of atomic positions represent unambiguous measures of the distance between the departure and the optimum geometries. On one hand, an unminimized conformer may be of high energy, although geometrically very similar to the next energy minimum of the potential energy surface, typically due to a short-range repulsive van der Waals contact that can be relaxed by means of a minimal atomic displacement of a few tenths of an angstrom. On the other hand, the minimization of a departure geometry already low in energy may be nevertheless accompanied by significant changes in molecular geometry if flat regions of the potential energy surface are crossed.

The geometrical criteria defined earlier can be viewed as an empirical measure of the shortest path length required to reach the neighborhood of conformers within Δ kcal/mol with respect to the next local minimum of the CVFF-based potential energy surface. In other words, they express a distance or dissimilarity between the departure conformer and the closest "reasonable" geometry of low strain energy, according to the currently applied molecular Hamiltonian. Table 1 shows that, on the average over all molecules, the atoms of the GDbM structures must be shifted by only 0.1 Å to reach the potential energy zone within $+20$ kcal/mol with respect to the next local optimum. While in the worst situation, reaching this potential energy zone was accompanied by geometrical rearrange-

ments implying interatomic distance changes of up to 5.7 Å, such events appear
to be rare: the corresponding average over all compounds of the largest fluctuation of interatomic distances within each molecule was only 0.6 Å. Geometries
obtained from catConf and UFF local optima appear to be only marginally closer
than GDbM structures to the low energy zones of the CVFF potential surface. On
the average over all compounds, the largest interatomic distance changes expected
in a molecule upon CVFF energy minimization are 2.7–3.8 Å. The starting geometries less affected by minimization are the ones stemming from catConf, while
the ones provided by GDbM do not behave significantly worse than the UFF optimal conformations, which were shown to be clearly different from CVFF optimal conformations. It can therefore be concluded that the GDbM geometries are
comparable to the ones obtained on the basis of low cost/low quality force fields.
Since MIDC scores monitor the *largest* fluctuation of interatomic distances seen
in a molecule, using CVFF to optimize GDbM geometries would not dramatically
alter the overall image of the interatomic distance distribution pattern on which
the FBPA are based.

B. Calibration of the FBPA-Based Similarity Metric

The optimal feature weights and fuzziness factor maximizing the ability of the
FBPA metric to discriminate between ligands displaying different types of biological activities are listed in Table 4. As expected, the weighing factors associated to the polar pharmacophoric features, which are generally believed to confer
the specificity of the ligands, were found to be consistently higher than the ones
of the hydrophobic and aromatic features. While hydrophobic site–ligand contacts
are thought to represent the most important contribution to the binding free energy, the same is basically true for most of the ligand–receptor interactions. Hy-

Table 4 Optimal Parameters of the FBPA
Metric, Maximizing Its Ability to Discriminate
Between Ligands Belonging to Different
Activity Families

Weighing factor	Value
Aliphatic	0.41
Aromatic	0.24
Hydrogen bond acceptor	1.00
Hydrogen bond donor	0.78
Cation	0.68
Anion	0.76
Fuzziness factor α	0.32

drophobic and aromatic groups are ubiquitous in all the druglike molecules, representing a less important differentiation criterion between different biological categories.

From the overall performance of the optimal FBPA metric with respect to the problem used to calibrate its parameters (Fig. 4), we saw that FBPA outperforms both our previously reported, less elaborate BPF-based metric (16) and the Euclidean metrics based on topological and shape descriptors and their most relevant five principal components. The quality of diversity selections performed on the basis of the FBPA metric is near optimal, whereas molecular topology and shape alone, ignoring the chemical nature of the included atoms, are basically unable to discriminate between classes of biological activity and appear to be no better than random selections.

C. Robustness of FBPA with Respect to Conformational Sampling Artifacts

Earlier (Fig. 5), we compared the expected fluctuations of the dissimilarity scores obtained from a molecule represented by its GDbM conformer set (< 40 conformers) and from alternative conformer families of the same compound. Ideally, these dissimilarity scores should be zero, independently of the sampling algorithms used, if these provide an exhaustive exploration of the conformational space. The FBPAs built on the basis of up to 40 catConf geometries actually compared fairly well with those obtained from GDbM structures (series A of Fig. 5), and the use of fuzzy scoring effectively "smoothed out" the differences due to sampling. Indeed, series B, comparing exactly the same two sets of conformers as series A, evidenced an increased probability of diverging dissimilarity scores when a "less fuzzy" scoring function was used. However, the failure to sample several significant conformers of the molecules may result in quite different FBPA fingerprints. Series C, comparing a single catConf geometry with the GDbM conformer set, evidenced an even larger rate of inconsistent scores than in series B, in spite of fuzzy scoring. Therefore, it may be concluded that, as expected, both the use of multiple conformers to build the FBPA and the application of fuzzy scoring functions to compare them significantly enhance the robustness of these 3D descriptors.

D. Validation of the Virtual Screening Algorithm

From a technical point of view, the virtual screening algorithm discussed in this chapter is a *pattern recognition* approach and should be judged in terms of its success in retrieving compounds with pharmacophoric patterns that are *similar* to the reference hypothesis. However, there is no absolute recipe to define "pharmacophoric pattern similarity" or any other type of *structural similarity*. In principle, any mathematical function fulfilling the conditions of a *metric* in a given vector space can be used as a similarity score.

1. Similarity-Based Search Tools as "Expert Systems"

Structural similarity is first of all a heuristic concept, and the possible mathematical definitions of structural similarity may agree more or less with a medicinal chemist's point of view about "similar molecules." A virtual screening algorithm acting as an "expert system" (e.g., selecting the analogs that a medicinal chemist would have picked out for synthesis and testing out of large databases) would already represent an extremely useful tool in drug design. However, it would obviously be difficult to give any quantitative estimate of its beneficial effects on the discovery of novel drugs. Furthermore, algorithms based on pharmacophoric pattern recognition may outperform the human operator in retrieving analogs that, from the latter's point of view, bear no obvious similarity to the lead at first sight (the two rightmost analogs among the "best matching candidates" in Fig. 6). However, their similarity to the lead can be easily understood and accepted by the medicinal chemist upon analysis of the superimposition model (upper right-hand corner, Fig. 6) showing how pharmacophorically related groups can be brought to overlap. The human view of molecular similarity is heavily biased toward topological "substructure" similarity, and, not astonishingly, most of the patents covering classes of therapeutically active compounds rely on a connectivity-centric approach featuring a basic scaffold and its substituents (Markush representations). Alternative, pharmacophore-based similarity metrics are excellent tools for retrieving compounds of similar pharmacophore group distributions of completely different connectivity with respect to a lead structure.

2. Structural Similarity and Property Similarity: The "Similarity Paradigm": Interpretation of the "Seeding" Experiments

From a practical point of view, however, the algorithm is expected to successfully select compounds that are likely to display similar *properties* with respect to the lead molecule. In other words, the "structural similarity" defined by the algorithm is expected to significantly correlate with the "property similarity" evidenced by experimental methods. Insofar as the descriptors entering the similarity metric are not drawn from exhaustive docking (36) simulations evidencing both the energetic and entropic aspects of the interaction of ligands with their targets, there is no fundamental reason to expect any absolute guarantees that a near neighbor of a given lead will certainly "inherit" the biological activity of the lead. The "similarity paradigm" relating the structural to the property similarity is actually a statistical law: the density of actives in a set of molecules that are similar to an active lead *must exceed* the density of actives in a set of "randomly chosen" molecules. This is clearly evidenced in Fig. 8, where more than 90% of the family of structurally distinct FPT inhibitors (Fig. 7B) are found among the 4% of compounds that are best ranked in terms of their FBPA similarity with respect to the reference ligand. In other words, a high throughput screening experiment conducted in the order of increasing dissimilarity with respect to the lead A and terminated after testing 4% of the compounds would nevertheless have discovered the large majority of a *completely original family B of FPT inhibitors.* By contrast, random screening of the same collection to achieve the same result would have required a roughly *20-fold* larger screening effort! Such an interpretation of a seeding experiment is however an oversimplified view of the much more complex reality of the drug discovery effort. Some of the other aspects that should be included in this analysis are enumerated as follows.

1. Many compounds *not* belonging to the "hidden" family to be retrieved may well be active FPT inhibitors, even though they were not counted as such in this particular in silico experiment. This is a general weakness of all the purely virtual validation studies of active analog retrieval methods making reference to electronic databases of *untested* compounds. Only the actual testing of the entire library would allow a realistic estimate of the overall enrichment factor achieved by the computational approach.

2. It is convenient to think about the "false positives"—inactives that were ranked higher than the truly active compounds—as being misclassified by the virtual screening software because of an "overestimated" similarity score with respect to the lead. In principle, an attempt to correct such problems by recalibration of the similarity metric could be envisaged. However, some of such "should-be actives" may actually represent "me too" compounds, strongly related to the reference lead. A forced reparameterization of the similarity metric in order to rank them in the list of candidates, if at all possible, would lead to highly counterintuitive results from a medicinal chemist's point of view.

3. Point 2 also evidences the nonnegligible impact of the choice of the "hiding" set on the result of a seeding experiment. In an ideal experiment, the set of presumed inactives against which the search of the hidden actives is conducted should in principle represent a homogeneous density coverage of the whole structural space occupied by the 10^{18} (37) "druglike" molecules. This is never the case in real drug discovery programs. It is relatively easy to achieve impressive enrichment factors when searching for the actives among diverse compounds. In the currently shown example, each of the many pharmacophore patterns is represented by a *small* number of examples. In such cases, it is highly improbable that one will encounter a massive set of inactive molecules also displaying the wanted patterns, and therefore erroneously selected among the "virtual hits." By contrast, if virtual screening were used to discriminate between active and inactive members of a *focused* library with a high degree of common pharmacophoric patterns, where activity/inactivity is controlled by local structural details of the compounds, the observed enrichment factors may be much lower.

4. The failure of the methodology to "discover" a set of structurally unrelated compounds does not necessarily invalidate the utility of the virtual screening approach. A failure can be considered as such only if the existence of a different metric, successfully evidencing the similarity of those apparently unrelated families, actually proves that the problem was *solvable in terms of the "similarity paradigm."* Indeed, ligand families of a given target may actually display different pharmacophore patterns corresponding to different anchoring modes to the site, or even bind to different sites. It is not to be expected that such families should ever be discovered on the basis of similarity considerations starting from a representative of another one. Therefore, "seeding" experiments expected to be successful without being trivial are not easy to design. In particular, experiments reporting statistics of retrieval of a heterogeneous set of actives against a single reference lead are always difficult to interpret. It is unclear how many of the "hidden" compounds are reasonably similar (in terms of action mechanism) to the lead, in order to count them among the molecules that can be reasonably expected to be virtual hits.

5. As previously discussed, a "seeding" experiment offers no hints of the number or structural class of the actives *not* selected among the best ranked and therefore lost if the systematic high throughput screening of libraries were to be replaced by a virtual-screening-based selection of compounds to screen. Obviously, the search for actives with novel pharmacophore patterns must be conducted by means of "blind" high throughput screening experiments or targeted "mutations" of the current pharmacophores (e.g., medicinal chemistry experiments involving insertions and/or deletions of functional groups). Knowledge of the 3D structure of the target would be of great help in this approach. The virtual screening tool may then be successfully used to extract all the possible chemotypes of the library of potentially synthetizable compounds fitting every given pharmacophore. The concerted application of the two approaches is required for the early discovery of drug candidates, sufficiently diverse to ensure that at least one of these starting points displays selectivity, bioavailability, and metabolic stability profiles promising enough to justify further optimization work.

3. Comparative Study of the Performances of Similarity Metrics in "Seeding" Experiments: Fuzziness vs. Informational Content

A comparative study of the performance of different similarity metrics with respect to the same retrieval problem of active FPT inhibitors of family B has been undertaken against a smaller "hiding" set of 629 drugs from the U.S. Pharmacopeia (see Fig. 9).

A first interesting observation is that in the current experimental setup, the use of the same FBPA metric reported in the previous seeding experiment would yield only a 10-fold enrichment—10% rather than 4% of the entire library must be screened before 90% of family B can be discovered. Enrichment factors are *not* invariant with respect to the size of the "hiding" library—they would be, though, if (1) the size of the hidden set would scale up proportionally to the size of the library and (2) the increase in library size was achieved not by populating novel regions of the structure space, but by increasing the compound density in the structure space zones already populated by the initial collection. In the current experiment, 35 members of family B are found among the first 52 best ranked compounds. The screening of this subset would have yielded a hit rate of 67.3%, leaving not much space for further improvement.

It can be seen that the Euclidean metric based on "classical" molecular descriptors completely failed to recognize the similarity between families A and B, actually biasing the selection *against* these molecules that appeared more *distant* from the reference than many of the supposedly FPT-inactive members of the "hiding" set. Although the electrotopological state keys conveyed information about the types and numbers of heteroatoms, none of these descriptors unambiguously described the relative arrangement of the functional groups in the molecules.

By contrast, both the FBPA and the three-point pharmacophore fingerprint metrics display a significant enrichment in FPT B-family members in the neighborhood of reference A. Intriguingly, the four-point pharmacophore fingerprints, encoding the relative position of quadruplets of groups rather than triplets or pairs, actually perform *significantly worse* in this test. It might be argued that four-point pharmacophores actually selected *other* actives from the "hiding" set—although to our knowledge none of the included drug molecules act against FPT, the possibility that some of them display a residual activity on this target cannot be excluded. However, an analysis of the four-point pharmacophore dissimilarity scores revealed that the *nearest* neighbor of the reference compound A had a dissimilarity of 0.98 on a scale between 0 (identical) and 1 (completely different). Even the most similar of the compounds from the screened collection has virtually no four-point pharmacophores in common with the reference molecule, compared to the wealth of pharmacophore bits highlighted by each compound! This is most likely a consequence of the markedly discontinuous nature of these fingerprints, where each combination of pharmacophore group types and the corresponding intergroup distances are associated with a particular bit in the overall fingerprint. Inevitably, such a four-point pharmacophore is prone to be very similar to *many* other four-point pharmacophores consisting of the same functional groups at roughly the same distances—except for the one distance that exceeds the cutoff value delimiting them. Highlighting a given three-point pharmacophore bit involves fewer constraints than hitting a given four-point pharmacophore. The probability that similar molecules fail to highlight exactly the same bit in the pharmacophore fingerprint, highlighting *related* pharmacophores instead, is significantly enhanced in four-point over three-point pharmacophore models. Since the relatedness of the pharmacophores associated with each bit is *not* accounted for by the scoring function, the existing elements of similarity are not recognized. The fuzziness of the FBPA scoring function, designed to avoid exactly this type of artifact, compensates for the more important informational contents stored in three-point pharmacophore fingerprints. With four-point pharmacophores, however, the absence of fuzzy scoring becomes the dominant effect and causes an overall loss of discriminant power.

4. Discovery of Active Dopamine Transporter Ligands by FBPA-Based Similarity Searching

Tables 2 and 3 regroup the experimental results obtained with the small combinatorial library, designed to encompass all the virtual hits produced by a similarity search around a lead from the literature, against a potential fingerprint library featuring 5 million combinatorial candidates. The most striking structural difference between the lead and its analogs is the replacement of the diphenylmethyl ether moiety with a diphenylketone fragment. It is difficult to evaluate the performance of the algorithm in terms of the enrichment rate of actives in the focused library. The discovery of four compounds with an inhibition percentage over 70% (hit rate of about 10%) may represent a significant enhancement over "blind" HTS results of lead-seeking libraries (no such data were available). But is the size of this focused library sufficient to ensure statistical relevance of these hit rates? None of the hits showed a stronger potency than the lead, the best of them being actually 10 times weaker—as pointed out earlier, nothing in the pattern recognition approach allows the actual prediction of the activities of analogs. However, the most interesting feature in these compounds is the presence of the amide bond, opening a second retrosynthetic pathway accessible by combinatorial chemistry. Therefore, a three-step focused library could now be easily designed for further optimization of the new lead, exploiting the available building block diversity of aromatic acids, cyclic diamines, and aromatic aldehydes. This enlargement of the scope of the synthesis effort may eventually prove to be more beneficial than the occasioned loss of one order of magnitude in terms of potency. Backed by a quite small synthesis and screening effort, the utilization of the FBPA-based virtual screening approach has led to the "discovery" of a novel strategy for the synthesis of DAT ligands. Certainly, the "merit" of the methodology does not reside in the purposeful "design" of the novel synthesis route, but in having showed that the amide derivatives are pharmacophorically compatible with the lead structure and should not be discarded (as would have probably happened with a substructure search procedure). This may be important, or it may prove to be just another dead end among other research directions, but nevertheless the potentiality of this similarity searching tool as an "idea generator" for medicinal and combinatorial chemists is illustrated.

5. Optimization of μ-Opiate Ligands

Some of the representative μ ligands and their respective IC_{50} values are given in Tables 5–8. As shown in Table 7, replacing the benzimidazolone fragment in the reference (Table 5) by the "spiro" moiety, shown by the virtual screening approach to be pharmacophorically equivalent, yields a class of topologically different carbamates that mostly conserve the low micromolar affinity of their parent compound. According to Tables 6 and 8, the exploration of alternative pharmacophore patterns by introduction of positive charges successfully led to the discovery of two distinct structural families of low nanomolar μ-opiate binders.

In this experiment, there is no meaningful way to *quantitatively* score the contribution of the FBPA-based methodology to the success of this real-life experiment, in which it was used as a tool among many others. Nevertheless, at least four broad conclusions with respect to the role of similarity searches in the drug discovery process can be drawn from this example.

1. A first general remark is that the FBPA-based similarity metric, an implicit function of both molecular size and number of polar groups in the two molecules being compared, is highly relevant with respect to the "druglikeness"

Table 5 Generic Template, Substitution Patterns, and Activities against Certain μ-Opiate Receptors[a]

R_1	R_2	R_3	R_4	n	p	IC_{50} (nM)	Template
NO_2	—CH_2—CH_2—		H	1	1	1050 *Reference	
Cl	—CH_2—CH_2—		H	1	1	147	
Cl	H	H	H	1	1	2480	
Cl	H	H	H	0	1	>10,000	
NO_2	—CH_2—CH=		H	1	1	>10,000	
Cl	—CH_2—CH_2—		H	1	2	2,590	
Cl	—CH_2—CH_2—		CH_3	1	1	1,235	

[a] These are μ-opiate receptors of "benzimidazolone" analogs that are *pharmacophorically similar* with respect to their "parent" reference lead. Largest pharmacophore difference in this series: replacement by the —NO_2 group in R_1 by a hydrophobic atom/group. Largest topological difference within this series: ring opening (R_2–R_3). Shaded rows represent compounds selected by the virtual screening approach.

Table 6 Benzimidazolone-Based Derivatives That Are Pharmacophorically Radically Different from the Parent Compound Owing to the Replacement of the Carbamate Linker by a Positive Ionizable Group

R_1	R_2	R_3	R_4	R_5	R_6	p	IC_{50} (nM)	Template
Cl	$-OCH_2O-$		$-CH_2-CH_2-$		H	1	1.2	
Cl	$-OCH_2O-$		H	H	H	1	235	
NO_2	$-OCH_2O-$		$-CH_2-CH_2-$		H	1	1.5	
NO_2	$-OCH_2O-$		$-CH_2-CH=$		H	1	19.3	
Cl	$-OCH_2O-$		$-CH_2-CH_2-$		H	2	6.3	
Cl	$-OCH_2O-$		$-CH_2-CH_2-$		CH_3	1	2.4	
Cl	OMe	H	$-CH_2-CH_2-$		H	1	6	

Table 7 Various Carbamates Featuring the "Spiro" Chemotype[a]

R_1	R_2	R_3	R_4	R_5	X	IC_{50}(nM)	Template
H	NO_2	$-OCH_2O-$		H	CH	>10,000	
H	Cl	$-OCH_2O-$		$-CH_2CONHMe$	CH	179	
H	Cl	$-OCH_2O-$		$-CH_2COOH$	CH	735	
H	Cl	$-OCH_2O-$		H	CH	1,730	
H	H	$-OCH_2O-$		H	CH	2,285	
H	H	$-$Thiadiazole$-$		H	CH	>10,000	
H	H	H	$-OC_6H_5$	H	CH	>10,000	
$-CH_2-CH_2-$		H	H	H	CH	1,470	
H	H	H	H	H	N	>10,000	

[a] Shaded compounds were "discovered" by the virtual screening approach because their pharmacophoric patterns were similar to the ones of the reference lead, despite differences in connectivities. The other two molecules present supplementary pharmacophore groups not present in the lead and are not similar to it according to the "strict" FBPA metric.

Table 8 "Spiro" Compounds Derived from the Virtual Screening Hits from Table[a]

R_1	R_2	R_3	R_4	IC_{50} (nM)	Template
Cl	—OCH₂O—		H	2.0	
Cl	—OCH₂O—	CH₂CONHCH₃	0.9		
H	—OCH₂O—		H	16.7	
H	Thiadiazole		H	24.9	
H	H	—OC₆H₅	H	83.5	

[a] These hits were derived as the result of a voluntary pharmacophore mutation replacing the carbamate linker with a positive ionizable tertiary amino group. The molecules differ with respect to the reference lead from both topological and pharmacophoric point of view.

of compounds. Any near neighbor of a druglike lead respecting the "rules of five" (38) will most likely also abide by these rules. Therefore, this metric can be successfully used for the design of "druglike" combinatorial libraries; that is, "druglikeness" may be seen as a valuable "by-product" of the FBPA-based selection mechanism.

2. The lead optimization scheme described here accumulates the strengths of both medicinal chemistry and combinatorial chemistry. The optimal pharmacophore was sought using search tools and electronic databases to select all the pharmacophoric patterns that were *easily* attainable by coupling of available building blocks. This may appear as a serious restriction with respect to classical medicinal chemistry, where labor-intensive chemistry is performed to achieve *any* desired substitutions thought to represent an "interesting" novel pharmacophore pattern. In this context, the ability of the FBPA-based searches to furnish alternative molecular topologies that are pharmacophorically equivalent significantly loosened this restriction. Indeed, a pharmacophoric pattern difficult to achieve on a given molecular skeleton may be brought within the scope of efficient resynthesis based on high throughput coupling chemistries when one is considering other potential ligand chemotypes. Two structurally different and highly potent μ-ligand families, involving the synthesis of a few hundred analogs, were discovered within a period of weeks.

3. The major result obtained from the FBPA-based similarity metric was the insight that the novel "spiro" chemotype may successfully replace the benzimidazolone moiety without significantly modifying the overall pharmacophore pattern of the molecule. The "breakthrough" in potency was *not* a result of this

substitution, but required a major step away from the original pattern in pharmacophore space—the introduction of a positive charge. However, the discovery of *two* potent, structurally different lead families offers a potentially capital backup option if one should prove to be incompatible with the pharmacokinetic constraints imposed to a drug molecule.

4.   A quite obvious conclusion pertaining to all pharmacophore-based similarity search methods, irrespectively of the metrics they use, appears nevertheless worth restating at this point: a *similarity* search around a weak binder with an obviously suboptimal pharmacophore pattern is expected to discover nearest neighbors with *suboptimal* pharmacophoric patterns—and therefore weak binders like the parent lead! To discover the optimal binding pattern based on a pharmacophore similarity search would imply selection of all neighbors within a similarity sphere of radius R_{opt}, representing the dissimilarity score of the current vs. optimal pharmacophore pattern (see Fig. 11). Obviously, R_{opt} is an unknown parameter at the moment the search is done, and the larger the cutoff taken, the larger the number of inactives selected among the quickly increasing set of neighbors. If starting and optimal pharmacophore patterns are different enough, then setting a low cutoff radius to select a set with a significant enrichment factor in de-

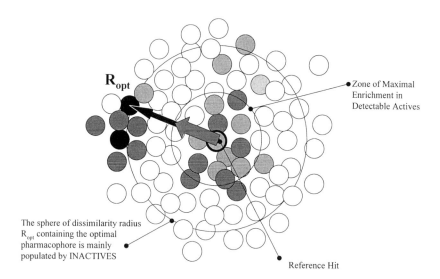

Figure 11   A possible distribution of actives and inactives in the "structure space" around a lead of average potency. The nearest neighborhood of a lead with a suboptimal pharmacophore pattern is enriched in analogs with similar potencies (compound activities are suggested by the intensity of gray shading; dark = potent). R_{opt} represents the difference between the reference and the optimal pharmacophore of a potent ligand according to the pharmacophore dissimilarity metric.

tectable actives is unlikely to yield any major improvement of affinity among the selected compounds. By contrast, setting a high cutoff value may well allow the selection of much more potent actives, but with no reduction at all of the experimental effort compared to "blind" high throughput experiments.

Pharmacophore query procedures (asymmetric similarity scores) may select analogs with novel features that enrich the original patterns and may *serendipitously* contribute to the improvement of their binding properties. However, the current experimental setup extended the scope of the search for the optimal pharmacophore pattern even more than a pharmacophore query approach would have done:

> It included analogs with *impoverished* pharmacophore patterns, which would have been discarded by a pharmacophore query approach for failing to match all the groups available in the reference compound.
>
> It included analogs based on commercial building blocks from the ACD that were not represented in the potential fingerprint library because they did not figure on the list of accessible starting materials at the moment the search was initiated.

IV. CONCLUSIONS

No attempt to reduce the heuristic concept of molecular similarity to a mathematical expression representing a distance metric in a descriptor space, no matter how ingenious, can ever compensate for the inherently imprecise working hypothesis behind the seemingly rigorous mathematical treatment. This central hypothesis, the "similarity paradigm," is best expressed as a statistical law stipulating that the probability of discovering actives of the same type as a reference lead is locally enhanced among the set of compounds occupying the *near neighborhood* of the reference, compared to the average chance of hitting such a molecule over the whole accessible volume of the structure space. The present work focused on the potentialities and caveats of various applications of this hypothesis in the drug design process. In an innovative technical context, the use of specific high throughput multiconformational modeling of combinatorial compounds and effective fingerprinting of average pharmacophoric patterns set new limits for the size of the primary pool of potentially synthetizable candidates against which various types of similarity query can be performed without compromising the relevance of the similarity metric.

Pharmacophore-based similarity metrics, and in particular the FBPA-based methodology presented here, are potent tools that allow the discovery of completely novel molecular connectivities, nevertheless conserving the original pharmacophoric patterns. This extremely useful feature displays an obvious synergy with the medicinal chemist's ability to imagine and synthesize compounds with alternative pharmacophoric "ornaments" around a given skeleton motif. The combination of the two approaches led to the discovery of two structurally distinct and very potent μ-ligand families, instead of only one such family likely to be discovered on the basis of a classical lead optimization approach not backed up by searches for novel chemotypes.

Real-life validation experiments of this type are also the ones most difficult to analyze in terms of optimality of their experimental design, in order to quantitatively score the advantages and misses triggered by the use of the similarity screening approach. In drug discovery, often hindsight is required to outline the single most important discovery that led to the success of a program.

Therefore, the drug discovery laboratory environment is too complex to serve as a playground for theoretical validation studies of hypotheses concerning the "optimal" mathematical definition of the molecular similarity metric. However, imperfect in silico validation simulations can nevertheless be valuable benchmark tools to compare the performances of different similarity metrics with respect to a given problem. This chapter has discussed the problem of the retrieval of structurally dissimilar analogs "hidden" within large sets of combinatorial and drug molecules. Interestingly, the use of fuzzy logic to score molecular dissimilarity appears to be a key factor in ensuring the robustness of the similarity metric with respect to geometric artifacts that may affect the 3D models of the compounds. The using of three- or four-point pharmacophore fingerprints does not improve the discriminant performance of the former and actually decreases that of the latter with respect to two-point FBPAs, which are less rich in information but greatly benefit from their "fuzziness."

ACKNOWLEDGMENTS

The research work described here was accomplished at the drug discovery company CEREP, between 1996 and 1999. Thanks to all the chemists and biologists of CEREP for insisting that molecular modeling tools should be close to chemical intuition, rather than to mathematically correct but obscure algorithms, to André Tartar and Benoit Deprez, who aroused my interest in and enthusiasm for the combinatorial way of thinking about chemistry, and to the workbench chemists and biologists from Lille and Poitiers for the validation studies—teams lead by Rebecca Poulain (μ receptor) and Isabelle Moully-Daffix (DAT).

REFERENCES

1. Blake, J.F. Cheminformatics—Predicting the physicochemical properties of drug-like molecules. Curr. Opin. Struct. Biol., 2000; 11:104–107.
2. Hopfinger, A.J., Duca, J.S. Extraction of pharmacophore information from high throughput screens. Curr. Opin. Struct. Biol., 2000; 11:97–103.

3. Chaiken, I.M., Kim, D.J., eds. Molecular Diversity and Combinatorial Chemistry: Libraries and Drug Discovery. American Chemical Society, Washington DC, 1996.

4. Gallop, M.A., Barret, R.W., Dower, W.J., Fodor, S.P.A., Gordon, E.M. Applications of combinatorial technologies to drug discovery. 1. Background and peptide combinatorial libraries. J. Med. Chem., 1994, 37:1233–1251.

5. Gordon, E.M., Barret, R.W., Dower, W.J., Fodor, S.P.A., Gallop, M.A. Applications of combinatorial technologies to drug discovery. 2. Combinatorial organic synthesis, library screening strategies and future directions. J. Med. Chem., 1994, 37:1385–1401.

6. Bollobas, B. Modern Graph Theory. Springer-Verlag, New York, 1998.

7. Menard, P.R., Mason, J.S., Morize, I., Bauerschmidt S. Chemistry space metrics in diversity analysis, library design and compound selection. J. Chem. Inf. Comput. Sci., 1998; 38:1204–1213.

8. Zheng, W.F., Cho, S.J., Tropsha, A. Rational combinatorial library design. 1. Focus-2D: A new approach to the design of targeted libraries. J. Chem. Inf. Comput. Sci., 1998; 38:251–258.

9. Cramer, R.D., Patterson, D.E., Clark, R.D., Soltanshahi, F., Lawless, M.S. Virtual compound libraries: A new approach to decision making in molecular discovery research. J. Chem. Inf. Comput. Sci., 1998; 38:1010–1023.

10. Matter, H. Selecting optimally diverse compounds from structure databases: A validation study of two-dimensional and three-dimensional molecular descriptors. J. Med. Chem., 1997; 40:1219–1229.

11. Brown, R.D., Martin, Y.C. The information content of 2D and 3D structural descriptors relevant to ligand–receptor binding. J. Chem. Inf. Comput. Sci., 1996; 36:572–584.

12. Cerius2 v 4.0. Molecular Simulations Inc., San Diego, CA.

13. Catalyst v 3.5. Molecular Simulations Inc., San Diego, CA.

14. The Daylight Theory Manual. http://www.daylight.com.

15. ISIS/Base, MDL Information Systems, San Leandro, CA.

16. Horvath, D., Deprez, B., Tartar, A.T. High throughput molecular modeling using "fast 3D" descriptors. Acta. Chim. Ther., 1997; 23:55–69.

17. Ermer, O. Calculation of molecular properties using force fields. Applications in organic chemistry. Struct. Bonding, 1976; 27:167–211.

18. Rappe, A.K., Casewit, C.J., Colwell, K.S., Goddard, W.A., Skiff, W.M. J. Am. Chem. Soc., 1992; 114:10024–10035.

19. Pickett, S.D., Mason, J.S., McLay, I.M. Diversity profiling and design using 3D pharmacophores: Pharmacophore-derived queries. J. Chem. Inf. Comput. Sci., 1996; 36:1214–1223.

20. Chen, L., Gasteiger, J. Knowledge discovery in reaction databases: Landscaping organic reactions by a self-organizing neural network. J. Am. Chem. Soc., 1997; 119:4033–4042.

21. Willett, P., Barnard, J.M., Downs, G.M. Chemical similarity searching. J. Chem. Inf. Comput. Sci., 1998; 38:983–996.

22. Kahn, S. Molecular Simulations, Inc. Personal communication.

23. Pötter, T., Matter, H. Random or rational design? Evaluation of diverse compound subsets from chemical structure databases. J. Med. Chem., 1998; 41:478–488.

24. This set can be downloaded from:
    http://www.msi.com/user/consortia/cchem/index.html
25. Dixon, S.L., Villar, H.O. Bioactive diversity and screening library selection via affinity fingerprints. J. Chem. Inf. Comput. Sci., 1998; 38:1192–1203.
26. Bradshaw, J. Introduction to the Tversky similarity measure. Presented at Daylight MUG Meeting, Laguna Beach, CA, February 1997. http://www.daylight.com/meetings/mug97/agenda97/Bradshaw/MUG9 7/tv tversky.html
27. Tversky, A. Features of similarity, Psychol. Rev., 1977; 84:327–352.
28. Patterson, D.E., Cramer, R.D., Ferguson, A.M., Clark, R.D., Weinberger, L.E. Neighborhood behavior, a useful concept in validation of molecular diversity descriptors. J. Med. Chem., 1996; 39:3049–3059.
29. Horvath, D. ComPharm—Automated comparative analysis of pharmacophoric patterns and derived QSAR approaches, novel tools in high-throughput drug discovery. In preparation.
30. Kaminski, J.J., Rane, D.F., Snow, M.E., Weber, L., Rothofsky, M.L., Anderson, S.D., Lin, S.L. Identification of novel farnesyl–protein transferase inhibitors using three-dimensional database searching methods. J. Med. Chem., 1997; 40:4103–4112.
31. Njoroge, F.G., Vibulbhan, B., Rane, D.F., Bishop, W.R., Petrin, J, Patton, R., Bryant, M.S., Chen, K.-J., Nomeir, A.A., Lin, C.-C., King, I., Liu, M., Chen, J., Lee, S., Yaremko, B., Dell, J., Lipari, P., Malkowski, M., Li, Z., Catino, J., Doll, R.J., Girijavallabhan, V., Ganguly, A.K. Structure–activity relationship of 3-substituted N-(pyridinylacetyl)-4-(8-chloro-5,6-dihydro-11H-benzo[5,6]cyclohepta[1,2-b]pyridin-11-ylidene)-piperidine inhibitors of farnesyl–protein transferase: Design and synthesis of in vivo active antitumor compounds. J. Med. Chem., 1997; 40: 4290–4301.
32. Hall, L.H., Kier, L.B. Electrotopological state indices for atom types: A novel combination of electronic, topologic and valence state information. J. Chem. Inf. Comput. Sci., 1995; 35:1039–1045.
33. Mason, J.S., Morize, I., Menard, P.R., Cheney, D.L., Hulme, C., Labaudiniere, R.F. New 4-point pharmacophore method for molecular similarity and diversity applications: Overview of the method and applications, including a novel approach to the design of combinatorial libraries containing privileged substructures. J. Med. Chem., 1998; 38:144–150.
34. Dutta, A.K., Xu, C., Reith, E.A. Tolerance in the replacement of the benzhydrylic O atom in 4-[2-(diphenylmethoxy)ethyl]-1-benzylpiperidine derivatives by an N atom: Development of new-generation potent and selective N-analogue molecules for the dopamine transporter. J. Med. Chem., 1998; 41:3293–3297.
35. Poulain, R.F., Horvath, D., Eckhof, C., Chapelain, B., Bodinier, M.C., Deprez, B. Structure–profile relationships in μ-opiate ligands. J. Med. Chem. Submitted.
36. Briem, D., Kuntz, I.D. Molecular similarity based on DOCK-generated fingerprints. J. Med. Chem., 1998; 38:144–150.
37. Clark, D.E., Pickett, S.D. Computational methods for the prediction of drug-likeness. Drug Discovery Today, 2000; 5:49–58.
38. Lipinski, C.A., Lombardo, F., Dominy, B.W., Feeney, P.J. Experimental and computational approaches to estimate solubility and permeability in drug discovery and development settings. Adv. Drug Delivery Rev., 1997; 23:3–25.

16

Applications of Cell-Based Diversity Methods to Combinatorial Library Design

Dora Schnur* and Prabha Venkatarangan†

Pharmacopeia, Inc.
Princeton, New Jersey

I. INTRODUCTION

Now that the methods for combinatorial synthesis have matured to a stage at which use of the resultant libraries is routine in the drug discovery process, the druglike character (1) of the molecules contained in libraries has become a critical area for design. Library quality is a key consideration for design whether the library is large or small. Over the last several years, many groups have developed methods and tools for designing diverse libraries (2–13) and various commercial tools have appeared on the market (14), but only recently has the computational emphasis shifted toward designing screening libraries that are druglike as well as diverse. Maximal diversity is no longer an adequate design criterion; library members must possess structural and/or property diversity relevant to the biological activity they are synthesized to explore. Much method development has focused on using commercial and corporate databases to develop diversity/similarity assessment and sampling tools (5–20). The next logical step is to find and define those

Portions of this work have been excerpted and/or adapted from J. Chem. Inf. Comput. Sci. 1999; 39:36–45. Copyright 1999 American Chemical Society.

* *Current affiliation*: Bristol-Myers Squibb, Princeton, New Jersey.
† *Current affiliation*: R. W. Johnson PRI, Raritan, New Jersey.

druglike (21–23) and activity-optimizing properties that will enable rational design of screening and other types of combinatorial library.

II. REAGENT BASED VS PRODUCT BASED DESIGN?

A key consideration in combinatorial library design is that small numbers of synthons yield large numbers of products; thus it becomes essential to find a means for selection of these small building blocks from the large superset of available reagents. This method should be flexible enough to achieve the desired level of diversity appropriate to the library's intended use. While it is intuitively reasonable and can be demonstrated (24) that analysis of product diversity is more desirable than that of synthon diversity, reagent/synthon selection is generally more feasible in terms of scale. Another important factor enters into the decision to design a library via building blocks rather than final products: synthetic feasibility. For the bench chemist, the cold hard reality of combinatorial synthesis lies, not in the actual library construction, but in developing practical reaction conditions for the entire set of selected reagents and over the entire set of conversions needed for that library. Often a set of reagents selected by an automated computational technique contains either obviously chemically unfeasible compounds or ones that owing to the vagaries of Mother Nature simply fail when tested on solid phase. Substitutions have to be made that may modify but must not destroy the diversity design. Since tight timelines are an integral part of pharmaceutical development, an iterative process for reagent selection must be fast and easy. Adherence to a very strict formal experimental design is difficult at best, and a means of finding replacement synthons for the design is essential. While optimal diversity may be an aesthetically desirable goal, it is far more pragmatic to strive for a library design whose diversity is defined by synthetic reality. It should also be based on understandable metrics that can be used later for a structure–activity relationship (SAR) hypothesis (or, perhaps, if synthesis is truly problematic, for a reactivity hypothesis).

One method (25) to achieve this is a fast, simple, property-based tool that groups synthons according to common property descriptors. The chemist chooses and adjusts the properties that he or she believes to be appropriate for the library design and can manually select the most desirable starting materials. (Automated subset selection is provided as well, if the chemist chooses to use it.) As a result, the library design method incorporates "medicinal chemistry intuition," synthetic feasibility, and the ability to make quick substitutions by selecting another compound from the same group. The emphasis is shifted from covering property space to mapping feasible parts of property space. Holes in the mapped property space either are understood to be synthetically unreachable or are consciously omitted. These property-based groupings also allow easy use of "druglike" filters or of filters based on ADME (adsorption, distribution, metabolism, excretion). Thus, some filled regions of the property space are omitted for sampling because they would produce undesirable products.

III. PROPERTY SPACE, LIBRARY SIZE, AND BIOLOGICAL ACTIVITY

Since library design by this method employs a kind of "hypothesis space," the intended use of the library is an important design factor. Is this a lead discovery library, a targeted discovery library, or an optimization library? The ultimate library size, amount of diversity/property space covered, sampling density of the space required to find actives, and the diversity metrics used to define the space are all determined by the intended use of the library. While the library should cover a region of diversity space, it is assumed that actives for a given target will be clustered together. Depending on the space and the nature of the target, the activity space may be broad or spiky. Clearly, the selection of appropriate metrics to deal with some of these issues poses a difficult problem.

Let us examine the requirements for libraries of different types. A discovery screening library for which the target assays are unknown or are themselves very diverse may be designed to cover a large region of diversity space; whether it does so sparsely or densely is a function of library size and the technology available. If lead finding is the only goal, the library can be smaller and sparser than if an initial SAR around possible leads is also desired. A targeted discovery library is most likely one that is intended to find leads for a target or family of targets about which some information is known, but for which most probably X-ray crystal structures are unavailable for structure-aided design. Alternatively, they may be used to supplement the latter design method. This type of combinatorial library generally contains fewer chemotypes (cores) than the screening library and is usually intended to provide some SAR. In this case, a smaller region of diversity space is sampled, and the issue is coverage of that region. A larger denser library may provide more SAR information depending on the nature of the biological activity. If the activity being sought is sensitive to small structural changes, dense sampling of the space may be essential. If the activity tolerates fairly broad structural changes, a much smaller, sparser library design will suffice to find the SAR. The same considerations regarding the nature of biological activity apply to optimization libraries. These are by definition the least diverse, since they concentrate on a fairly small region of diversity space and their purpose is specifically to find the most potent druglike molecules. The quest at this stage may actually be similarity rather than diversity if bioisosteric substitutions are required.

How does one select the metrics that define a diversity space (also called chemistry space)? Clearly, the diversity one achieves will be dependent on the metrics chosen to define the diversity space hypothesis. The best set of metrics may well differ depending on the type of library or target. Ideally those metrics should be orthogonal to each other and, from an ideal mathematical perspective, continuous. This last is a difficult requirement to fill in chemistry space, so we need to modify it in practice. Gaps or voids in property space are acceptable if it is possible to determine whether they can be filled. If at all possible, the metrics used should be interpretable by the chemist and usable for follow-up library design and SAR exploration.

It is clear that whatever diversity space is used for library design and analysis, it must have some validation with regard to biological activity. Obviously, it is also necessary to analyze the product diversity in libraries designed using a synthon approach. Since the ability to compare sets of large combinatorial libraries for relative coverage of diversity space was considered to be essential for the application examples discussed, a cell-based diversity analysis method was chosen as the most appropriate. In contrast to that of clustering methods, the property space in cell-based methods can be independent of the molecules studied. As a result, molecules can be added or deleted from the set and different sets of molecules compared in the same space. Pearlman's "Diverse Solutions" (26) can easily handle individual analyses of the libraries under consideration, which ranged from 20,000 to 170,000 members and can evaluate merged sets of hundreds of thousands of compounds. The BCUT metrics (26) used by this program are based on simple physical properties such as partial charge, hydrogen bonding, and polarizability, and therefore should relate to biological activity. We have found this to be the case (25). In addition, we sought to find the means of performing comparable analyses with alternative descriptors within the MSI Cerius2 (27) tool set. Such methods and descriptors sets were found.

IV. METHODS

A. Synthon Selection/Activity Clustering Tool

The synthon selection tool has been discussed at length elsewhere (25), and the method has been implemented in Cerius2 (28). The binning method has an additional use, however, in the analysis of activity clusters, so a brief description is needed.

Our diversity tool assists in choosing, from among a list of synthons characterized by various molecular properties, a subset in which all feasible combinations of these properties are represented. The basis of this approach is the QSAR assumption that, in the case of potential drug molecules, there is a reasonable number of simple whole-molecule properties that may discriminate, alone or in combination, between actives and inactives for a given target. The tool also may be used to bin active neighbors in diversity space based on their distances from each other in that property space.

To make the problem manageable, the range of possible values for each property is subdivided into two or three ranges. For instance, log P could be subdivided into low, intermediate, and high ranges, which we will denote by "–," "0," and "+," respectively. A property such as the presence or absence of a hydrogen bond donor on the molecule can be denoted by just "+" or "–". Any synthon that we might consider using for a given step, if it meets all synthetic criteria, is thus assigned to a particular "bin." This bin is identified by the property pattern that describes it, that is, "+" for lipophilic, "–" for not having a hydrogen bond donor, "0" for not having an overall charge, or "+ – 0." These "bins" correspond to the rows of a full factorial design (29) of either three levels or mixed two and three levels.

Choosing a diverse set of synthons relative to a set of parameters now simply reduces to taking at least one out of each of the filled bins: one from "+ + +," one from "+ + –," one from "+ 0 +," one from "+ 0 –," and so on. When a bin is empty, the particular combination of parameters may be physically difficult or impossible to realize. Most real-world parameters are not completely independent, so finding a small, charged, lipophilic molecule, for instance, would be difficult. Another more interesting possibility is that the set from which we are choosing our fragments lacks some perfectly reasonable cases and so we should go look for other possible sources—possibly custom synthesis—from which we can fill at least some of these bins.

The analysis of activity clusters is a matter of adjusting the "+ – 0" threshold ranges of the metrics for a set of products until the actives of interest appear in one bin, which may be a very nonsymmetrical multidimensional space. (Activity is used as one of the parameters.) Some of the metrics from the original diversity space may turn out to be irrelevant for activity and others may form relevant subspaces (25,30). The tool thus provides an easy means of visualizing (via multiproperty histograms) n-dimensional activity space. Examples of this will be provided.

B. Diverse Solutions Cell-Based Analyses

For the analyses discussed, various versions of Pearlman's "Diverse Solutions" were used. The methodology and parameters used by this program are discussed elsewhere (26), but a brief overview may be useful.

The descriptors used in Diverse Solutions are extensions of parameters originally developed by Frank Burden (31). The Burden parameters were based on a combination of the atomic number for each atom and a description of the bond types for adjacent and nonadjacent atoms. Pearlman expanded the types of atomic property considered, expanded the measures for connectivity relationships, and introduced various weighting schemes. The resultant whole-molecule descriptors, called "BCUTs," correspond to the highest and lowest eigenvalues of matrices that contain the property of interest on the diagonal and some form of interatomic distance relationship on the off-diagonal. Since there are many choices for connectivity and atomic information and many possible scaling factors to weight the relationship between these two types of information, a large number of BCUT metrics may be generated. A need arises then for an algorithm to select which ones should actually be chosen as the axes for a chemistry space. Pearlman and Smith developed a chi-squared-based "auto-choose" algorithm to accomplish this (26). The algorithm examines all possible combinations of metrics and evaluates both their orthogonality and the ability of the resultant chemical spaces to distinguish the structural differences in a compound population.

The specific set of descriptors used for the analyses discussed were the 3D hydrogen-suppressed BCUTs. The default properties used for the matrix diagonals included Gasteiger charges, tabulated polarizability, hydrogen bond donors, and hydrogen bond acceptors. All the default distance relationships and weighting factors were included. Although Pearlman recommends removing 10% of the molecules in libraries analyzed as outliers, all molecules that could be successfully converted from Daylight (32) SMILES to 3D structures (SYBYL mol2 files or MACCS sd) via CONCORD3.2.4 (33) were used. While this yields larger diversity or chemistry spaces with more voids, it was felt that all molecules that were actually synthesized and screened should be included in the analysis. Visualization of the resultant chemistry spaces was performed by selecting subsets from the entire library that could be viewed in either SYBYL6.3 (or 6.4) (33) or in Cerius2 (27). In general, subset selection was done by proportional cell-based sampling with a sample size corresponding to roughly 10% of the actual library. Since Diverse Solutions chemistry spaces range from four to six parameters depending on the size and diversity of the combinatorial library being studied, it is easiest to look at all possible 3D subspaces using the selected subset of molecules. (This assumes the ability to rotate the subspace on a computer screen, an obvious impossibility in a 2D document.) This, in fact, proves interesting when the bioactive structures are added to the subset. (There is no guarantee that any subset picking method will find any of the actives, much less all of them, so the picked set is manually supplemented.) It now becomes possible to find receptor-relevant subspaces visually. [Since this work was done, Pearlman has implemented algorithms for finding receptor relevant subspaces in Diverse Solutions (30).]

C. Cerius2 Cell-Based Analyses

The molecular descriptors and, alternatively, their principal components can be used as input to the Cerius2 cell-based diversity analysis. The choice of the molecular descriptors to be used for diversity analysis depends on a number of factors. These include the speed with which the descriptors can be calculated for practical situations and the goal of the diversity analysis. The 2D structure captures many of the physical properties and much of the reactivity of a molecule (34). In addition, practical considerations make 2D descriptors an obvious choice. On the other hand, if the purpose of the diversity analysis is to cluster biologically active molecules, the 3D descriptors may be considered. A number of 2D and 3D descriptors can be calculated in Cerius2 (35).

The choice of descriptors for a particular target depends on the identification of the descriptors that are able to separate the active molecules in diversity space from the inactives. One such approach (36) is to identify the descriptors that contribute to activity by using 3D-QSAR analysis on a subset of active molecules. The analysis used genetic function approximation (GFA) with partial least squares (PLS) to select the descriptors (37,38). The next step was the calculation of the significant descriptor set identified from the 3D-QSAR analysis for the combinatorial library. The dimensionality of these descriptors was further reduced using principal component analysis (PCA) (39). All descriptors were assigned the same weight by setting the mean to zero and the variance to unity. Usually a variance of 90% of the information determines the number of principal components. This number is usually 5–10 components. All the members of the combinatorial library were included in the analysis. The subset picking was done using cell-based density to cover about 10% of the library size. Different metrics available as a target function to evaluate the coverage of the selected subsets include cell-based fraction, cell-based chi-squared, cell-based entropy, and cell-based density (40). A Monte Carlo algorithm was used to optimize the diversity function (8).

V. APPLICATIONS OF CELL-BASED DIVERSITY ANALYSIS

Comparisons of the library being designed with preexisting libraries or other structural databases with cell-based methods (26) are straightforward and need not be addressed in this chapter.

Since diversity spaces are ultimately useful only if they actually cluster actives, we have looked at numerous examples of our combinatorial libraries and their corresponding activity data for various targets. A very early study (25) using Diverse Solutions BCUT-based chemistry spaces investigated this with a series of actives from a library designed to inhibit carbonic anhydrase. As shown in Fig. 1, there are two major clusters and one minor cluster in this representative 3D subset, which was based on charge, polarizability, and H-bond donation.

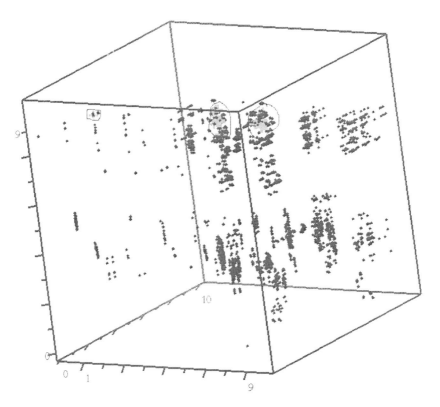

Figure 1   Clusters of carbonic anhydrase actives (gray, circled) in diversity space. (Reproduced/adapted with permission from J Chem Inf Comput Sci 1999; 39:36–45. Copyright 1999 American Chemical Society.)

This result is reasonable based on structure classes studied. Experimentally they fall into three types. Another study (25) included an analysis of eight actives found for CXCR2 receptor from a 170,000-compound library. As shown in Fig. 2, it was possible to find a 3D subspace that clustered the actives. The other possible subspaces did not cluster them. The worrisome aspect of this analysis is that it is unlikely that any subset picking method on this library would have found these compounds. Assuming that it is reasonable to use diversity spaces as hypothesis spaces for designing active compounds, it is clear from this example that optimal sampling density of a space for a particular target must be considered. It also illustrates the usefulness of large discovery libraries for targets with "spiky" activity profiles.

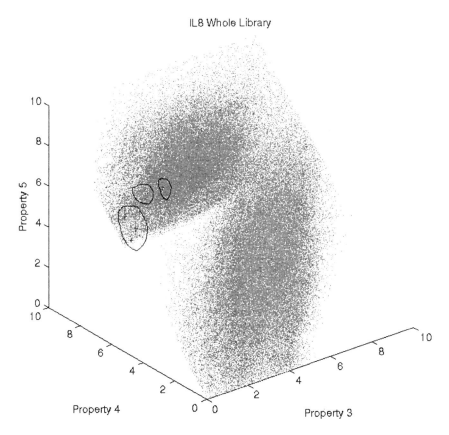

Figure 2   Eight CXCR2 receptor actives (circled) in 150,000 compound library. The axes correspond to BCUT charge, polarizability, and hydrogen bond descriptors. (Reproduced with permission from J Chem Inf Comput Sci 1999; 39:36–45. Copyright 1999 American Chemical Society.)

In another study of BCUT-derived diversity spaces, five libraries, including three discovery and two focused ones, and containing multiple core structures, were compared in a common 6D diversity space with their actives for a particular target. Not only was it found (25) that the actives do cluster better in some subspaces than in others, but it also became apparent that the more active compounds clustered more closely. Activity data ranged from ~100 μM to less than 10 nM. To provide further insights into the activity clusters, the BCUTS for some 90 actives were imported into our factorial design tool and the parameter thresholds were set so that all actives less than 1 μM were in one bin [–000000]:

[- 000000]	33:###
[+-0--+-]	1:#
[+00-0+0]	1:#
[+00-+00]	3:#####
[+000-0-]	1:#
[+0000-0]	1:#
[+00000-]	1:#
[+000000]	30:###
[+00000+]	2:###
[+000+00]	2:###
[+000+0+]	1:#
[+00+0-0]	1:#
[+0+0000]	3:#####
[+0+000+]	1:#
[++00000]	4:#######
[++00+0-]	1:#
[++00+00]	1:#
[++00+0+]	1:#
[++0++0+]	2:###

The qualitative message here is clear: while the more active compounds cluster together, there are also many compounds that are less active (bold) in the same region of cell-based BCUT diversity space. Clearly, in this instance, the parameters function well for diversity and lead finding/optimization but are not appropriate as QSAR (41) parameters. Actives do cluster, but less active compounds will be found in the same cells.

Another test of diversity spaces with regard to activity clustering is the examination of actives for two related targets for which one wants to find selectivity. An example is shown in Figs. 3–8. The four-dimensional chemistry space for these analyses is described by the following BCUT 3D hydrogen-suppressed descriptors:

1. Lowest eigenvalue of BCUT Gasteiger charge with inverse distance squared and scale factor 1.00
2. Highest eigenvalue of BCUT Gasteiger charge with inverse distance to the sixth and scale factor 1.00
3. Highest eigenvalue of BCUT hydrogen acceptor with inverse distance and scale factor 1.00
4. Lowest eigenvalue of BCUT polarizability with inverse distance squared and scale factor of 1.00

The actives for the two targets (plasmepsin II and cathepsin D) share common regions of the diversity space but also have nonoverlapping regions. Figure 3 illustrates two libraries that were designed through a combination of structure-aided and diversity approaches. Figure 4 shows that the decodes (structures identified from their tags) from the plasmepsin II screen cluster in BCUT-derived diversity subspaces. Figure 5 illustrates that decodes from the two libraries, which have different cores, cluster in some subspaces. Figure 6 shows that in some of the subspaces, the decodes for plasmepsin II and cathepsin D cluster selectively. Figure 7 highlights the most selective cathepsin compounds and illustrates that these compounds cluster tightly in at least one subspace and that in three of the subspaces shown they are not located in the same regions of diversity space as the plasmepsin II hits. Figure 8 shows a less happy result for the most selective plasmepsin II compounds: it was not possible within this diversity space to find a 3D subspace that selectively clustered these hits. They are all fairly near the regions of cathepsin D activity. The implication here is that another description is needed to understand plasmepsin II selectivity, whereas the descriptor set does seem to explain the cathepsin D selectivity.

Since it was of interest to our group at Pharmacopeia to find out whether the descriptor sets available in Cerius2 could similarly show activity and selectivity clustering, the same data set was analyzed using a PCA-based descriptor space derived from the default (35) descriptor set. The result is shown in Fig. 9. Clearly, both plasmepsin II and cathepsin D decodes cluster in this space, although not as selectively as might be desired. Figure 10 illustrates the use of the prototype (pre-Cerius2 integrated version) factorial design tool to visually examine the activity clusters. The top third of the figure focuses on the plasmepsin II hits, and the center highlights the cathepsin D hits. In the latter case, it is possible to adjust the property thresholds so that all the cathepsin decodes are in the [000] bin (along with many other compounds from the library). While the bin may contain inactive compounds (or ones that for some reason were not chosen for decoding), it certainly represents a relevant region of diversity space to use for subsequent design. In the case of plasmepsin II, the top third of the figure

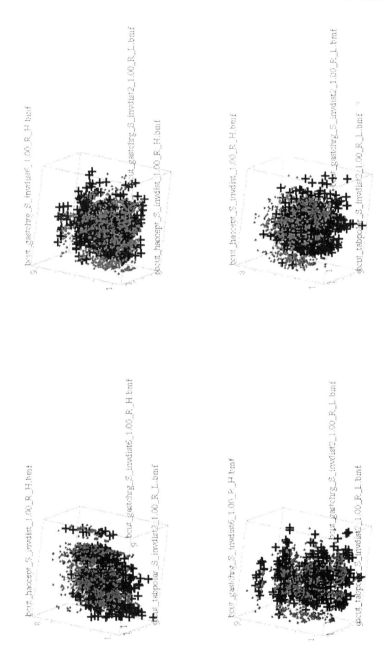

Figure 3 Two libraries (hydroxystatin in black and hydroxyphosphinate in gray) designed to be active for plasmepsin II and cathepsin D in 3D subsets of BCUT-based diversity space. The axes for these plots are as follows: top right, x, y, x = chemistry space descriptors 1, 2, 3 as listed in text; top left, x, y, z = 2, 3, 4; lower left, x, y, z = 1, 2, 4, and lower right, x, y, z = 1, 3, 4.

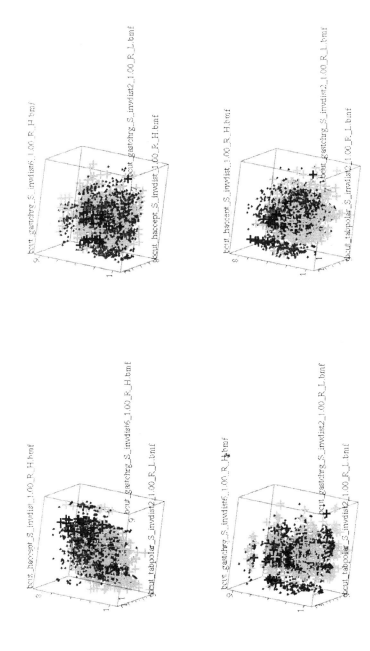

Figure 4 Plasmepsin II decodes (black) in 3D subsets of BCUT-based diversity space. The axes are described in Fig. 3.

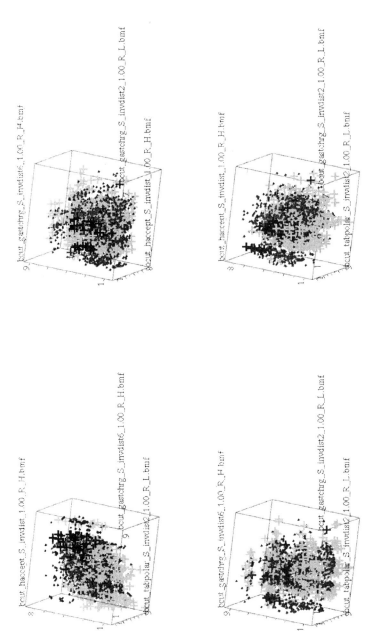

Figure 5 Plasmepsin II decodes in 3D subsets of BCUT-based diversity space with actives from the two libraries in black and dark gray + signs. The hydroxystatin library is represented by gray dots. The hydroxyphosphinate library is represented by light gray + signs. The axes are described in the legend to Fig. 3.

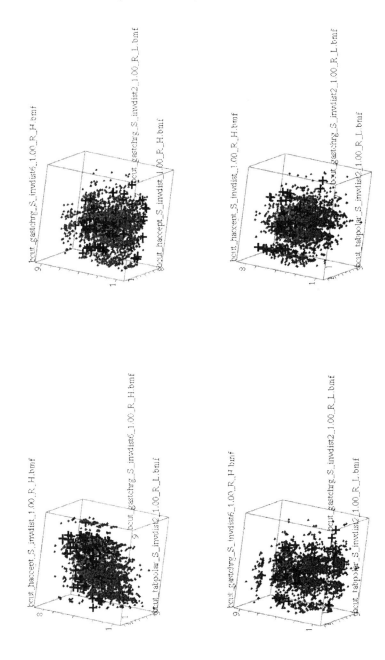

Figure 6 Plasmepsin II (black +) and cathepsin D (gray +) decodes in 3D subsets of BCUT-based diversity space. The axes are described in the legend to Fig. 3.

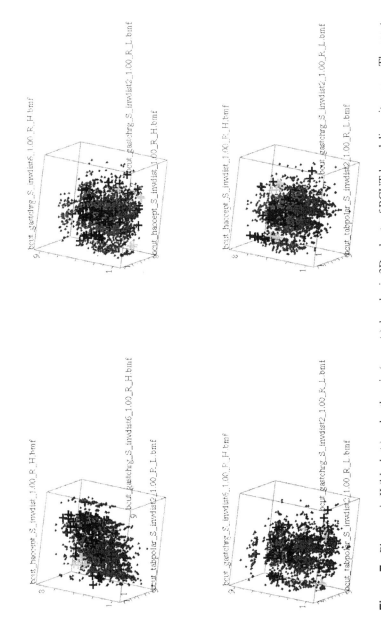

Figure 7   Plasmepsin II (black +) and cathepsin (gray +) decodes in 3D subsets of BCUT-based diversity space. The most selective cathepsin D decodes are in light gray. The axes are described in the legend to Fig. 3.

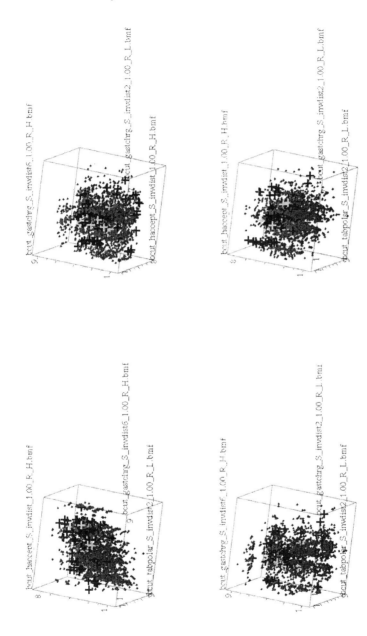

Figure 8 Plasmepsin II (black +) and cathepsin D (gray +) decodes in 3D subsets of BCUT-based diversity space. The most selective plasmepsin decodes are very light gray + signs. The axes are described in the legend to Fig. 3.

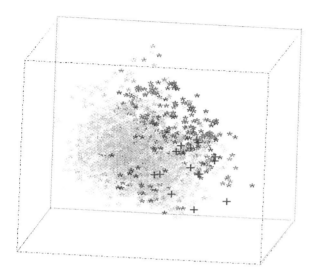

Figure 9 Plasmepsin II (*) and cathepsin D (+) decodes in Cerius2 PCA space derived using default descriptors. The axes correspond to the first three principal components.

shows that the decodes are spread out across a number of bins. There is some concentration with regard to PCA1—no hits were found in the range below the [–] threshold. These results are thus qualitatively similar to those from Diverse Solutions for this data set. It is certainly true that either the BCUTs or the PCAs or both could be used for these targets to select additional compounds for screening or for resynthesis and follow-up assays.

Another example of diversity spaces that are validated through clustering of actives involves a set of combinatorial libraries diverse enough to yield hits across a variety of targets and yet similar enough to hit the same set of targets. In this particular case, selectivity in diversity of space between hits for CXCR2 receptor and B1 receptor will be examined. Figures 11 and 12 show the some interesting 3D subspaces from Diverse Solutions with and without a 10% subset of the combinatorial libraries from whence the highlighted actives came. Again, it is possible to find subspaces that cluster the activities with some amount of selectivity. The six-dimensional chemistry space for these libraries is described by the following hydrogen-suppressed, nonlinearly scaled BCUT descriptors:

    1.   Lowest eigenvalue of BCUT Gasteiger charge with inverse distance to the sixth with scale factor 0.5

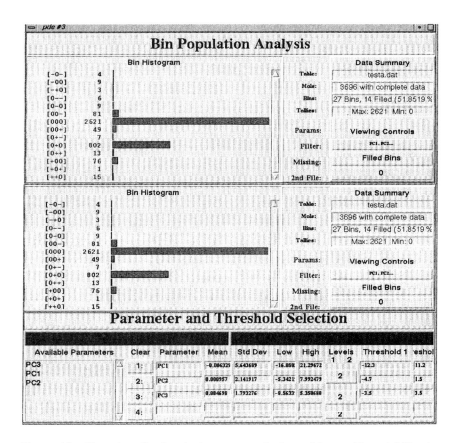

Figure 10 Clustering of actives in the presence of subset of the combinatorial libraries using Cerius 2 PCA-based diversity space. The upper third shows attempts to cluster plasmepsin II selectively (light gray bins). The middle shows clustering of the cathepsin D actives into the [000] bin.

2. Lowest eigenvalue of BCUT Gasteiger charge with inverse distance squared with scale factor 2.00
3. Highest eigenvalue of BCUT hydrogen acceptor with inverse distance with scale factor 0.5
4. Highest eigenvalue of BCUT hydrogen donor with inverse distance with scale factor 0.3
5. Highest eigenvalue of BCUT-tabulated polarizability with inverse distance squared with scale factor 2.00
6. Lowest eigenvalue of BCUT-tabulated polarizability with inverse distance to the sixth with scale factor 2.00

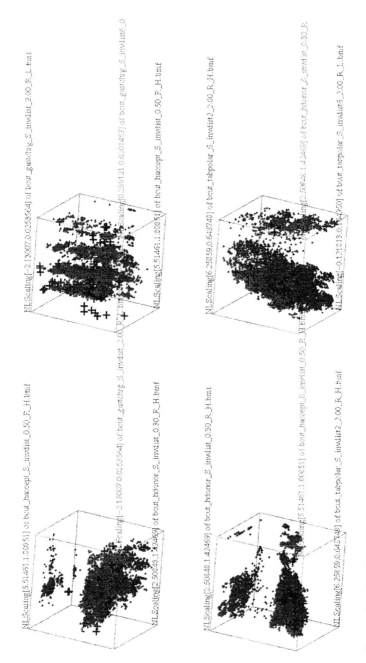

Figure 11 CXCR2 (black +) and B1 receptor (gray +) actives with a subset of the combinatorial library in 3D subspaces from Diverse Solutions. The axes for these plots are as follows: top right, x, y, x = chemistry space descriptors 1, 2, 3 as listed in text; top left, x, y, z = 2, 3, 4; lower left, x, y, z = 3, 4, 5 and lower right, x, y, z = 4, 5, 6.

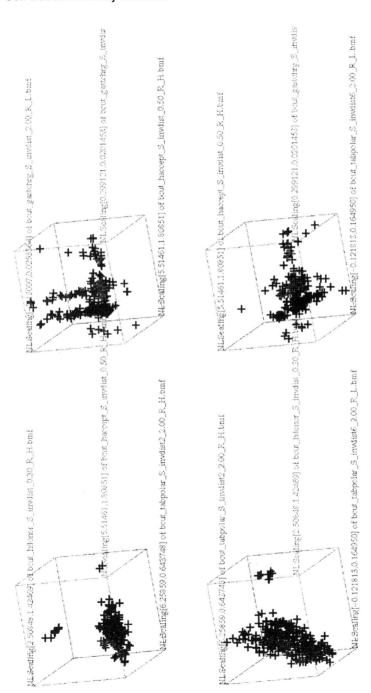

Figure 12 CXCR2 (black) and B1 receptor (gray) actives without the subset of the combinatorial library in 3D subspaces from Diverse Solutions. The axes for these plots are as follows: top right, x, y, x = chemistry space descriptors 1, 2, 3 as listed in text; top left, x, y, z = 2, 3, 4; lower left, x, y, z = 3, 4, 5; and lower right, x, y, z = 4, 5, 6.

Figure 13 illustrates the use of CXCR2 receptor actives to find other possible hits for screening. The left-hand side shows the distribution of the active compounds from several libraries. The eight most active compounds of 146 clustered in the same bin and 36 additional actives were found in that same region of diversity space. The bins showing the region where the most active compounds are found are shown in lighter gray. The right-hand side illustrates results obtained when the same threshold values were used for purposes of virtual screening. The bin containing the most active compounds has been opened to examine the list of additional compounds that are found in that region of diversity space. There are 67 compounds in that bin—44 are known actives from the training set, and 23 more are recommended for further screening.

The separation of the CXCR2 receptor and B1 receptor actives in diversity space was also examined using a Cerius2 cell-based method. The GFA/PLS (37,38) analysis on a subset of actives for both the targets yielded a set of descriptors (36). The descriptors selected for diversity analysis include the E-state sum indices (42) for the elements C, N, O, and S, charged partial surface area (43) descriptors, A log P (44,45), number of rotatable bonds, number of hydrogen bond donors, number of hydrogen bond acceptors, molecular weight, and the desolvation free energy in water/octanol (46). The first three principal components (PCs) of these descriptors were used as input for subset picking. As shown in Fig. 14, it was possible to cluster the actives of both the targets separately in the PC-based descriptor space. This suggests that although derived from a common core, the two targets have different pharmacophores (36). The more active molecules were more tightly clustered. It was seen that the less active compounds fell in bins adjacent to the more active compounds. Although the information in the descriptors used in this study was not adequate to discriminate between the more active and less active molecules for a given target, the analysis was able to identify activity space for CXCR2 receptor and B1 receptor for further lead optimization and focused library design. Since the descriptor set includes "druglike" properties, focused library can be designed with ADME considerations in mind. Another observation made was that some of the less active molecules for the two targets fell in a common area in diversity space between the two target active clusters. This may mean that some of these less active molecules are less discriminant or have somewhat nonspecific binding with respect to the two targets. An interesting follow-up analysis would be to keep track of active space and druglike space for a virtual focused library design.

Further analysis of the PC-derived active space for CXCR2 receptor and B1 receptor was done using the factorial design tool (25). Ninety-eight compounds were selected for the two targets ranging in activity of ~ 12 μM to less than 10 nM. The first three PCs for the subset were used in the factorial design tool. The thresholds were adjusted such that the activity value of less than 300 nM fell in the bin [***0], with the * representing the PCs. One bin (**bold**) represents the B1 recep-

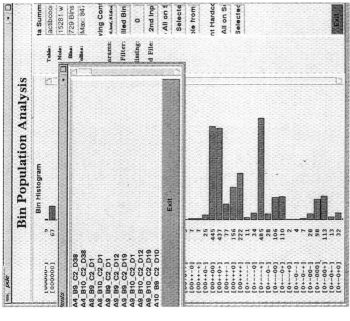

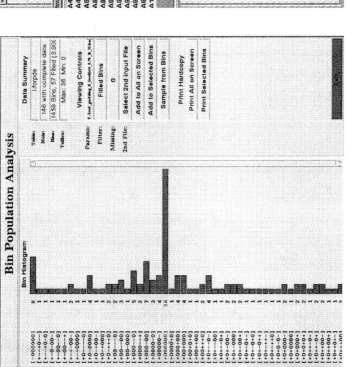

Figure 13 Use of CXCR2 actives to find other possible hits for screening. (Left) Distribution of the active compounds from several libraries. The bins showing the region where the most active compounds are found are in lighter gray. (Right) The same threshold values were used for purposes of virtual screening. The bin containing the most active compounds has been opened to examine the list of additional compounds that are found in that region of diversity space.

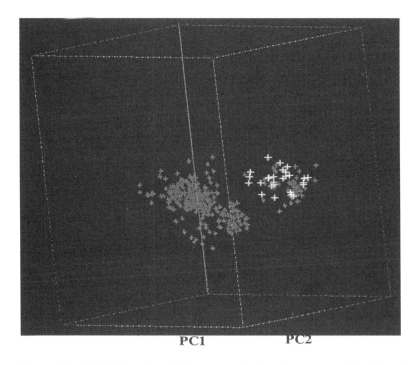

PC1   PC2

Figure 14 CXCR2 (gray) and B1 receptor (black) actives separated in PC-derived descriptor space using Cerius2.

tor set and the other bin (**bold**) represents the CXCR2 receptor set. The actives of the two targets were separated based on the third principal component, but the inactives [***+] were in adjacent bins. Based on the highly weighted variables in the third PC and the threshold value separation of the two targets, it may be possible to identify the significant separate descriptors that separate the actives of the two targets in descriptor space. Although this exercise was not done, a QSAR analysis will probably provide the same information for the same set of descriptors and data set.

[——0]	1:#
[–0–0]	2:###
[–0++]	2:###
[0—+]	1:#
[00–0]	**32:###**
[00–+]	**9:###############**
[0000]	1:#
[00+0]	**25:##**
[00++]	**18:##############################**

[0++0]   3:#####
[+0−+]   4:#######

VI. CONCLUSIONS

We have discussed the utility of a simple property-based reagent/synthon selection tool intended for the bench chemist. This tool is designed to bin reagents according to patterns based on the ranges of a set of user-selected properties that form a diversity hypothesis. Since the chemist may select synthons manually from the bin sets, choices may be biased to allow for synthetic feasibility and medicinal chemistry knowledge/intuition. Simple whole-molecule property sets have been found that are in concert with the medicinal chemist's intuition and knowledge of molecular similarity. The same tool is also useful in the analysis and visualization of activity clusters from n-dimensional diversity spaces.

In addition, we have found it useful to analyze these synthon-designed libraries as full product libraries by cell-based analyses using Diverse Solutions, both as individual libraries and for library comparisons. The size of the individual libraries analyzed ranged from several thousand to 170,000 compounds.

Moreover, active molecules have been found to cluster in various 3D subspaces of higher dimensional diversity spaces. In particular, for an analysis based on five libraries, it was found that actives clustered better in some subspaces than others and that more active compounds were more tightly clustered. Since multiple scaffolds were present in the libraries, their proximity in diversity space suggests the possibility of a common pharmacophore for this target. An important observation about the BCUT parameters used for this analysis is that they function well as SAR indicators by clustering actives, but they are not QSAR descriptors. Less active compounds were found in proximity to more active ones. In a second type of analysis, compounds active for related targets were found to cluster selectively using both BCUT-based descriptors or PCA-based ones. Additionally, different targets could be found to cluster actives selectively in a set of combinatorial libraries containing related structures.

With regard to MSI Cerius2 cell-based diversity tools, we have also been able to find descriptors sets that cluster active molecules in PCA space. Many of the same conclusions regarding BCUT descriptor–based analyses apply to these as well. In particular, they function well as SAR indicators by clustering actives, but they are not QSAR descriptors for these targets. In the case of a common core set of libraries with different targets, selectivity was seen among the clusters, suggesting different pharmacophores for these targets in the same chemotype.

Since the ultimate goal of diversity analysis tools is library design, the ideal tools and descriptor sets would provide obvious, intuitive direction to a synthetic chemist. Whether this is a realistic expectation is another issue. Unfortunately, neither of these sets of descriptors is sufficiently intuitive that a chemist can decide on the basis of inspection which molecule to make next. The descriptors may, however, be used to screen large virtual libraries for a variety of purposes. The catch at this stage is that virtual molecules must be built and tested by the current tools. Once built, they may be filtered for ADME properties and examined relative to the cell-based design space to determine their potential activity with regard to multiple targets and even, in some cases, their selectivity/specificity. Structural databases may be virtually screened to find potential leads with enhanced probability of activity. Follow-up libraries may be designed with potentially enhanced activity using these descriptor spaces.

ACKNOWLEDGMENTS

Many individuals contributed to this work, both through helpful discussions, recent or long past, and through programming. Among them are Bob Pearlman and Karl Smith (University of Texas at Austin); Jack Baldwin, Peter Gund, George Lauri, Diane Lynch, Drew Leamon, Evert Homan, Janet Cohen, and I-Ping Cheng (Pharmacopeia), Dave Rogers, Moises Hassan, and Marvin Waldman (MSI), Mark Grieshaber (Monsanto), and Bob Clark (Tripos). In addition, we would be remiss if we omitted mentioning the many, many biologists and chemists of Pharmacopeia whose experimental work provided the data, inspiration, and motivation for our analyses. To name just a few: Carolyn DiIanni Caroll, Rob Horlick, Jim Inglese, Kirk McMillan, Ilana Stroke (biology), Ron Dolle, Mike Ohlmeyer, Tim Herpin, Kevin Moriarty, Yvonne Shimshock, and Shawn Erickson (chemistry).

REFERENCES

1. Lipinski, CA, Lombardo, F, Dominy, BW, Feeney, PJ. Experimental and computational approaches to estimate solubility and permeability in drug discovery and development settings. Adv. Drug Delivery Rev. 1997; 23(1):3–25.
2. Brown, RD, Martin, YC. Designing combinatorial library mixtures using a genetic algorithm. J. Med. Chem. 1997; 40:2304–2313.
3. Chapman, D. The measurement of molecular diversity: A three-dimensional approach. J. Comput-Aided Mol. Design 1996; 10:501–512.
4. Blaney, JM, Martin, EJ. Combinatorial approaches for library design and molecular diversity. Curr. Opin. Chem. Biol. 1997; 1:54–59.
5. Martin, EJ, Blaney, JM, Siani, MA, Spellmeyer, DC, Wong A, Moos, WH. Measuring diversity: Experimental design of combinatorial libraries for drug discovery. J. Med. Chem. 1995; 38:1431–1436.

6. Good, AC, Lewis, RA. New methodology for profiling combinatorial libraries and screening sets: Cleaning up the design process with HARPick. J. Med. Chem. 1997; 40:3926–3936.

7. Pickett, SD, Mason, JS, McLay, IM. Diversity profiling and Design using 3D pharmacophores: Pharmacophore-derived queries (PDQ). J. Chem. Inf. Comput. Sci. 1996; 36:1214–1223.

8. Hassan, M, Bielawski, JP, Hempel, JC, Waldman, M. Optimization and visualization of molecular diversity of combinatorial libraries. Mol Diversity. 1996; 2:64–74.

9. Patterson, DE, Ferguson, AM, Cramer, RD, Garr, CD Underiner, TL, Peterson, JR. Design of a diverse screening library. In Devlin, JP, High Throughput Screening. New York, Marcel Dekker, 1997, pp. 243–250.

10. Agrafiotis, DM. Stochastic algorithms for maximizing molecular diversity. J. Chem. Inf. Comput. Sci. 1997; 37:841–851.

11. Pickett, SD, Luttmann, C, Guerin, V, Laoui, A, James, E. DIVSEL and COMPLIB—Strategies for the design and comparison of combinatorial libraries using pharmacophoric descriptors. J. Chem. Inf. Comput. Sci. 1998; 38:144–150.

12. Lewis, RA, Mason, JA, McLay, IM. Similarity measures for rational set selection and analysis of combinatorial libraries: The diverse property-derived (DPD) approach. J. Chem. Inf. Comput. Sci. 1997; 37:599–614.

13. Zheng, W, Cho, SJ, Waller, CL, Tropsha, A. Rational combinatorial library design. 3. Simulated annealing guided evaluation (SAGE) of molecular diversity: A novel computational tool for universal library design and data mining. J. Chem. Inf. Comput. Sci 1999; 39:738–746.

14. For example, Molecular Simulations: Cerius2 Diversity Manager. Tripos Associates: Selector, Chemical Design: Chem-Diverse.

15. Lewell, XQ, Smith, R. Drug-motif based diverse monomer selection: Method and application in combinatorial chemistry. J. Mol. Graphics Modelling 1997; 15:43–48.

16. Young, SS, Sheffield, CF, Farmen, M. Optimum utilization of a compound collection or chemical library for drug discovery. J. Chem. Inf. Comput. Sci. 1997; 37:892–899.

17. Higgs, RE, Bemis, KG, Watson, IA, Wikel, JH. Experimental designs for selecting molecules from large chemical databases. J. Chem. Inf. Comput. Sci. 1997; 37:861–870.

18. Matter, H. Selecting optimally diverse compounds from structure databases: A validation study of two-dimensional and three-dimensional molecular descriptors. J. Med. Chem. 1997; 40:1219–1229.

19. Kearsley, SK, Sallamack, S, Fluder, EM, Andose, JD, Mosley, T, Sheridan, RP. Chemical similarity using physicochemical property descriptors. J. Chem. Inf. Comput. Sci. 1996; 36:118–127.

20. Bayada, DM, Hamersma, H, van Geerestein, VJ. Molecular diversity and representativity in chemical databases. J. Chem. Inf. Comput. Sci. 1999; 39:1–10.

21. Ajay, Walters, WP, Murcko, MA. Can we learn to distinguish between "drug-like and "non-drug-like" molecules? J. Med. Chem. 1998; 41:3314–3324.

22. Martin, EJ, Critchlow, RE. Beyond mere diversity: Tailoring combinatorial libraries for drug discovery. J. Comb. Chem. 1999; 1:32–45.

23. Ghose, AK, Viswanadhan, VN, Wendoloski. A knowledge-based approach in designing combinatorial or medicinal chemistry libraries for drug discovery. 1. A qualitative and quantitative characterization of known databases. J. Comb. Chem. 1999; 1:55–68.

24. Gillet, VJ, Willett, P, Bradshaw, J. The effectiveness of reactant pools for generating structurally diverse combinatorial libraries. J. Chem. Inf. Comput. Sci. 1997; 37:731–740.

25. Schnur, DM. Design and diversity analysis of large combinatorial libraries using cell-based methods. J. Chem. Inf. Comput. Sci. 1999; 39:36–45.

26. Pearlman, RS. Novel software tools for addressing chemical diversity. Network Science 1996.
http://www.awod.com/netsci/Science/combichem/feature08.html

27. Cerius2 v 3.0. Molecular Simulations, Inc., San Diego, CA.

28. Cerius2 Combinatorial Chemistry Consortium release (fall 1999) is distributed by Molecular Simulations, Inc., San Diego, CA.

29. Austel V. Experimental design in synthesis planning and structure–property correlations. Methods Princ. Med. Chem. 1995; 2:49–62.

30. Pearlman, RS, Smith, KM. Metric validation and the receptor-relevant subspace concept. J. Chem. Inf. Comput. Sci. 1999; 39:28–35.

31. Burden, FR. Molecular identification number for substructure searches. J. Chem. Inf. Comput. Sci. 1989; 29:225–227.

32. Daylight Chemical Information Software, v 4.51. Daylight Chemical Information, Inc., Irvine, CA.

33. Syby16.3 and Concord3.2.4. Tripos Associates, St. Louis, MO.

34. Brown, RD. Descriptors for diversity analysis. Perspect. Drug Discovery Design 1997; 7/8:31–49.

35. Cerius2 reference manual, v 3.0. Molecular Simulations, Inc., San Diego, CA.

36. Venkatarangan, P, Schnur DM. In preparation

37. Rogers, DG. Splines: A hybrid of Friedman's multivariate adaptive regression splines (MARS) algorithm with Holland's genetic algorithm. Proceedings of the Fourth International Conference on Genetic Algorithms, San Diego, CA, July 1991.

38. Rogers, D, Hopfinger, AJ. Applications of genetic function approximation to quantitative structure–activity relationships and quantitative structure–property relationships. J. Chem. Inf. Comput. Sci. 1994; 34:854–866.

39. Glen, WG, Dunn WJ III, Scott, DR. Principal components analysis and partial least squares. Tetrahedron Comput. Methods 1989; 2:349–354.

40. Cerius2 online documentation 1999.
http://www.msi.com/doc/cerius40/cchem

41. For case in which BCUTs were in fact used for QSAR, see Stanton, DT. J. Chem. Inf. Comput. Sci. 1999; 39:11–20.

42. Hall, LH, Kier LB. Electrotopological state indices for atom types: A novel combination of electronic, topological, and valence state information. J. Chem. Inf. Comput. Sci. 1995; 35:1039–1045.

43. Stanton, DT, Jurs, PC. Development and use of charged partial surface area structural descriptors in computer-assisted quantitative structure–property relationship studies. Anal. Chem. 1990; 62:2323.

44. Viswanadhan, VN, Ghose, AK, Revankar, GR, Robins, RK. Atomic physicochemical parameters for three-dimensional structure directed quantitative structure–activity relationships. 4. Additional parameters for hydrophobic and dispersive interactions and their application for an automated superposition of certain naturally occurring nucleoside antibiotics. J. Chem. Inf. Comput. Sci. 1989; 29:2080.

45. Ghose, AK, Crippen, GM, Atomic physicochemical parameters for three-dimensional structure directed quantitative structure-activity relationships: Partition coefficient as a measure of hydrophobicity. J. Comput. Chem. 1986; 7:565.

46. Hopfinger, AJ. Hydration shell model. Conformational Properties of Macromolecules. New York: Academic Press, 1973, pp. 71–85.

17

Structure-Based Combinatorial Library Design and Screening: Application of the Multiple Copy Simultaneous Search Method

Diane Joseph-McCarthy

Genetics Institute
Wyeth Research
Cambridge, Massachusetts

I. INTRODUCTION TO STRUCTURE-BASED COMBINATORIAL LIBRARY DESIGN

Recent advances in combinatorial chemistry have dramatically increased the number of new compounds that can be synthesized (1,2). While high throughput robotic methods (3) have accelerated the process of screening large corporate databases and combinatorial libraries, it is still not possible to experimentally screen all available compounds. As a result, there is a critical need for new computational methods for designing, constructing, and screening virtual libraries. This includes methods for designing diverse libraries as well as those for designing focused libraries for specific targets or for computationally screening diverse libraries for binding to a specific target.

Most of the drugs currently on the market today have been found through large-scale random screening of compounds for activity against a target, for which no 3D structural information was available. Genome sequencing projects have caused the number of known sequences to increase at a rapid rate (4). Many people are, therefore, working on ways to try to predict the structure of a protein from its one-dimensional amino acid sequence (5–7). There is also a worldwide effort in functional genomics to determine as many 3D structures of proteins as possible or to develop computational approaches to cluster sequences into families of related proteins and then select and solve the 3D structure of a representative sequence from each family (8). Thus, the numbers of homology models and known structures for medically relevant targets are increasingly large. Structure-based computational methods that utilize the information contained in the three-dimensional structure of a macromolecular target can, therefore, increasingly be used to focus a combinatorial library or screen a diverse virtual library. Hits from the virtual library can then be screened experimentally or used to guide the design and synthesis of a second-generation library that spans less of the available chemical space and is more likely to contain compounds that will bind to the target in question.

As the first step in structure-based drug design (Fig. 1), the 3D structure of the target macromolecule (protein or nucleic acid) is determined by X-ray crystallography or NMR techniques. In a few instances a homology model (9) has been used as the starting point. In general, the more accurate the structural infor-

Structure-Based Drug Design

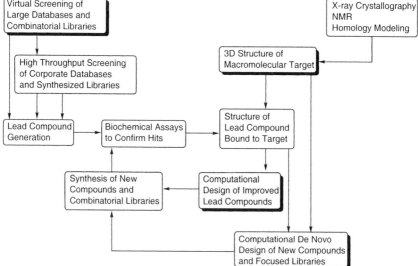

Figure 1 Overview of the structure-based drug design process. (From D. Joseph-McCarthy, Computational approaches to structure-based ligand design. Pharmacol. Ther. 1999; 84:179–191.)

mation, the more predictive the computational results will be. Once a lead compound has been found by some means, an iterative process begins that involves solving the 3D structure of the lead compound bound to the target, examining that structure and characterizing the types of interaction the bound ligand makes, and using computational methods to design improvements to the compound. This last step—designing improvements to existing lead compounds—is the point at which computational methods have played an important role in the drug discovery process during the last 5–10 years. A small subset of the most promising proposed compounds is then synthesized and tested. For compounds with improved activity, one determines their 3D structure bound to the target. One problem with using screening to find an initial lead compound is that the compound must already exist.

In combinatorial chemistry, libraries or mixtures of compounds are simultaneously synthesized from all possible combinations of up to hundreds of molecular fragments. Combinatorial computational methods that mirror this experimental approach can be used to design structure-based libraries that take into account more of the available experimental information on the specific target. These large virtual combinatorial libraries of compounds can be constructed on the computer and screened computationally before one goes to the effort and expense of actually synthesizing and testing them. Fragment positioning methods, such as the multiple copy simultaneous search (MCSS) method, optimally place functional or chemical groups in the binding site of a macromolecular structure. They can be used to place fragments likely to be monomers incorporated into the synthesis of a ligand library or to position potential library scaffolds into the binding site. In addition, functional group maps for a given target can be used to search large databases and virtual libraries for compounds likely to bind to that target.

On average it can take 15 years and $350–$500 million for a drug to reach the market (10). The structure-based combinatorial computational methods described in this chapter are expected to accelerate and reduce the cost of the drug discovery process. This computationally guided approach is now feasible owing to dramatic increases in computer power (11,12), developments in the computational methodologies, and improvements in the accuracy of the empirical energy functions (13–17) used to model atomic interactions in large biological systems.

The following areas of computational library design and screening are discussed: fragment positioning methods and their use in combinatorial library design, the specifics of the MCSS approach, the use of MCSS to design structure-based peptide libraries, small organic compound libraries and large focused libraries, and the use of MCSS-generated theoretical pharmacophores for database and virtual library screening.

II. FRAGMENT POSITIONING METHODS AND THEIR USE IN LIBRARY DESIGN

There are three basic classes of computational methods for the de novo design of structure-based ligands: fragment positioning methods, fragment methods coupled to database searches, and molecule growth methods (18). In molecule growth methods, a seed atom (or fragment) is first placed in the binding site of the target structure. A ligand molecule is successively built by bonding another atom (or fragment) to it. There are a number of molecule growth methods available including SMoG (19,20), GrowMol (21,22), GenStar (23), GroupBuild (24), and GROW (25). Of the fragment positioning methods, two well-known programs are GRID (26) and MCSS (27,28). These methods determine energetically favorable binding site positions for various functional group types or chemical fragments. Methods that couple fragment positioning to database searching include HOOK (29) and LUDI (30,31).

A. Fragment Positioning Methods

The program GRID calculates protein interaction energies for functional groups represented as single-sphere probes on a grid surrounding the target structure. The GRID nonbonded interaction energy includes an explicit hydrogen-bonding term (32–34) in addition to electrostatic and van der Waals terms. The resulting grid contour map for a given probe looks like electron density into which fragments of that probe type can be built; therefore GRID should be fairly intuitive for a crystallographer to use and is particularly useful for designing modifications to existing lead compounds. For example, GRID was used to suggest the replacement of a single hydroxyl by an amino group in an existing inhibitor of influenza virus sialidase (2-deoxy-2,3-didehydro-N-acetylneuraminic acid) that led to an inhibitor (4-aminoNeu5Ac2en) with dramatically improved binding affinity (two orders of magnitude improvement in K_i) (35). The newer versions of GRID, offer the ability to create multisphere probes, but at least three atoms in the multisphere probe must be capable of making hydrogen bonds and must not be in a linear arrangement (so a multisphere phenol group, e.g., cannot be created). In contrast, with the MCSS program, the probes are fully flexible and individual atoms are represented using the CHARMM (36) potential energy function. GRID, in its standard single-atom probe mode, is fast but gives much less detailed information than MCSS. A detailed comparison of the two methods (37) has shown that the time required for a typical MCSS calculation for methane, for example, is approximately 2.5 times that required for the corresponding GRID calculation, although neither time is prohibitive and the results are similar. For larger functional groups (such as phenol), the MCSS calculation takes significantly longer than the corresponding GRID single-sphere probe calculation (as for an aromatic hydroxyl), but the results are effective at indicating

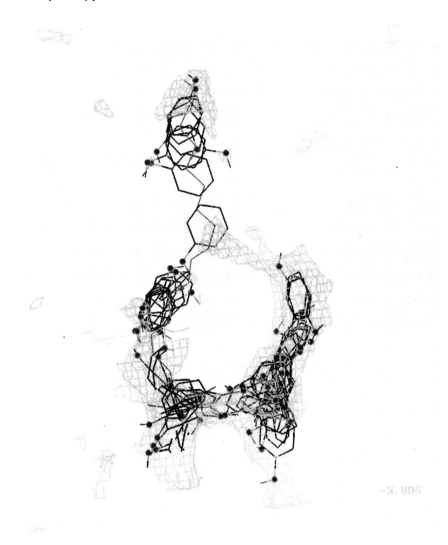

Figure 2 Comparison of an MCSS functional group map for phenol and a GRID map for the aromatic hydroxyl probe, both calculated for the poliovirus capsid protein. MCSS phenol minima with $E \leq -13$ kcal/mol are shown in black with spheres indicating the hydroxyl oxygens. GRID density contoured at $E \leq -3.8$ kcal/mol is in light gray. The natural ligand for poliovirus (sphingosine) is also shown in dark gray and indicates the binding pocket.

where in the binding site the group can be accommodated (Fig. 2). The resulting MCSS maps are more analogous to experimental mapping of a protein surface by determining its 3D structure in various organic solvents (38–40). MCSS has been used to suggest improvements to HIV-1 protease inhibitors (41) and thrombin inhibitors (42), and to design novel picornavirus capsid binding ligands (43), as described later (Sect. V). Related methods include the fragment positioning mode of LUDI (30).

B. A Three-Step Design Process

Fragment positioning methods can be considered as the first step in a three-step approach to de novo drug design. The second step in the process involves clustering and connecting the optimally placed molecular fragments to form chemically sensible candidate ligands. The third step involves estimating how well the proposed compounds should bind relative to one another and to existing drugs; this third step can also be done experimentally by using combinatorial synthesis to form molecules of all combinations of the computationally selected monomers and then screening the library for binding to the target. Several different approaches can be employed for the second step, including MC minimizations using a pseudo–potential energy function (41) and a link procedure involving the optimization of linker carbon positions and their connectivity to selected functional group minima (44). As another example, the newer program OLIGO (E. Evensen and M. Karplus, unpublished) can also construct peptide backbones by using a simulated annealing MC minimization procedure and a pseudopotential. In this case, however, each MC move is the substitution of one backbone monomer fragment (an N-methylacetamide minimum position) for another in the chain. Allowed side chains (in their optimal positions in the binding site) are then automatically and exhaustively added to these backbones. Two related dynamical approaches are DLD (dynamic ligand design) (45) and CONCERTS (creation of novel compounds by evaluation of residues at target sites) (46). DLD saturates the target binding site with sp^3 carbons, which can use a pseudo-energy function to connect to each other or to functional group minima (as determined by MCSS or a related method) to form molecules with the correct stereochemistry. This potential function depends on the Cartesian coordinates of the atoms as well as their occupancies and types. In the present implementation, it is sampled and optimized using MC-simulated annealing. CONCERTS saturates the binding site with multiple copies of various molecular fragments and does both the fragment positioning and connection by means of molecular dynamics (MD) with the AMBER potential energy function. The fragments are fully flexible during the minimization, and only connected fragments interact with each other. Connections can occur along user-specified bonds to hydrogen in each fragment; when an interfragment bond is formed, two hydrogens (one belonging to each fragment) are deleted. During the optimization procedure, bonds can break as well as form if the result lowers the overall energy of the molecule or macrofragment. With both DLD and CONCERTS, multiple molecules are simultaneously formed and scored.

C. Fragment Methods Coupled to Database Searches

Fragment positioning methods can also be coupled to database searching techniques, either to extract existing molecules from a database that can be docked into the binding site with the desired fragments in their optimal positions or for *de novo* design. HOOK (29) and LUDI (30,31) can be used either for *de novo* design or database searching. For *de novo* design, LUDI uses either statistical data from small-molecule crystal structures, geometric rules, or output from the program GRID to identify interaction sites in the target binding site. Molecular fragments (taken from a library of hundreds) are then placed in binding site positions such that atoms of the appropriate type superimpose with up to four of these favorable hydrogen-bonding or hydrophobic interaction sites. Smaller linker groups such as CH_2 can be used interactively to connect these larger, optimally placed fragments into candidate ligands. LUDI's empirical scoring function takes into account hydrogen bonds, ionic interactions, the lipophilic protein–ligand contact surface, and the number of rotatable bonds in a ligand. It was calibrated by fitting to experimental binding affinities for 45 protein–ligand complexes to obtain the individual energy contributions for an ideal neutral hydrogen bond (-4.7 kJ/mol), an ideal ionic hydrogen bond (-8.3 kJ/mol), a lipophilic contact (-0.17 kJ/mol), and one rotatable bond in the ligand ($+1.4$ kJ/mol). Deviations from ideal geometry reduce these contributions, and the sum of all interactions gives an estimate of the free energy of binding for a given protein–ligand complex. Since its scoring function is based solely on geometric considerations, LUDI is very fast and can be used interactively to predict protein–ligand complex structures, but it may sometimes miss optimal positions that are due to more delocalized electrostatic and van der Waals interactions. Instead of docking molecular fragments from a library, LUDI can similarly be used to dock and score molecules from a large database or virtual library.

In its *de novo* design mode, HOOK (29) first creates a database of molecular skeletons by stripping off all the functional groups on the database molecules and then searches this database for those molecular skeletons that can be fit into the target binding site in such a way that two MCSS functional group minima can be attached or hooked onto them. After the initial docking by geometrical superposition (of two designated hooks—methyl groups and attached atoms—in the skeletal molecule onto two functional group minima), the fit of the skeleton in the binding site is scored using a simplified, inverted Lennard-Jones type of contact potential. If the fit is acceptable, secondary searches are carried out to attach additional MCSS minima to the skeleton possibly through an extra carbon. CAVEAT (47) is similar in that it searches a database of 3D structures of small molecules (often cyclic molecules) to use as molecular frameworks to connect fragments already optimally placed in the binding site.

Virtual Combinatorial Library Design and Screening

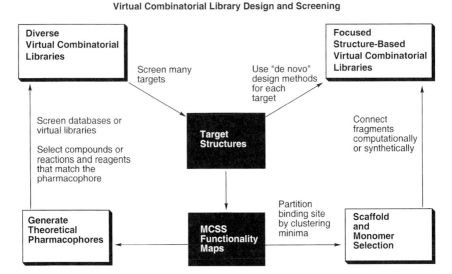

Figure 3 Schematic of the use of combinatorial computational ligand design methods for the construction and screening of virtual libraries.

D. Virtual Library Construction and Screening

The *de novo* design methods just described can be used to construct large virtual combinatorial libraries of compounds that can be screened computationally (Fig. 3). MCSS functional group maps, for example, have been used to design small directed libraries of poliovirus capsid binding ligands (43) and large structure-based libraries for class II major histocompatibility complex (MHC) molecules (48) as described shortly. An automated method for generating combinatorial libraries by iteratively and exhaustively connecting MCSS minima has also been developed (49). Starting with the MCSS minimum with the lowest binding free energy, small linker units (from 0 to 3 covalent bonds) are used to add additional fragment minima. The calculation is fast because lists of mutually excluding (overlapping) fragment pairs and of possible bonding fragments pairs are precomputed. Also, ligand growth is stopped if the average value of the approximated binding free energy of its fragments exceeds a specified cutoff. In addition, HOOK can be used with a database of all allowed conformations of a scaffold or a set of scaffolds with only positions that can be combinatorialized designated as hooks, or to search a virtual library for molecules containing an MCSS-generated theoretical pharmacophore for the target structure. The flexible docking program FLO99 (50) can also be used to generate and score combinatorial libraries in an automated manner. Furthermore, the program DOCK can be used with MCSS-generated site points to screen a conformationally expanded virtual library, as discussed in Section VII.

III. THE MULTIPLE COPY SIMULTANEOUS SEARCH APPROACH

The original MCSS method (27) has been completely reimplemented and improved in several significant ways (28) (Fig. 4). The enhanced MCSS method (48) is implemented as an Expect script and is more efficient in the initial dis-

Multiple Copy Simultaneous Search

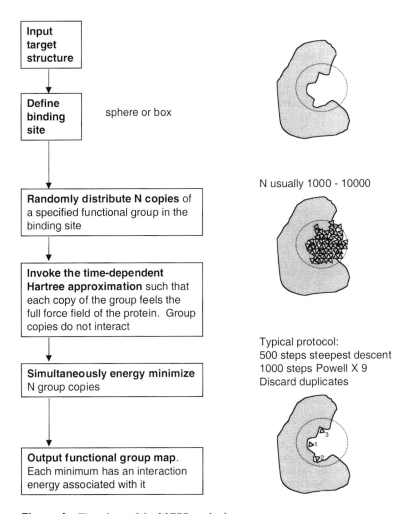

Figure 4   Flowchart of the MCSS method.

tribution of the group copies. In addition, it has the ability to generate box- as well as sphere-shaped distributions and allows user specification of minimization protocols, constraints, and replication selections. The latter feature allows a cycle of minimization to stop once it has converged; it also permits parts of the target structure to be flexible and target side chains to be replicated along with the functional groups. The newer version of MCSS also runs standard CHARMM (36), has greatly simplified the creation of new groups, and can use either a polar hydrogen (51), an all-hydrogen (13), or a hybrid (polar for the group and all-hydrogen for the target structure) representation for the calculations. The validity of the hybrid representation has been extensively tested (52). All the features of this version (2.1) are now available in a C program version (2.5) distributed by Harvard University and Molecular Simulations, Inc. The use of protocols other than minimization is under development. Alternate protocols involving Monte Carlo (MC) (37) or molecular dynamics (53) have also been implemented for annealing the functional group positions in the binding site. These other protocols are still being fully tested and refined. It has been shown that when using molecular dynamics and allowing parts of the target structure to move, every local energy minimum on the LES [locally enhanced sampling by replicating a small subsystem of the overall system (54)] potential energy surface is also a local minimum on the real potential surface, but the converse is not true (55). MC sampling should enable the use of MCSS for the placement of larger ligands in the binding site because it affords greater conformational sampling of the ligand and MC dynamics avoids problems with disproportional increases in the kinetic energy of the system when the target is also flexible. The use of MCSS with a force field that includes an implicit solvation correction is also being explored.

IV. DESIGN OF PEPTIDE LIBRARIES

Multiple copy simultaneous search methods can be used to design peptide libraries for specific macromolecular targets (Fig. 5). N-Methylacetamide (NMA) minima can be used to construct peptide backbone chains in the target binding site, and side chain minima can subsequently be added to the selected backbone chain. Conversely, side chain minima can be used to partition the binding site into distinct "interaction sites" or clusters of functional group minima, and database searches of existing peptide or protein structures can be carried out to identify backbone structures capable of connecting selected side chain minima. In an application of the first type, MC minimizations were performed using a pseudo–potential energy function to connect NMA minima in the binding site of HIV-1 protease to form backbones for candidate peptide inhibitors (41). When two NMA minima are connected, the acetyl carbon of one NMA and the N-methyl carbon of

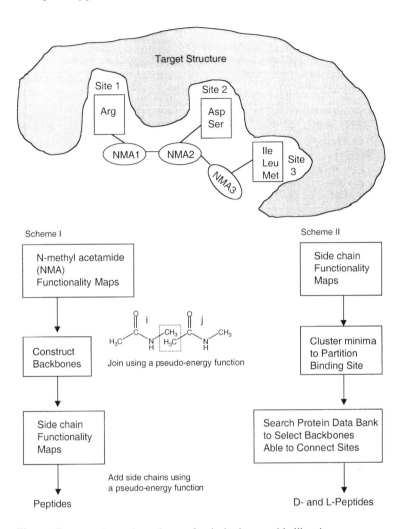

Figure 5 Two alternative schemes for designing peptide libraries.

an adjacent NMA are superposed and the ideal distance between them r_o is 0 (Fig. 5). Side chains were added to the backbone chains by using the same pseudo-energy function. Two alternate binding modes of a known peptide inhibitor were predicted and subsequently confirmed experimentally, and modifications to improve the inhibitor were proposed. Zeng and Treutlein used a related approach to design potential peptide inhibitors for Ras that should block its interaction with Raf (56). They first modeled the interaction of Ras with Raf based upon the X-ray structure of the Ras-binding domain (RBD) of Raf with Rap, a protein highly homologous to Ras. A helix of the RBD modeled to interact with Ras was used as a minimal template for potential inhibitors. NMA minima in the binding site were clustered and randomly selected to form backbones capable of extending the helix to a peptide predicted to bind to Ras with high affinity. The bound backbone chains were subjected to conjugate gradient minimization. Maps of the binding site were also calculated for functional groups representing side chains; minima for all these groups were clustered, and representatives from each cluster were assembled into a "computational combinatorial library" of monomers. Relatively low energy minima for each side chain group type (~150 each) were selected and attached to a given backbone chain, at each backbone position, whenever the distance to the specific backbone α-carbon was within a certain tolerance. The resulting candidate sequences were aligned and the optimal peptide inhibitor was chosen based on the amino acid preference at each peptide position. This approach is similar to the experimental derivation of a consensus peptide from the effectors of a target protein. The Ras computational results reproduced the known consensus sequence of Ras effectors and predicted the sequences of new candidate peptides of inhibitors of Ras. In contrast, Caflisch and coworkers have selected candidate peptides based on a scoring function or an estimated binding free energy. This approximated binding free energy is the sum of an electrostatic solvation energy calculated by the numerical solution of the linearized Poisson–Boltzmann equation and a nonpolar term that is proportional to the buried solvent-accessible surface area (49).

An example of the second type of application involves the design of candidate D-peptide inhibitors of hepatitis delta antigen (Dag) dimerization (57). MCSS maps of the monomer structure of the dimerization region of Dag were calculated for all side chain groups. Functional group minima were clustered and used to identify six "critical interaction sites." The Protein Data Bank (PDB) was searched for backbone templates that could be used to connect side chain minima at these various sites. The database search specified the sites to match, the protein side chains allowed at each site, and the allowed distance and length of the chain between the sites; it did not specify the relative orientation of the sites. Therefore about half the hits (segments of protein structures in the PDB) retrieved would be expected to bind to the actual target and half to the mirror image of the target. A 14-residue helix expected to bind to the D-isomer of the target structure was selected as the template for the subsequent inhibitor design. Over 200,000 mutant sequences of this peptide were generated based on the MCSS results. A secondary structure prediction algorithm (58,59) was used to screen all sequences, and in general only those that were predicted to be highly helical were retained. Approximately 100 of these 14-mers were built as D-peptides and docked with the L-isomer of the target monomer. Each peptide was energy-minimized in the presence of the fixed target structure using the program CHARMM (36). Based on calculated interaction energies, predicted helicity, and intrahelical salt bridge patterns, a small number of peptides were selected as the most promising candidates. This computational strategy for the design of D-peptide ligands is the analog of experimental mirror-image phage display (60).

V. DESIGN OF SMALL FOCUSED ORGANIC COMPOUND LIBRARIES

In one of its first applications to the design of nonpeptide compounds, MCSS was used to design a series of capsid binding ligands for poliovirus and the related rhinovirus (44). In all picornaviruses, a ligand binding pocket exists in the β-barrel of VP1, one of the viral capsid proteins (61–63). When a candidate drug, or high affinity ligand, is bound in this pocket, it inhibits replication by stabilizing the virus particle so that it cannot undergo conformational changes necessary later in the viral life cycle. MCSS maps of the binding region were calculated for many diverse groups. Functional group minima were clustered, and selected minima were connected by linker CH_2 groups; the connectivity and the position of the linker carbons were optimized by energy minimization and molecular dynamics in the presence of the fixed target protein. Normalized interaction energies (the interaction energy of the ligand with the fixed target protein minus the internal energy of the ligand in a vacuum minimum energy conformation) for the candidate ligands were compared to those for known ligands. Based on the results of this initial study, several small combinatorial libraries of on the order of 100 compounds were designed and synthesized (43). Each library ligand consists of three linked monomers. The functionalities allowed at each monomer position and the overall architecture of the libraries was derived from the MCSS results. A novel mass spectrometry assay as well as a functional immunoprecipitation assay were developed to screen the libraries. Several hits were identified and confirmed by X-ray crystallography. In one case, correct prediction of the binding mode of the ligand was dependent on enabling conformational flexibility of a few side chains of the protein.

VI. DESIGN OF LARGE COMBINATORIAL LIBRARIES

Recent advances in combinatorial chemistry and high throughput screening have increased the demand for computational methods for designing, constructing, and screening large virtual libraries. There are two types of large virtual library one might want to construct. A large virtual library of several million compounds could be designed for screening a wide range of biological targets. In this case, it would ideally include "druglike" (64–66) molecules or computationally modeled compounds that constitute a diverse set capable of representing many 3D pharmacophores. In the other case, an equally large "focused" library can be designed that consists of compounds likely to fit a specific 3D pharmacophore. Often a library of this type is designed around a common scaffold or template with attached R groups that are varied. For a virtual library that is significantly larger than what can be examined experimentally in a high throughput screen, computational screening could be done in two steps as described by Ghose et al. (67). A high throughput virtual screen (e.g., using 1D- and 2D-QSAR techniques) can be followed by a medium throughput virtual screen (e.g., a flexible pharmacophore search, 3D-QSAR, molecular mechanics energy minimization of the protein–ligand complex) on the remaining compounds (67).

Ghose et al. developed a reaction-based combinatorial library design approach using a 3D pharmacophore hypothesis. Given the 3D structure of SHPTP2 phophotyrosine, MCSS maps were calculated to identify binding pockets and generate a pharmacophore hypothesis for the target SH2 domain (67). Starting with this MCSS-generated 3D pharmacophore, a systematic approach for reaction and reagent selection was employed that led to the design of a combinatorial library satisfying the pharmacophore geometry. In general MCSS and other fragment positioning methods have proved useful in selecting scaffolds and corresponding reagents. In this application, the MCSS interaction energy indicated that the phosphate binding pocket is probably the best binding region. The pharmacophore obtained for this target is consistent with the known binding affinities for several synthetic peptide ligands. Their library design approach involves identifying the most important pharmacophoric feature, choosing one or two other nearby pharmacophoric features for the initial reagent search (one of which must contain a functionality that can be used to add the remaining pharmacophoric features to the compound), and then using the pharmacophoric distance constraints to screen a database for reagents; if there are too many hits from the reagent search, the physicochemical properties of the ligands should be diversified while imposing tighter pharmacophoric distance constraints.

In another application, MCSS was used in a more de novo manner to design a structure-based combinatorial library for the class II MHC molecule HLA-DR4 (48). MCSS maps were calculated for a variety of different functional groups. The scaffold and overall design of the library were determined by the functional group map for propane. A library of six-monomer branched ligands was designed. Two of the six monomers were anchor positions whose functionality was determined by the specificity of the DR molecule. The functionalities allowed at the other four monomer positions were chosen based on the MCSS results as well synthesis considerations. The library, which was synthesized by a split-and-pool approach with one compound per bead (or solid support resin), consists of approximately 250,000 unique compounds. An antibody-based assay was developed and used to screen the library for compounds that bind specifically to HLA-DR4. One of the consensus hits was modeled by superimposing the ligand "side chains" with selected functional group minima in the binding site and then using simulated annealing to optimize its position. This individual ligand was subsequently synthesized and tested, using another assay to show that it competes with the native peptide ligand for binding to HLA-DR4 and therefore may bind in a similar mode as predicted. Retrospectively, clustering and diversity techniques were developed to analyze the composite of the MCSS results for systematic incorporation of the data in library design. Since typically several thousand functional group minima are determined for a given target, selection of minima can be difficult and time-consuming; this type of analysis greatly reduces the number of individual minima that need to be examined and therefore should aid in the design of future libraries. The initial virtual screen of the target binding site using MCSS can be thought of as analogous to the experimental screen of a large exploratory library to identify the area of chemical space to be searched by a second-generation library.

VII. STRUCTURE-BASED COMPUTATIONAL SCREENING OF VIRTUAL LIBRARIES AND SUBSEQUENT SEARCHES OF EXISTING DATABASES FOR SIMILAR COMPOUNDS

Fragment positioning methods can also be used to determine or combinatorially generate possible structure-based pharmacophores. Traditionally, a pharmacophore is the set of features common to a series of active molecules. A 3D pharmacophore specifies the spatial relationship between the groups or features, often defining distances or distance ranges between groups, angles between groups or planes, and exclusion spheres. Programs like Catalyst (68), ISIS (MDL Information Systems Inc., 1997), and UNITY (Tripos Associates, 1995) can use a pharmacophore to search a database for new molecules that possess the pharmacophore. The ability to rapidly and accurately dock large numbers of small molecules into the binding site of a target macromolecule such that the compounds are rank-ordered with respect to their goodness of fit is key to lead generation and virtual library screening in structure-based drug design (69). One of the older and more widely used computational docking programs is DOCK (70–72). DOCK systematically attempts to fit each compound from a database into the binding site of the target structure such that three or more of the atoms in the database molecule overlap with a set of predefined site points (or a clique) in the target binding site.

A. MCSS to Generate Pharmacophoric Site Points

The default method for site point generation involves creating an inverse surface of the binding site. This is defined by the set of all overlapping spheres that fill the binding site and are tangent to at least two protein atoms. The sphere centers are used as site points. Crystallographic water molecules or experimental positions of known ligand atoms are also often taken as site points. A site point can be "chemically labeled" to indicate the type of atom that it is allowed to match, and it can be required that at least one site point from a subset, or a critical cluster, be matched. Chemically labeled site points can be generated in an automated fashion using the script MCSS2SPTS (73). This script runs a series of MCSS calculations on the macromolecular structure and extracts the chemically labeled site points from the resulting functional group maps. The MCSS2SPTS script calculates maps for acetic acid, methyl ammonium, N-methylacetamide, methanol, water, cyclohexane, oxazole, and benzene and then clusters minima based on type, position, and energy to generate a set of chemically labeled site points (see Fig. 6 for an example).

B. DOCK to Score Library Ligands

A virtual ligand library can be generated in a format that is searchable by DOCK. The 3D structures of compounds can be generated by the program CONCORD (Tripos Associates, 1995) (74), which uses a combination of geometry rules and optimization procedures to select the lowest energy conformer of the molecule for inclusion in the database. Each match or docking of a molecule is scored on a grid throughout the binding site of the macromolecular target using precalculated values for the protein part of the interaction energy. A number of different energy functions can be employed: molecular mechanics force fields such as Amber (15) or CHARMM (13,51), contact scoring functions, or Delphi electrostatic potential maps (75–77). In customized versions of DOCK, a solvation correction for the database compound can be added to the score (78). DOCK has been used to generate lead compounds for a number of important biological targets including HIV-1 protease (79,80), dihydrofolate reductase (81), B-form DNA (82), RNA (83), hemagglutinin (84), a malaria protease (85), and thymidylate synthase (72).

C. Docking Conformationally Expanded Databases or Libraries

In an attempt to account for ligand flexibility, DOCK databases have been constructed with multiple conformations for each molecule, or ensembles of superimposed conformations. In the first case each, conformation of a molecule is docked separately. In the other case, either the largest rigid fragment of a molecule

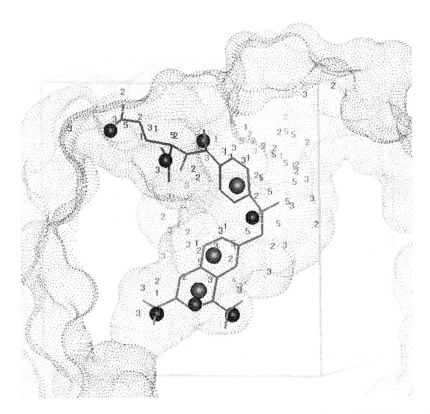

Figure 6 Set of approximately 100 site points generated with MCSS2SPTS for the DHFR structure. Acceptor sites are labeled "1", donor sites "2," dual sites (which can match both acceptors and donors) "3," and ring centroids "5." Spheres indicate site points that overlay with similar features in MTX bound to DHFR. MTX is in stick representation, molecular surface for DHFR is drawn with small dots, and the box used for the MCSS and subsequent DOCK calculations is shown.

(86) or its largest three-dimensional pharmacophore (87) can be used to dock the ensemble of conformations. The speedup with either ensemble method is that the ensemble is docked instead of the individual conformers (orientation matrices are determined simultaneously for all the associated conformers), and far fewer atoms are used to match to the DOCK site points. With the pharmacophore-based ensemble docking, only the pharmacophoric feature points of each ensemble are used to match to the DOCK site points (not the individual atoms). In addition, with this method chemical labeling is fully functional and is expected to further decrease the search time without sacrificing sampling (88). The latter should partic-

ularly be true when the DOCK site points have been derived from MCSS-generated theoretical pharmacophores.

D. Flexible Docking Methods

The newest version of the program, DOCK 4.0, can be run in a flexible ligand mode (89), although the latter is still very CPU intensive. Other methods for flexible ligand docking include FLO99 (50), AUTODOCK (90,91), GOLD (92), and FlexX (93,94). The FLO99 algorithm involves Monte Carlo (MC) perturbation (wide-angle torsional Metropolis perturbation as well as translation and rotation of ligand atoms) followed by energy minimization in Cartesian space for flexible ligand binding to a target structure; therefore, there is full flexibility for cyclic and acyclic molecules. AUTODOCK (90,91) employs simulated annealing in torsion space and therefore is best suited for ligands with only a few rotatable bonds. GOLD (92) also searches torsion space but uses a genetic algorithm approach. FlexX (94) is more distinct from the other docking methods in that it first decomposes the ligand into fragments by breaking all single acyclic and nonterminal bonds. A hashing pattern recognition technique is then used to dock a set of base fragments into the binding site. Base fragments are docked by matching three ligand interaction centers to three interaction points on the receptor surface. The ligand is incrementally built up starting from the position of a base fragment. The set of allowed interaction types or physicochemical properties and the empirical scoring function are defined as in the program LUDI (see Sect. II-C for a more detailed description of this method) (95), with slight modifications. This model of discrete conformational flexibility for the ligand, with finite sets of allowed torsional angles for single acyclic bonds and precomputed conformations for ring systems, allows the docking to be fast. Flexible docking methods can reasonably be used to screen virtual libraries of on the order of only a few thousand compounds.

E. Ligand Binding Scoring Functions

The success of docking molecules into a target site, designing ligands *de novo,* or constructing and screening large virtual combinatorial libraries is ultimately dependent on the accuracy of the scoring function that ranks the compounds or how well the corresponding relative binding affinities can be predicted. Ligand binding is governed by kinetic and thermodynamic principles. Factors that contribute to ligand binding include the hydrophobic effect, van der Waals and dispersion interactions, hydrogen bonding, other electrostatic interactions, and solvation effects (96). If the change in free energy associated with complex formation is negative, the association will be favorable.

In order of increasing complexity, the various approaches for estimating binding affinities include scoring functions based on the statistical analysis of known structures of protein–ligand complexes (97) and those based on physicochemical properties (98), molecular mechanics force field calculations, force field calculations with added solvation corrections, and free energy perturbation calculations (99). The simplest physicochemical scoring functions include those that count the number of receptor atom contacts within specified distances or, like HOOK, scale these counts depending on the distance from the ligand. More complicated ones include the LUDI energy function and similar empirical scoring functions (100,101). Molecular mechanics force field calculations attempt to model explicitly the atomic interactions in the system. The resulting interaction energies represent the enthalpic contribution to the free energy. The simplest force field calculations are performed with the ligand–target complex in vacuum using truncation schemes for the nonbonded interactions. The calculated ligand–target interaction energies include electrostatic and van der Waals interactions between the ligand and target and often also include the internal energy (bond, angle, and torsion terms) of the ligand or a ligand strain term (the internal energy of the ligand in its bound conformation minus a reference energy for the ligand in an unbound conformation). In a number of cases of sets of related compounds, a reasonable correlation exists between the van der Waals interaction energy alone and binding affinities (18,44,102–105).

A mean force field approximation or continuum representation for solvent can be used to calculate an electrostatic term that is substituted for the molecular mechanics Coulombic term to estimate the electrostatic contribution to the free energy. This continuum treatment of long-range electrostatic interactions involves first calculating the electrostatic potential for the final state and the individual reference states, using a finite difference approach to solve the linearized Poisson–Boltzmann equation as implemented in the University of Houston Brownian Dynamics Program (106,107) or Delphi (75,77). Calculation of the electrostatic energy from the electrostatic potential is trivial, and for ligand binding the difference in the electrostatic energy approximates the difference in the electrostatic contribution to the free energy (i.e., for the binding of ligand L to protein P, $\Delta G_{elec} \approx \Delta U = U_{PL} - U_P - U_L$). To account further for solvation, the solvent-accessible surface area can be calculated for the ligand, the protein, and the ligand–protein complex. The surface area buried upon complex formation can be related to the free energy of nonpolar solvation, or the hydrophobic effect associated with ligand binding (108,109). A number of groups have had some success in using a weighted sum of a continuum electrostatic term and a buried surface area term, sometimes with the addition of a ligand internal energy term, to predict binding affinities (49,110–112). Another approach is to incorporate an implicit solvation term directly into the molecular mechanics force field [e.g., an excluded-volume implicit solvation model (113)].

VIII. SUMMATION AND FUTURE OUTLOOK

The first structure-based computational drug design methods came into existence in the early 1980s, and the greatest successes to date are the various HIV-1 protease inhibitors (114). With the development of new computer-aided drug design technologies and their use in connection with combinatorial chemistry, there promise to be many more successes. Improved scoring functions, faster computers, and better database storage methods will facilitate progress. Expert systems for organic synthesis, such as LHASA (115,116) or WODCA (117), may be used to assess the ease of synthesis for a set of compounds. Even once a compound with a high affinity for the target has been developed, factors such as toxicity, bioavailability, and resistance often determine its fate. There is now a greater emphasis on trying to account for some of these factors in the initial screening and optimization process. The "druglike" character of monomers or functional groups can be assessed before they are incorporated into the library design. Large virtual libraries will be constructed based on available chemistry or on a set of existing combinatorial scaffolds, as well as on known drug properties. In the future, as structural information become more readily available, the use of the target structure to design focused libraries as well as to screen virtual libraries will be increasingly important in the drug discovery process.

ACKNOWLEDGMENTS

The author thanks Martin Karplus, Erik Evensen, Ryan Putzer, Juan C. Alvarez, and Bert Thomas for helpful discussions.

REFERENCES

1. Wilson EK. Combinatorial chemistry. Chem. Eng. News 1997; 75:24–25.
2. Borman S. Combinatorial chemistry. Chem. Eng. News 1997; 75:43–53.
3. Houston JG, Banks M. The chemical–biological interface: Developments in automated and miniaturised screening technology. Curr. Opin. Biotechnol. 1997; 8:734–740.
4. Andrade MA, Sander C. Bioinformatics: From genome data to biological knowledge. Curr. Opin. Biotechnol. 1997; 8:675–683.
5. Westhead DR, Thornton JM. Protein structure prediction. Curr. Opin. Biotechnol. 1998; 9:383–389.
6. Onuchic JN, Luthey Schulten Z, Wolynes PG. Theory of protein folding: The energy landscape perspective. Annu. Rev. Phys. Chem. 1997; 48:545–600.
7. Dunbrack RL, Gerloff DL, Bower M, Chen XW, Lichtarge O, Cohen FE. Meeting review: Second Meeting on the Critical Assessment of Techniques for Protein Struc-

ture Prediction (CASP2), Asilomar, CA, December 13–16, 1996. Fold. Design 1997; 2:R27–R42.

8. Rost B. Marrying structure and genomics. Structure 1998; 6:259–263.

9. Ring C, Sun E, McKerrow J, Lee G, Rosenthal P, Kuntz I, Cohen F. Structure-based inhibitor design by using protein models for the development of antiparasitic agents. Proc Natl Acad Sci USA 1993; 90:3583–3587.

10. Petsko GA. For medicinal purposes. Nature 1996; 384:7–9.

11. Couzin J. Supercomputing—Computer experts urge new federal initiative. Science 1998; 281:762–762.

12. Buzbee B. Workstation clusters rise and shine (Computing in Science: Perspective). Science 1993; 261:852–853.

13. MacKerell AD, Bashford D, Bellott M, Dunbrack RL, Evanseck JD, Field MJ, Fischer S, Gao J, Guo H, Ha S, Joseph-McCarthy D, Kuchnir L, Kuczera K, Lau FTK, Mattos C, Michnick S, Ngo T, Nguyen DT, Prodhom B, Reiher WE, Roux B, Schlenkrich M, Smith JC, Stote R, Straub J, Watanabe M, Wiorkiewicz-Kuczera J, Yin D, Karplus M. All-atom empirical potential for molecular modeling and dynamics studies of proteins. J. Phys. Chem. B 1998; 102:3586–3616.

14. Halgren TA. Merck molecular force field. I. Basis, form, scope, parameterization, and performance of MMFF94. J. Comput. Chem. 1996; 17:490–519.

15. Cornell WD, Cieplak P, Bayly CI, Gould IR, Merz KM, Ferguson DM, Spellmeyer DC, Fox T, Caldwell JW, Kollman PA. A 2nd generation force-field for the simulation of proteins, nucleic-acids, and organic-molecules. J. Am. Chem. Soc. 1995; 117:5179–5197.

16. Maxwell DS, Tiradorives J, Jorgensen WL. A comprehensive study of the rotational energy profiles of organic-systems by ab-initio MO theory, forming a basis for peptide torsional parameters. J. Comput. Chem. 1995; 16:984–1010.

17. Lii JH, Allinger NL. The MM3 force-field for amides, polypeptides and proteins. J. Comput. Chem. 1991; 12:186–199.

18. Caflisch A, Karplus M. Computational combinatorial chemistry for de novo ligand design: Review and assessment. Perspect. Drug Discovery Design 1995; 3:51–84.

19. DeWitte R, Shakhnovich E. SMoG: De novo design method based on simple, fast, and accurate free energy estimates. 1. Methodology and supporting evidence. J. Am. Chem. Soc. 1996; 118:11733–11744.

20. DeWitte R, Ishchenko A, Shakhnovich E. SMoG: De novo design method based on simple, fast, and accurate free energy estimates. 2. Case studies on molecular design. J. Am. Chem. Soc. 1997; 119:4608–4617.

21. Bohacek RS, McMartin C. Multiple highly diverse structures complementary to enzyme binding-sites—results of extensive application of a de-novo design method incorporating combinatorial growth. J. Am. Chem. Soc. 1994; 116:5560–5571.

22. Bohacek RS, McMartin C. De-novo design of highly diverse structures complementary to enzyme binding-sites—Application to thermolysin. Comput.-Aided Mol. Design 1995; 589:82–97.

23. Rotstein SH, Murcko MA. Genstar—A method for de novo drug design. J. Comput.-Aided Mol. Design 1993; 7:23–43.

24. Rotstein SH, Murcko MA. Groupbuild—A fragment-based method for de novo drug design. J. Med. Chem. 1993; 36:1700–1710.

25. Moon J, Howe W. Computer design of bioactive molecules: A method for receptor-based de novo ligand design. Proteins 1991; 11:314–328.

26. Goodford P. A computational procedure for determining energetically favorable binding sites on biologically important macromolecules. J. Med. Chem. 1985; 28:849–857.

27. Miranker A, Karplus M. Functionality maps of binding sites: A multiple copy simultaneous search method. Proteins 1991; 11:29–34.

28. Evensen E, Joseph-McCarthy D, Karplus M. MCSS v2. Cambridge, MA: Harvard University, 1997.

29. Eisen MB, Wiley DC, Karplus M, Hubbard RE. HOOK: A program for finding novel molecular architectures that satisfy the chemical and steric requirements of a macromolecule binding sites. Proteins 1994; 19:199–221.

30. Bohm HJ. LUDI—Rule-based automatic design of new substituents for enzyme–inhibitor leads. J. Comput.-Aided Mol. Design 1992; 6:593–606.

31. Bohm HJ. On the use of LUDI to search the fine chemicals directory for ligands of proteins of known 3-dimensional structure. J. Comput.-Aided Mol. Design 1994b; 8:623–632.

32. Boobbyer D, Goodford P, McWhinnie P, Wade R. New hydrogen-bond potentials for use in determining energetically favorable binding sites on molecules of known structure. J. Med. Chem. 1989; 32:1083–1094.

33. Wade R, Goodford P. Further development of hydrogen bond functions for use in determining energetically favorable binding sites on molecules of known structure. 2. Ligand probe groups with the ability to form more than two hydrogen bonds. J. Med. Chem. 1993; 36:148–156.

34. Wade R, Clark K, Goodford P. Further development of hydrogen bond functions for use in determining energetically favorable binding sites on molecules of known structure. 1. Ligand probe groups with the ability to form two hydrogen bonds. J. Med. Chem. 1993; 36:140–147.

35. von Itzstein M, Dyason JC, Oliver SW, White HF, Wu WY, Kok GB, Pegg MS. A study of the active site of influenza virus sialidase: An approach to the rational design of novel anti-influenza drugs. J. Med. Chem. 1996; 39:388–391.

36. Brooks BR, Bruccoleri RE, Olafson BD, States DJ, Swaminathan S, Karplus M. CHARM: A program for macromolecular energy, minimization, and dynamics calculations. J. Comput. Chem. 1983; 4:187–217.

37. Putzer R, Joseph-McCarthy D, Hogle JM, Karplus M. Optimal placement of functional groups in a binding site: A comparison of the MCSS and GRID methods. J. Med. Chem. Submitted, 2000.

38. Shuker SB, Hajduk PJ, Meadows RP, Fesik SW. Discovering high-affinity ligands for proteins: SAR by NMR. Science 1996; 274:1531–1534.

39. Joseph-McCarthy D, Fedorov AA, Almo SC. Comparison of experimental and computational functional group mapping of an RNase A structure: Implications for computer-aided drug design. Protein Eng. 1996; 9:773–780.

40. Allen KN, Bellamacina CR, Ding XC, Jeffery CJ, Mattos C, Petsko GA, Ringe D. An experimental approach to mapping the binding surfaces of crystalline proteins. J. Phys. Chem. 1996; 100:2605–2611.

41. Caflisch A, Miranker A, Karplus M. Multiple copy simultaneous search and construction of ligands in binding sites. J. Med. Chem. 1993; 36:2142–2167.

42. Grootenhuis PDJ, Karplus M. Functionality map analysis of the active site cleft of human thrombin. J. Comput.-Aided Mol. Design 1996; 10:1–10.

43. Joseph-McCarthy D, Tsang SK, Hogle JM, Karplus M. Use of MCSS to design small libraries of picornavirus ligands: Successful application of the computational methodology. J. Am. Chem. Soc. Submitted, 2000.

44. Joseph-McCarthy D, Hogle JM, Karplus M. Use of the multiple copy simultaneous search (MCSS) method to design a new class of picornavirus capsid binding drugs. Proteins 1997; 29:32–58.

45. Miranker A, Karplus M. An automated method for dynamic ligand design. Proteins 1995; 23:472–490.

46. Pearlman D, Murcko M. CONCERTS: Dynamic connection of fragments as an approach to de novo ligand design. J. Med. Chem. 1996; 39:1651–1663.

47. Lauri G, Bartlett PA. CAVEAT—A program to facilitate the design of organic molecules. J. Comput.-Aided Mol. Design 1994; 8:51–66.

48. Evensen E, Joseph-McCarthy D, Weiss GA, Schreiber SL, Karplus M. A combinatorial approach to ligand design: Application to HLA-DR4. J. Comput.-Aided Mol. Design. Submitted, 2000.

49. Caflisch A. Computational combinatorial ligand design: Application to human α-thrombin. J. Comput.-Aided Mol. Design 1996; 10:372–396.

50. McMartin C, Bohacek RS. QXP: Powerful, rapid computer algorithms for structure-based drug design. J. Comput.-Aided Mol. Design 1997; 11:333–344.

51. Neria E, Fischer S, Karplus M. Simulation of activation energies in molecular systems. J. Chem. Phys. 1996; 105:1902–1921.

52. Elkin CD, Joseph-McCarthy D, Hogle JM. EVAn: Application of visual data mining to ligand design with MCSS. J. Mol. Graphics. Submitted, 2000.

53. Stultz CM, Karplus M. MCSS functionality maps for a flexible protein. Proteins 1999; 37:512–529.

54. Elber R, Karplus M. Enhanced sampling in molecular dynamics: Use of the time-dependent Hartree approximation for a simulation of carbon monoxide diffusion through myoglobin. J. Am. Chem. Soc. 1990; 112:9161–9175.

55. Stultz CM, Karplus M. On the potential surface of the locally enhanced sampling approximation. J. Chem. Phys. 1998; 109:8809–8815.

56. Zeng J, Treutlein HR. A method for computational combinatorial peptide design of inhibitors of Ras protein. Protein Eng. 1999; 12:457–468.

57. Elkin CD, Zuccola H, Hogle JM, Joseph-McCarthy D. Computational design of D-peptide inhibitors of hepatitis delta antigen dimerization. J. Comput. Aided Mol. Design. 2000; 14:705–718.

58. Chandonia JM, Karplus M. Neural networks for secondary structure and structural class predictions. Protein Sci. 1995; 4:275–285.

59. Chandonia J, Karplus M. Secondary structure/class prediction program. Cambridge, MA: Harvard University; 5 1997.
    http://yuri.med.harvard.edu/~jmc/2ary.html

60. Schumacher T, Mayr L, Minor D, Milhollen M, Burgess M, Kim P. Identification of D-peptide ligands through mirror-image phage display. Science 1996; 271: 1854–1857.

61. Hogle JM, Chow M, Filman DJ. Three-dimensional structure of poliovirus at 2.9 Å resolution. Science 1985; 229:1358–1365.

62. Filman DJ, Syed R, Chow M, Macadam AJ, Minor PD, Hogle JM. Structural factors that control conformational transitions and serotype specificity in type 3 poliovirus. EMBO J. 1989; 8:1567–1579.

63. Badger J, Minor I, Oliveira MA, Smith TJ, Rossmann MG. Structural analysis of antiviral agents that interact with the capsid of human rhinoviruses. Proteins 1989; 6:1–19.

64. Walters WP, Murcko A, Murcko MA. Recognizing molecules with drug-like properties. Curr. Opin. Chem. Biol. 1999; 3:384–387.

65. Ghose AK, Viswanadhan VN, Wendoloski JJ. A knowledge-based approach in designing combinatorial or medicinal chemistry libraries for drug discovery. 1. A qualitative and quantitative characterization of known drug databases. J. Comb. Chem. 1999; 1:55–68.

66. Lipinski CA, Lombardo F, Dominy BW, Feeney PJ. Experimental and computational approaches to estimate solubility and permeability in drug discovery and development settings. Adv. Drug Delivery Rev. 1997; 23:3–25.

67. Ghose AK, Viswanadhan VN, Wendoloski JJ. Adapting structure-based drug design in the paradigm of combinatorial chemistry and high-throughput screening. In Parrill AL, Reddy MR, eds. Rational Drug Design: Novel Methodology and Practical Applications. ACS Symposium Series 719. Washington, DC: American Chemical Society, 1999, pp. 226–238.

68. Sprague PW. Automated chemical hypothesis generation and database searching with Catalyst(R). Perspect. Drug Discovery Design 1995; 3:1–20.

69. Kuntz I. Structure-based strategies for drug design and discovery. Science 1992; 257:1078–1082.

70. Kuntz ID, Blaney JM, Oarley SJ, Langridge R, Ferrin TE. A geometric approach to macromolecule-ligand interactions. J. Mol. Biol. 1982; 161:269–288.

71. Meng EC, Shoichet BK, Kuntz ID. Automated docking with grid-based energy evaluation. J. Comput. Chem. 1992; 13:505–524.

72. Shoichet BK, Stroud RM, Santi DV, Kuntz ID, Perry KM. Structure-based discovery of inhibitors of thymidylate synthase. Science 1993; 259:1445–1450.

73. Joseph-McCarthy D, Thomas BE, IV, Alvarez JC. Automated generation of MCSS-derived pharmacophore-based DOCK site points for searching multi-conformation databases. Proteins. Submitted, 2000.

74. Pearlman RS. Rapid generation of high quality approximate 3D molecular structures. Chem. Design Aut News 1987; 2:1–6.

75. Nicholls A, Honig B. A rapid finite-difference algorithm, utilizing successive over-relaxation to solve the Poisson–Boltzmann equation. J. Comput. Chem. 1991; 12:435–445.

76. Sharp KA, Honig B. Electrostatic interactions in macromolecules—Theory and applications. Annu. Rev. Biophys. Biophys. Chem. 1990; 19:301–332.

77. Gilson MK, Sharp KA, Honig BH. Calculating the electrostatic potential of molecules in solution: Method and error assessment. J. Comput. Chem. 1988; 9: 327–335.

78. Shoichet BK, Leach AR, Kuntz ID. Ligand solvation in molecular docking. Proteins 1999; 34:4–16.

79. Friedman SH, Ganapathi PS, Rubin Y, Kenyon GL. Optimizing the binding of fullerene inhibitors of the HIV-1 protease through predicted increases in hydrophobic desolvation. J. Med. Chem. 1998; 41:2424–2429.

80. Rose JR, Craik CS. Structure-assisted design of nonpeptide human immunodeficiency virus-1 protease inhibitors. Am. J. Respir. Crit. Care Med. 1994; 150: S176–S182.

81. Gschwend DA, Sirawaraporn W, Santi DV, Kuntz ID. Specificity in structure-based drug design: Identification of a novel, selective inhibitor of *Pneumocystis carinii* dihydrofolate reductase. Proteins 1997; 29:59–67.

82. Grootenhuis PDJ, Roe DC, Kollman PA, Kuntz ID. Finding potential DNA-binding compounds by using molecular shape. J. Comput.-Aided Mol. Design 1994; 8:731–750.

83. Chen Q, Shafer RH, Kuntz ID. Structure-based discovery of ligands targeted to the RNA double helix. Biochemistry 1997; 36:11402–11407.

84. Hoffman LR, Kuntz ID, White JM. Structure-based identification of an inducer of the low-pH conformational change in the influenza virus hemagglutinin: Irreversible inhibition of infectivity. J. Virol. 1997; 71:8808–8820.

85. Li RS, Chen XW, Gong BQ, Selzer PM, Li Z, Davidson E, Kurzban G, Miller RE, Nuzum EO, McKerrow JH, Fletterick RJ, Gillmor SA, Craik CS, Kuntz ID, Cohen FE, Kenyon GL. Structure-based design of parasitic protease inhibitors. Bioorg. Med. Chem. 1996; 4:1421–1427.

86. Lorber DM, Shoichet BK. Flexible ligand docking using conformational ensembles. Protein Sci. 1998; 7:938–950.

87. Thomas BE, IV, Joseph-McCarthy D, Alvarez JC. Pharmacophore-based molecular docking. In Guner OF, ed. Pharmacophore Perception, Development, and Use in Drug Design. La Jolla, CA: International University Press, 2000, pp. 351–367.

88. Joseph-McCarthy D, Thomas BE, IV, Belmarsh M, Moustakas D, Alvarez JC. Pharmacophore-based molecular docking. Proteins. Submitted, 2000.

89. Makino S, Kuntz ID. Automated flexible ligand docking method and its application for database search. J. Comput. Chem. 1997; 18:1812–1825.

90. Morris GM, Goodsell DS, Huey R, Olson AJ. Distributed automated docking of flexible ligands to proteins: Parallel applications of AutoDock 2.4. J. Comput.-Aided Mol. Design 1996; 10:293–304.

91. Goodsell DS, Morris GM, Olson AJ. Automated docking of flexible ligands: Applications of AutoDock. J. Mol. Recognit. 1996; 9:1–5.

92. Jones G, Willett P, Glen RC, Leach AR, Taylor R. Development and validation of a genetic algorithm for flexible docking. J. Mol. Biol. 1997; 267:727–748.

93. Kramer B, Rarey M, Lengauer T. CASP2 experiences with docking flexible ligands using FLExX. Proteins 1997:221–225.

94. Rarey M, Wefing S, Lengauer T. Placement of medium-sized molecular fragments into active sites of proteins. J. Comput.-Aided Mol. Design 1996; 10:41–54.

95. Bohm HJ. The development of a simple empirical scoring function to estimate the binding constant for a protein ligand complex of known 3-dimensional structure. J. Comput.-Aided Mol. Design 1994; 8:243–256.

96. Ajay, Murcko MA. Computational methods to predict binding free energy in ligand–receptor complexes. J. Med. Chem. 1995; 38:4953–4967.

97. Koppensteiner WA, Sippl MJ. Knowledge-based potentials—Back to the roots. Biochemistry-Moscow 1998; 63:247–252.

98. Bohm HJ, Klebe G. What can we learn from molecular recognition in protein–ligand complexes for the design of new drugs? Angew. Chem. Int. Ed. Engl. 1996; 35:2588–2614.

99. Gilson MK, Given JA, Bush BL, McCammon JA. The statistical-thermodynamic basis for computation of binding affinities: A critical review. Biophys. J. 1997; 72:1047–1069.

100. Eldridge MD, Murray CW, Auton TR, Paolini GV, Mee RP. Empirical scoring functions .1. The development of a fast empirical scoring function to estimate the binding affinity of ligands in receptor complexes. J. Comput.-Aided Mol. Design 1997; 11:425–445.

101. Jain AN. Scoring noncovalent protein–ligand interactions: A continuous differentiable function tuned to compute binding affinities. J. Comput.-Aided Mol. Design 1996; 10:427–440.

102. Grootenhuis PDJ, Van Helden SP. Rational approaches towards protease inhibition: Predicting the binding of thrombin inhibitors. In Wipff G, ed. Computational Approaches in Supramolecular Chemistry. Dordrecht: Kluwer Academic Press, 1994, pp. 137–149.

103. Kurinov IV, Harrison RW. Prediction of new serine proteinase inhibitors. Nature Struct. Biol. 1994; 1:735–743.

104. Grootenhuis PDJ, van Galen PJM. Correlation of binding affinities with nonbonded interaction energies of thrombin–inhibitor complexes. Acta Crystallogr. 1995; D51: 560–566.

105. Holloway MK, Wai JM, Halgren TA, Fitzgerald PMD, Vacca JP, Dorsey BD, Levin RB, Thompson WJ, Chen LJ, deSolms SJ, Gaffin N, Ghosh AK, Giuliani EA, Graham SL, Guare JP, Hungate RW, Lyle TA, Sanders WM, Tucker TJ, Wiggins M, Wiscount CM, Woltersdorf OW, Young SD, Darke PL, Zugay JA. A priori prediction of activity for HIV-1 protease inhibitors employing energy minimization in the active site. J. Med. Chem. 1995; 38:305–317.

106. Davis ME, Madura JD, Luty BA, McCammon JA. Electrostatics and diffusion of molecules in solution—simulations with the University-of-Houston-Brownian Dynamics Program. Comput. Phys. Commun. 1991; 62:187–197.

107. Madura JD, Briggs JM, Wade RC, Davis ME, Luty BA, Ilin A, Antosiewicz J, Gilson MK, Bagheri B, Scott LR, McCammon JA. Electrostatics and diffusion of molecules in solution—Simulations with the University-of-Houston Brownian Dynamics Program. Comput. Phys. Commun. 1995; 91:57–95.

108. Ooi W, Oobataki M, Nemethy G, Scheraga HA. Accessible surface areas as a measure of the thermodynamic parameters of hydration of peptides. Proc. Natl. Acad. Sci. USA 1987; 84:3086–3090.

109. Eisenberg D, McLachlan AD. Solvation energy in protein folding and binding. Nature 1986; 319:199–203.

110. Froloff N, Windemuth A, Honig B. On the calculation of binding free energies using continuum methods: Application to MHC class I protein–peptide interactions. Protein Sci. 1997; 6:1293–1301.

111. Simonson T, Archontis G, Karplus M. Continuum treatment of long-range interactions in free energy calculations. Application to protein–ligand binding. J. Phys. Chem. B 1997; 101:8349–8362.

112. Novotny J, Bruccoleri RE, Davis M, Sharp KA. Empirical free energy calculations: A blind test and further improvements to the method. J. Mol. Biol. 1997; 268:401–411.

113. Lazaridis T, Karplus M. Effective energy function for proteins in solution. Proteins. 1999; 35:133–152.

114. Wlodawer A, Vondrasek J. Inhibitors of HIV-1 protease: A major success of structure-assisted drug design. Annu. Rev. Biophys. Biomol. Struct. 1998; 27:249–284.

115. Corey EJ, Long AK, Lotto GI, Rubenstein SD. Computer-assisted synthetic analysis—Quantitative assessment of transform utilities. Recl. Trav. Chim. Pays-Bas-J. R. Neth. Chem. Soc. 1992; 111:304–309.

116. Long AK, Kappos JC. Computer-Assisted Synthetic Analysis—Performance of tactical combinations of transforms. J. Chem. Inf. Comput. Sci. 1994; 34:915–921.

117. Fick R, Ihlenfeldt WD, Gasteiger J. Computer-assisted design of syntheses for heterocyclic compounds. Heterocycles 1995; 40:993–1007.

18

Genetic Algorithm-Directed Lead Generation

Jasbir Singh

Cephalon, Inc.
West Chester, Pennsylvania

Adi M. Treasurywala

Mississauga, Ontario, Canada

I. INTRODUCTION

Advances in molecular biology and functional genomics have made important strides in the last few years, and new proteins are being identified at a much faster pace than ever before. As a result, the number of proteins that will serve as important therapeutic targets is expected to dramatically increase in the near future. Usually, however, the proof that these newly discovered proteins are causatively linked to a pathology that is treatable by somehow interfering with the protein's function must await the discovery of potent, selective small molecules that interfere with these events. This will mean that scientists involved in early lead discovery will face a new challenge: to rapidly provide leads (ligands, antagonists, inhibitors, etc.), that will be suitable for early proof-of-principle (POP) studies for these newly discovered proteins and thus help prioritize newly discovered targets for drug discovery programs. The leads thus generated will be optimized to identify development candidates and provide new therapeutically useful agents for clinical trials and ultimately for commercialization.

The last decade witnessed a shift to a new paradigm in drug discovery via the implementation of combinatorial/parallel synthesis of both oligomeric and nonoligomeric (low molecular weight "druglike" compounds) libraries of diverse compounds, and high throughput (HT) screening has provided a format for the identification of new lead compounds for various molecular targets (1). This new paradigm for the exploration of the chemical diversity space for rapid lead identification is very appealing, especially when one has very limited three-dimensional (3D) structural information on the molecular target(s)—which is likely to be the case for newly discovered targets. In any given template, however, the number of possible compounds that can be synthesize in combinatorial or permutational (2) libraries is enormous, often in the millions to billions of compounds (1,2). Under these circumstances, computational tools are utilized to evaluate the molecular diversity contained in a given library prior to synthesis and finally to prepare a subpopulation that maximizes dissimilarity (3) among the selected members. The biological evaluation of this subpopulation, rather than the entire library, is a more manageable task.

One of the important goals in building combinatorial libraries is to maximize diversity, which is based on dissimilarity among the library members, whereas identification of compounds that are active versus a *given* molecular target (a property-dependent criteria) is based on similarity measures. It is important to note that a nonactive molecule, by definition, is dissimilar to an active molecule even though the two may or may not be (chemically) diverse. Therefore, diversity in biological space does not automatically imply large dissimilarity in the chemical space (4). Therefore, to ensure a greater probability of finding "actives" (i.e., to enhance the likelihood of lead generation) for a *given* biological target (a problem-dependent scenario), adaptive learning algorithms such as genetic algorithms are particularly suited.

Genetic algorithms have been successfully utilized to find solutions to problems involving a relatively large search space where traditional methods are not feasible (5,6,7). In fact, genetic algorithms (GA) are distinguished for their powerful optimization characteristics, enabling them to find a set of very good (but not necessarily the best) solutions rapidly from among an astronomically larger number of potential possibilities (8,9). Like combinatorial libraries, genetic algorithms, a machine learning technique originated by Holland (10) and referred to as the computational analogs of Darwinian evolution, are based on a building-block concept. The ability to find good solutions to the problem at hand by efficient combination of a given set of building blocks is a key source of the GAs' strength. The basic ideas—selection, crossover, and mutations—are common to all GA applications; however, the "chromosome" representation and fitness function that codes information are specific to the problem under investigation. GA involves starting with a random population (coded by "genes"). A *fitness value* for each member is determined, *which becomes the basis for selection of the fit members to produce the next generation,* based on the natural process of selection, mating, (crossover), and mutation. The definition of the fitness function is a large part of the GA search paradigm, which measures the extent to which the genome provides a solution for the problem under evaluation. These terms are explained later in greater detail.

This chapter describes some examples for use of GA-based lead generation. These examples are subdivided into two main sections: fitness functions based on *experimentally determined* value(s) and fitness functions based on *computed properties.* Since the design principles for use of GA for structure-based problems were covered in detail in the preceding chapter, our description is restricted and focused to relate to lead generation.

II. GENETIC ALGORITHMS

This section briefly describes the basic ideas of the GA method, some issues involved in its use as a tool for the selection and representation of chemical structures, some of the details and variations in its implementation, and finally our view for its future applications and potential for lead optimization and development.

GA optimization methods are based on several strategies from Darwinian theories of evolution, where "adaptivity" to the constantly changing environment is a key for the survival of any given species. "Adaptivity" is what best distinguishes GA from other "stochastic" or "heuristic" algorithms. We will briefly touch on how one may be able to take advantage of this uniqueness of GA in the drug discovery realm. In the normal survival and evolution of the species, new genetic variants constantly arise, and their survival and "dominance" is based on their ability to find food, to reproduce, and to resist "assault" on their existence. These would be classified in the language of genetic algorithms as the "objective" function (or *fitness function*). The GA method seeks to quickly find "individuals" that have very good scores on the fitness function, using the process modeled on natural selection, or "survival of the fittest." For example, for a living and evolving population made up of compounds instead of species of organisms, with the evolutionary pressure being applied being their biological activity, one could envisage exactly the same process occurring. The GA utilized here is based on three basic strategies (11): *selection, crossover, and mutation* (though the order and relative rate for crossover and mutation may vary).*

The first strategy—*selection,*—is the use of a breeding population in which the individuals that are more "fit" in some sense (e.g., higher biological response/activity) have a higher chance of producing offspring and passing on their "genetic" information. The second strategy is the use of *crossover* (mating), in which a child's genetic material is a mixture of that of its parents. In the final strategy, *mutation,* the genetic material is occasionally "changed or corrupted" to maintain a certain level of spontaneous and random genetic mutation in the population. The solutions obtained will differ according to the relative rates of crossover and mutations. Generally, when the probability of crossover is much larger than mutation, it is possible to tune GA toward simple optimization, since information used in producing the next generation relies largely or wholly on information contained in the previous generation. This leads to very little exploration of "fringe" possibilities and usually provides a "local optimum" solution to the problem. When the reverse is the case, GA may be driven toward a more purely stochastic exploration of search space (i.e., Monte Carlo–like diversity exploration). Generally some balance of these two (crossover and mutation) operators is sought in GA-driven searches.

One GA paradigm (see later, Sect. III.B), a modified version of the Genesis GA (12 code) is outlined in Scheme 1.

One generally works with a population of individuals, which interact through their genetic operators to carry out an optimization process. An individual is specified by a chromosome. The first step in using GA for analysis of chemical space is to use a suitable coding strategy for the representation of chemical structures—building blocks. The most common strategy is to use binary bit strings (i.e., a sequence of 1's and 0's) to code building blocks. Each unique binary bit string corresponds to only one building block; however, a building block may be encoded by more then one unique binary bit string. An individual is completely specified by a chromosome, a bit string in this case. Initially, a set of n_{pop} individuals is formed by choosing a set of N_b bit strings at random, and each member is evaluated for fitness. In the case in which "individuals" are compounds and the fitness is evaluated by, for example, biological activity in an assay, each compound is synthesized and screened.

Let us assume that a hexapeptide is to be represented by a bit string of 30 bits (or digits). Each amino acid is represented by 5 bits, the first amino acid being coded into bits 1–5, the second being coded into 2–10, and so on. Each 5-bit code can essentially code for 2^5 or 32 unique amino acids. Since there are only 20

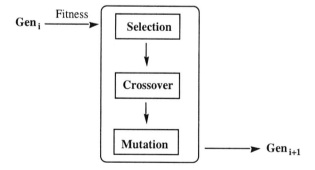

Scheme 1   Summary of variables used for genetic algorithms. Adopted with permission from Ref. 13. Copyright [1996] American Chemical Society.

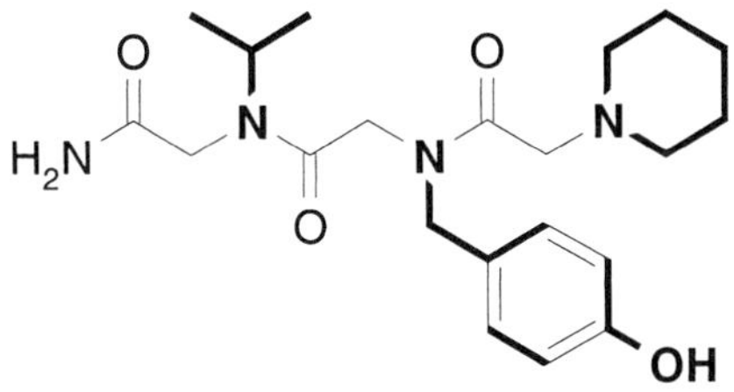

SMILES: NC(=O)C **N(C(C)C)**C(=O)C **N(Cc1ccc(O)cc1)** C(=O)C **N2CCCCC2**

Decimal   1 **2** 1 **88** 1 **101**

Binary   1001 1110 1111

Cyclic   -1001 1110 1111 01-

Figure 1   Various encoding strategies used to represent a chemical structure's "genome." A tripeptoid and various encoding systems to code for genome members for genetic algorithms are shown. The building blocks are separated by alternate bold substructures and are also designated in bold in the coding systems shown. (From Ref. 14.)

amino acids, this 5-bit codon can easily accommodate a unique pattern for each amino acid (13). Therefore, the 30-bit string can be translated into a unique hexapeptide. Organic compounds may be coded by alternate coding systems and some of these are shown in Fig. 1 (14).

A *fitness function,* also called the objective function (see above), is used to rank the individual chromosome. The optimization proceeds because a population selects individuals that have increasingly higher fitness. Initially, members of a set of N_{pop} individuals, which have been synthesized (or generated), are evaluated for fitness. The exact nature or origin of the fitness function is, in a sense, of no consequence as long as it is reasonably consistent and objective, hence the name. [*Note:* For the examples in Sec. III, the fitness functions are derived from experimentally determined values, whereas for the examples in Sec. IV, the fitness functions are computed corresponding to a given set of properties and are described later.] A roulette wheel is conceptually created, and the "slice" on the wheel for any given individual is proportional to the value of the fitness for that individual. A compound with greater biological activity gets a larger slice on the wheel and an inactive compound get a smaller slice. In the selection process, one may imagine mating pairs to be selected by spinning this wheel. (*All* individuals have a place on the wheel and therefore have a finite chance to be selected.) This produces a list of pairs for mating.

Subsequent generations are formed as follows. In one particular implementation of the process each member of the first generation (15) is ranked by *fitness*, and the fittest individual is placed into the next generation with no change. Next, pairs of individuals (from the selection step above) are crossed over to form the next generation. The *crossover* step may be visualized as shown in Scheme 2 and

Genomes: (Gen$_i$)	Fitness (Score)
10011 11010 10100 10010 00100	70
11000 10011 10110 01010 11111	48
10111 01010 00100 11010 11100	46
10110 10101 10110 01110 00001	42
10000 11011 10110 01010 11110	32
10110 01101 00011 10000 10011	27
11011 01110 10110 01010 10101	19
10001 11111 00010 01110 00001	12
00100 11001 01111 10110 01011	07
11010 11010 11100 01011 10011	02

Selection (via (replication and roulette wheel)
10011 11010 10100 10010 00100
10011 11010 10100 10010 00100
11000 10011 10110 01010 11111
10111 01010 00100 11010 11100
10110 10101 10110 01110 00001
10001 11111 00010 01110 00001
10011 11010 10100 10010 00100
10110 10101 10110 01110 00001
10011 11010 10100 10010 00100
10110 01101 00011 10000 10011

↓

Genome after crossover (with bits crossed over underlined)
10011 11010 10100 10010 00100
10011 11010 10110 01010 11111
11000 10011 10100 10010 00100
10111 01010 10110 01110 00001
10110 10101 00100 11010 11100
10001 11111 10100 10010 00100
10011 11010 00010 01110 00001
10110 10101 10110 01110 00001
10011 11010 10100 10010 00100
10110 01101 00011 10000 10011

← CROSSOVER

Parents for Crossover
10011 11010 I 10100 10010 00100
11000 10011 I 10110 01010 11111
10111 01010 I 00100 11010 11100
10110 10101 I 10110 01110 00001
10001 11111 I 00010 01110 00001
10011 11010 I 10100 10010 00100

↓

Site of Mutation (highlighted)
10011 11010 10100 10010 00100
10011 11010 10110 01010 11111
11000 10011 10100 10010 00100
10111 01010 10110 01110 00001
10110 10101 00100 11010 11100
10001 11111 10100 10010 00100
10011 11010 00010 01110 00001
10110 10101 10110 01110 00001
10011 11010 10100 10010 00100
10110 01101 00011 10000 10011

→ MUTATION

Genome (Gen$_{i+1}$)
10011 11010 10100 10010 00100
10011 11010 10110 01010 11111
11000 10011 10100 10010 00100
10111 01010 10110 01110 00001
10110 10101 00100 11010 11100
10001 11111 10100 10010 00100
10011 11010 00010 01110 00001
10110 10101 10110 01110 00001
10011 11010 10100 10110 00100
10110 01101 00011 10000 10011

Scheme 2 Overview of GA process: selection, crossover, and mutation. This example is for a genome size of 10 individuals, replication of a single best member, crossover rate of 60% with duplicates allowed, and a very low mutation rate. Point of crossover is shown by 'I' and a bit selected for mutation is highlighted in bold and underlined.

described as follows: ARBITRARILY making a "cut" at a randomly chosen spot in the genomes (represented by fixed-length bit strings) of two individuals selected as parents and then recombining the first part of the first genome with the second part of the second and vice versa, to generate two "new" offspring individuals with new genomic information (16). It is important to note that this is a completely random process and may transect the "allele" or part of the bit string used to code a particular pair. The process serves thus to change that particular part it describes radically sometimes.

Normally, since a genome is composed of linearly disposed bit strings, one may generally expect lower probability of crossover for building blocks near the terminus. To address this issue, one can envision linking up the individual genome to form a cyclic genome for each of the two members of a mating pair and then ARBITRARILY making two "cuts" at randomly chosen spot in both the genomes, to open the ring and generate two parts of the resulting string. Recombining the first part of the first genome with the second part of the second and vice versa generates two "new" offspring individuals. These two scenarios are shown schematically in Fig. 2. The scenario in which the genome is cyclized to form a circular genome and subsequently subjected to crossover process (Fig. 2b) offers a potential advantage-crossover takes place with the same probability for each bit (building block), whereas for linear bit strings the crossover may be lower near the ends of the genome. This entire process is called the *crossover* step.

It is important to note that the total number of individuals N_{pop} selected for the subsequent generations (Gen_{i+1}) remain identical to the initial random population (Gen_i), since each pair of parents produces exactly two offspring. It is worth noting that the crossover process using a "circular genome" in essence allows one to code the genome as a linear bit string and does not require the use of cyclic coding, as described in Fig. 1. The crossover rate corresponds to the percentage of individuals in Gen_{i+1} that are formed by pure crossover of individuals in the Gen_i. For example, in a population of 100 individuals, if 10 are copied over directly to the next generation, the maximum crossover rate (assuming no mutations, etc.) will be 90%. The rate of crossover varies, and is indicated in the examples (Sects. III and IV). In several reported examples in the literature, 100% has been used for the crossover rate.

After applying the selection and crossover steps as outlined above and thus producing a population of "new" individuals for the next generation, one applies the *mutation* operator. This simply consists in our case of "flipping" a bit (from 0 to 1 or vice versa). The frequency of this mutation is usually preset and held constant throughout the run. The choice of which individual to mutate and which bit in that individuals' genome to mutate is purely random (Scheme 2). One constraint that may be applied in the binary encoding process is know as Bensen *gray-scaling*, which assures that no resulting mutated individual or

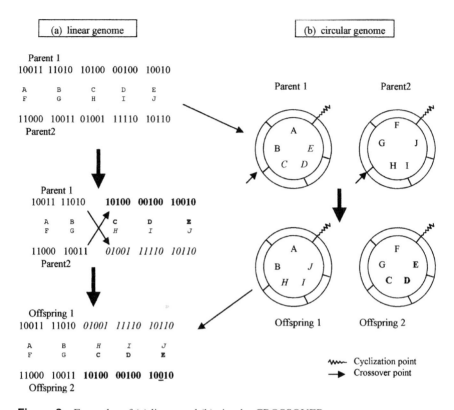

Figure 2 Examples of (a) linear and (b) circular CROSSOVER.

mutated molecular compound differs significantly from the corresponding molecule that underwent mutation. For example, in a peptide-based library, it may be desirable to mutate the initial building block Ala to Leu, but not to Arg. The gray-scaling method assures that point mutation does not significantly alter the character of the encoded trait. A variation on this theme, called *neighbor mutation,* is described briefly in Section IV.B. A number of variants have already been discussed, and Scheme 2 captures the essence of the GA process for going from Gen_i to Gen_{i+1}.

In addition, an elitist GA can be used, where a set number of best individuals from generation Gen_i are copied to Gen_{i+1} without allowing these to undergo crossover or mutation. (A variation referred to as the "best-third method" is described in Sect. IV). Also, duplicate checking procedures may be used to ensure that an identical parent is not present in any given population and subsequent generations. In these cases, mechanisms are provided to keep the population size fixed (see Sect. III.B for an example).

III. EXAMPLES WITH EXPERIMENTALLY DERIVED FITNESS FUNCTION

A. Identification of Thrombin Inhibitor

One of the earliest published reports for combining a combinatorial synthesis with a genetic algorithm utilizing an experimentally derived fitness value was provided by Weber et al. (17). A well-known Ugi reaction based combinatorial library with $10 \times 40 \times 10 \times 10$ building blocks representing a total of 160,000 possible compounds was explored. An initial population size of 20, with a crossover probability of 100% and a mutation rate of 1% in about 50 generations, was explored with an elitist GA. The crude products from each generation were assayed in a thrombin chromogenic assay for inhibitory activity. In this GA experiment, no duplicates were allowed. The data and the population members from run Gen_1 and Gen_2 were stored in the same database, and the best 20 new compounds were selected from this database for synthesis and biological evaluation. This process was repeated, and after the production of 20 unique Ugi products in 20 generations, the most active product was found in generation 18. This reaction was scaled up and the products were purified, which revealed the intended racemic four-component product **2** with $K_i = 1.4$ μM, along with a three-component by-product (**3**) with $K_i = 0.24$ μM (Fig. 3). Thus the most active compound, **3,** was not the one identified through GA optimization, it was product **2.** For comparison, the best product from the first generation was very weakly active, with $K_i = \sim 300$ μM. It is worth noting that by synthesis and screening of merely 400 compounds from a

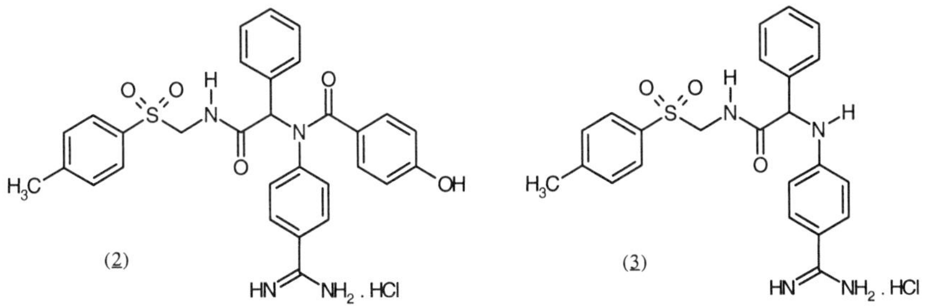

Ki = 1.4 uM   Ki = 0.24 uM

Figure 3 GA-driven identification of thrombin inhibitors. (From Ref. 17.)

possible 160,000-member library, GA-derived lead generation identified a sub-micromolar thrombin inhibitor.

B. Identification of Stromelysin Substrates

In an independent study, Singh et. al. (18) described a GA-based approach for rapid exploration of protease substrate specificity and selectivity leading to the identification of a unique substrate for stromelysin. The total diversity space for this problem represented a 64-million-hexapeptide library, represented by $X_1PX_3X_4X_5X_6$, where each X represents all 20 natural amino acids, (except Cys, which was replaced with S-Me-Cys, denoted U). An initial population size of 60 randomly formed hexapeptides was used. A constraint of $X_2 = P$ (proline) was invoked only for the initial population, since our previous work had shown that proline at this site is preferred, and it was felt useful to start with all available knowledge in a real-life example. All subsequent generations were free to choose nonproline amino acids for position X_2. The peptides were synthesized using controlled-pore glass as a solid support (19). Each peptide was capped with a fluorescent tag (coumarin propanoic acid, COP). A glass-bound peptide sample used for biological assays (20) could be generically represented as COP-A-$X_1X_2X_3X_4X_5X_6$-$(Acp)_5$-βAla-AMP-CPG. Automated biological assays were performed on a small sample (typically, 4.0 ± 0.3 mg) of the glass-bound peptides, using the protease solution (stromelysin or collagenase) added using a

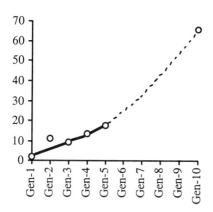

Figure 4   Performance evaluation of GA-based lead identification. plot of average activity (fluorescence x 10^3) versus GA-based generation. Solid lines represent actual experimental data, the dotted line shows a projected plot if one assumes that G_{10} represents a termination point, such that each individual of G_{10} is a good substrate with at least 20% substrate processing. Adapted with permission from Ref. 13. Copyright [1996] American Chemical Society.

Packard PROBE, in a 96-well format. The samples were mixed on a variable-speed vortexer, the solution was allowed to settle by gravity, and a fixed amount of the supernatant from each sample, containing the processed substrate carrying the fluorescence tag (COP group), was transferred to the appropriate position of a 96-well plate. Fluorescence was then read by means of a fluorescence plate reader. Thus, the amount of fluorescence for each sample was proportional to the extent of the corresponding substrate processed by the protease (i.e., proportional to the biological activity).

Samples from each generation were screened for substrate activity versus stromelysin (mSl-t) and/or collagenase (mCl-t) (21). The negative value of the observed fluorescence of given samples of generation Gen_i served as a measure of the sample's potency as a substrate, as input for optimization in the GA to provide sequences for the subsequent generation, Gen_{i+1}. A plot of the average fluorescence (22) (activity) versus generation is shown in Fig. 4. Even in this relatively small number of generations evaluated, one can clearly see the trend to greater activity, as well as the variability, as the program automatically explores the space it is optimizing over. Not only did the average activity per generation improve (as highlighted in Fig. 4) but each generation also identified new sequences with greater activity than displayed by the previous generations (see Fig. 5). The graph-

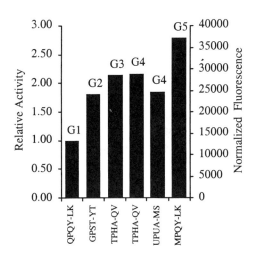

Figure 5 The most active sequence from each generation. The site of processing by stromelysin is indicated by a hyphen for these sequences listed. Relative activity is plotted with best sequence for $G_1 = 1.0$. Two samples are shown for G_4. Even though the most active sample for G_4 and G_3 had an identical sequence, the next most active sample for G_4 represented a different sequence and distinct P_2 and P_1 residues than those observed for G_2. Adopted with permission from Ref. 13. Copyright [1996] American Chemical Society.

ical representation of the data in Fig. 4 provides evidence for GA's capability to explore diversity space. A single assay per sample was carried out and the observed fluorescence, "raw data," had been used as a fitness function to drive GA-based optimization. This underscores the tolerability of this method of optimizing even over a "noisy" fitness function. The Gen_{i+1} sequences are directly derived from the biological data obtained for the preceding generation, Gen_i.

It is also important to realize that, as part of the normal GA process, a given sequence was identified several times and these multiples were accepted. (In this regard, this GA differs from most of the other examples in the literature, both experimentally and computationally driven, where duplicates are excluded, and this difference explains why a noisy fitness function was still acceptable.) As part of the experimental design for this GA-driven study, a fundamental decision was made to conduct biological evaluations for that sample equal to the number of times the GA demanded it and to use the individual biological responses as fitness function for the subsequent cycle. This decision was made to address the inherent variance in any biological evaluation, and it was argued that this experimental design will over time (generations) take care of the noise inherent in the biological screening. If a decision were reached to use, for example, chemical samples of less than 100% purity (as often occurs with high throughput synthesis), this same strategy could be extended to resynthesized samples. This was not necessary in this case because the samples were virtually pure.

C. Identification of Trypsin Inhibitors (23)

A population of 24 randomly chosen hexapeptides was optimized: Ac-XXXXXX-NH$_2$, where X denotes one of the four building blocks [phenylalanine (F), isoleucine (I), lysine (K), and theronine (T)] exhibiting increased inhibitory activity for trypsin. In this example, the strings coding the genome sequence were the single-letter codes of the amino acids (building blocks). The initial, randomly chosen 24 peptides (generation 1) were synthesized, and trypsin inhibitory activity for each member was determined. For the *selection* step, all 24 peptides from the first generation were ranked for trypsin inhibitory activity; the best 12 peptides (1st–12th) were kept and the worst 12 (13th–24th) were discarded. For assigned peptides 1–6, three copies were made (providing 18 peptides) and for peptides 7–12 a single copy of each was saved. The resulting 24 new peptide sequences were randomly divided into 12 pairs, which were then subjected to the random, single-point *crossover* process (i.e., a crossover rate of 100% was used), producing 24 new members. Finally these sequences were subjected to *mutation,* in which any given amino acid (building block) can be replaced by any one of the four possible building blocks with a 3% probability. Finally, the resulting 24 peptide sequences, representing the second generation, were synthesized and assayed. The process was repeated until a total of six generations, or 144 peptides, had been evaluated. The analysis revealed that the average inhibitory activity increased from 16% for the initial randomly chosen population to 50% for the sixth generation. Also, 17 of the 22 discrete peptides from the sixth generation belonged to the two predominant patterns, Ac-XXKIXX-NH$_2$ and Ac-XKIXXX-NH$_2$, whereas only two peptides belonged to these patterns in the first generation. It is interesting to note that the most active peptide, Ac-TIKIFT-NH$_2$, with 89% inhibitory activity, was identified in the sixth generation. This sequence had been identified earlier by Eichler and Houghten (24), who started, with Ac-XXKIXX-NH$_2$ peptide libraries representing about 3.2 million peptides. Therefore, from the synthesis of 144 (3%) of 4096 (4^6) possible sequences (from F, I, K, and T building blocks), a peptide with 90% inhibitory activity was successfully identified by means of a GA.

IV. EXAMPLES WITH COMPUTATIONALLY DERIVED FITNESS FUNCTION

We know that when the number of possible compounds that may be synthesized/generated in combinatorial or permutational libraries (2) is enormous, computational tools can be utilized to identify subpopulations that meet a predefined paradigm, for subsequent synthesis and biological screening. Some of these computational tools include computer-based structure–activity relationship (SAR) methods (dissimilarity or similarity measures) and generation of molecular structure with constraints (pharmacophore, active site modeling/fitting, especially when a 3D model or a crystallographic or NMR-based 3D structure is known). The demand for the judicious and timely use of proprietary corporate compound collections and the increasing availability of commercial compound collections also fuel the development of heuristic algorithms for the selection of subpopulations with an increased probability of generating leads in target-driven biological screening. Whether these are the existing compound collections or the combinatorial virtual libraries, an array of computational tools have been used for property-based compound selections. Next we briefly describe some examples in which selected properties were computed and their match to optimal values used as a fitness function for GA-guided lead generation. Rather than presenting an exhaustive listing, we highlight diverse examples taken from published primary literature.

A. Virtual Library Screening: Peptoid-Based Lead Generation

GA has been developed and described by Sheridan and Kearsley (25), to select molecules (and identify a preferred set of building blocks) from a large virtual library. This GA utilized a measure of similarity to an active probe structure(s) as a measure of fitness. A virtual peptoid library representing over 20 billion members, from a set of $1 \times 2507 \times 1 \times 2507 \times 3312$ building blocks was constructed and stored "in silico." A population of 300 molecules was selected at random, and each molecule was scored using a topological descriptor using atom pairs separated by a specific number of bonds as a selection criteria of the chemical similarity against a reference probe. A number of GA parameter strategies—"third-best" method and stochastic search, with random or neighbor mutation (see below)—were used. For all methods, duplicates were excluded.

The "*third-best*" method is briefly described as follows. Three hundred molecules, selected at random (generation Gen$_i$), were scored, the best 100 were saved, and the remaining 200 were deleted. Three copies of the best 100 were made. The first copy was saved unchanged to the next generation, Gen$_{i+1}$. For each individual of the second copy, a single residue was chosen at random and that residue was mutated with another residue from the list of building blocks for that position; the resulting 100 molecules were added to the generation Gen$_{i+1}$. The third set of copies of 100 molecules was crossed over with another copy from this set at a randomly chosen position, and the resulting molecules were added to the generation Gen$_{i+1}$. The generation (Gen$_{i+1}$) was subsequently scored as above, and the process was repeated.

In the *stochastic method*, 100 molecules from Gen$_i$ were randomly selected to survive unchanged, and these became part of Gen$_{i+1}$. The next 100 molecules from Gen$_i$ were randomly selected, and two parents were selected at random from the remaining 100 molecules of Gen$_i$ to produce 100 new molecule for Gen$_{i+1}$. The relative probability for selection of any individual molecule "*i*" depended on

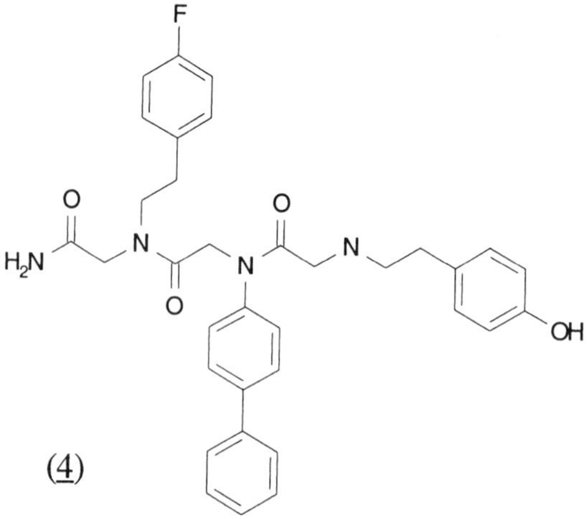

(4)

Figure 6   A predetermined tripeptoid used as test case. (From Ref. 25.)

a linear function of its score. The rel prob$_i$ = (score$_1$ − minscore)/(maxscore − minscore), where maxscore and minscore, respectively, refer to the maximum and minimum score for Gen$_i$. The neighbor mutation, similar to the gray-scaling mentioned in Section II, assures that a molecule mutates to a closely related molecule.

Thus starting with a population size of 300 molecules, and utilizing the two selection methods just described, GA was run for 25 generations. The "third-best" method was found to be much better than the stochastic method at increasing the average score more quickly with either of the two mutation choices. To study these GAs' optimization concept for this combinatorial problem, identification of the predetermined tripeptoid *4* (Fig. 6) was chosen as the test case. Amazingly, with the evaluation of just 3600 molecules (<0.00002%) from a virtual combinatorial population of over 20 billion, this GA-driven method found the "right answer" after 12 generations.

Another example explored in this work was the chemical similarity to the cholecystokinin antagonist tetrapeptides. In this case the score of each molecule for the GA run was taken as the similarity to the average value of the descriptors for the two probes **5** and **6** (Fig. 7). Again, a GA using the "third-best" method with neighbor mutation was run for 25 generations. In this case, peptoid-based molecules with similarity score of 0.90–1.00 were observed between generations

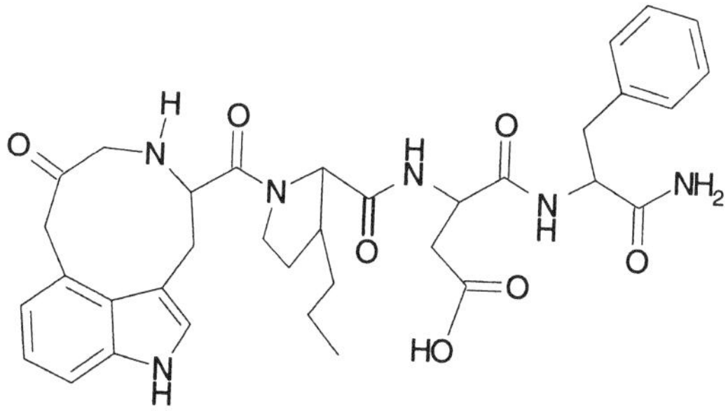

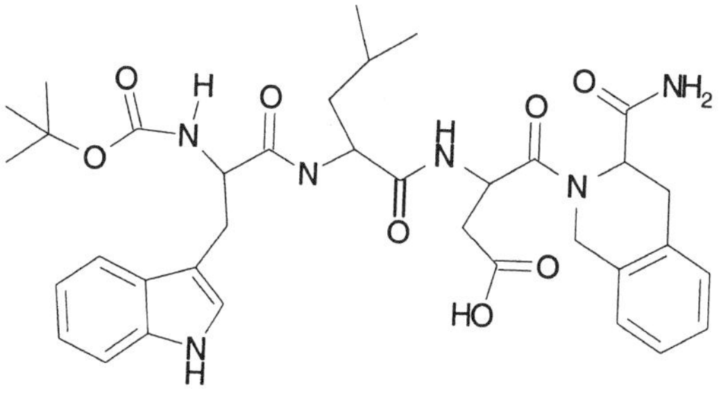

Figure 7 Two probes used in GA run. (From Ref. 25.)

15 and 23. Unfortunately, the biological data for these molecules were not reported.

B. Virtual Library Screening and Ligand Design for Lead Generation

Genetic algorithms are among the most versatile and efficient means of solving a given problem quickly. Their applications in the area of drug design have only begun to be tapped. Frequently in the domain of high speed analoging one wishes to produce libraries of compounds in chemical series to quickly and efficiently sample the SAR possibilities. The aim in such exercises is to produce a fraction of the total member of combinatorial possibilities of analogs while at the same time covering maximally the diversity space they represent.

A virtual library of compounds derived from a core template with a number of groups replaced can be constructed based on 3D grid properties, using a genetic algorithm protocol. An example of this with the goal of achieving maximum molecular diversity, for a benzodiazepine-based library, is described by DongXiang et. al. (26). The core molecule is labeled with dummy atoms "Du" at sites where substituents are to be attached. Only one putative atom on each fragment was allowed to be involved in covalent bond formation with the core molecule. When every "Du" atom on the core molecule had been replaced with a fragment chosen from the fragment list, a molecule was generated *in silico* and saved in SYBYL MOL2 format. Molecular dynamics/mechanics techniques were used to optimized the structures of the virtual compounds in SYBYL, the optimized structures were aligned onto a template structure ("*j*"), and this orientation of the optimized structure was saved in the database containing the virtual library members. In this example the size of the virtual library was set to 100 compounds. A C (sp^3) atom with +1 charge was used as a probe atom, and the interaction energy matrix [comparative **m**olecular **f**ield **a**nalysis (CoMFA) descriptors] obtained was subjected to (principal component analysis) (PCA). The CoMFA matrix was analyzed with SYBYI QSAR module (v. 6.1) to extract 28 principal components, which were used as a measure of molecular dissimilarity MS_{ij} (i.e., a measure of molecular dissimilarity to the template structure "*j*"; the bigger the MS_{ij}, the more different is compound "*i*" from the reference compound "*j*"). Rather than using the dissimilarity to a reference compound, these authors considered the overall dissimilarity of compound "*i*" to all other compounds in the library. If all compounds in the library are said to be "*n*," then the overall diversity measure "MS_i" is described by the following formula.

$$MS_i = \frac{\sum_{j=1}^{n} MS_{ij}}{n}$$

The value of MS_i was maximized as a fitness function (i.e., $Score_i = MS_i$) for this GA-guided virtual library construction. An inheritance of 40% was used; that is, the first 40 compounds (from Gen_i, with the highest fitness score, were saved unchanged to Gen_{i+1}. Crossover and mutation were randomly selected according to a Gaussian function that was driven by a computer's random number generator. For crossover, fragments linked onto the same randomly selected connection site on the core molecule were exchanged between the two parents. For mutation, two alternate protocols—"mutation-in" and "mutation-out" were chosen at random. For "mutation-in," the two fragments at two different sites on the same molecule were swapped, whereas for "mutation-out" a randomly selected fragment on a molecule is replaced with fragment chosen at random from the fragment list. In this GA-directed study, duplicate compounds were disallowed. For the generation of this virtual library, the same set of 72 diverse fragments for each of the three connection sites, represented by R_1, R_2, and R_3 in the generic structure (Fig. 8), were chosen.

A total of 20 generations were evaluated, and it was reported that the maximum score was reached by generation 10 [i.e., with evaluation of 1000 members (0.26 %) out of the possible 373,248 members] and remained unchanged after that. In addition, following the performance as the generations evolve, a list of fragments was tracked, as well as their respective frequencies at each of the sites. This list provided useful information of the building blocks, which could then be used for the synthesis of a targeted benzodiazepine library with maximum molecular diversity. This analysis provided the number of diverse fragments with higher frequency at sites: R_1 ($= 6$), R_2 ($= 5$), and R_3 ($= 12$) with very little overlap among these fragments. Only one fragment was identical for R_1 and R_3; two fragments were identical for R_2 and R_3. Unfortunately, the actual synthesis of benzodiazapine library members derived from these preferred fragments was not reported.

Another interesting case that was considered in this report was the construction of a virtual library of huperzine analogs (for the generic structure, see Fig. 9). The important difference in this case was that coordinates of the (–)-huperzine A in the complex of huperzine and acetylcholinesterase (AChE) (molecular target) at 2.5 Å were used as the reference structure for library compound overlap. The fitness function for this GA-driven study was the ratio of MS_i with

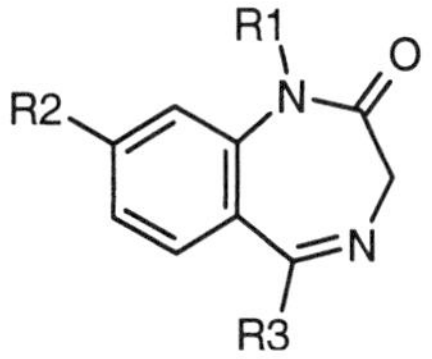

Figure 8   Three connection sites, R_1, R_2, and R_3 in the generic structure.

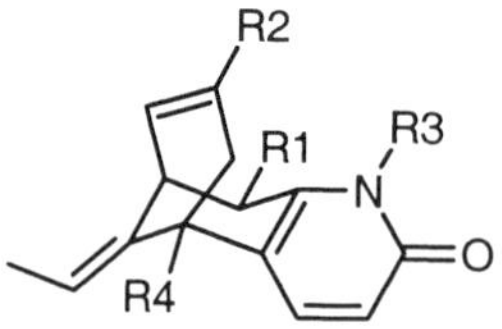

Figure 9

MS_{ir}, where MS_{ir} refers to the molecular dissimilarity of the library compound "i" to the reference compound:

$$Score_i = \frac{MS_i}{MS_{ir}}$$

Thus the compounds with high dissimilarity to other library members and similarity to the reference are given higher probability for being selected as parents for the next generation. For the score for compound "i" to be higher on this scoring system, the value of MS_{ir} must be smaller and the value of MS_i larger. For this virtual library generation, nine fragments (H, CH_3, CH_2CH_3, CH_2OH, CH_2CH_2OH, OH, NH_2, CH_2NH_2, and COOH) were selected. The GA run, using the foregoing conditions, produced a total of 30 generations. The maximum score at the equilibrium state was found to be 0.73. After ninth generation the maximum score fluctuated between 0.73 and 1.0. Also, examinations of the molecular structures revealed that the (−)-huperzine was identified in the ninth generation and was inherited in the subsequent generations. In addition, the virtual library members were docked in the AChE receptor and their binding energies were also evaluated. It was observed that most of the library members had computed binding energies near that of the (−)-huperzine. In fact, one library member, compound **7** (see below: Fig. 10), has been reported (27) to be biologically active (IC_{50} = 35 nM). This experimental data point was claimed to provide indirect verification and support of this GA-driven study.

C. Genetically Evolved Receptor Model

The goal of this example, described here as reported by Walter et al. (28), was to produce an atomic-level model of receptor (active) site based on a small set of known structure–activity relationships. A number of explicit model atoms (40–60 atoms) were placed at points in space around a series of ligands, and intermolecular interactions between the ligand and receptor model atoms were calculated. By using, say, 60 atoms and changing the atom types from among eight possible atom types, more than 10^{54} models could be produced. Since this large multidimensional search space obviously could not be evaluated via an exhaustive, systematic approach, a GA was employed. That is, by changing the atom types at various positions, one would produce models that should optimally have high correlation between the calculated binding energy and bioactivity, with the assumption that the observed bioactivity is proportional to the ligand–receptor interaction energy. This study involved 22 sweet-tasting compounds representing three structural classes: (L)-aspartate derivatives, aryl ureas, and guanidine-aliphatic acids. Interestingly, the bioactivity for these analogs spans 3.5 orders of magnitude. Using the (L)-aspartic acid derivatives (Fig. 10) as an example, four sets of models were generated. Each set had six compounds, which were used as templates to calculate bioactivity for eight analogs each. The (L)-aspartic acid derivatives were identified by compound numbers as follows:

Set 1:   **8, 9, 10, 12, 13, 14**
Set 2:   **7, 9, 10, 11, 13, 14**
Set 3:   **7, 8, 10, 11, 12, 14**
Set 4:   **7, 8, 9, 11, 12, 13**

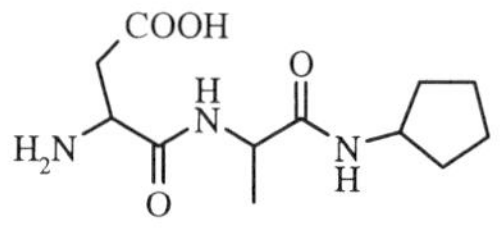

(7) log (potency) = 1.70 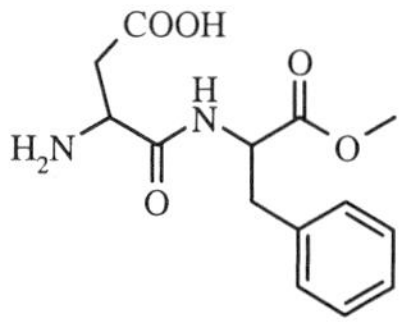   (8) [Aspartame] log (potency) = 2.26 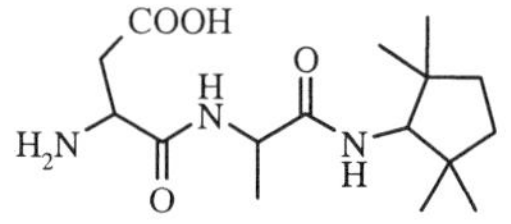   (9) log (potency) = 2.90

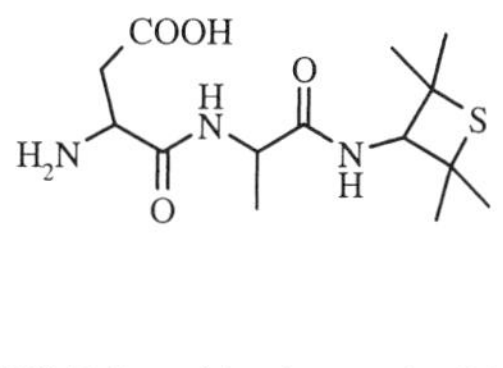

(10) [Alitame] log (potency) = 3.30 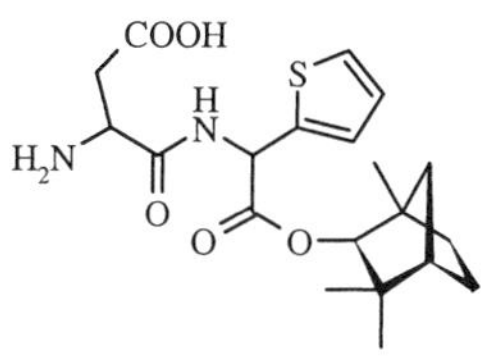   (11) log (potency) = 3.38 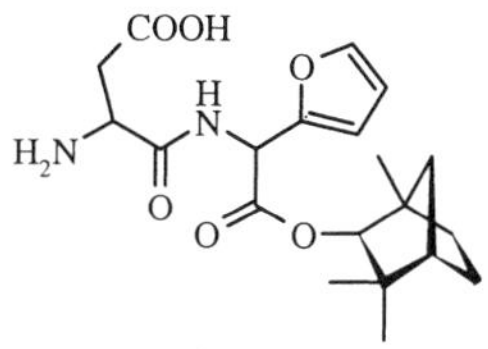   (13) log (potency) = 4.22

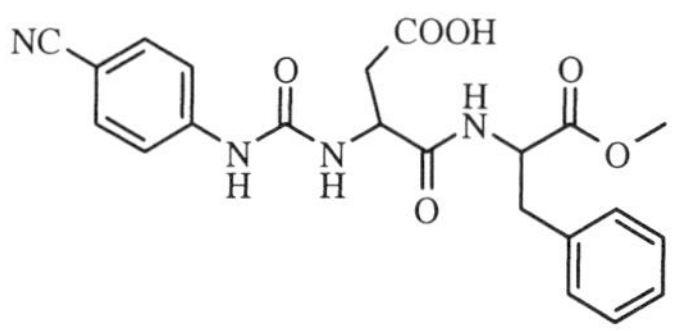

(12) [Superaspartame] log (potency) = 4.00 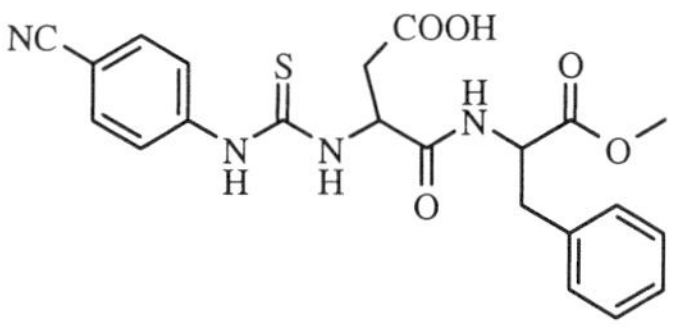   (14) log (potency) = 4.70

Figure 10   Structures and potencies of aspartic acid derivatives. Adopted with permission from Ref. 23. Copyright [1994] American Chemical Society.

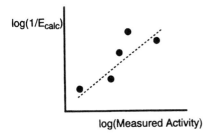

Figure 11 Fitness score for a model is calculated by first calculating an interaction energy with each ligand, then calculating the regression coefficient for 1/exp(energy) versus log(bioactivity). Reprinted with permission from Ref. 23. Copyright [1994] American Chemical Society.

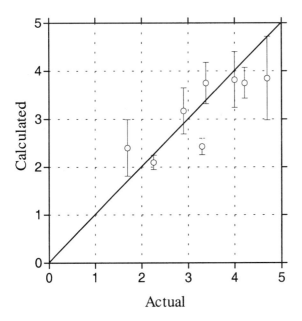

Figure 12 Composite calculated values for all eight aspartic acid derivatives, each taken from the set in which it was not part of the model-building process. Data points are calculated as average over the first 100 models in the population and the error bars indicate standard deviation. Adopted with permission from Ref. 23. Copyright [1994] American Chemical Society.

The "fitness function" score for each member was produced by first calculating the intermolecular van der Waals and electrostatic energies between the model and each individual ligand, in the series of superimposed ligands; the correlation coefficient for 1/exp-(energy) versus log(bioactivity) was used as the fitness score (Fig. 11). The parameters used in the GA run include the following: atoms/model; 60; population size, 2000; mutation rate, 1.0/generation; and number of generations, 10,000. A Poisson distribution is applied to the mutation rate, so that with the overall mutation rate of 1.0/generation, the probability of having 0, 1, 3, 4, or 5 mutations in a given gene is 0.368, 0.368, 0.184, 0.062, 0.016, and 0.004, respectively. This GA-driven model-building procedure generated models with excellent correlation (r = 0.98–0.99) between calculated energy and bioactivity. The results are shown graphically in Fig. 12.

In an extension of this study, all 22 diverse sweet-tasting compounds were used simultaneously. Only 11 of these were used for model building. This model was then used to predict the activity of all 22 compounds. The results are shown graphically in Fig. 13. As anticipated, larger errors were observed for compounds not included for the model building, as is the case with other methods based on in-

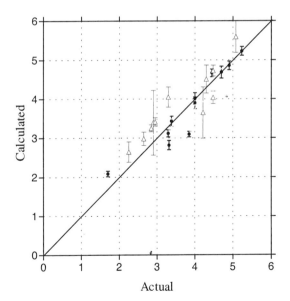

Figure 13   Calculated bioactivities for all 22 compounds. Filled circles indicate the 11 compounds used for evaluation of the model, and open triangles are calculated values for the 11 compounds that were not included in the model generation. All values are averages calculated for the first 100 models in the population. Adopted with permission from Ref. 23. Copyright [1994] American Chemical Society.

terpolation of data. Moreover, such model building with limited SAR information may provide useful constraints on the diversity space to be explored. The results are expected to be useful for designing and evaluating virtual library members in order to arrive at potentially active sets for targeted screening and lead generation. The analogy between the use of these virtual active site models and the use of active sites via X-ray crystallographic measurements of enzymes is obvious.

V. DISCUSSION

The examples described in this chapter were chosen to illustrate the versatility of GA-driven processes for lead generation. In Section III, we focused on examples of GA driven by experimentally determined fitness score values. These three examples provided significant variations for the principal GA operators: inheritance (replication), crossover, and mutation rates (values). These examples also differed in the level of input based on the existing data/information (a problem-dependent variable) or "bias" for selection of the initial "random" generation.

All the examples in Section IV utilized computationally generated fitness score values to drive GA-based lead generation. The examples were arranged to illustrate fitness score evaluation in comparison to probe molecule/molecules utilizing fitness scores based on dissimilarity measures (1) derived from a topological descriptor using atom pairs separated by a specific number of bonds, (2) based on 3D grid properties, and (3) utilizing the coordinates of a probe based on its complex with its molecular target. Finally (4) an example was given of generation of the 3D receptor model from a known SAR data set. In addition to the information already discussed, we feel that a number of other points need to be addressed, which are touched upon next.

For all the examples given in Section III, a single, simple fitness function was used. In addition to the GA-derived optimization process for the stromelysin (example in Sect. IIIB), the samples were screened for collagenase to obtain selectivity information among these closely related matrix–metalloproteinase targets. Even though, a GA-derived utilization of this selectivity information was presented, it is obvious that one may well use the biological data from screening with collagenase for every sample as a penalty value for determining the final fitness score, which will be a sum of the partial score versus stromelysin and partial "penalty" score versus collagenase. Thus, such a GA could be easily utilized to provide optimized and selective leads (solutions) for the target of choice. In general the most complex and discontinuous fitness functions can be used with *no* increase in the complexity of the GA approach.

One may choose to incorporate a number of interdependent parameters or molecular properties as fitness functions, or one may employ penalty functions for some variables as a part of the fitness function paradigm, as described earlier. For example one may construct complex fitness functions incorporating the molecular weight of each compound, its selectivity for the target enzyme over the related enzymes, its degree of overall charge, solubility in water, and so on. *The only requirement of this overall fitness function is that a value for each member be determinable.* These values may be a combined aggregate of several experimental and computed values as well.

Many optimization strategies require a continuous, mathematically definable fitness function. Real-life SAR optimization problems rarely afford this situation. This is not the case with GA, which readily optimizes discontinuous fitness functions by its very nature. The method also readily accommodates variability in the fitness value (as exemplified in Sect. IIIB). This is an essential part of SAR work and is also not readily accommodated by other methods of optimization. In fact, it is stated that the GAs succeed best where the fitness functions are highly complex, discontinuous and "noisy" (10). GAs could easily be used to optimize a number of parameters simultaneously, since they are known to provide solutions to the kinds of problem not suited to other methods. The combinatorial chemistry based diversity assessment/solution is such a problem.

Screening formats are continually being changed (e.g., from the 96-well format all the way to microarray formats providing roughly half a million samples per run) to generate larger amounts of screening data points per unit time. As a consequence, meaningful data mining (informatics) is becoming rate limiting. Therefore, emphasis is being placed on constructing smaller focused libraries representing structurally diverse compounds rather than aiming for exhaustive chemical sample pools. The fitness function in a GA run ranks individuals at the product level (i.e., fully enumerated products as opposed to reagents or pathways). Thus, one would expect generation of more diverse libraries. The work of Gillet et al. (29) presents a very important analysis providing the upper and lower limits on diversity for a given library. This work points out that the reactant-based selection, which attempts to maximize diversity at the building-block level, results in libraries noticeably less diverse than those obtained when the selection is at the product level. Interestingly, their work also demonstrates that the GA-driven process provides combinatorial libraries that are more diverse than those generated by means of reagent-based selections.

An important aspect of lead generation for which GA has been increasingly used is the virtual library screening and identification of a focused library, some of whose members may then be synthesized and screened. An important emphasis for these studies has been to augment (amplify) the corporate or compound collection library for screening against diverse biological targets. An important piece generally missing from the published literature to date has been experimental validation or support for these computationally designed libraries.

In most of the computational examples with GA, it is generally acceptable to carry the run through a relatively large numbers of generations before reaching convergence. However, for the cases involving chemical synthesis, it is essential that the GA converge relatively quickly (in <15 generations). It is worth noting that in well-designed GA runs, this was in fact the case regardless of whether the fitness function was experimentally or computationally derived.

Increasingly, as even the limited SAR data are gathered for newly discovered targets, the GA-based approaches are potentially a very powerful way to quickly generate pharmacophore models, from limited data set. The work of Walters et al., described earlier, is an excellent example of this GA-derived approach. Of course, a set of known sweet-tasting compounds was used to exemplify and provide a critically needed validation. We hope that such studies will be carried out much more frequently, to provide a large database of experience in this area. A pharmacophore generated by a GA-derived process may be further utilized for GA-derived ligand design as described by DongXiang et al. (27), as highlighted earlier with the huperzine example.

Additionally, in any given hit identification process from HT screening, one generally encounters issues related to the *dynamic range of the biological data.* One has to choose optimum concentration range to ensure a hit rate from the samples at hand of less than 1%. Thus, the screening process is simply a hit identification process, not necessarily lead generation. An important contrast to this is the GA approach, where potentially one can deal with a large dynamic range of concentrations, from millimolar to nanomolar. This is because the fitness function in reality has meaning for a given generation Gen_i under consideration, and this information (from Gen_i) is utilized only for the production of the very next generation, Gen_{i+1}. Thus, at the start of the GA, one may use a biological assay at high concentration (say, 10–100 μM), such that differential activity is identified. Note: that the best member of the Weber's first generation had $K_i = 300$ μM, as mentioned in Section III.A. Even so, as the GA proceeds, one may easily modify the biological assay (provided one is still optimizing the same property space) and run the assay, say, to identify submicromolar to nanomolar range hits. This iterative GA-driven process for finding good solutions thus allows accommodation of huge dynamic ranges of values within one experiment. This is in contrast to the screening of huge combinatorial libraries, where one is restricted to a much narrower range of biological data (e.g., inhibition or antagonism of screened compounds at 1 μM, followed by determination of IC_{50} or K_i values for the hit members only).

Genetic algorithms have recently been applied to a variety of applications related to drug design. These examples are not discussed here, but a number of excellent papers in the published literature provide detailed discussions of the use of genetic algorithms for de novo ligand and library design (30), flexible structure docking (31), and molecular recognition (32).

VI. LEAD OPTIMIZATION

Finally, we emphasize that the entire process of iterative optimization of some SARs has an inherent potential advantage over that of making large libraries before screening them. That advantage is one of "data digestion." In the drug discovery process, as it is generally run in project team settings in the pharmaceutical industry, the project team has an opportunity to assess results from one generation of compounds and to "spin off" other avenues of investigation related to the project goals. When GA is applied to the optimization process, it fits this natural rhythm quite well. We have found this fit to offer a great advantage in our hands. An example of such a "spinoff" bonus, *identification of selective substrates— "unique" hits,—*was provided in our work described Section III.B, where samples from Gen$_1$ were utilized for determination not only of potency as a substrate of stromelysin, but also of selectivity over collagenase.

The goal in the stromelysin project was to identify not only good but also selective substrates. This substrate selectivity information would be subsequently translated into a selective inhibitor. In view of this objective, we have also assayed the initial set (generation 1) vs. collagenase (mCl-t). The results for the collagenase vs. stromelysin assays are shown in Fig. 14.

The site of processing, in the sequences highlighted, is indicated by a hyphen. However, translation of these substrates to a known class of matrix metalloprotease inhibitors (33) would only involve the P$_1'$ and P$_2'$ portion of the information. It is known that the P$_1'$ residue imparts a greater selectivity among matrix metalloproteases. Thus, a sequence GPST-YT, which is selectively processed by stromelysin, as shown in Fig. 14, was identified. *This processing between Thr and Tyr represents a unique selectivity between these two metalloproteases.* Therefore, this stromelysin-selective substrate (34) was utilized, and a focused set of sequences that explores variations at the Y (Tyr) position was prepared in an independent, "spinoff" study while the GA optimization proceeded on a parallel track. A number of substituents were selected to evaluate a variety of electronic and steric properties for their effects on the relative hydrolysis of the substrates and thus their relative importance on the overall binding energy in the active site of

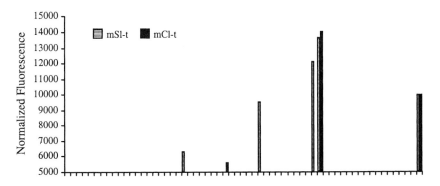

Figure 14   Assay results for collagenase (mCl-t) and stromelysin (mSl-t) for G$_1$ samples. Adopted with permission from Ref. 13. Copyright [1996] American Chemical Society.

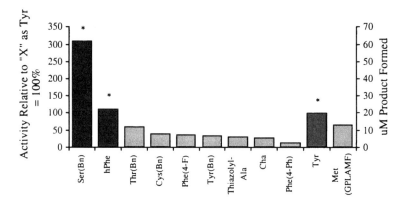

Figure 15 Assay results of samples COP-AGPST;"X"T-(Acp)5-bAla-AMP-CPG. All of the samples showed processing by stromelysin only, and the data is shown above. Asterisks (*) represent that these samples were confirmed to show a single site of cleavage as confirmed by Edman sequencing of the post-enzymology solid (glass-bound) sample and by amino acid analysis and LC/MS of the supernatant [for details, see reference 24(b) as cited in the Ref. 13 in this chapter]. Two columns are shaded differently to highlight the groups that show activity greater than the reference: Tyr. Adopted with permission from Ref. 13. Copyright [1996] American Chemical Society.

stromelysin. The outcome of this study also provides some additional SAR information. The results of this study, summarized in Fig. 15, show a preference of oxygen vs. sulfur [by comparison of Ser(Bn) vs. Cys(Bn)] and indicates β-branching to be deleterious [comparison of Ser(Bn) and Thr(Bn)]. In conjunction with a 3D model of stromelysin, we were able to rationalize the data based on the steric and electronic effects and to suggest additional novel P_1' groups.

This process, as already described, illustrates the flexibility of this GA-based approach. However, it is important to emphasize that this activity *must* proceed independently of the GA optimization and must not be used to influence its convergence. Thus one *should not* change the variables for the GA experiment on the basis of the outcome of information from a given generation.

In summary we learned three lessons from this experience. First, it is usually better to start a GA runoff with as much of the current thinking about the problem as possible built into the *first* generation—but not required to be carried over to subsequent generations unless it truly deserves to be. In this way one can circumvent several generations of the "first principles" type of learning as the run evolves. Second, it is advisable—even desirable—sometimes to deal with noisy data, provided the data point is remeasured each time it is called for in the GA run. We firmly believe that this ability to deal with and accommodate raw, noisy data in the context of the optimization is a unique feature of GA that has not been well exploited to date. Finally, as the run progresses it is important not to interfere with its progress by yielding to the normal human inclination of micromanaging the course of the evolution. At the same time it is also profitable to monitor the progression from one generation to the next. It has been our experience that knowledge gained by watching the progress of such GA optimizations can lead to insights, and side projects that may have great value in their own rights and can be independently exploited as the run progresses on a parallel path.

One of the goals of the recent efforts from virtual library generation and GA-based screening is to derive important building-block information. For example, in addition to looking for new molecules that meet the profile of desired molecules (i.e., come close to the desired properties), one may determine the frequency at a given site on the template of building blocks that lead to these desired properties. This building-block information may then be used to construct "biased" libraries. This has been described in some detail by Sheridan and Kearsley (25) in their work for GA-driven analysis of the "virtual" tripeptoid library described earlier.

VII. SUMMARY

From the experimentally (or computationally) driven examples provided here, it is clear that GA-based lead generation/optimization can proceed after synthesis (or analysis) of a small fraction of the combinatorial population. This approach provides an alternative strategy for the effective exploration of diversity space without the construction and assay of large libraries to identify lead candidate(s). These algorithms further facilitate the process of lead discovery/optimization by reducing user bias and should provide a powerful tool to help focus drug discovery.

VIII. FUTURE DIRECTIONS

These early examples of the demonstration of GA as an efficient tool to guide chemical synthesis-based problems should provide an incentive for further exploration of genetic algorithms for increasingly important problems based on chemical diversity. The current information for experimentally driven GA-based lead generation or optimization indicates that all the population members must be synthesized and screened (i.e., scored). However, as computational tools become more robust, and as members of a given generations are produced before one moves on to the execution phase (chemical synthesis), it may become possible to incorporate algorithms that rank the compounds in order of synthetic feasibility. Such an "ease of synthesis" penalty function then could be fluidly built into the overall fitness function during GA optimization. Other such terms might include bias against toxicophores or metabolic instability if those can be quantified and objectively computed. The possibilities are limitless.

In addition, we are increasingly witnessing creative applications of high throughout principles to the obtaining of physicochemical data: (aqueous solubility (35), log P determinations (36)), pharmacokinetic parameters via "N-in-1" cassette dosing strategies (37), and even toxicity screening (38) from milligram amount of samples in parallel. These experimentally determined values may be used as "fitness functions" to permit the use of GA-based lead development. Inasmuch as any "drug" possesses a diverse set of optimal parameters, we believe it is the discontinuous nature of these parameters that makes the GA-driven process particularly appealing for the lead development phase of drug discovery, which calls for the simultaneous optimization of several discontinuous functions.

In the last several decades, the pharmaceutical arena has witnessed the increasing need for new and effective antibiotics to combat resistant strains of microorganisms; and the area of AIDS is plagued by rapidly evolving mutations in the HIV protease in response to early inhibitors. Drug discovery must constantly deal with and respond to evolving biological systems. It is intriguing that genetic algorithms, based as they are on the principles of "genetics," may be particularly suited to finding solutions when the definition of "optimal" is constantly changing. We believe that this "adaptivity" of GA to respond to changes in the environment may allow us to use this tool to find solutions to these problems. For example, a GA-driven lead generation aimed at finding optimal solutions (identifying inhibitors) not just to one but to a host of HIV proteases with different mutations may be worth exploring. The area of receptor antagonists, has evolved from selectivity among the various classes (5HT vs. dopaminergic vs. histaminergic) to the discovery of subtypes and pursuit of subtype-selective drugs. This adaptation is trivial for an advanced GA optimizer. We are now poised perhaps on an age of seeking particular profiles ($5HT_2/D_2$, etc.), where combination of binding activity at a pluralities of receptor subtypes may be sought. Again, this is a simple task for an advanced GA optimizer. This line of thought may well be extended to finding an optimal solution for cancer therapy, where combinations of drugs that are diverse in mechanism/mode of actions, may provide an optimal outcome.

REFERENCES

1.	For excellent reviews see: (a) Gallop MA, Barrett RW, Dower WJ, Fodor SPA, Gordon EM. Application of combinatorial technologies to drug discovery. 1. Background and peptide combinatorial libraries. J Med Chem 1994; 37:1233–1251. (b) Gordon EM, Barrett RW, Dower WJ, Fodor SPA, Gallop MA. Application of combinatorial technologies to drug discovery. 2. Combinatorial organic synthesis, library screening strategies, and future directions. J. Med. Chem. 1994; 37:1385–1401, and references cited therein.

2. Pirrung MC, Jrlung C. J. Am. Chem. Soc. 1995; 117:1240–1245.

3. Martin EJ, Blaney JM, Siani MA, Spellmeyer DC, Wong AK, Moose WH. Measuring diversity: Experimental design of combinatorial libraries for drug discovery. J Med Chem 1995; 38:1431–1436.

4. There is tremendous effort in producing and screening chemically diverse compound libraries. However, the real interest for the pharmaceutical industry is the biological diversity that is embodied in it (i.e., biodiversity, not necessarily chemodiversity). The key interest for later is primarily from the point of view of intellectual property, that is, to assure that the new chemical entities (NCEs) remain proprietary.

5. Davis L. *Handbook of Genetic Algorithms.* New York: Van Nostrand Reinhold, 1991.

6. Forrest, S. Genetic algorithms: Principle of natural selection applied to computation. Science. 1993; 261:872–878.

7. Goldberg D. *Genetic Algorithms in Search, Optimization and Machine Learning.* Reading, MA: Adison-Wesley, 1989.

8. (a) Wagener M, Gasteiger J. The determination of maximum common substrate by a genetic algorithm: Application in synthesis design and for the structural analysis of biological activity. Angew. Chem. Int. Ed. Eng 1994; 33:1189–1192. (b) Wehrens R., Lucasius C., Buyden L., Kateman G. HIPS, a hybrid self-adapting expert system for nuclear magnetic resonance spectrum interpretation using genetic algorithms. Anal Chim Acta 1993; 277:313–324.

9. (a) Judson RS, Jaeger EP, Treasurywala AM, Peterson ML. Conformational searching methods for small molecules. II. Genetic algorithm approach. J Comput Chem 1993; 14:1407–1414. (b) Judson RS, Jaeger EP, Treasurywala AM. A genetic algorithm based method for docking flexible molecules. J. Mol. Struct. (THEOCHEM) 1994; 308:191–206.

10. Holland JH. Genetic algorithms. Sci. Am. 1992; 66–72.

11. Generally, in the computational paradigm utilizing GA, deletion and insertion are not used; one does not want to change the overall size of the chromosome and therefore, overall length of the bit strings.

12. Genesis v 1.2. Available by file transfer protocol:
ftp:aic.nrl.navy.mil

13. Since binary bit strings (0's and 1's) are being used, one need 2^5 (i.e., 32) bits to represent 20 amino acids. Therefore, some of the 20 amino acids are represented by more than one string. The choice of this bit degeneracy was selected at random, but once selected it was kept constant throughout the experiment.

14. Weber L. Evolutionary combinatorial chemistry: Application of genetic algorithms. Drug Discovery Today. 1998; 3:379–385.

15. The generations are referred to as Gen_i, where "*i*" is an interger. But in a true sense this initial randomly generation should be referred to as generation 0 (zero) as far as GAs are concerned, since there are no fitness functions that need to be evaluated by GA to provide the initial population of members.

16. Crossover is the single most important aspect that provides for optimum assurance to explore the gene population for selecting a set of more fit members.

17. Weber L, Wallbaum S, Broger C, Gubernator K. Optimization of biological activity of combinatorial compound libraries by a genetic algorithm. Angew. Chem. Int. Ed. Eng. 1995; 34:2280–2282.

18. Singh J, Ator MA, Jaeger EP, Allen MP, Whipple DA, Soloweij JE, Chowdhary S, Treasurywala AM. Application of genetic algorithms to combinatorial synthesis: A computational approach to lead identification and lead optimization. J Am Chem Soc 1996; 118:1669–1676.

19. Ator M, Beigel S, Dankanich T, Echols M, Gainor J, Gilliam C, Gordon T, Koch D, Kruse L, Morgan B, Olsen R, Siahaan T. Immobilized peptide arrays: A new and powerful tool in protease characterization. In R Hodges and J Smith, eds. *Peptides: Chemistry Structure and Biology. Proceedings of the 13th American Peptide Symposium.* 1994, pp. 1012–1016.

20. Singh J, Allen MA, Ator MA, Gainor JA, Whipple DA, Soloweij JE, Treasurywala AM, Morgan BA, Gordon TD, Upson DA. Validation of screening immobilized peptide libraries for discovery of protease substrates. J. Med. Chem. 1995; 38: 217–219.

21. mCl-t: a recombinant form of the human fibroblast collagenase and mSl-t: a recombinant form of the human fibroblast stromelysin. For details, see (a) Chowdhury SK, Vavra, KJ, Brake PG, Banks T, Falvo J, Wahl R, Eshraghi J, Gonyea G, Chait BT, Vestal CH. Examination of recombinant truncated mature human fibroblast collagenase by mass spectroscopy: Identification of differences with published sequence and determination of stable isotope incorporation. Rapid Commun Mass Spectrom 1995; 9(7): 563–569. (b) Brownell, J, Earley W, Kunec E, Morgan BA, Olyslager B, Wahl RC, Houck DR. Comparison of native matrix metalloproteinases and their recombinant catalytic domains using a novel radiometric assay. Arch Biochem Biophys. 1994; 314:120–125.

22. Average fluorescence is simply the sum of normalized fluorescence value for all samples divided by 60 (total number of samples N_{pop} in any generation).

23. Yokobayashi Y, Ikebukuro K, McNiven S, Karube I. Directed evolution of trypsin inhibiting peptides using a genetic algorithm. J. Chem. Soc. Perkin Trans. 1 1996; 2435–2437.

24. Eichler J, Houghten RA. Identification of substrate–analog trypsin inhibitors through the screening of synthetic peptide combinatorial libraries. Biochemistry 1193; 32:11035.

25. Sheridan RP, Kearsley SK. Using genetic algorithms to suggest combinatorial libraries. J. Chem. Inf. Comput. Sci 1995; 35:310–320.

26. DongXiang L, HuaLiang J, KaiXian C, RuYun J. A new approach to design virtual combinatorial library with genetic algorithm based on 3D grid property. J. Chem. Inf. Comput. Sci. 1998; 38:232–242.

27. (a) Raves ML, Harel M, Pang YP, Silman I, Kozikowski, AP, Sussman JL. Structure of acetylcholinesterase complex with the nootropic alkaloid, (–)-huperzine A. Nat Struct. Biol 1997; 4:57–63. (b) Vrland G. WHAT IF: A molecular modeling and drug design program. J. Mol. Graphics 1990; 8:52–56.

28. Walter DE, Hind RM. Genetically evolved receptor models: A computational approach to construction of receptor models. J Med Chem 1994; 37:2527–2536.

29. Gillet VJ, Willet P, Bradshaw. The effectiveness of reactant pools for generating structurally-diverse combinatorial libraries. J Chem Inf Comput Sci 1997; 37: 731–740.

30. (a) Glen RC, Payne AWR. A genetic algorithm for the automated generation of molecules with constraints. J Comput-Aided Mol Design 1995; 9:181–202. (b) Gillet VJ, Willet P, Bradshaw J, Green DV. Selecting combinatorial libraries to optimize diversity and physical properties. J Chem Inf Comput Sci 1999; 39:169–173. (c) Brown RD, Martin YC. Designing combinatorial library mixtures using a genetic algorithm. J Med Chem 1997; 40:2304–2314.

31. (a) Oshiro CM, Kuntz ID, Dixon JS. Flexible ligand docking using genetic algorithm. J Comput-Aided Mol Design 1995; 9:113–130. (b) Jones G, Willet P, Glen RC, Leach AR, Taylor R. Development and validation of a genetic algorithm for flexible docking. J Mol Biol 1997; 267:727–748. (c) Jones G, Willet P, Glen RC, Leach AR, Taylor R. Further development of a genetic algorithm for ligand docking and its applications to screening combinatorial libraries. In AL, Parrill and MR, Reddy, eds. *Rational Drug Design: Novel Methodology and Practical Applications,* ACS Symposium series 719. Washington, DC: American Chemical Society, 1999, pp. 271–291. (d) Jones G, Willet P, Glen RC. Genetic algorithms and their applications to problems in chemical structure handling. In Harry C, ed. Proc Int Chem Inf Conf 1994, pp. 135–148.

32. (a) Nelen MI, Eliseev AV. Use of molecular recognition to drive chemical evolution: mechanism of an automated genetic algorithm implementation. Chem Eur J 1998; 4(5):825–834. (b) Willet P. Genetic algorithms in molecular recognition and design. Trends Biol Technol 1995; 13:516–225. (c) Clark DE, Westhead DR. Evolutionary algorithms in computer-aided molecular design. J Comput-Aided Mol Design 1996; 10:337–358. (d) Gehlhaar DK, Verkhivker GM, Rejto PA, Sherman CJ, Fogel DB, Fogel LJ, Freer ST. Molecular recognition of the inhibitor AG-1343 by HIV-1 protease: Conformationally flexible docking by evolutionary programming. Chem Biol 1995; 2:317–324.

33. Various classes of MMP inhibitors have been reported in the literature. (a) For hydroxamate series, see: Singh J, Conzentino P, Cundy K, Gainor J, Gordon T, Johnson J, Morgan B, Whipple D, Gilliam C, Schneider E, Wahl R. Relationship between structure and bioavailability in a series of hydroxamate based metalloprotease inhibitors. BioMed Chem Lett 1995; 5:537–542, and Johnson WH, Roberts NA, Borkakoti NJ. Collagenase inhibitors: Their design and potential therapeutic use. Enzyme Inhib. 1987; 2:1–22. (b) For *N*-carboxyalkyl series, see: Chapman KT, Kopka IE, Durette PL, Esser CK, Lanza TJ, Izquierdo-Martin M, Neidzwiecki L, Change B, Harrison RK, Kuo DW, Lin T, Stein RL, Hagmann WK. Inhibition of matrix metalloproteinases by *N*-carboxyalkyl peptides. J Med Chem 1994; 36:4293–4301. (c) For phosphonate series, see Bartlett PA, Marlowe CK. Possible role for water dissociation in the slow binding of phosphorus-containing transition-state analogue inhibitors of thermolysin. Biochemistry 1987; 26:8553–8561 and Bird J, DeMallo RC, Harper GP, Hunter DJ, Karran EH, Maekwell RE, Miles-William AJ, Rahman SS, Ward RW. Synthesis of novel *N*-phosphonylalkyl dipeptide inhibitors of human collagenase. J Med Chem 1994; *37*:158–169.

34. The sequence: GPST-YT, chosen for P'_1 variations, was observed in all five generations, as described in Section III.B. This sequence is also processed selectively by stromelysin, as shown in Fig. 14.

35. Lipinski, CA. Solubility screening in an early drug discovery setting. Presentation at the IBC sponsored meeting, Early ADME and Toxicology in Drug Discovery: Techniques for Accelerating and Optimizing Drug Candidate Selection, Berkeley, CA, November 12–13, 1998.

36. Singh J, et al. Unpublished work.

37. Berman J, Halm K, Adkinson K, Shaffer J. Simultaneous pharmacokinetic screening of a mixture of compounds in dog using API LC/MS/MS analysis for increased throughput. J Med Chem 1997; 40:827–829.

38. Todd MD, Xiaodong L, Stankowski LF, Desai M, Wolfgang GHI. Toxicity screening of a combinatorial library: Correlation of cytotoxicity and gene induction to compound structure. J Biomol Screening 1999; 4:259–268.

19

Enhancing the Drug Discovery Process by Integration of Structure-Based Drug Design and Combinatorial Synthesis

Donatella Tondi and Maria Paola Costi

Università degli Studi di Modena e Reggio Emilia
Modena, Italy

I. INTRODUCTION

In the last decade, drug discovery has changed radically. This change has been driven by technological advances in biological science, rapid follow-up activities in chemistry, and powerful computer techniques. Of these, structure-based drug design (SBDD) and combinatorial chemistry (CC) have proved to be successful methodologies. Knowledge of the three-dimensional (3D) structure of a biological target can help to identify a lead compound and subsequently guide its optimization. Combinatorial chemistry, on the other hand, is a synthetic strategy whereby molecular diversity can be rapidly reproduced in the laboratory with the specific aim of targeting a biological system.

To accelerate the identification of hits suitable for lead development and then lead optimization, the whole discovery process should be as fast as possible; technologies with the necessary features are grouped under the term HTT (high throughput technology). This approach involves the rapid design, synthesis, and testing of large numbers of molecules in an iterative fashion. Critical to the whole process is the design step, which should be consistent in order to avoid waste of time, money and human effort (Fig. 1).

The identification of leads take place in the "combinatorial section," where library design and chemistry optimization are geared to synthesis. The final step, that of biological evaluation, is high throughput screening (HTS), and from here the highest scoring compounds are promoted to the following cycle. The process is iterative and tends toward convergence when the hits constitute a good lead for optimization in the "medicinal chemistry section." In this phase, the approach is similar to that adopted previously, but on a smaller scale: a reduced number of molecules is processed at a higher level of refinement, the compounds being studied sequentially, one by one, following classical precepts of design and chemistry. Detailed computational methods and traditional medicinal synthetic chemistry are applied.

The integration of combinatorial chemistry with the structure-based methods (1) is achieved by including the combinatorial synthetic approach in the structure-based drug design cycle. Studies from several groups suggest that structural information can usefully impose constraints on the diversity available in combinatorial exploration, leading more rapidly to more potent inhibitors. In some cases, the 3D structure of the biological target can be used as a guide to direct the combinatorial strategy; in other cases, the 3D structure can be used as the final template into which the best inhibitor can be fitted. In the latter case, the structural basis for the molecular recognition of the compounds is taken into account and accelerates library design. Rational library design provides focused molecular state depiction, and only the highest scoring compounds are synthesized. This approach is possible, provided the preliminary structural information on the addressed target is to hand. When this information is not available, the library design is based on structure–activity relationship (SAR) analysis from previous work; the X-ray structure is part of the final cycle and serves to explain the basis for inhibitor–target interaction and eventually to continue the process in an iterative fashion. In both cases, the library design is the bridging point between the CC and SBDD.

II. COMBINATORIAL CHEMISTRY

Combinatorial chemistry is the most important recent advance in medicinal chemistry, in view of its great potential for speeding up the drug discovery process. Combinatorial library methods have been successfully used in the past few years not only to generate new leads for a specific target but also as a powerful tool for the optimization of an already known lead.

The initial development of combinatorial chemistry derived from solid phase synthetic techniques (2) and peptide synthesis (3). In moving from peptide synthesis to the synthesis of small organic molecules, a wide range of chemistry

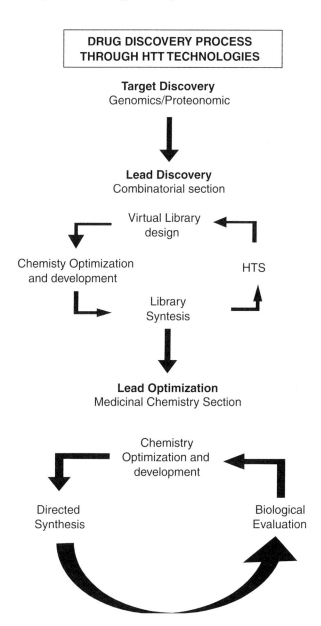

Figure 1 Drug discovery process adopting high throughput technologies.

could be adopted by applying retrosynthetic strategy principles, in which the desired final structure is dissected via several steps suitable for in-parallel/serial combinatorial synthesis, while the core structure (shared scaffold) is synthesized off-line by traditional organic chemistry (4). The retrosynthetic combinatorial analysis finds its software application in the program RECAP (5).

At first combinatorial chemistry involved the design and synthesis of a huge number of molecules to cover the synthetic space available for a specific molecular scaffold set and to generate as much diversity as possible. However, the need for a rational strategy in the development of the drug discovery process finally arose: the approach turned out to be uneconomical owing to frequent redundancy in molecular state description and the poor analytical control of the products. New techniques for the design, synthesis, and purification of molecules continue to be reported, and there are a number of excellent reviews providing background information on combinatorial chemistry (6–15).

Combinatorial library methods, applicable to a small-molecule library, include parallel solid-phase or solution phase synthetic library methods requiring deconvolution, the one-bed/ one-compound library method, and synthetic library methods using affinity chromatography selection.

Of these, the in-parallel synthesis of a library of spatially separated compounds has the advantage that the chemical structure of each compound is predetermined. Each molecule, analog of a core structure, is in fact synthesized on a solid support in a spatially addressable format. Synthesis and screening of the libraries are performed in parallel through binding assays or solution assay, depending on whether the screening is performed on the final molecule still attached to the solid support.

The techniques that allow for synthesis in parallel are multipin technology (16), SPOTS membrane (17), light-directed peptide synthesis on chips (18), and Diversomer technology (19). In view of the benefits of such synthetic techniques, new combinatorial chemistry methods are generally oriented toward the in-parallel synthesis of a restricted number of separate compounds, rather than a mixture of thousands of molecules, with obvious screening limitations. The major limitation of the in-parallel approach is that only a relatively small number of compounds can be synthesized in parallel, resulting in a limited library. It is therefore crucial to identify the optimum subset of a library for synthesis. However, even though it is relatively easy, given a starting template, to design a large library by simply combining all the sets of commercially available starting compounds, there are practical reasons for not synthesizing every compound: it is wasteful of time and money, and the information coming from a such library may not be proportional to the size of the entire set.

Of the numerous methods developed to select the right subset of molecules to synthesize, the integration of modeling and structure-based drug design with combinatorial chemistry techniques has proved to be successful. Combinatorial chemistry methodology has in fact reached such a level of productivity that it is increasingly incorporated into the iterative drug design and discovery cycle. Many examples of combinatorial libraries applied to small molecules are known in the literature and refer to benzodiazepines, diketopiperazines, isoquinolinones, 1,4-dihydropyridines, pyrrolidines, hydantoins, imidazoles, triazines, acylpiperidines, β-lactams, and many others (11).

Despite the availability of several excellent reviews providing a clear background to combinatorial chemistry and solid phase synthesis, however, only recently has it been shown how the combinatorial and in-parallel synthesis of small libraries can be positively combined with computational and molecular modeling methods to assist in the discovery of new lead candidates. In the last couple of years this approach was applied to various targets, as shown by the examples discussed in this chapter.

III. STRUCTURE-BASED DRUG DESIGN

Structure-based drug design takes advantage of the structural information about a biological target to design drugs that can fit into the target. The structural information can derive from the experimental determination of an X-ray crystal structure or an NMR structure, from a homology modeling approach, or from a calculated structure. Many reviews have been published on this topic (20–27). The technology that paved the way for structure-based methodology was the development of recombinant DNA techniques, which made large amount of proteins available for crystallographic studies. The number of available 3D structures of unbound and ligand binding proteins has since grown exponentially; nowadays, over 10,000 X-ray crystals are available in the Protein Data Bank.

The reliable prediction of the energy of interaction between a ligand and its biological target is the first, challenging step in the design of a new drug, and a large number of variables need to be taken into account. One weak point is our limited understanding of the basic principles of the molecular recognition process that guides the dynamic reactions within the biological systems, and much research in this area is currently underway (28).

Another issue is the accurate description of the conformational change that occurs in the target macromolecule upon ligand binding. The computation of this change is limited to the closest residues surrounding the ligand binding site, with the rest of the protein kept rigid (29,30). This limitation is due to computational restrictions that will be overcome in the future with more powerful algorithms and computers. Increasingly complex biological systems are being computationally studied to address the issue of protein–protein interface interactions and protein surface–ligand interactions.

Closely related to the conformational problem is the influence of solvation effects on specific and nonspecific binding. In fact, the process of solvation and desolvation upon ligand binding affects the exposed accessible surface area. This aspect is very important and provides key elements for further drug development. Moreover, thermodynamic description of the drug–protein interactions suggests the relative importance of the energy terms contributing to binding affinity and strongly supports the structural data (31,32).

The formation of protein–ligand interactions competes with the rupture of interactions with water molecules. Both protein and ligand are solvated before complex formation. They lose part of their solvation shell upon binding, a process that involves enthalpic and entropic contributions. Hydrogen bonds are broken and new ones are formed. Neither lipophilicity nor hydrogen bond network formation, considered as separate contributions, can explain why a molecule displays high affinity toward a target (32). A combination of both, plus other properties, accounts for the overall result.

The *modular approach to drug design* is based on the supposition that free binding energy can be considered to be the sum of the binding effects of the fragments deriving from the dissection of the ligand. This principle has had a strong impact on structure-based studies: for example, modeling programs like LUDI are based on it (33). Also, substructure dissection on a lead and subsequent combination with different fragments is a widely used method that has found rational application in library design and combinatorial chemistry.

Two distinct approaches to structure-based drug design can be recognized: the virtual screening of small-molecule databases and the refined detailed study of the drug–target interactions. Virtual screening affords the rapid and automatic analysis of the interaction of a large number of molecules of a database evaluated through a scoring function that assigns a score (not a true energy value) to the database components; this is important in order to accelerate the screening process. The model of the selected target plays a critical role: a target in its bound conformation is usually the best choice, but it leaves the subsites of the available conformational space of the target only partly accessible. This means that the exploration of the inner available binding surface is not complete unless more conformations of the target are considered (multiple docking). This issue is very important when the structure-based approach is combined with the combinatorial technique in the HTT process (34,35). Combinatorial libraries can be designed and selected on a structural template, and the selection process is therefore highly dependent on the available inner surface: compounds potentially capable of binding upon conformational change can be lost, in principle. Careful attention should be paid to the selection criteria. When unpredictable information about the target is lost, diverse approaches remain valid, since they offer the advantage of a wider range of lead candidates.

IV. LIBRARY DESIGN AS A BRIDGING POINT BETWEEN CC AND SBDD

A combinatorial chemical library is as an ensemble of molecules whose preparation and screening determine the so-called hit rate, hence the success or failure of the design (36). Preliminary information is required to develop the whole library in a rational manner. Structure-related properties based on SAR analysis or on a 3D model of the biological target are necessary to constrain the huge number of possible molecules (Fig. 2).

It is important to direct the construction of the library along the most accessible and rapid synthetic pathway. The analysis of an ensemble of molecules whose available synthetic space is very large (i.e., suitable for structural and functional modulation at more than one site) is essential to the selection of the basic scaffold. This scaffold must be referred to not as a true pharmacophore, but as a structure that precedes pharmacophore generation.

Other considerations are crucial to library construction, such as the distribution of the molecules within the library, which is a topological issue. Related to this is the neighborhood concept, which serves to avoid redundancy and to keep the number of molecules to a minimum (37). In principle, a library can be an ensemble of all the molecules designed from a starting scaffold, with a harmonic modulation of the different molecular states representing the bioactive ligand; a huge number of molecules can therefore be candidates. A critical analysis of these molecular states should be performed, with only some highly representative molecules kept in the library. On this basis, it is possible to select a molecule from among a large number of analogs because it is the best compound that samples a particular molecular state. This leads back to the homogeneous distribution of the molecules in the library: a molecule chosen to be part of a library must represent a family of molecular, steric, and electronic states. If the molecule is well chosen, it is not necessary to include analogs, thereby avoiding the unnecessary repetition of similar compounds.

Molecular similarity/diversity concepts must be considered. There is a subtle dividing line between similarity and diversity properties, which are strongly

Figure 2 Library design as a bridging point between CC and SBDD.

dependent on the particular context and cannot be defined outside it: two molecules that one target considers diverse may be considered similar by another target. The macromolecule, on the molecular recognition basis, "decides" which molecules are similar and which are diverse. The ones that can bind with similar binding energy and in the same binding site are supposed to be very similar. The degeneracy of the binding energies (i.e., same target can bind two different molecules with the same binding energy) is not in itself a sufficient criterion for the similarity concept, and a topological description of the binding site is necessary. An example of this issue is the structures of three thrombin inhibitors: their crystallographically determined binding geometry has been superimposed, and it is clear that they share a common binding orientation and molecular recognition properties, despite having dissimilar structures and very different molecular descriptors (34) (Fig. 3).

Why should we consider testing the same molecules for two different targets? Because of the screening concept on which the initial development of a drug discovery project is based. At the outset, a whole, roughly selected database is tested virtually, and databases of commercially available compounds are increas-

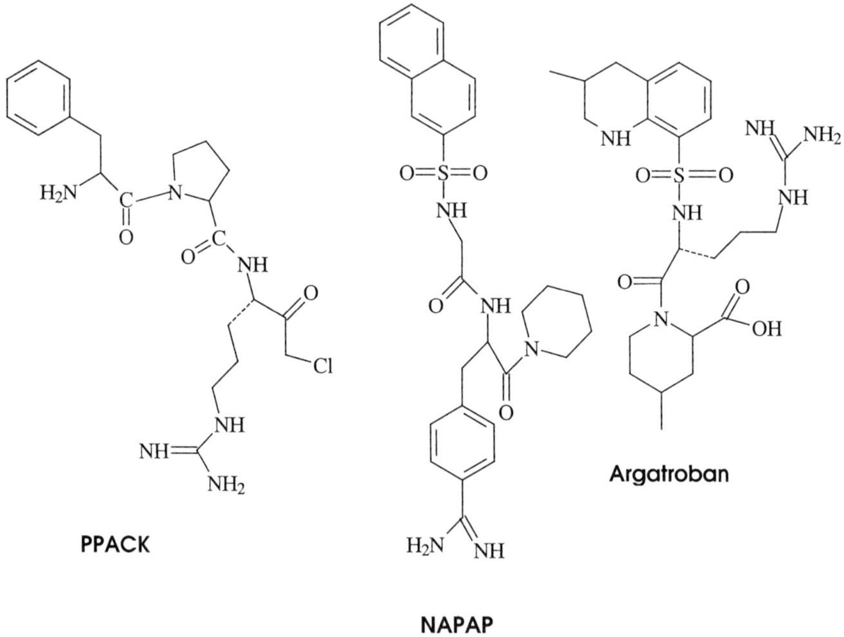

Figure 3 The chemical structure of three thrombin inhibitors. Despite their differing chemical structures and molecular properties, the molecules share a common mode of molecular recognition.

ingly accessible to many users. Therefore, in principle, the same molecules can be tested for different targets. This is interesting because the results of this "natural" selection can demonstrate that a molecule is just a probe for the enzyme, and specificity is a relative property of a molecule but an intrinsic property of a target macromolecule (38).

It can be very complicated to design a library rationally, since all the foregoing variables must be taken into account. Powerful programs and computer technology combined afford a solution to this problem (1,39,40). Library design is an iterative process within the drug discovery program, and the library dimension and the design methodology can change during the process, depending on the knowledge acquired and the subsequent goal to be achieved (Fig. 4). The iteration of library design is a common feature of each logical process aimed at the discovery of a new drug. At the beginning of the drug discovery process, large numbers of compounds are part of both a virtual and a practical library. The screening methodologies are applied in this initial step, and, depending on the biological activity results, only a few molecules can be promoted to the second step. This second step requires a new library to be constructed, but with a lower, and more focused, number of compounds (41,42).

The final selection step relies on the strong relationship between the library design based on 3D structure information and the *molecular recognition principles* (Fig. 5). We cannot refer to a particular bioactive electronic structure or conformation; rather, we must indicate a bioactive molecular state. The different levels of complexity of the problem have been described (28). A receptor or an enzyme does not recognize particular atoms or groups; instead, it interacts with the (electrostatic- and orbital-based) electronic properties projected by a certain geometric arrangement of these atoms. Bioisosterism affords an idea of how similar two groups are in terms of their receptor binding properties.

Similarity/diversity measures are widely applied for rational set selection in the analysis of a combinatorial library. Approaching the problem from the small-molecule ligand viewpoint is relatively easy in terms of molecular properties, but it does not give a complete picture. The macromolecular receptor must be described through its available molecular states; despite continuing efforts to study the behavior of the macromolecules in solution by means of computational models, such a description is at present problematic.

The aspect of the library design concerning the high affinity lead/drug like molecule design deserves particular attention. Structure-based drug design has proved to be extremely successful and, if it is applied in iterative fashion, tightly bound ligands can be generated. Apart from potency and chemical novelty, additional factors usually requiring attention are the so-called ADME factors (absorption, distribution, metabolism, and excretion). Experience indicates that any single feature of a molecule can be optimized, but this often requires the preparation of many hundreds of potential candidates. ADME considerations are rarely used

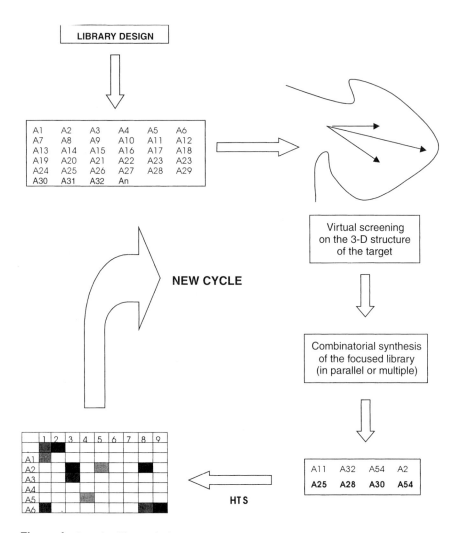

Figure 4 Iterative library design.

to design a combinatorial library. Discouraging results are often shown in the pharmacokinetic properties of lead compounds. Such molecules are suitable for defining molecular ligand–target interactions (maybe drugs for parenteral administration to treat acute, resolvable diseases if they are not toxic), but they are not useful for many common purposes (1,43). Consequently, medicinal chemists' efforts are centered on discovering methods for the synthesis of libraries of more druglike compounds.

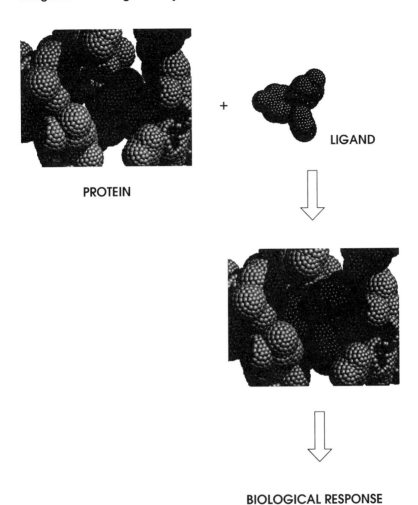

PROTEIN

LIGAND

BIOLOGICAL RESPONSE

Figure 5   Molecular recognition: protein–ligand interaction leads to biological response.

How can libraries be designed that are not only structurally diverse but also contain molecules with good bioavailability and metabolism, and possess a suitable pharmacokinetic and toxicological profile? The difficulties surrounding this question arise from our basic inability to understand clearly why a tightly binding ligand is a drug, while a close analog is not. We still do not know the laws underlying this complex process, namely, the pharmacokinetic behavior of a molecule. There are some well-recognized structures that look more like drugs, such as the heterocyclic compounds. There are also certain ranges of physicochemical and

other properties that are preferred in a lead. Many strategies, from the database filtering of commercial chemical databases to complex algorithms based on selected
molecular descriptors, are being developed to resolve this problem (44).

V. EXAMPLES FROM THE LITERATURE

Several cases in which combinatorial library synthesis successfully complements
structure-based drug design techniques in drug discovery are presented and discussed in the subsections that follow.

A. Discovery of Potent Nonpeptide Inhibitors of Cathepsin D

To discover potent, nonpeptide inhibitors of the aspartyl protease cathepsin D, Ellman and collaborators at the University of California at Berkeley applied SBDD
in conjunction with combinatorial chemistry (41).

   The aspartyl proteases are ubiquitous endopeptidases that use the aspartic
acid residues to catalyze the hydrolysis of amide bonds through tetrahedral intermediates. They have been implicated in a number of pathologies and represent important therapeutic targets. Ellman and coworkers have been conducting detailed
studies on three aspartyl proteases: cathepsin D (Cat D), plasmepsin I (Plm I), and
plasmepsin II (Plm II). Mimetic isosteres of the tetrahedral intermediates of these
enzymes are statin, hydroxyethylamine, and hydroxyethylene (Fig. 6), and drug

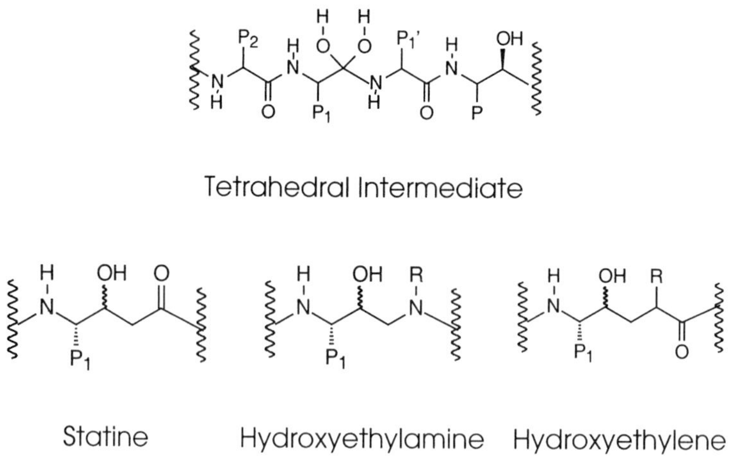

Tetrahedral Intermediate

Statine   Hydroxyethylamine   Hydroxyethylene

Figure 6   Isosteres that mimic the tetrahedral intermediate of the amide bond hydrolysis
catalyzed by aspartyl proteases.

discovery has been mainly directed at developing new nonpeptide inhibitors of the enzymes.

More particularly, Cat D is a lysosomal aspartyl protease, with well-determined structural and chemical properties, but a less clearly defined biological role. In stratified epithelia, the chronology of cathepsin D activation and degradation can be connected with stages of cellular differentiation. Cat D is also believed to be associated with proteolytic processes leading to local invasion and seeding of tumor cells in breast cancer. Moreover, it has been reported that Cat D may be involved in the proteolytic cleavage of amyloid precursor protein in the brains of individuals with Alzheimer's disease (44–47).

In their work the authors started from the mimetic isostere of the transition state analog, hydroxyethylamine, which has several positions capable of coupling to many different chemical groups, allowing for the introduction of a large number of variations in the starting molecule (Fig. 6). They identified three specific sites suitable for elaboration. Primary amines were used to introduce diversity at one site, while carboxylic acid and sulfonyl chloride introduced various groups at the second and third sites (Fig. 7). The combinations of all amine and acylating agents commercially available would have generated a library of 10 billion molecules, whose synthesis and subsequent evaluation would have involved considerable cost and effort.

The authors opted for the design of two smaller libraries, using two different approaches in the choice of building blocks. The first method generated a *diverse* library, since the building blocks necessary for synthesis were selected in terms of maximum diversity. The second method referred to the available crystal structure of Cat D–pepstatin to select building blocks for a *directed* library. For the generation of both libraries, the authors used commercially available building blocks (amines, carboxylic acid, sulfonyl chloride, and isocyanates, MW < 275 Da), selected from the Available Chemicals Directory (ACD). Others factors such as cost and availability were taken into consideration. The building blocks were also chosen for their reactivity, to ensure high yields of coupling at each position in the library. The final list consisted of 2600 compounds that would have provided over a billion molecules, and the authors turned to computational methods to reduce this number.

For the *diverse* library, the list of suitable building blocks was reduced by clustering all the compounds on the basis of their similarity through the Jarvis–Patrick algorithm, using the Daylight connectivity measure of similarity. In the case of the *directed* library, the molecular modeling program CombiBuild was used to run a conformational search. The scaffold was first oriented in the binding site, assuming it could adopt a binding orientation similar to that adopted by pepstatin in the X-ray complex with Cat D. An initial conformational search found four possible families of orientation for the scaffold. CombiBuild performed a conformational search on all possible components at each variable position on

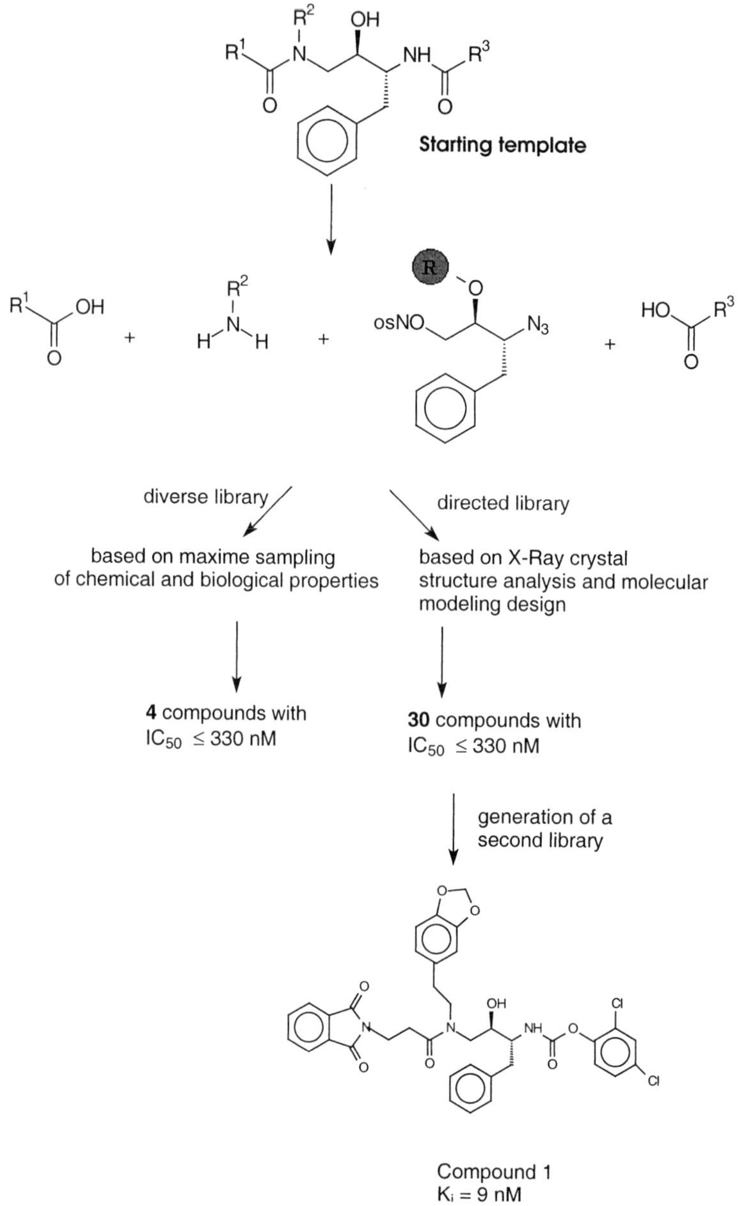

Figure 7 Discovery of potent inhibitors of cathepsin D through iterative structure-based drug design and in-parallel synthesis. Comparison of the diverse and direct library design approaches.

each family of orientation and scored the components according to their potential interaction with Cat D. Based on their AMBER score, the 50 best components at each position and for each family were selected and then merged for all three variable positions.

The synthesis of each library was accomplished through solid phase, in-parallel synthesis of a total of 1000 compounds for each library. The scaffold was anchored to polystyrene beads and the libraries synthesized in a spatially separated array using a 96-well filter apparatus. Each compound in the libraries was screened for inhibitory activity against Cat D in high throughput fluorometric assay tests performed in the reaction vessels.

In the first round of synthesis the directed library yielded a higher number of active compounds, the best of which had a K_i of 78 nM; the diverse library provided inhibitors four times less potent. The most potent compound from the directed library was then rapidly optimized through the generation and the screening of a second library. This second effort led to compound **1** with a K_i of 9 nM (Fig. 7).

In their project, Ellman et al. set out to address the problem of how to maximize the impact of chemical library design on the drug discovery process. In particular, their paper addresses the question of whether a library based on target structure information can be comparable with a library based only on diversity criteria. The *directed* structure-based combinatorial library allowed a larger area of the inner target surface to be explored, producing an increased number of active inhibitors with respect to the *diverse* library. The ability to combine structure-based drug design with combinatorial chemistry overcomes many structure-based design limitations, such as poor structural information and unpredictable conformational change, and focuses the combinatorial synthesis process.

B. Inhibitors of Malarial Aspartyl Protease Plasmepsin II

After their study of the Cat D inhibitors, Ellman and coworkers reported the development of a general method of using solid phase synthesis to incorporate diverse functionalities at all variable sites of the intermediate hydroxyethylamine with a view to further optimization of the inhibitors for Cat D, Plm I, and Plm II. The secondary alcohol represents the only invariant part of the inhibitor structure and serves as the linkage to the resin (Fig. 6) (48).

The identification of novel inhibitors of Cat D with an optimized side chain at the P_1 pocket and the development of appropriate solid phase synthetic methodologies were the basis for a subsequent outstanding paper in which the coapplication of SBDD and combinatorial chemistry guided the discovery of selective plasmepsin II aspartyl protease (Plm II) inhibitors (30). The authors deal with the discovery of potent, low molecular weight, nonpeptide inhibitors of malarial aspartyl protease Plm II. Interest in inhibitors of plasmepsin aspartyl proteases of

Plasmodium falciparum with potential chemotherapeutic activity has increased because of the alarming increase in the resistance of the malaria parasite. Plm II complexed with pepstatin A, a peptide mimetic inhibitor of the enzyme from *P. falciparum,* was the starting crystal structure used in the modeling work.

Iterative library design was the central topic of the work. To constrain the number of possible compounds in each library in a rational manner and also to obtain molecules with optimized requisite properties, the authors focused on iterative library design. They paid great attention to the improvement of binding affinity and pharmacokinetic properties, taking into account the general rules on bioavailability such as low molecular weight, good hydrophobicity range, and low serum albumin binding. The paper is rigorous, and step-by-step control of the modeling is achieved by using the available X-ray structure to address the iterative library design. The library synthesis was run in solid phase fashion.

First of all, authors took advantage of the close homology of Plm II, for which no inhibitor has so far been identified, with Cat D, for which a library of a total of 1039 compounds had been synthesized (41). These libraries were screened against Plm II and led to the identification of two leads: compounds **2** and **3** were validated as submicromolar inhibitors as well as being potent inhibitors of Cat D (Fig. 8). Their common structure represented the starting core suitable for optimization through the library design of more active and selective Plm II inhibitors. To reduce the size of the possible libraries, the authors chose three sites in the starting scaffold, R^A, R^B, and R^C, and derivatized one site at a time. The P_1 site was also included in a final step (Fig. 9).

Library R^A was constructed by truncating the pepstatin A molecule and introducing a benzyl moiety into the structure. Using AMBER, the scaffold thus modeled was minimized in the Plm II active site and used for R^A library generation. The selection of possible reagents for each library was carried out using toolkits from a Web-based program, UC Select, and Daylight. A variety of acylating agents for introduction into R^A were chosen from the ACD on the basis of their low molecular weight, chemical reactivity, and cost. The final lists consisted of 4093 acylating agents. Cartesian coordinates were generated, and each building block was oriented in the active site by attaching it to the amide (R^A side chain) of the scaffold. The other radicals, R^B and R^C, were kept constant. A total of 4086 different side chain compounds were generated by means of the "anchor and grow" algorithm from DOCK 4.0. Each compound was then scored on the basis of inter- and intramolecular van der Waals and electrostatic terms. The highest scoring compounds were chosen visually for their conformation and number of hydrogen bonds with the enzyme, while the selected molecules were clustered according to their metric similarity. The 74 compounds selected for synthesis were screened through enzymatic assays. Compound **4** bearing a 4-(benzyloxy)-3.5-dimethoxybenzoic acid side chain, was chosen as the highest scored compound (K_i 100 μM). A second 12-member library (library 2) focusing on compound **4**

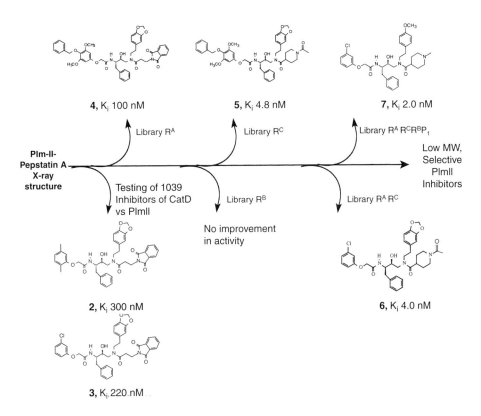

Figure 8 Plm II inhibitor optimization. Combinatorial elaboration at multiple sites leads to potent molecules with desired characteristics.

was prepared by synthesizing the neighborhood molecules, but no improvement in the biological properties was obtained (Fig. 8).

The scaffolds for the R^B and R^C libraries were generated as for R^A by truncating pepstatin A. A hydroxyethylamine scaffold was then built, and leucine and serine side chains were attached to R^B and R^C to give the starting scaffold for libraries R^B and R^C. Reagents for R^B were chosen on the basis of their diversity: 473 primary amines were selected for incorporation in the R^B site, while R^C was kept fixed as a 3-phthalimidoproprionic acid side chain. The docking and the clustering procedures for side chains R^B were similar to those in R^A library design. None of the synthesized compounds showed any increase in inhibitory activity with respect to the starting R^B side chain scaffold, and the 3,4-(methylenedioxy)phenethylamine group was kept in the following libraries.

Difficulties were encountered in trying to model side chains into the S1' and S2' pockets (for the R^B and the R^C side chains, respectively) of Plm II in the pep-

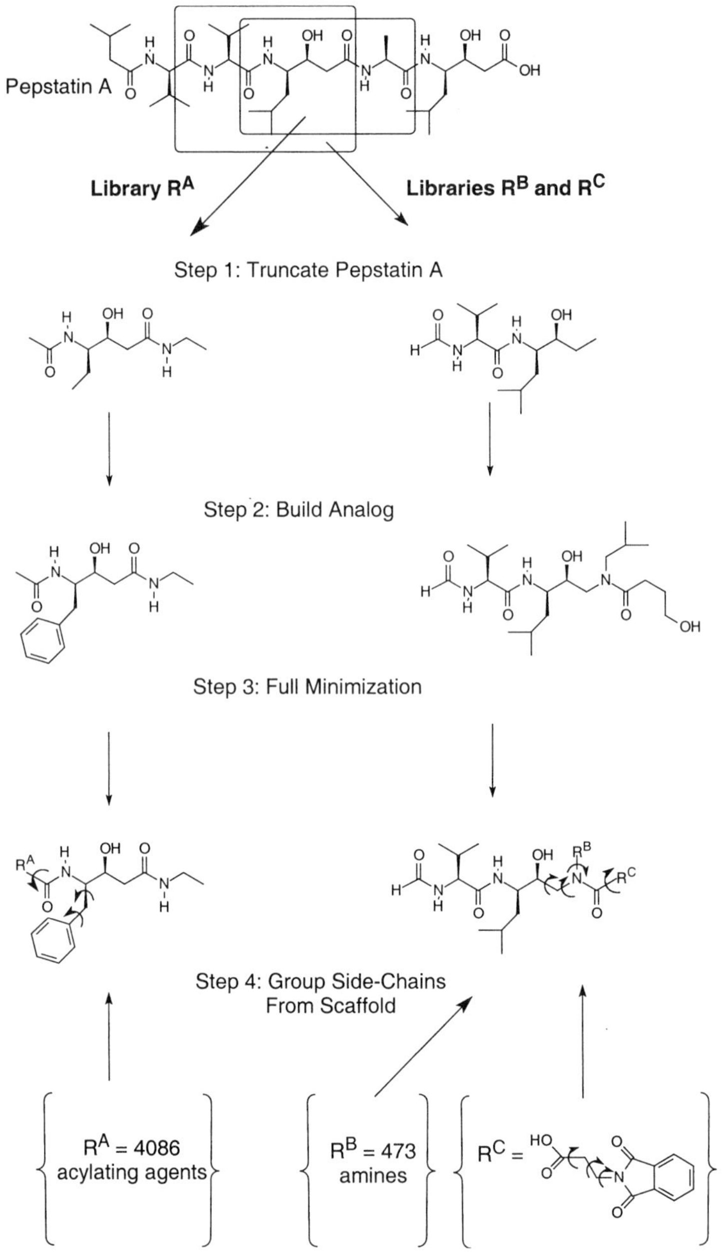

Figure 9 Structure-based design strategies in the development of Plm II inhibitors. (From Ref. 30, fig. 9, p. 1437. Copyright November 30, 1999, American Chemical Society.)

statin A–Plm II crystal structure. The X-ray structure of Plm II–pepstatin A is conformationally unable to accommodate easily new chemical groups at the S1' and S2' levels. The X-ray crystal structure of the Plm II–compound **4** binary complex was obtained next, and it confirms the enzyme's plasticity and ability to relax and bind the whole molecule.

Library R^C (library 4) consisted of 44 compounds, and, owing to the difficulty in the modeling, only one diverse set was incorporated. Upon screening, compound **5**, which incorporated an *N*-acetyl-nipecotic acid side chain, displayed a K_i of 4.8 nM, 20 times more potent than compound **4** (Fig. 8).

The latter two libraries were based on multiple site modifications of the initial scaffold. The first one, consisting of 80 compounds, varied R^A and R^C simultaneously and kept the overall molecular weight rigorously low. Elements of the R^A library and the R^C library were combined to investigate cooperativity effects between the two sites. Compound **6** was identified as a molecule slightly more potent than **5** but with a significantly reduced molecular weight (650 Da vs. 752 Da).

A final, sixth library examined all three sites of the hydroxyethylamine scaffold simultaneously. In addition, the benzyl P_1 side chain was converted into an isobutyl moiety. However, this modification did not improve affinity, since, with respect to the starting benzyl P_1 side chain, the isobutyl group lost three interactions with residues F111, F120, and the β-C of S79. Of the several compounds synthesized in the final, most focused, library, compound **7** was a highly potent and selective Plm II inhibitor. Moreover, this and other analogs had low molecular weight and low human serum albumin binding affinities; in cultured parasite-infected human erythrocytes, they proved to be poor micromolar inhibitors of parasite growth.

Upon applying an interesting iterative small-molecule library design with synthesis and evaluation sequence, several nanomolar inhibitors of Plm II were identified. Moreover, the approach applied allowed the identification of selective and active molecules. The integration of SBDD with combinatorial in-parallel chemistry was successful. Finally, we stress the importance of taking into account previously synthesized libraries: if appropriate, they afford useful starting information. In the present study, the application of SBDD methodologies in the design of a druglike molecule, not only a more potent enzyme inhibitor, was successful.

C. Thrombin Inhibitors

Thrombin, a serine proteinase, is one of the most actively studied enzymes. The interest in thrombin arises from its central role in blood coagulation: it catalyzes the conversion of fibrinogen into fibrin and also converts factor XIII into factor XIIIa, which crosses binding fibrin. It is known that thrombin inhibitors have anticoagulant effects, which are important in the treatment of pathological states such thrombosis and stroke. Although many potent peptidomimetic inhibitors of thrombin are known, members of this class of known molecules always present poor pharmacokinetic properties such as problematic oral absorption and short plasma half-life; the development of active thrombin inhibitors has consequently remained a major goal for medicinal chemists.

Researchers at Merck have been working on the development of an orally potent thrombin inhibitor. Brady et al., starting from two previously developed peptidomimetic structures displaying poor oral absorption (Fig. 10, compounds **8** and **9**) set out to discover new thrombin inhibitors with improved pharmacokinetic properties (49). These authors based their studies on the analysis of the X-ray crystal structure of compound **9,** a 1-(N-methyl-D-phenylalanyl)-pyrrolidine-2-carboxylic acid *trans*(4-aminocyclohexyl)methyl amide derivative, bound to thrombin. The crystal structure showed how the aminocyclohexyl group is located in the S$_1$ subsite and the pyrrolidine in the S$_2$ subsite, while the N-methyl-D-phenylalanyl residues bind into a lipophilic pocket (S$_3$) of the enzyme. The binding orientation thus suggested the possibility of introducing molecular diversity at the P$_3$ site of the starting compound through several hydrophobic groups that could be accommodated in the S$_3$ subsite, thereby increasing the binding affinity.

A brief SAR generated a more active compound, the diphenyl derivative of compound **9** (compound **10,** Fig. 10) on which molecular modeling studies were also run. The results showed one phenyl ring, the one in common with compound **9,** binding in the same position as compound **9,** while the second phenyl was oriented at the beginning of site S$_3$, facing the solvent interface but close enough to the side chains Ile174 and Glu217 for apolar interactions. However, the pharma-

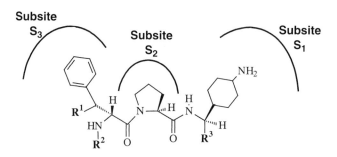

Compound 8. R^1=H; R^2= CH$_3$; R^3= CO-CONHCH$_3$ K$_i$= 0.09 nM
Compound 9. R^1=H; R^2= CH$_3$; R^3= H   K$_i$= 4.9 nM
Compound 10. R^1=C$_6$H$_5$; R^2= H; R^3= H   K$_i$= 0.1 nM

Figure 10 Lead compounds for the design of new thrombin inhibitors.

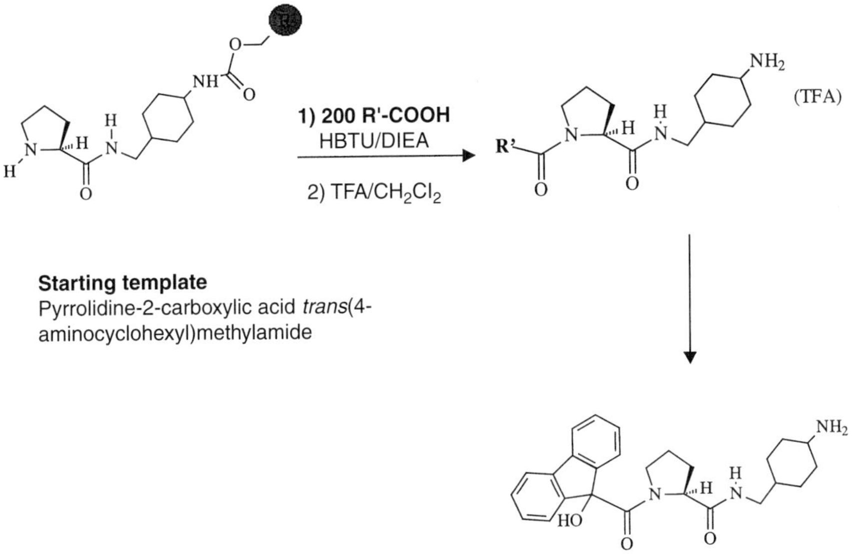

Starting template
Pyrrolidine-2-carboxylic acid *trans*(4-aminocyclohexyl)methylamide

Compound 11:
K_i 1.5 nM vs. thrombin
K_i 860 nM vs. trypsin

Figure 11 Introducing diversity at the P_3 site of a given template through solid phase, in-parallel synthesis leads to a potent and selective thrombin inhibitor.

cokinetic properties of compound **10** were not desirable, since in the rat and dog models it was impossible to register significant blood level.

To develop a more diverse library, the authors devised a solid phase approach where the starting template (Fig. 11) was linked to a polystyrene support via an acid-labile carbamate and then coupled in parallel with 200 different lipophilic carboxylic acids. This procedure allowed the rapid elaboration of a small, rational library of derivatives of compound **9** wherein the N-Me-D-Phe (P_3) residue was replaced with 200 different moieties, chosen on the basis of diversity and potential ability to interact with the S_3 subsite.

In vitro determinations were carried out through fast-binding (reversible-noncovalent) thrombin inhibitors. Among the synthesized compounds was the potent fluorenyl derivative compound **11** (K_i 1.5 nM, Fig. 11), which displayed good selectivity with respect to trypsin and no effects versus several other serine proteases. Moreover, compound **11**, tested in vivo, retained significant oral bioavailability in the rat and dog models.

Molecular modeling studies suggested a possible binding orientation for this molecule. The fluorenyl system partially overlaps the phenyl ring of com-

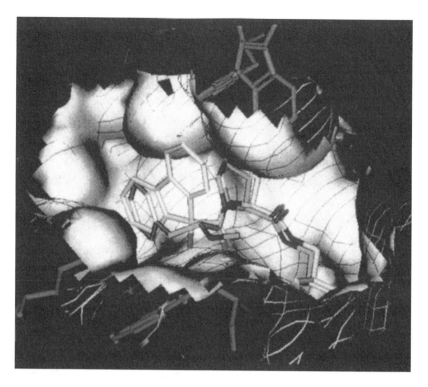

Figure 12 Superimposition of X-ray coordinates of the complex **9**–thrombin and the predicted orientation of **11** (fluorenyl derivative). (From Ref. 49, fig. 4, p. 404. Copyright February 3, 2000, American Chemical Society.)

pound **9** and takes advantage of hydrophobic interaction with residues Tyr60A and Trp60D. Hydroxyl group hydrogen bonds with the carbonyl of Gly216. The presence of the hydroxyl group is also important in the improvement of oral bioavailability (Fig. 12).

The integration of structure-based design with the in-parallel synthetic elaboration of a 200-compound library led to a new, highly selective inhibitor of thrombin. The access to a wide range of functionality and, consequently, of physicochemical properties allowed the development of an orally available compound. Crucial to the improved efficiency of the design cycle was also the availability of an enzyme assay that allowed for high throughput screening.

The validity of an approach in which structure-based design is combined with a rapid synthetic elaboration was confirmed in a paper published shortly thereafter by the same Merck research group, continuing their research into thrombin inhibitors. The study, aimed at orally available thrombin inhibitors, was reported by Lumma et al. (50) starting from the visual analysis of a X-ray crystal structure of the inhibitor, compound **10** (see the previous work by Brady, Figure 10) binding to thrombin. However, whereas Brady's work focused on the P_3 site of the molecule, keeping the P_1 and P_2 sites constant, interest is now wholly concentrated on the P_1 part of the molecule and, consequently, its counterpart on the enzyme S_1 site. The surface representation of the inhibitor–thrombin P_1 pocket revealed the presence of a lipophilic cavity. None of the known molecules binding to the enzyme takes advantage of this hydrophobic hole buried on the P_1 pocket.

The potential importance of the lipophilic cavity in terms of increasing the binding affinity was investigated through the rapid solid phase synthesis of diverse benzylamide derivatives of the starting lead. To increase oral availability and reduce the known side effects of basic thrombin inhibitors, the authors also opted for the deletion of the basic P_1 amino group and for the substitution of the cyclohexyl ring with a phenyl ring (compound **12**, Fig. 13). Eighteen derivatives were then synthesized by means of solid phase, in-parallel synthesis.

Of the compounds thus synthesized, compound **13** (the 2,5-dichlorobenzylamide derivative) was the most active; several others showed K_i between 10 and 50 nM. The SAR information provided by the analysis of this family of compounds showed how the most active compounds were those carrying a *meta*-hydrophobic substituent. Docking studies with compound **13** suggested that the *meta* substituent is oriented toward the lipophilic corner and takes advantage of it (Fig. 14).

As an illustration of the power of the spatial analysis of X-ray crystal structure to direct the solid phase synthesis of small focused libraries, the authors suggest that an extension of this work may already have led to low-picomolar thrombin inhibitors (51).

Another example from the literature that deserves comment is the work on thrombin inhibitors conducted at Hoffmann-La Roche. The authors report the

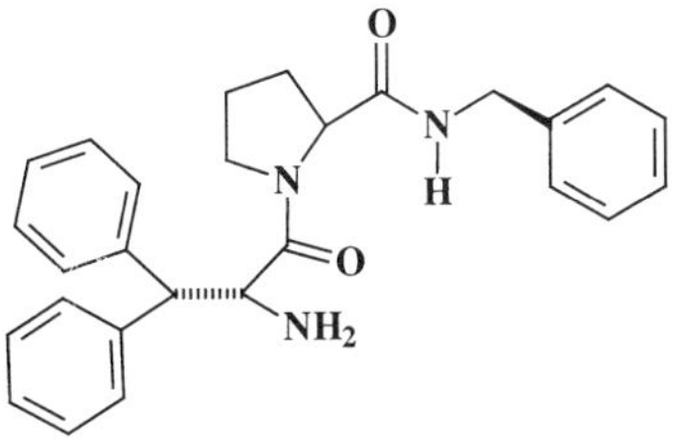

Compound 12 K_i 110 nM 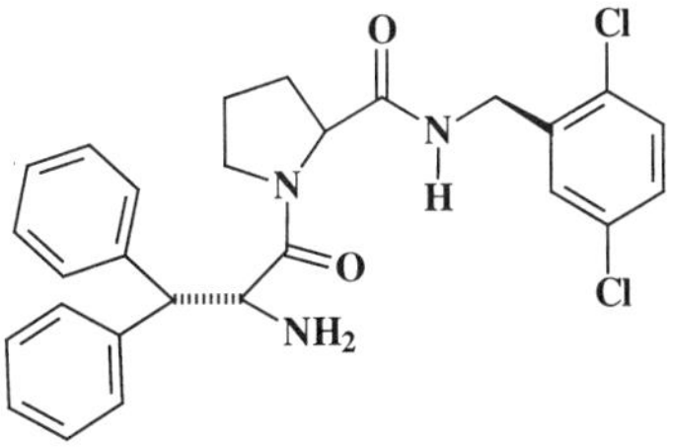   **Compound 13** K_i 3 nM

Figure 13   Follow-up derivatives of compound **10**. The introduction of lipophilic moiety at the P_1 site of the lead compound allows the exploration of an additional binding site (see Fig. 10).

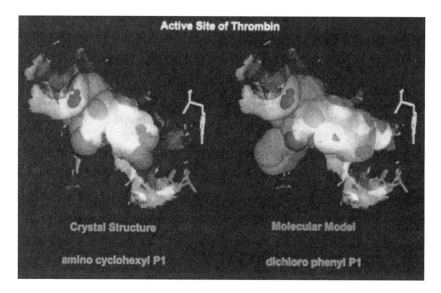

Figure 14   Surface representation of the P_1 pocket of thrombin binding compounds **10** (left) and **13** (right). The 2,5-dichlorophenyl group of **13** extends into the lipophilic cavity of the S_1 subsite, interacting with it. (From Ref. 50, Fig. 1, p. 1012. Copyright February 3, 2000, American Chemical Society.)

coapplications of a novel computational algorithm with synthesis in parallel for the de novo design of nonpeptide thrombin inhibitors. The authors took into account the synthetic accessibility of the designed compound: they focused on chemical reactions amenable to parallel or combinatorial synthesis (52).

The computer program LUDI was used to generate suitable positions for functional groups in the thrombin binding site. The interactions with the enzyme were evaluated in terms of possible hydrogen bond and hydrophobic interactions. Three dimensional coordinates of fragments from the ACD, a total of 5300 primary amines, were docked into the thrombin binding site and chosen on the basis of their ability to form at least one hydrogen bond with the side chain of Asp189, located in the S_1 thrombin site. The first round of calculus found that of the total number of primary amines, the highest scoring compound was *p*-aminobenzamidine, a known inhibitor of thrombin (K_i 34 µM). The computer program CORINA was used to connect the top-scoring amine with different benzaldehyde fragments onto the amine group in the binding site, thus generating a virtual library of 98 hits. Only hits with a predicted K_i smaller then 1 mM were accepted for the subsequent synthetic optimization step. Of the initial 98 hits, only 10 benzaldehydes, based on their predicted score and their commercial availability, were selected and bought for synthesis.

The 10 synthesized compounds showed an at least 10-fold increase in inhibitory activity toward thrombin. In particular, the most interesting compounds were those carrying a lipophilic substituent in the para position, able to take advantage of the P_3 pocket of thrombin. Moreover, synthetic optimization improved selectivity toward thrombin with respect to trypsin (compound 14, Fig. 15).

In the approach followed by the authors, an important role was played by molecular modeling studies. The design of a starting lead was carried out in a straightforward way, allowing the authors, through the synthesis of a very small subset of compounds, to discover a very active compound. Once the chemical characteristics essential to the activity in a lead have been well determined, the work of synthetic elaboration can be considerably reduced.

The crystal structure determination of compound 14 (K_i 95 nM) was subsequently solved. Small, unpredictable conformational changes occurred in the crystal with respect to the modeled orientation: the S_2 subsite of thrombin was markedly reduced in its site, causing a shift in the orientation of the central part of the molecule (52). However, most of the features predicted from the modeling were confirmed (Fig. 16).

The results achieved were possible thanks to a few favorable factors: the well-defined S_1 pocket in the thrombin structure suitable for detailed modeling (otherwise it would have been difficult to find a small molecule such as *p*-

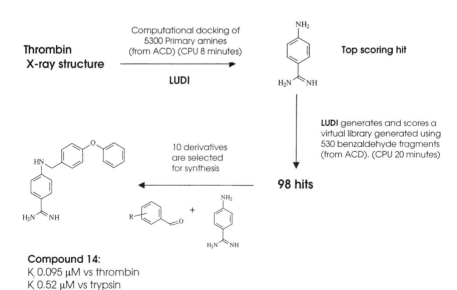

Figure 15 Combinatorial, de novo design and in-parallel synthesis of non-peptide thrombin inhibitors.

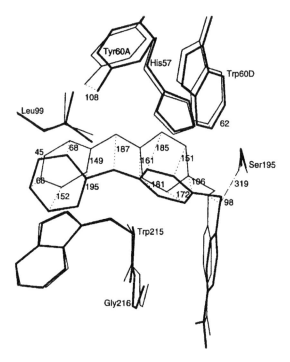

Figure 16 Comparison of the X-ray (bold line) and the predicted (light line) binding orientation for compound **14**. (From Ref. 52, Fig. 2, p. 54; Copyright February 4, 2000, Kluwer Academic Publishers.)

aminobenzamidine with measurable binding affinity); the single-step reaction that generated a potent inhibitor; and finally, a really sensitive biological assay used for testing.

Nowadays, new methodology combining elements of structure-based drug design and combinatorial chemistry are being increasingly developed, as in the work just discussed (52).

The generation of virtual libraries will be discussed in greater detail elsewhere, but the applications of new and sophisticated algorithms that have been developed deserve a brief comment (53–56). These programs generally start from a synthetically accessible template, chemically complementary to the target active site. A database of potential substituents for each derivatizable position of the starting template is then sought. The substituents are selected on the basis of their ability not only to interact with certain residues in the active site but also to couple with the template through known and accessible synthetic routes. This generated list of virtual and synthetizable ligands is then computationally screened against the active site and ranked on the basis of the binding free energy calculated for each of them.

D. Novel Inhibitors of Matrix Metalloproteinases

The report from Rockwell and coworkers at Dupont–Merck in 1996 may be the first example of the complementarity of SBDD and combinatorial chemistry (29). This work was directed toward the identification of novel inhibitors of matrix metalloproteinases (MMPs) human stromelysin (MMP-3), and human neutrophil collagenase (MMP-8) with possible cartilage protection activity. MMPs are, in fact, a family of zinc-dependent enzymes involved in the degradation and reconstruction of the extracellular matrix. These enzymes are thus targets in autoimmune diseases, arthritis, and cardiovascular diseases.

The researchers at Dupont Merck started from N-carboxyalkyl peptides and hydroxamic acids already known as MMP inhibitors and for which X-ray structures with the enzyme were known. The major problem with these molecules is their peptidic structure, which is responsible for limited bioavailability and reduced plasma half-life. A starting template, consisting of 2-(1-carboxyethylamino)-4-phenylbutyric acid (Fig. 17), was chosen as common scaffold based on preliminary crystallographic and enzymatic results, which showed that the presence of the methyl and phenyl rings enhances inhibitory activity against MMP-3. However to gain longer in vivo half-life, the authors attempted to replace the peptidic part of the known inhibitors, responsible for molecular recognition with subsite S_2' and S_3', with mimetic substituents.

The three-step synthetic elaboration was accomplished in solid phase by attaching the starting template to a solid support and coupling it with several amines (Fig. 17). All synthesized compounds were tested without previous purification.

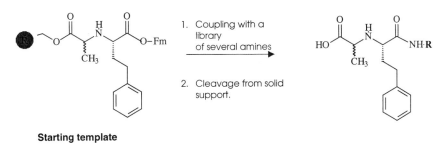

Starting template
2-(1-carboxy-ethylamino)-4-phenyl-butyric acid

Figure 17   Solid phase elaboration of novel matrix metalloproteinase (MMP) inhibitors.

Of the several molecules, compound **15,** which carries a methyl group, proved to be particularly interesting: at 100 μM it displayed a 33% inhibition vs. MMP-3. For this molecule, a binding orientation was calculated suggesting that the terminal methyl group could bind the S_2' pocket of the enzyme. The authors then introduced into the molecule a phenyl ring, known to be a good P_2' substituent, thereby obtaining compound **16,** which displayed 72% inhibition at 200 μM vs. MM-3 (Fig. 18).

Finally the benzhydrylamine group was introduced into the more active hydroxamic template to yield compound **17.** Further modeling studies of compound **17** binding to the enzyme were carried out but failed to predicting the binding orientation correctly. The rigid fit of compound **17** into the active site of a known X-ray complex was unsatisfactory: no well-defined hydrophobic pocket for the additional phenyl ring was shown.

The crystal structure of compound **17** solved at 1.8 Å resolution showed how the loop made by residues 222–231 undergoes a conformational change al-

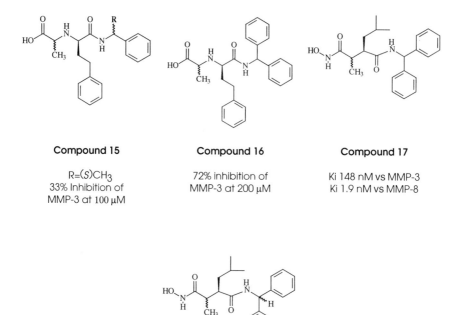

Compound 15

R=(*S*)CH₃
33% Inhibition of
MMP-3 at 100 μM

Compound 16

72% inhibition of
MMP-3 at 200 μM

Compound 17

Ki 148 nM vs MMP-3
Ki 1.9 nM vs MMP-8

Compound 18 Ki 9 nM
Compound 19 Ki 150 nM
(vs MMP-3)

Figure 18   Novel matrix metalloproteinase (MMP) inhibitors.

lowing the accommodation of the benzhydryl group. Such a conformational modification in the protein, necessary for a better stabilization of the enzyme–inhibitor complex, was completely unexpected from the analysis of known MMP-3 structures. However, as predicted from the design, the pro-S-phenyl group is oriented in the S_2' side chain.

The analysis of this complex directed the design and the synthesis of a final molecule with enhanced activity. Since the crystal suggested the introduction in the phenyl ring of a hydrogen bond acceptor, compound **18** and its isomer **19,** carrying a pyridine instead of phenyl, were synthesized: the K_i vs. MMP-3 was 9 and 150 nM, for the two isomers, respectively, indicating stereospecific binding affinity.

In this work the integration of SBDD with combinatorial chemistry was crucial to the discovery of a novel potent MMP inhibitor. The generation of a combinatorial library was important at the beginning of this work in that it suggested as leads susceptible of further optimization molecules otherwise not supported by modeling studies, which had always failed to predict conformational changes at the MMP-3 binding site. When compound **17** and its complex with MMP-3 were available, a synthetic optimization in an iterative fashion was possible. The data derived from the crystallographic complex explained the binding affinity of compound **17** by showing the conformational changes occurring at the binding site upon accommodation of larger entities and successfully suggesting the modification necessary to enhance binding.

It should be emphasized that structure-based ligand design can fail to predict alternative binding modes of new molecules as they are obtained through combinatorial chemistry methods. Molecular diversity cannot fully complement the enzyme active site, which is kept rigid, while deep and unpredicted conformational change often occurs as a result of enzyme plasticity. In such cases the availability of 3D coordinates of the target–inhibitor complexes is a crucial integrating tool in the drug discovery process.

The crystal structure of peptidylhydroxamates bound to matrix metalloproteinases (MMPs) was the starting point of a study at Affimax, where rational design in conjunction with combinatorial techniques led to the discovery of diketopiperazine (DKP) inhibitors (57). Of the works reviewed here, this is the only one in which the authors used split-and-mix synthetic methodologies rather than in-parallel, spatially addressed synthesis. Thus deconvolution methodologies were essential to identify the active compounds.

On the basis of known crystal structures of various MMPs binding succinylhydroxamate inhibitors, a pharmacophore model was generated that replaced the peptidyl-succinate portion and incorporated the P_1 and P_2 side chains and the zinc ligand. The 2,5-diketopiperazine scaffold was chosen as starting template: it retained the appropriate spatial relationship between ligand and protein and had the potential to interact through the H-bond with the protein. Moreover new interactions would have been possible with the introduction of new functionalities. In the designed molecules, the thiol function of a cysteine served as a zinc ligand.

Pharmacophore model generation was based on X-ray crystal structure analysis, but no precise details are given regarding the approach used to generate the 2,5-diketopiperazine scaffold (DKP) and the iterative diverse libraries based on its structural modifications. SAR analysis was largely responsible for guiding the rational design development of the libraries.

Synthesis through split-and-mix methodology was accomplished in a completely automatic fashion, allowing the rapid generation of large combinatorial libraries. Two different DKP scaffolds, DKP-I and DKP-II, varying in the spatial arrangement of the side chains relative to one another, were investigated: in the first, cysteine was represented by the R^1 group, in the second, by R^3. Both L- and D-cysteine were included (Fig. 19). Two 648-member libraries were constructed: cysteine served as the zinc ligand, while diversity was introduced at two different sites; in the first library, 19 different aldehydes were introduced at R^2 and 18 amino acids at R^3. In the second library, 19 amino acids were introduced at R^1 and 18 aldehydes at R^3.

A sensitive fluorogenic assay compatible with collagenase-1 (MMP-1), gelatinase-B (MMP-9), stromelysin-1 (MMP-3), and matrilysin was developed. From the DKP-I library, two compounds with low micromolar activity vs. collagenase-1 (compound **20**, Fig. 19) were identified, while the DKP-II library provided submicromolar MMP inhibitors (vs. MMP-1 and MMP-9). The anisaldehyde pool was in fact deconvoluted and supported important SAR observations for the development of a nanomolar inhibitor: first, a wider range of functionality at R^2 position was tolerated with respect to the R^1 site, where small variations were more critical for activity; second, hydrophobic substituents enhanced affinity; third, the L-cysteine configuration is better for activity. Compound **21** (Fig. 19) was the most active isolated compound. Its activity toward MMP-1 is comparable to that of succinylhydroxamate inhibitors, described previously. However, it represents a novel class of inhibitors.

This novel class of MMP inhibitors was generated starting from a scaffold rationally chosen from structural information and onto which diverse sets of building blocks were incorporated in a combinatorial fashion. SBDD was determinant in design of the starting scaffold, while synthesis guided the introduction of diversity. Libraries incorporating rational design and diversity led rapidly to the identification of nanomolar inhibitors.

E. Thymidylate Synthase Inhibitors

Thymidylate synthase (TS) catalyzes the final step on the biosynthetic pathway to thymidylate; it is consequently a target for anticancer drugs in human cells and also for antimicrobial chemotherapy. X-ray crystal structures of TS from several

Structure-based pharmacophore generation

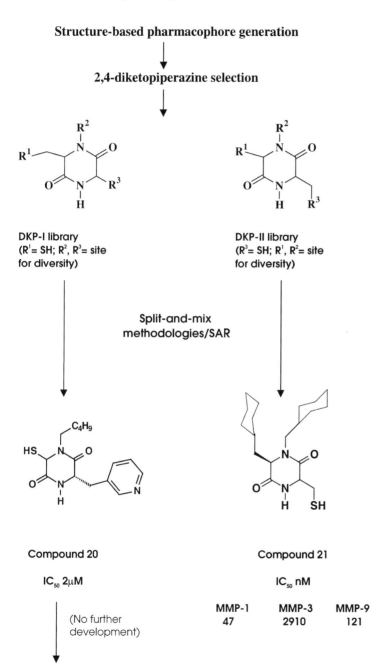

Figure 19   Design of MMP inhibitor scaffold and subsequent elaboration through combinatorial chemistry techniques.

species bound to numerous ligands has increased our understanding of the enzyme's catalytic mechanism. The enzyme has also become a well-studied system for the application of structure-based method to the design of antiproliferative drugs.

In an effort to discover novel inhibitors that bind to the folate site of *Lactobacillus casei* thymidylate synthase (*Lc*TS) but do not structurally resemble the cofactor and elaborate it rapidly, we combined structure-based screening with in-parallel synthetic elaboration (42). First we used the molecular docking computer program DOCK 3.5 to screen the Available Chemicals Database (ACD) for compounds that complemented the three-dimensional structure of the active site of the enzyme. The crystal structure of *Lc*TS in ternary complex with the uridine monophosphate dUMP and CB3717 (an *N*-10-propargylquinazoline derivative of folate) determined to 2.5 Å resolution was used. Potential binding sites for ligand atoms were defined by a set of 64 spheres corresponding to the atom coordinates of the folate analog CB3717 and phenolphthalein, a known nonfolate analog inhibitor of *Lc*TS. DOCK scored each molecule on the basis of the electrostatic and van der Waals interaction energies, with the enzyme correcting for ligand solvation. The 500 highest scoring compounds were saved.

On the basis of the DOCK score and the number of specific interactions with the protein, five hits (IC$_{50}$ between 300 μM and 2 mM) were tested for the ability to inhibit *Lc*TS. To allow for rapid elaboration, we also insisted that the DOCK hits have chemical characteristics suitable for optimization through solid phase techniques. Of the selected molecules dansylhydrazine (IC$_{50}$ 439 μM; K_i 176 μM) seemed to offer a good point of departure for synthetic elaboration (Fig. 20).

Through further docking, liquid phase synthesis and enzymatic studies on different dansyl derivatives we discovered dansyltyrosine: it was not only a better inhibitor of *Lc*TS (K_i 65 μM), but it was more suitable for solid phase synthetic elaboration. The molecule contained in its structure both an anchor site and a diversity-derivatizable group (the carboxyl group and the amino group, respectively). We introduced diversity on dansyltyrosine by synthesizing a series of analogs carrying different substituents on the amino group (Fig. 21). To have as much diversity as possible, we chose different building blocks, all commercially available, and we obtained aliphatic, aromatic, and heterocyclic derivatives. The carboxyl group served as anchor to the TentaGel Wang Resin-OH, which has the ability to restore the free carboxyl functionality (in our case crucial to the activity) after cleaving with trifluoroacetic acid (TFA). Thirty-three dansyl derivatives of this lead were then synthesized in parallel on the automatic synthesizer Advanced Chem Tech 357.

After synthesis, the compounds were tested against the enzyme *Lc*TS. The assays were performed spectrophotometrically, following the increasing absorbance at 340 nm due to the oxidation reaction of N^5,N^{10}-methylenetetrahydrofolate to dihydrofolate. The most active compounds were also tested vs. chy-

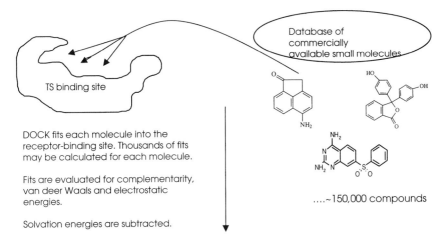

DOCK fits each molecule into the
receptor-binding site. Thousands of fits
may be calculated for each molecule.

Fits are evaluated for complementarity,
van deer Waals and electrostatic
energies.

Solvation energies are subtracted.

....~150,000 compounds

Visual analysis of 400 saved molecules

Inhibition constant of the best fitting molecules
established in enzyme assays.

Analog search in ACD

$IC_{50} = 439 \ \mu M$

Insoluble, discarded

$IC_{50} = 163 \ \mu M$

A lead is chosen for solid-phase
in parallel elaboration.

Figure 20   Docking flow in the de novo design of thymidylate synthase inhibitors.

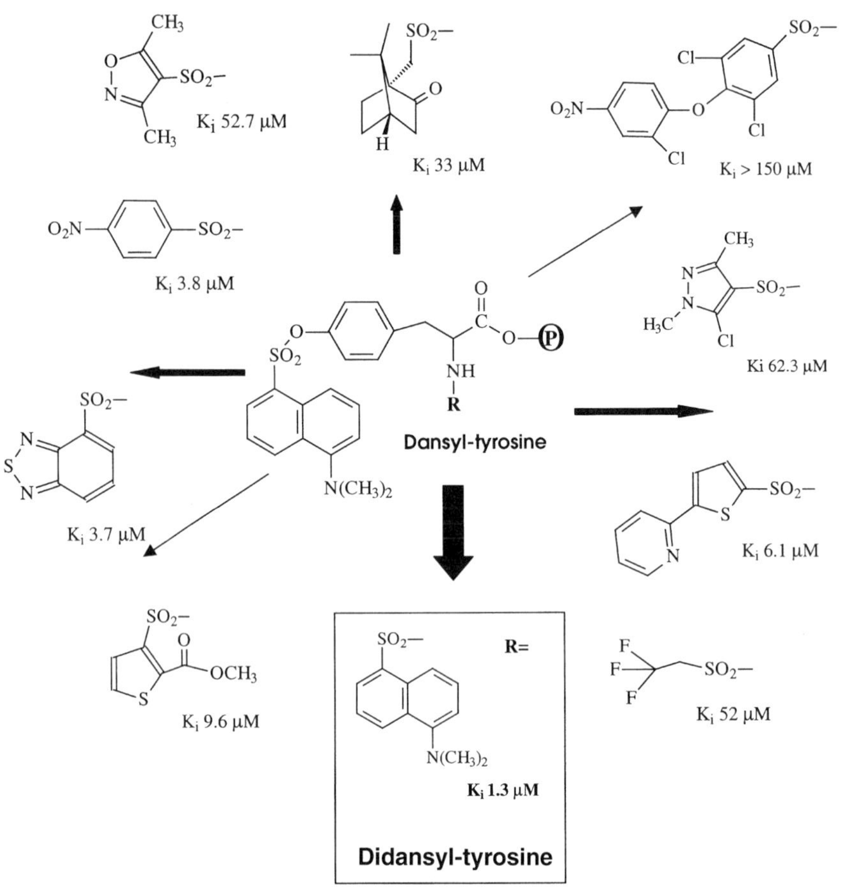

Figure 21 In-parallel elaboration of dansyltyrosine leads to a 1.3 μM non-folate analog competitive inhibitor. The **R** variable in the dansyltyrosine is represented from one of the surrounding groups. K_i values are reported. Only part of the total library is represented.

motrypsin, dihydrofolate reductase (DHFR), and human TS, to investigate their species-specificity and selectivity.

The best of the synthesized analogs, didansyltyrosine, had a K_i of 1.3 μM, while several others had K_i values below 10 μM. These analogs are structurally dissimilar to the substrate but bind competitively with it (Fig. 22). The tighter binding inhibitor was also the most specific for LcTS vs. the related human TS and bovine DHFR enzymes.

Further calculations were performed to understand the basis for DDT affinity: 500 multiple conformations were generated by rotating all single, nonterminal bonds in increments of 120°. Each conformer was independently docked and scored into the *Lc*TS binding site. In the best scoring orientation shown (Fig. 22), the anchor dansyl ring forms nonpolar interactions with the pyrimidine ring of dUMP. The dansyl ring also makes nonpolar contacts with folate binding residues, including Ile81 4.0, Å, Arg23, 4.0 Å, Ala315, 3.9 Å; and Leu195, 4.5 Å. As the inhibitor extends out of the quinazoline-binding site of *Lc*TS, the active site cleft opens up, leaving room for more orientations of the flexible N-dansyltyrosine group. In most orientations, the tyrosine ring interacts with hydrophobic residues such as Phe228 and Leu224. In the orientation represented in Fig. 22, for instance, the tyrosine ring atoms interact with Phe228 and with Leu224. In this orientation, the tyrosine carboxylate maintains its interaction with Lys50 (O22—N$_Z$ of Lys50, 2.8 Å), while the sulfonamide groups interacts with the backbone of Ala312 (O14—N of Ala312, 3.2 Å). In other high-scoring orientations this interaction is lost, and the carboxylate extends out toward the solvent.

It must be noted that while DOCK oriented the dansylsulfonate in the folate binding site in most of the calculated orientations, the conformation of the second dansyl group varied most in the docked orientations, since it was bound to the most and least defined open parts of the TS site. The most energetically favorable configuration is that which places this ring close to residues Ala309 (ligand C12—Cβ, 3.5 Å, and ligand C6—Cβ, 3.7 Å) and Ile310 (ligand, C8—O, 3.4 Å) and with Lys51 (ligand C17—Nζ 4.1, Å), at the mouth of the active site. The binding orientation we calculated roughly supported the enzymatic finding: these molecules bind into the active site of TS, preventing folate binding and therefore acting as competitive inhibitors.

This work demonstrates that the integration of computational chemistry and in-parallel synthesis in the quest for a novel lead really does accelerate the design

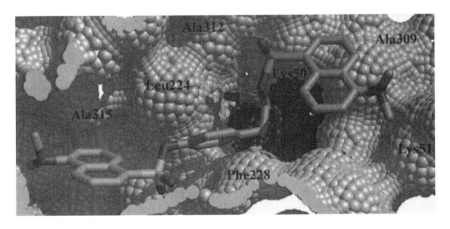

Figure 22 Predicted binding orientation for didansyltyrosine in the *Lc*TS binding site

of new inhibitors for a given target. In particular, by applying in-parallel synthesis to compounds structurally correlated to an active scaffold discovered through modeling studies, one is more likely to discover a new and more active compound. Analysis of the structure–activity relationship is also accelerated thereby.

More information will be forthcoming from the resolution of the crystal structure of compound DDT bound to *Lc*TS, at present under study; it is expected that the crystal structure will confirm the predicted orientation for didansyltyrosine. Moreover the X-ray structure will significantly integrate the iterative cycle of design, synthesis, and enzymatic activity so far applied.

Didansyltyrosine was the starting point for a new project focusing on the development of dansyl derivatives with exalted species-specificity vs. bacterial TSs. It is expected that these molecules, which are able to discriminate between the several TS species, will have antimicrobial activity (58).

Combining structure-based discovery with in-parallel synthetic techniques allowed the rapid discovery of novel competitive inhibitors that have no obvious similarity to the enzyme substrate.

The approach followed exploits the complementarity between the two techniques. Structural data can constrain the diversity available in the combinatorial elaboration of known lead compounds. The example drew on the ability of molecular docking to discover a new scaffold and then used in-parallel techniques to elaborate upon it rapidly.

VI. CONCLUSIONS

The design and synthesis of active and selective molecules is an important goal in medicinal chemistry. Two powerful tools applicable successfully to drug discovery are structure-based drug design and combinatorial chemistry. SBDD uses information from the 3D structure of a macromolecule to design new inhibitors; combinatorial chemistry is used to rapidly synthesize libraries of related compounds. A design protocol combining the two techniques can reduce the number of compounds in a library and increase the production of active molecule against specific targets.

Moreover the combination of SBDD and CC is used not only to improve ligand affinity for a target but more and more often applied for the optimization of the pharmacokinetic properties of the leads when basic elements for a drug-like action are considered.

To highlight this aspect, Table 1 summarizes the progress in terms of ligand affinity taken from the examples given earlier. The optimized ligand does not always show a better binding affinity with respect to the starting lead; however the second-generation compound can have better pharmacokinetic properties with respect to the first-generation ligand, as in the cases reported by Brady

Table 1 Coapplication of SBDD and CC: Summary of Results

| Target | Starting lead | Affinity, K_i | | Ref. |
		First generation	Second generation	
Cathepsin D (Cat D)	Reaction intermediate	78 nM	9 nM	Kick et al. (41)
Plasmepsin II (Plm II)	220 nM	100 nM,	2.0 nM	Haque et al. (30)
Thrombin	4.9 nM	0.1 nM	1.5 nM	Brady et al (49)
Thrombin	0.1 nM	110 nM	3 nM	Lumma et al (50)
Thrombin	de novo	34 μM	95 nM	Bohm et al (52)
Matrix metalloproteinases MMP-3	3 nM	33% Inhibition at 100 μM	9 nM	Rockwell et al (29)
Matrix Metalloproteinases MMP-1	de novo	2 μM vs	47 nM vs	Szardenings et al (57)
Lactobacillus casei thymidylate synthase (LcTS)	de novo	1.3 μM	0.13 μM	Tondi et al (42)

(49), Lumma (50), Böhm (52), and Rockwell (29). This is mainly due to the selection criteria applied along with the library design and to the taking into account of other features such as pharmacokinetic properties, in addition to the binding affinity.

In this chapter we have collected a few representative examples of the combination of the two technical aspects of drug discovery, highlighting the most important aspects of this new approach. The field has the opportunity to advance rapidly owing to more efficient combinatorial chemistry techniques, reliable HTS screening, and more accurate computational methods.

REFERENCES

1. Salemme, FR, Spurlino J, Bone R. Serendipity meets precision: The integration of structure-based drug design and combinatorial chemistry for efficient drug discovery. Structure. 1997; 5:319–324.
2. Merrifield, RB. J. Am. Chem. Soc. 1963; 85:2149.
3. Notariell Beglaubigtes. Document 362371982. Budapest, Hungary, 1982.
4. Corey EJ, Cheng XM. *The Logic of the Chemical Synthesis*. Wiley, New York, 1989.
5. Lewel XQ, Judd DB, Watson SP, Hann MM. Retrosynthetic combinatorial analysis procedure: A powerful new technique for identifying privileged molecular fragments with useful application in combinatorial chemistry. J. Chem. Inf. Comp. Sci. 1988; 38:511–522.

6. Kundu B, Khare SK, Rastogi SK. Combinatorial chemistry: Polymer supported syn thesis of peptide and nonpeptide libraries. Prog. Drug. Res. 1999; 53:89–156.

7. Gallopp MA, Barrett RW, Dower WJ, Fodor SPA, Gordon EM. Application of com-binatorial technologies to drug discovery. 1. Background and peptide combinatorial libraries. J. Med. Chem. 1994; 37(9):1233–1251.

8. Gordon EM, Barret RW, Dower WJ, Fodor SPA, Gallop MA. Application of combi-natorial technologies to drug discovery. 2. Combinatorial organic synthesis, library screening strategies and future directions. J. Med. Chem. 1994; 37(10):1385–1401.

9. Terret NK, Gardner M, Gordon DW, Kobylecki RJ, and Steele J. Combinatorial syn-thesis—The design of compound libraries and their application to drug discovery. Tetrahedron, 1995, 51(30):8135–8173.

10. Ellman JA. Design, synthesis and evaluation of small-molecule libraries. Acc. Chem. Res. 1996; 29:132–143.

11. Thompson LA, Ellman JA. Synthesis and application of small molecule libraries. Chem. Rev. 1996; 96:555–600.

12. Lam KS. Application of combinatorial library methods in cancer research and drug discovery. Anticancer Drug Des. 1997; 12(3):145–167.

13. Armstrong RW, Combs AP, Tempest EA, Brown SD, Keating TA. Multiple compo-nent condensation strategies for combinatorial library synthesis. Acc. Chem. Res. 1996; 29:123–131.

14. Combs AP, Kapoor TM, Feng S, Chen JK, Daudé-Snow LF, Schreiber SL. Protein structure-based combinatorial chemistry: Discovery of non-peptide binding elements to Src SH3 domain. J. Am. Chem. Soc. 1996; 118:287–288.

15. Van Hijfte L, Marciniak G, Froloff N. Combinatorial chemistry, automation and molecular diversity: New trends in the pharmaceutical industry. J. Chromatogr. B. 1999; 725:3–15.

16. Geysen, HM, Melven RH, Barteling SJ. Use of peptide synthesis to probe viral anti-gens for epitopes to a resolution of a single amino acid. Proc. Natl. Acad. Sci. USA. 1984; 81(13):3998–4002.

17. Frank R. SPOT-synthesis: An easy technique for the positionally addressable, paral-lel chemical synthesis on a membrane support. Tetrahedron. 1992; 48:9217.

18. Fodor SPA, Read JL, Pirrung MC, Stryer L, Lu AT, Solas D. Light-directed, spatially addressable parallel chemical synthesis. Science. 1991; 251:767–773.

19. DeWitt SH, Kiely JK, Stankovic CJ, Schroeder MC, Cody DMR, Pavia MR. Diver-somers: An approach to nonpeptide, nonoligomeric chemical diversity. Proc. Natl. Acad. Sci. USA. 1993; 90(15):6909–6913. DeWitt SH, Czarnik AW. Combinatorial organic synthesis using Parke-Davis's DIVERSOMER method. Acc. Chem. Res. 1996; 29:114–122.

20. Kuntz, ID, Meng EC, Shoichet BK. Challenges in structure-based drug design. In PM Dean, G Jolles, and CG Newton, eds. *New Perspectives in Drug Design.* Academic Press, San Diego, CA, 1995, pp. 137–154.

21. Bohacek RS, McMartin C, Guida WC. The art and practice of structure-based drug design: A molecular modeling perspective. Med. Res. Rev. 1996; 16:3–50.

22. Purshkar NK. Drug discovery: Past, present and future. Prog. Drug. Res. 1998; 550:12–105.

23. Kubinyi H. Structure-based design of enzyme inhibitors and receptor ligands. Curr. Opin. Drug Discovery Dev. 1998; 1:4–15.

24. McCarthy JD. Computational approaches to structure-based ligand design. Pharmacol. Ther. 1999; 84(2):179–191.

25. Kirkpatrick DL, Watson S, Ulhaq S. Structure-based drug design: Combinatorial chemistry and molecular modeling. Comb. Chem. High Throughput Screen 1999; 2(4):211–221.

26. Apostolakis J, Caflisch A. Computational ligand design. Comb. Chem. High Throughput Screen 1999; 2(2):91–104.

27. Roberts GC. NMR spectroscopy in structure-based drug design. Curr. Opin. Biotechnol 1999; 10(1):42–47.

28. Testa B, Kier LB, Carrupt PA. A system approach to molecular structure, intermolecular recognition, and emergence-dissolvence in medicinal chemistry. Med. Res. Rev. 1997; 17:303–326.

29. Rockwell A, Melden M, Copeland RA, Hardman K, Decicco CP, DeGrado WF. Complementarity of combinatorial chemistry and structure-based drug design application to the discovery of novel inhibitors of matrix metalloproteinases. J. Am. Chem. Soc. 1996; 118:10337–10338.

30. Haque TS, Skillman AG, Lee CE, Habshita H, Gluzman IY, Ewing TJA, Goldberg DE, Kuntz ID, Ellman JA. Potent, low-molecular-weight non-peptide inhibitors of malarial aspartyl protease plasmepsin II. J. Med. Chem. 1999; 42:1428–1440.

31. Jelesavor I, Bosshard HR. Isothermal titration calorimetry and differential scanning calorimetry as complementary tools to investigate the energetics of biomolecular recognition. J. Mol. Recognit. 1999; 12:3–18.

32. Böhm HJ, Klebe G. What can we learn from molecular recognition in protein–ligand complexes for the design of new drugs? Angew Chem Int Ed Engl 1996, 35:2588–2614.

33. Böhm HJ. The computer program LUDI: A new method for the de novo design of enzyme inhibitors. J. Comput-Aided Mol. Design 1992; 6(1):61–78.

34. Li J, Murray CW, Waszkowycz, Young SC. Targeted molecular diversity in drug discovery: Integration of structure-based design and combinatorial chemistry. Drug Discovery Today 1998; 3:105–112.

35. Antel J. Integration of combinatorial chemistry and structure-based drug design. Curr. Opin. Drug Discovery Dev. 1999; 2:224–233.

36. Janda KD. Tagged versus untagged libraries: Methods for the generation and screening of combinatorial libraries. Proc. Natl. Acad. Sci. USA 1994; 91:10779–10785.

37. Patterson DE, Ferguso AM, Weienberger LE. Neighborhood behavior: A useful concept for validation of "molecular diversity descriptors. J. Med. Chem. 1996; 39:3049–3059.

38. Dean PM, Barakat MT, Todorov NP. Optimization of combinatoric problems in structure generation for drug design. In PM Dean, G Jolles, and CG Newton, eds. *New Perspectives in Drug Design.* Academic Press, San Diego, CA, 1995, pp. 155–184.

39. Singh J, Ator MA, Jaeger EP, Allen MP, Whipple DA, Soloweij JE, Chowdhary S, Treasurywala AM. Application of genetic algorithm to combinatorial synthesis: A computational approach to lead identification and lead optimization. J. Am. Chem. Soc. 1996; 118:1669–1676.

40. Lewis RA, Mason JS, McLay IM. Similarity measures for a rational set selection and analysis of combinatorial libraries: The diverse property-derived (DPD) approach. J. Chem. Inf. Comput. Sci. 1997; 37:599–614.

41. Kick EK, Roe DC, Skillman AG, Liu G, Ewing, TJA, Sun Y, Kuntz I, Ellman JA. Structure-based drug design and combinatorial chemistry yield low nanomolar inhibitors of cathepsin D. Chem Biol 1997, 4(4):297–307.

42. Tondi D, Slomiczynska U, Costi MP, Watterson DM, Ghelli S, Shoichet BK. Structure-based discovery and in-parallel optimization of novel competitive inhibitors of thymidylate synthase. Chem. Biol. 1999; 6(5):319–331.

43. Fecik RA, Frank KE, Gentry EJ, Menon SR, Mitscher LA, Telikepalli H. The search of orally active medications through combinatorial chemistry. Med. Res. Rev. 1998; 3:149–185.

44. Polinsky A. Curr. Opin. Drug Discovery Dev. 1999; 2:197–203.

45. Horikoshi T, Arany I, Rajaraman S, Chen SH, Brysk H, Lei G, Tyring SK, Brysk MM. Isoforms of cathepsin D and human epidermal differentiation. Biochimie. 1998; 80(7):605–612.

46. Strojan P, Budihna M, Smid L, Vrhovec I, Skrk J. Cathepsin D in tissue and serum of patients with squamous cell carcinoma of the head and neck. Cancer Lett. 1998; 14;130(1–2):49–56.

47. Sadik G, Kaji H, Takeda K, Yamagata F, Kameoka Y, Hashimoto K, Miyanaga K, Shinoda T. In vitro processing of amyloid precursor protein by cathepsin D. Int. J. Biochem. Cell. Biol. 1999; 31(11):1327–37.

48. Lee CE, Kick EK, Ellman JA. General solid-phase synthesis approach to prepare mechanism-based aspartyl protease inhibitor libraries. Identification of potent cathepsin D inhibitors. J. Am. Chem. Soc. 1998; 120:9735–9747.

49. Brady, SF, Stauffer KJ, Lumma WC, Smith GM, Ramjit, HG, Lewis SD, Lucas BJ, Gardell SJ, Lyle EA, Appleby SD, Cook JJ, Holahan MA, Stranieri MT, Lynch JJ, Lin JH, Chen IW, Vastag K, Naylor-Olsen AM, Vacca JP. Discovery and development of the novel potent orally active thrombin inhibitor N-(9-hydroxy-9-fluorenecarboxy)prolyl *trans*-4-aminocyclohexylmethyl amide (L-372,460): Co-application of structure-based design and rapid multiple analogue synthesis on solid support. J Med Chem 1998; 41:401–406.

50. Lumma WC, Witherup KM, Tucker TJ, Brady SF, Sisko JT, Naylor-Olsen AM, Lewis SD, Lucas BJ, Vacca JF. Design of novel, potent, noncovalent inhibitors of thrombin with nonbasic P-1 substructures: Rapid structure-activity studies by solid-phase synthesis. J. Med. Chem. 1998; 41:1011–1013.

51. Tucker TJ, Brady SF, Lumma WC, Lewis SD, Gardell SJ, Naylor-Olsen AM, Yan Y, Sisko JT, Stauffer KJ, Lucas BJ, Lynch JJ, Cook JJ, Stranieri MT, Holahan MA, Lyle EA, Baskin EP, Chen IW, Dancheck KB, Krueger JA, Cooper CM, Vacca JP. Design and synthesis of a series of potent and orally bioavailable noncovalent thrombin inhibitors that utilize nonbasic groups in the P_1 position. J. Med. Chem. 1998; 13:41 (17):3210–3219.

52. Böhm, HJ, Banner DW, Weber L. Combinatorial docking and combinatorial chemistry: Design of potent non-peptide thrombin inhibitors. J. Comput-Aid Mol. Design 1999; 13:51–56.

53. Murray CW, Clark DE, Auton TR, Firth MA, Li Jin, Sykes RA, Waszkowycz B, Westhead DR, Young SC. PRO_SELECT: Combining structure-based drug design and combinatorial chemistry for rapid lead discovery. 1. Technology. J. Comput-Aid Mol. Design 1997; 11:193–207.
54. De Julian-Ortiz JV, Galvez J, Munoz-Collado C, Garcia-Domenech R, Gimeno-Cardona C. Virtual combinatorial syntheses and computational screening of new potential anti-herpes compounds. J. Med. Chem. 1999; 42:3308–3314.
55. Makino S, Ewing TJ, Kuntz ID. DREAM++: Flexible docking program for virtual combinatorial libraries. J. Comput-Aid Mol. Design 1999; 13(5):513–532.
56. Sun Y, Ewing TJA, Skillman AG, Kuntz ID. CombiDOCK: Structure-based combinatorial docking and library design. J. Comput-Aid Mol. Design 1998; 12:597–604.
57. Szardenings AK, Harris D, Lam S, Shi L, Tien D, Wang Y, Patel DV, Navre M, Campbell DA. Rational design and combinatorial evaluation of enzyme inhibitor scaffold: Identification of novel inhibitors of matrix metalloproteinases. J. Med. Chem. 1998; 41:2194–2200.
58. Tondi D, Ghelli S, DeCol L, Rinaldi M, Pecorari PG, Costi, MP. Modeling and biological activity of dansyl derivatives as thymidylate synthase inhibitors. Poster communication, XV EFMC International Symposium on Medicinal Chemistry, Edinburgh, September 6–10, 1998, p 89.

20

Design of Structural Combinatorial Libraries That Mimic Biologic Motifs

Hanoch Senderowitz and Rakefet Rosenfeld*

Peptor, Ltd.
Rehovot, Israel

I. INTRODUCTION

Numerous bioactive peptides and their peptidomimetic analogs have been synthesized, and some are used in the clinic (1). Linear peptides composed of the natural amino acids are limited in their therapeutic value because their high degree of flexibility commonly leads to low binding affinity and a fast rate of degradation by proteases. Strategies to overcome these problems have led to a variety of molecules displaying very broad chemical diversity—ranging from constrained, cyclic peptide analogs through molecules having various nonpeptide links between amino acids all the way to molecules with chemical scaffolds that bear no resemblance to peptidic backbones (2). Yet success with protein-mimetic analogs (analogs that can disrupt protein–protein interactions) has not followed. This may be due, in part, to the limited conformation space that standard cyclization methods (disulfide, head-to-tail, or side chain-to-side chain) confer to the peptidic backbone. Peptor's backbone cyclization technology (3) enables the creation of large ensembles of conformationally constrained peptidomimetic analogs by bridging any two positions along their backbones through bridges of varying sizes and chemical compositions. Peptor uses this technology to generate large ensembles of structurally shaped compounds termed SCAPLs—small cyclic analogs of protein loops—which have the potential to disrupt protein–protein interactions. An example of a SCAPL is depicted in Fig. 1.

---

* *Current affiliation:* QBI Enterprises, Ltd., Nes Ziona, Israel

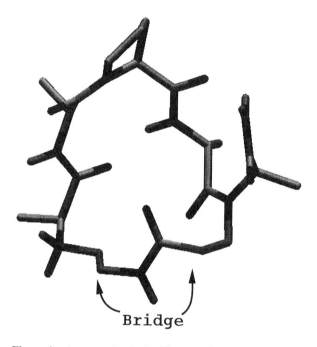

Figure 1   An example of a backbone cyclic peptide. The sequence Ala-Ala-Pro-(D)Ala-(D)Ala was cyclized by a bridge connecting the backbone nitrogens of Ala(1) and (D)Ala(4) through a lactam bond.

The uniqueness of this technology lies in its inherent ability to separate the effects of chemical changes from those of structural changes. Using the backbone cyclization technology, one can generate both *conformational libraries* in which many diverse amino acid sequences share similar structures (thereby enabling the optimization of a known three-dimensional biological motif) and libraries in which a single sequence is entrapped in a large variety of conformations (thereby identifying the active conformation of a biologically active sequence). This chapter describes the design process of libraries of the former type, through an example of the discovery of a lead protein-mimetic SCAPL that disrupts the interaction of tumor necrosis factor (TNF) with its membrane-bound receptor.

The main idea behind conformational SCAPL libraries based on biological motifs is to extract from its natural protein environment a structural motif that may be responsible for binding or recognition and stabilize it in its bioactive conformation. The stabilized structure can then serve as a scaffold on which a combinatorial library can be built. The structural motif we aimed to mimic is a hypervariable loop of a neutralizing anti-TNF antibody. TNF is a cytokine implicated in several normal physiological conditions and, when in excess, in pathological conditions such as rheumatoid arthritis, Chron's disease, and multiple sclerosis (4). Inhibition of the interaction of TNF with its membrane-bound receptor is expected to be of considerable therapeutic value.

The immune system can be viewed as a protein engineering system designed to generate an endless repertoire of molecular surfaces (antibodies) capable of recognizing any given molecular entity (antigen). Antigen recognition is conferred to the antibody by a limited number of hypervariable surface loops, differing in sequence and in length between different antibodies and connected to a conserved framework structure. Several reports in the literature have presented both linear and cyclic peptides derived from a single hypervariable loop of an antibody that can compete with the antibody for target binding (5). Döring et al. (6) have demonstrated that a 19-residue linear peptide derived from the first hypervariable loop of the anti-TNF antibody light chain (termed CDR-L1) can prevent TNF-induced cytotoxicity in vitro. Although the affinity of this peptide was very low (and in fact in our assays undetectable), we aimed to design SCAPLs that *structurally* resemble this loop and therefore stand a good chance of mimicking the antibody.

Several cycles of design resulted in lead compounds with affinities ranging between 30 and 50 μM. This chapter discusses the considerations involved in the design process and demonstrates that based on simple one- and two-dimensional molecular properties (descriptors) of our SCAPLs, one can construct a reliable model that can be used to optimize these leads. Additionally, we show that the predictive ability of the model can be enhanced by individually calculating the descriptors for each amino acid in the sequence rather than for the entire SCAPL and further enhanced by augmenting this set of descriptors with a simple structural descriptor.

II. DESIGN OF INITIAL CONFORMATIONAL LIBRARY

During the early 1990s, the notion was raised that screening large enough numbers of molecules against targets of interest will ultimately result in lead discovery. However, research has shown that sheer numbers are not enough to crack the lead discovery problem. Consequently today more effort is being put into rational design of combinatorial libraries (7).

Augmenting sequence information with structural information, we designed an initial library based on the first hypervariable loop of a neutralizing anti-TNF antibody light chain, in which all molecules stand a good chance of adopting the conformation of the CDR-L1 hypervariable loop in the intact antibody. We started by modeling the three-dimensional structure of the hypervariable loop, stabilized its conformation in isolation (i.e., outside the antibody), and designed a combinatorial library that is structurally focused around this conformation.

Initially, a three-dimensional model of the CDR-L1 loop was constructed. The antibody loop has 13 residues, and its conformation was constructed by homology to the sequentially closest CDR-L1 found in the Protein Data Bank (PDB) (8), namely, 1mbl (PDB code). Figure 2 depicts the structure of the modeled CDR-L1. The positions depicted in the solid boxes differ between the two antibodies, yet point outward and do not affect the conformation of the loop. Therefore these positions were modeled based on the most common rotamer found for these residues when they are solvent-exposed, and in any case their exact positions did not affect our design process. The positions depicted in the dashed boxes differ between the two sequences, and point inward, thus contributing to conformational stabilization. However, since the changes in sequence in these two positions are highly conserved (Ile-Val and Leu-Val), it was possible to optimize their positions

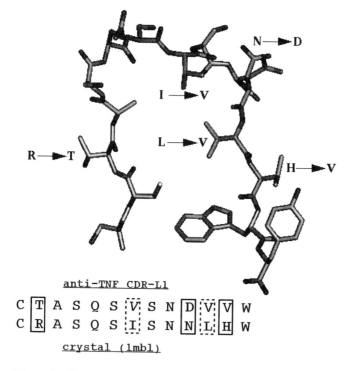

Figure 2   The model of the anti-TNF CDR-L1 is depicted, along with a comparison of its sequence to that of the closest crystallographically determined antibody. The positions in which the two sequences differ are boxed in the single letter sequence representation below: solid box - positions pointing outwards, and potentially involved in antigen binding; dashed box - positions pointing inwards, involving conformational stabilization. Note that the changes in the positions marked by a dashed box are highly conserved.

by a local minimization, thus achieving a degree of packing of the aliphatic residues in the modeled loop similar to that observed in the crystal structure of 1mbl. The final model consists of 13 residues:

Cys-Thr-Ala-Ser-Gln-Ser-Val-Ser-Asn-Asp-Val-Val-Trp

in which positions Cys^1, Ala^3, Val^7, Val^{11} and Trp^{13} interact, and can be viewed as forming a "hydrophobic core" that stabilizes the conformation.

In the intact antibody, the structure depicted in Fig. 2 is stabilized by interloop interactions, as well as by interactions between the loop and the antibody framework and by intraloop interactions. To design libraries of loop mimetics, one must stabilize the conformation of an isolated loop to compensate for the stabilizing interactions lost upon isolation. Additionally, synthetic considerations encouraged us to attempt to shorten the 13-residue loop with a minimal reduction on the chance of finding active mimetics. A subsequence of 9 residues, beginning with Ala^3, was chosen, based on its solvent exposure, as a shorter candidate for mimicking the 13-residues loop (the linear sequence described in the literature and reported to mimic this antibody contains 19 residues) (6). A process of rational core design was applied to stabilize the 9-mer sequence in its native conformation, so that the stabilized, synthetically feasible structure could be subsequently used as a scaffold for a combinatorial library.

The first step in the stabilization process was to identify appropriate bridging points and bridge lengths for the SCAPL scaffold. Peptor bridges have specific sizes and conformations, enumerated in a virtual library of approximately 10^4 bridges and conformations. For each bridge length, all conformations within a predefined energy cutoff were enumerated and ranked by their relative energies. Since not all sequence positions can be cyclized without severe distortions to the backbone, we have developed a geometrical test that compares four geometrical parameters for the peptide in its desired conformation and for all bridges in the virtual bridge library (Fig. 3). Only bridges that match the sequence in all four parameters are used in the library design.

Based on this test we identified all sequence positions that can be bridged and the appropriate bridge size for each. Five candidate bridges were chosen, based on their energy ranking, for our initial library.

To further stabilize the bioactive conformation, we divided the sequence positions into two groups, the first primarily responsible for structure stabilization, the other primarily responsible for binding/recognition. Our goal then was to find, for the structure determining positions, a set of amino acids that, together with the scaffold (the peptidic backbone and one of the above chosen bridges), would stabilize the desired conformation. It would then be possible to choose a set of diverse or similar side chains—based on the knowledge accumulated in previous optimization cycles on the importance of each position for binding and recognition—to adorn remaining positions.

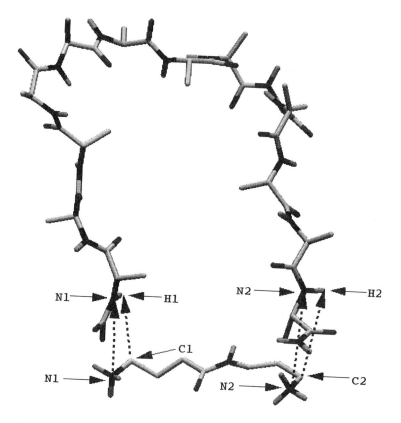

Figure 3 Structural parameters of backbone structure and potential bridges compared by the geometric test. The features compared are: distances: $N(1)\text{-}N(2)_{sequence}$ vs. $N(1)\text{-}N(2)_{bridge}$; bond angles: $H(1)\text{-}N(1)\text{-}N(2)_{sequence}$ vs. $C(1)\text{-}N(1)\text{-}N(2)_{bridge}$ and $N(1)\text{-}N(2)\text{-}H(2)_{sequence}$ vs $N(1)\text{-}N(2)\text{-}C(2)_{bridge}$; torsional angles: $H(1)\text{-}N(1)\text{-}N(2)\text{-}H(2)_{sequence}$ vs. $C(1)\text{-}N(1)\text{-}N(2)\text{-}C(2)_{bridge}$. Only bridges that match the sequence in all four parameters are considered for cyclization.

Visual inspection of the CDR-L1 conformation clearly demonstrates that among the 9 residues constituting our scaffold Ala[3], Val[7], and Val[11] point inward (Fig. 2), and properly chosen side chains at these positions can potentially stabilize the structure by forming a "hydrophobic core." All the other positions point outward and do not have a direct influence on the conformation of the loop.

Stabilizing a given backbone conformation by selection of appropriate side chains has become an important research topic in the last few years. One of the state-of-the art methods in this field uses the dead end elimination (DEE) algorithm (9). This algorithm, given a defined backbone conformation, rapidly scans all combinations of amino acids at each position and identifies the lowest energy one (i.e., the combination of amino acids that should best stabilize the desired backbone conformation). To be able to test the enormous number of possible combinations, the algorithm uses a virtual library of side chain rotamers (i.e., a discrete representation of the conformational space available to each side chain). Yet this discretization, while enabling a highly efficient search, sacrifices accuracy. Since our stabilization process involves only three sequence positions, we could avoid the usage of rotamer libraries and carry out a more elaborate procedure, albeit with a compromise on the number of side chain combinations tested. To decide which side chain combinations to test, we relied on chemical intuition.

Considering for each of the three structure-determining positions, a set of eight hydrophobic amino acids (Ala, Phe, Leu, Ile, Met, Trp, Tyr, Val) in combination with a set of five different bridges results in more than 2500 combinations, a number that is beyond our computational abilities. From inspection of the size of the loop, however, it is clear that large residues can enable good side chain packing. Thus by choosing only combinations in which at least two positions are occupied by large amino acids, we reduced the number of combinations to several hundred. For each such combination we performed a rigorous conformational search using the Monte Carlo minimization algorithm of Li and Scheraga (10) and the Merck molecular force field (MMFF) (11) [as implemented in the Macromodel program (12)]. In each search we varied all side chains of the hydrophobic core as well as the bridge, while keeping the rest of the sequence fixed in its native conformation. The stability of each combination of residues was evaluated by the nonbonded energy of the entire system, since this term is closely related to the degree of packing of each structure. To test our working assumption that at least two large amino acids are necessary for stabilization, we included several combinations with a larger number of small amino acids in the calculations. Not unexpectedly, we found that replacing a large amino acid with a small one in any of the three positions always resulted in poorer packing (i.e., increased the nonbonded energy). Figure 4 shows the best core for a 9-mer anti TNF CDR-L1 mimetic, clearly demonstrating its good packing.

A library of 10^7 9-residue CDR-L1 mimetics was designed around the scaffolds chosen as best stabilizing the "hydrophobic core." This core completely defines the identity of amino acids at positions 1, 5, and 9 (corresponding to positions 3, 7, and 11 of the original CDR-L1). However, to account for possible errors in our calculations, we introduced limited variability in these positions by including a small number of amino acids that displayed higher energy core structures. By choosing very limited variability at these three positions, we were able to introduce more variability in the remaining six. The amino acids in these positions were chosen based on their similarity to the original antibody sequence.

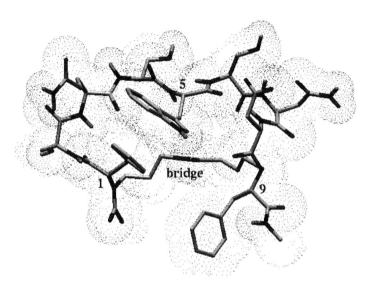

Figure 4   VdW surface for the amino acids combination that best stabilizes a 9-mer CDR-L1 mimetic. The tight packing of the hydrophobic core comprising the amino acids at sequence positions 1, 5 and 9 and the selected bridge is clearly visible.

Choosing similar amino acids requires a definition of a multidimensional space in which each amino acid is characterized by a set of molecular properties (descriptors) and methods for selecting similar (or diverse) sets of amino acids from this space. Yet this process is not trivial, since in this context "similar" implies that small differences in amino acids translate into small differences in activity—a requirement that depends on an atomic resolution understanding of the factors that govern biological activity (13). Several sets of descriptors aimed at meeting this requirement have been proposed, and in general these can be divided into two groups: experimentally derived physicochemical properties (14) and computational descriptors (15), usually involving intensive calculations that depend on the three-dimensional conformations of the side chains. Because of the large resources required for the measurement or calculations of these descriptors, they have been determined for limited sets of amino acids. Since we use a database of about 300 amino acids, we tested the ability of a set of easy-to-calculate one- and two-dimensional descriptors to measure amino acid similarity. This set of descriptors is listed in Table 1.

Following the synthesis of the designed library and its biological assay, a set of active sequences emerged that was characterized by a subset of amino acids, more focused than in the original design, at almost each sequence position.

Table 1 The 49 One- and Two-Dimensional Descriptors Used in the Present Work[a]

Cerius^2 name	Description	Number
Charge	Sum of partial charges	1
Fcharge	Sum of formal charges	1
Apol	Sum of atomic polarizabilities	1
Information content		9
MW	Molecular weight	1
Rotbonds	Number of rotatable bonds	1
Hbond acceptor	Number of H-bond acceptors	1
Hbond donor	Number of H-bond donors	1
AlogP/AlogP98	Octanol/water partition coefficient	2
Fh2o	Desolvation free energy for water	1
Foct	Desolvation free energy for octanol	1
MolRef	Molecular refractivity	1
Topological descriptors		28
Total		49

[a] All descriptors were calculated by the QSAR$^+$ module of Cerius^2 (Molecular Simulations, Inc., San Diego, CA, 1999).

III. DESIGN OF FOCUSED LIBRARIES

The consensus obtained from the initial library was subjected to further focusing by means of the position-by-position optimization method, in which a set of analogs, each differing from the original consensus sequence in one position, is synthesized. Similarity considerations, based on the descriptors listed in Table 1, were used to replace each amino acid in the active consensus by a set of amino acids similar to it. This method has the advantage of allowing the introduction of rather large sets of amino acids at each position, thereby leading to a relatively exhaustive position-by-position optimization. For example, in a 96-well plate, each position in a sequence of 9 amino acids can be represented by a set of 10 or 11 similar amino acids.

Two consecutive position-by-position optimization cycles, each consisting of the synthesis of a single 96-well plate, resulted in two potent TNF inhibitor leads. To test the validity of our descriptors as measurements of similarity, we chose 32 SCAPLs that were well characterized both biologically and analytically (i.e., their biological activity was clear-cut—either active or inactive—and mass spectrometric analysis demonstrated the existence of the desired product). Our 49 descriptors (Table 1) were reduced to a smaller set of principal components by means of principal component analysis (PCA), and when the set of 32 SCAPLs was viewed in the space of the first 3 PCs (corresponding to 93% of the variance in the original data set), an almost perfect separation between actives and inactives was obtained along PC1 (Fig. 5). This demonstrates that simple one- and two-dimensional descriptors, when calculated on whole molecules, meet the stringent "similarity" requirement and can be used to optimize our leads.

The position-by-position optimization method suffers from several drawbacks, however. First the method completely disregards possible favorable interactions between the different positions. Additionally, this optimization method tends to provide a poor and biased coverage of property space (Fig. 6). Criticism along these lines has also been brought up by Hellberg et al. (14a). Consequently we looked for more global optimization models.

A more global optimization can be performed by designing a virtual library (VL) focused in some sense around an active sequence and screening it against computational models derived from the biological data in order to select appropriate candidates for synthesis. The VL can be designed based either on whole-molecule or on fragment properties (16). Fragment-based design assumes that using similar fragments in a combinatorial synthesis will ultimately lead to similar final compounds. The obvious risk of working with fragments is, as noted above, that such designs disregard the interactions between the fragments in the final

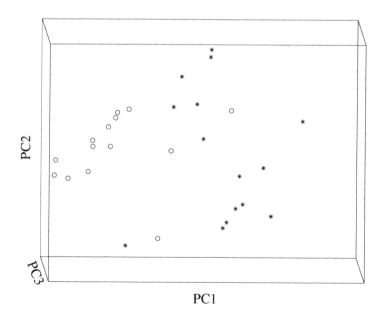

Figure 5   PCA of 32 CDR-L1 mimetics classified as inactive (17, circle) and active (15, star). Shown are the first 3 PC's corresponding to 93% of the variance in the original data set. An almost perfect separation between inactives and actives is obtained along PC1.

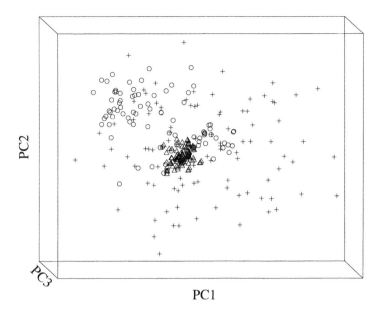

Figure 6 A comparison of coverage of property space between three sets of SCAPLs: (1) 96 diverse SACPLs selected from the virtual library (+); (2) 96 SCAPLs designed in the second position-by-position optimization cycle (○); (3) 96 SCAPLs similar to the best lead from the first position-by-position optimization cycle (△). See text for discussion.

molecule. A related study performed at Molecular Simulations Inc. (MSI) showed that better coverage of property space (i.e., better diversity) is obtained when a diverse set of molecules is selected from a completely enumerated library than when the molecules are assembled from a previously selected diverse set of fragments (17). The whole-molecule approach, however, is limited by computer resources. If each of the 9 positions in the foregoing active consensus were represented by a set of 10 amino acids (as in the position-by-position approach), an enumerated library of all resulting combinations would contain 10^9 molecules. Clearly, enumeration of a library of this size, if at all possible, could use only the most primitive descriptors and selection methods. It may therefore be more dangerous to assume that the use of simplistic descriptors and selection procedure will lead to effective libraries than it is to employ the fragment-based approach. Furthermore, inclusion of the entire backbone in the properties calculation may disguise information about the substituents (16).

Following these arguments we used the fragment-based approach and our in-house database of about 300 amino acids to design several virtual libraries of varying sizes (i.e., containing 1000, 10,000, 100,000 and 150,000 SCAPLs), each focused on the active sequences obtained from the position-by-position optimization cycles. Figure 7 presents profiles of several molecular properties of interest as calculated for each library. Clearly 10,000 SCAPLs suffice to produce stable, convergent profiles for these properties, so we are confident that a VL of 150,000 SCAPLs accurately describes the property space around our leads. Such a VL was screened in the PCA space defined earlier.

Figure 6 compares a set of 96 diverse SCAPLs selected from the VL, with the 96 SCAPLs designed in the second position-by-position optimization cycle and a set of 96 SCAPLs chosen according to their similarity to the most active SCAPL obtained in the first optimization cycle (which served as a starting point for the second optimization cycle). Two conclusions are obvious: (1) a diverse set of SCAPLs obtained from a focused library provides a better coverage of the property space than a set of SCAPLs obtained from a position-by-position optimiza-

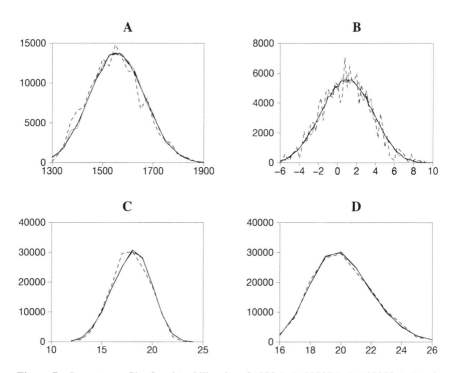

Figure 7   Property profiles for virtual libraries of 1000 (---), 10000 (—), 100000 (—) and 150000 (···) SCAPLs. Profiles are given for molecular weight (A), AlogP (B), number of H-bond acceptor (C) and number of H-bond donors (D). Note the almost perfect overlap for the three libraries of 10,000 and greater SCAPLs, indicating convergence in the measured properties.

tion procedure, and (2) replacing each amino acid of a target sequence by a set of amino acids most similar to it does not necessarily lead to a set of sequences most similar to the target sequence. Thus fragment-based similarity is inferior to whole-molecule-based similarity, as was shown in the case of diversity in previous work (17).

IV. STRUCTURE–ACTIVITY MODELS

Using standard peptide chemistry to synthesize SCAPL-based combinatorial libraries requires multiple steps, many more than are typically used in small-molecule combinatorial library synthesis. Consequently synthesis products are impure, and quantitative activity analysis is difficult during the initial optimization cycles. Thus, library focusing should employ computational tools capable of handling large amounts of qualitative high throughput screening (HTS) data where biological activity is expressed, for example, as inactive, slightly active, or active. We tested two such tools for their ability to provide reasonable computational models: a diversity model in the space of the first three principal components and recursive partitioning (RP) via classification trees.

A. Diversity Model in PCA Space

Our entire set of tested compounds contained 296 SCAPLs classified as 230 inactive (78%), 39 moderately active (13%), and 27 active (9%). Mapping this set onto the PCA space (Fig. 8) resulted in a picture similar to that obtained for the previous set of 32 SCAPLs (Fig. 5) in the sense that regions of space characterized by high density of actives and inactives could be observed. For the entire set, the first three PCs covered 82% of the variance in the original data set. This result lends further support to the conclusion that our set of descriptors provides a valid representation of molecular properties responsible for the differentiation between actives and inactives for the current set of SCAPLs

Thus a reasonable procedure for choosing candidates for synthesis is to project the VL onto the PCA space and select a diverse set of SCAPLs from among those that fall in the active region of the space. The major drawback of this model lies in its nonintuitive nature—the principal components are not readily interpretable in physical terms.

B. Recursive Partitioning (RP) Model

A more readily interpretable model can be constructed by using recursive partitioning (18). This method enables fast derivation of classification models for the prediction of activities or other properties. Given a set of molecules X whose

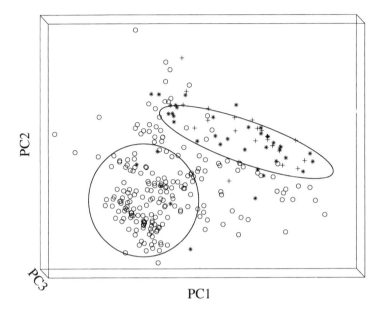

Figure 8   PCA of 296 CDR-L1 mimetics classified as inactive (230, ○), moderately active (39, *) and active (27, +). Shown is the space defined by the first 3 PCs corresponding to 82% of the variance in the original data set. Note the high concentration of inactives in the large circle and of moderately active and actives in the ellipsoid.

members are characterized by a set of descriptors, RP constructs a binary tree by repeated splits of subsets of X into two descendant subsets, beginning with X itself. In the simplest implementation, descriptors are tested one at a time and splits are chosen so that each of the descendant subsets is "purer" than the data in the parent subset. The method can rapidly provide accurate classifiers for the properties of interest, as well as intuitive insight into the factors underlying them, while handling, simultaneously, very large sets of quantitative and qualitative (including alphanumeric) data. The resulting tree can subsequently be used to predict the activity of members of a virtual library.

RP models were built using the CART module of Cerius^2 (MSI) (19). All trees were constructed to a maximal depth of 5, and cross-validation was used for evaluating the statistical parameters and predictive power of the resulting trees. For this process the data points were divided into four groups; then four trees were constructed, each leaving out one group, and tested using the group left out. The final statistics are an average of performance for all four trees. The results were assessed by means of three statistical parameters: enrichment (defined as fraction of the actives in the active bin divided by their fraction in the original data set), false negative rate, and false positive rate.

The data set for the RP models consisted of our entire set of 296 compounds. When the tree was constructed using the 49 descriptors of Table 1, calculated on the entire SCAPLs, we obtained active enrichment of 3.3, a false negative rate of 19% and a false positive rate of 20%.

We then investigated the performance of a model developed by describing each sequence by means of positional descriptors, (i.e., descriptors of the properties of the individual amino acid side chains). As noted earlier, several such sets appear in the literature and are sometimes referred to as principal properties. Principal properties were first derived for the 20 natural amino acids by applying principal component analysis to a matrix of physicochemical observables (14a) and later expanded to include a set of 87 natural and nonnatural amino acids (14b). Additional sets have been developed and in general can be divided into experimentally (14) and computationally (15) derived descriptors. Assuming a linear correlation between differences in descriptor values and activities for a pair of molecules, Matter, who performed a validation study of some of these descriptor sets, concluded that experimentally derived descriptors are in general more acceptable than computationally derived ones (13).

Since both the experimental and computational characterization of our amino acids database is demanding, we wanted to test the performance of easy-to-calculate one- and two-dimensional descriptors. Characterizing each of the amino acids comprising our SCAPLs by the initial set of 49 descriptors results in a set of descriptors so large that if used for constructing an RP model, it would most probably lead to data overfitting. Since the previous PCA demonstrated that the original variables are highly correlated, we attempted to choose smaller sets of variables displaying less correlation. Sets of minimally correlated descriptors (comprising 4, 6, 8, 9, 10, 11, 12, 13, 14, and 16 descriptors) were obtained by applying multidimensional scaling (MDS) to the discriptor correlation matrix followed by cluster analysis.

Figure 9 presents the RP results for the different descriptor sets. As the number of descriptors increases, the enrichment improves until it levels off at 5.1–5.2 for sets of 10–14 descriptors and decreases to 4.8 as the number of positional descriptors reaches 16. While the false positive rate remains approximately constant at 21–27%, the false negative rate covers a much larger range (26–45%, where in general larger descriptor sets result in a smaller false negative rate). Since for all models, the maximal tree depth was limited to 5, these results indicate that the improvements in the model performance are not merely a consequence of increasing the number of descriptors but result from a better description of the factors underlying the behavior of the system. Based on the results presented in Fig. 9, a model in which each amino acid is represented by a set of 12 positional descriptors provides the best RP model.

Describing each amino acid by its first three principal components obtained by the PCA of the original 49 descriptors as calculated for our entire database of

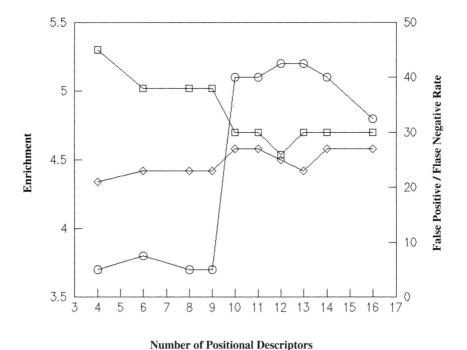

Number of Positional Descriptors

Figure 9   Comparison of statistical parameters (enrichment ($\bigcirc$), false negative rate ($\square$), false positive rate ($\diamondsuit$)) for RP models derived for different numbers of positional descriptors for the entire data set of 296 SCAPLs. Representing each amino acid by 12 descriptors yields the overall best results.

about 300 amino acids gave a reasonably good RP model (enrichment, 4.9; false negative rate, 26%; false positive rate, 24%). However, the ease of interpretation of the results inherent in RP models is lost when the descriptors take on the complicated form of principal components.

    Next we wanted to test whether augmenting the foregoing descriptors with structural ones could improve the predictive power of the model. Since the initial design of the SCAPLs was based on the assumption that an appropriate choice of side chains for the amino acids at sequence positions 1, 5, and 9 (the structural positions) will help to stabilize the CDR-L1-like conformation, we added four structure-related descriptors to the foregoing set: the volume of the side chain at each of the core-stabilizing positions (V1, V5, and V9) and the sum of these volumes (Vt). The resulting RP model provided an enrichment of 5.6, a false negative rate of 26%, and a false positive rate of 24% (vs. 5.2, 26%, and 25% for the original 12 positional descriptors set) and comprises the best RP model we could obtain for this data set (Fig. 10). In this tree, a split occurred on Vt, dividing the SCAPLs into those with total volumes less than 305 Å^3 (58 cases) and those with Vt above this value (5 cases). The outcome of this split resulted in a node having 41 out of 58 (70%) inactive SCAPLs. This result points to the importance of the total core volume in differentiating between actives and inactives, rather than the volume of the side chain at any one of these positions, and supports our initial assumption that bioactive conformations can be stabilized by a tightly packed hydrophobic core.

To investigate our activity classification of SCAPLs, two new classifications were generated. Moderately active SCAPLs were allocated to the inactive group in the first and to the active group in the second. RP models generated with the set of 12 positional descriptors augmented with Vt gave enrichment of 4.8 and 2.4, false negative rate of 26 and 23, and false positive rate of 10 and 26% for the first and second classification, respectively. These results lend credence to our original classification (since alternative classifications produced overall poorer models) and caution against oversimplified active/inactive schemes.

Finally, RP models were obtained for five randomized activity data sets, each having the original fraction of inactive, moderately active, and active SCAPLs. When the 12 positional descriptors + Vt were used, the results, reported as mean ± SD for the three parameters, were enrichment, 0.9 ± 0.2; false negative rate; 72

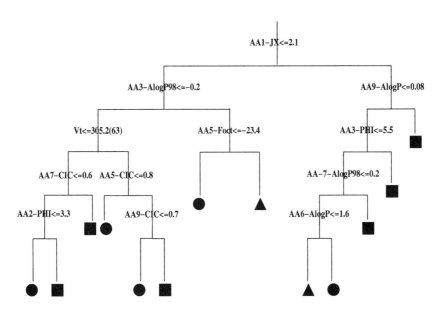

Figure 10 Classification tree obtained for the best RP model where each amino acid is described by 12 positional descriptors augmented by structural (volume) parameters. Terminal nodes are marked by activity: ▲-inactive, ■-moderately active, ●-active.

$\pm$ 14%; and false positive rate; 47 $\pm$ 15%. These results clearly demonstrate that a significant RP model could not be derived from random activity data.

V. CONCLUSIONS

In this work we presented a rational procedure for designing combinatorial libraries of small cyclic analogs of protein loops—SCAPLs—that mimic biological motifs. The method consists of identifying a biological motif of interest, stabilizing its conformation in isolation by backbone cyclization and by the selection of appropriate substituents for sequence positions identified as structure determining, and using the stabilized scaffold as a combinatorial library core. The resulting library was subjected to repeated optimization cycles, which led to potent TNF inhibitor leads in only a few steps of position-by-position optimization cycles. Sets of synthesis candidates obtained by a method of global optimization based on the design of a virtual library focused on an active motif showed better coverage of property space than did compounds obtained by position-by-position optimization. High throughout screening data were shown to be sufficient for the derivation of computational models with reasonable differentiation between actives and inactives using standard, easy-to-calculate one- and two-dimensional molecular descriptors. For the PCA model, reasonable separation of actives from inactives could be obtained in the space defined by the first three principal components. For the RP model, larger enrichments were obtained when positional rather than global descriptors were used and when a volume-related descriptor, specifically derived for the molecular model in hand, was included. The predictive power of this improved model is awaiting synthetic verification.

ACKNOWLEDGMENTS

We gratefully acknowledge the help of Roberto Olender for developing the geometric test and of Chen Keasar for helpful discussions. The synthetic work was performed by Yoseph Salitra, Tamar Yechezkel, Anna Balanov, Ella Nadler, and Eran Hadas from Peptor's chemistry group, the analytical work by Manuela Lazarov and Dvira Shohat from Peptor's analytical group, and the biological assays by Mical Gozi, Pninit Litman, Moshe Bracha, and Amnon Wolf from Peptor's biology group. The help of all these people is gratefully acknowledged.

REFERENCES

1. (a) Marraud M, Aubry A. Biopoly (Peptide Sci) 1996, 40, 45. (b) Selfer M A, He JX, Sawyer TK, Holub KE, Omecinsky DO, Reily MD, Thanabal V, Akunne HC, Cody WL. J Med Chem 1995; 38:249. (c) Jirácek J, Yiotakis A, Vincent B, Lecoq A, Checler F, Dive V. J Biol Chem 1995; 270:21701.

2. (a) Blondelle SE, Takahashi E, Weber PA, Houghten RA. Antimicrob. Agents Chemother. 1994; 38:2280. (b) Giannis A, Kolter T. Angew Chem Int Ed Engl 1993; 32:1244. (c) Gante J. Angew Chem Int Ed Engl 1994, 33, 1699. (d) Soth MJ, Nowck JS. Curr. Opin. Chem. Bio. 1997; 1: 120. (e) Hruby VJ, Ahn J-M, Liao S. Curr Opin Chem Biol 1997; 1:114. (f) Kirshenbaum K, Zuckermann RN, Dill KA. Curr Opin Struc Biol 1999; 9:530.

3. (a) Gilon C, Halle D, Chorev M, Selinger Z, Byk G. Biopolymers. 1991; 31:745. (b) Gilon C, Huenges M, Mathä B, Gellerman G, Hornik V, Afargan M, Amitay O, Ziv O, Feller E, Gamliel A, Shohat D, Wanger M, Arad O, Kessler H. J Med Chem 1998; 41:919.

4. (a) Aggarwal BB, Natarajan K. Eur. Cytokine Network. 1996; 7:93. (b) Raine, CS. Nat Med 1995; 1:211.

5. (a) Williams WV, Kieber-Emmons T, VonFeldt J, Greene MI, Weiner DB. J Biol Chem 1991; 266:5182. (b) Monnet C, Laune D, Laroche-Traineau J, Biard-Piechaczky M, Briant L, Bès C, Pugnière M, Mani JC, Pau B, Cerutti M, Devauchelle G, Devaux C, Granier C, Chardès T. J Biol Chem 1999; 274:3789.

6. Döring E, Stigler R, Grütz G, Von Baehr R, Schneider-Mergener J Mol Immun 1994; 31:1059.

7. Walters WP, Ajay, Murcko MA. Curr Opin Chem Biol 1999; 3:384.

8. Berenstein FC, Koetzle TF, Williams GJB, Meyer EF, Brice MD, Rodgers JR, Kennard O, Shimanouchi T, Tasumi M. J Mol Biol 1977; 112:535.

9. (a) Desment J, DeMaeyer M, Hazes B, Lasters I. Nature 1992; 356:539. (b) Goldstein RF. Biophys. J. 1994; 66:1335. (c) Dahiyat BI, Mayo S. Science 1997; 278:82.

10. Li Z, Scheraga HA. Proc Natl Acad Sci USA. 1987; 84:6611.

11. (a) Halgren TA. J. Comput Chem 1996; 17:490. (b) Halgren TA. J. Comput Chem 1996; 17:520. (c) Halgren TA. J Comput Chem 1996; 17:587. (d) Halgren TA, Nachbar R. J. Comput Chem 1996; 17:616. (e) Halgren TA. J. Comput Chem 1999; 20:720. (f) Halgren TA. J Comput Chem 1996; 20:738.

12. Mohamadi F, Richards NGJ, Guida WC, Liskamp R, Lipton M, Caufield C, Chang G, Hendrickson T, Still WCJ. J Comput Chem 1990; 11:440.

13. Matter H. J. Peptides Res 1998; 52:305.

14. (a) Helberg S, Sjöström M, Kagerberg B, Wold S. J Med Chem 1987; 30:1126. (b) Sandberg M, Ericksson L, Jonsson J, Sjöström M, Wold S. J Med Chem 1998; 41:2481. (c) Fauchère J-L, Charton M, Kier LB, Verloop A, Pliska V. J Peptide Protein Res 1988; 32:269.

15. (a) Norinder W. Peptides. 1991; 12:1223. (b) Collantes ER, Dunn WJ III, J Med Chem 1995; 38:2705. (c) Zaliani A, Gancia E. J. Chem Inf Comput Sci 1999; 39:525.

16. Martin EJ, Critchlow RE. J Comb Chem 1999; 1:32.

17. Jamois EA, Hassan M, Waldman M. J Chem Inf Comput Sci 2000; 40:63.

18. Breiman L, Friedman JH, Olshen RA, Stone CJ. *Classification and Regression Trees.* Wadsworth, Belmont, CA, 1984.

19. Cerius^2 ccI. Molecular Simulations, Inc. San Diego, CA, 1999.

Index

3D QSAR, 131, 479
  CoMFA, 132, 146
  Compass, 132, 150
  Distance Geometry, 134
  Egsets, 132, 139
  EGSITE, 132, 140
  Ligand Paradigm, 143
  Molecular Similarity, 132, 149
  MSA, 132, 143
  MTD, 132, 144
  Pharmacophore, 133
  REMOTEDISC, 132, 144
  Receptor Paradigm, 133
  Voronoi Models, 137
  YAK, 132, 134
ACD (Available Chemical Directory),
  282

Acetyl cholinesterase, 547
ADME (Absorption, distribution,
  metabolism and excretion), 2,
  474, 494, 571
Antigen recognition, 607
AlogP, 269, 613
  AlogP98, 452, 613
Aromaticity, 440
Autocorrelograms, 439, 440
  Bipolar, 439
  FBPA, 439, 442
  Fuzzy, 439
  Pharmacophore, 439
  Feature pair partial dissimilarity
    score, 442
AUTODOCK, 520
Activity Clustering, 476

Binding Affinity, 238, 240, 598
  3D QSAR, 241
  Scoring Functions, 245
  VALIDATE, 239
  VolSurf, 252
Binding Energy, 157, 202
  Landscapes, 157
  Orientation, 597
Bioactive Conformation, 236, 609
Biological Activity, 475

Carbonic Anhydrase, 480
CASP2, 249
Cathepsin D, 483, 490, 574
CAVEAT, 200
CCLD (Computational Combinatorial
    Library Design), 200
CHARMM, 506
CHEM-DIVERSE, 404
Chemoinformatics, 429
Chemical languages, 432
Cholecystokinin, 545
CMC (Comprehensive Medicinal
    Chemistry), 282
CoMFA, 546
CONCORD, 478, 518
Combinatorial Chemistry, 1
  Analog Builder, 432
  Backbone cyclization technology,
    605
  Building blocks, 432, 436
  Chemoinformatics, 429
  Combinatorial Products, 430, 432
    Conformer families, 434
    Reactive center, 433
    Substitution, 433
  Combinatorial Synthesis, 563
  Discovery Libraries, 25
  Enumeration, 432
  Library Design Concept, 1
  Library enumeration, 436
  Molecular Modeling, 432, 583
  Molecular Fragments, 505
  Parallel elaboration, 595
  Parallel synthesis, 11, 566, 597

[Combinatorial Chemistry]
  Peptide Libraries, 512, 513
  Peptide Synthesis, 564
  Pooled synthesis, 14
  Solid phase synthesis, Solid phase
    organic synthesis (SPOS), 4,
    564, 595
  Solution phase synthesis, 8
  Split and mix methodology, 592
  Synthons, 432, 474
Combinatorial Library Design, 19, 429,
    473, 503, 564, 569
  Augmenting Existing Libraries, 326
  Biological motifs, 605
  Bradykinin Potentiating Activity, 369
  Cell-Based Metrics, 329
  CHARMM, 506
  CHEM-DIVERSE, 404
    Deficiencies, 406
  Cluster-Based Compound Selection,
    380
  Clustering and diversity techniques,
    516
  Combinatorial Analysis, 566
  Combinatorial Constraint, 303
  Comparing Whole Libraries, 326
  Comparison of Spanning-Tree, 329
  Compound Selection, 281
  Conformational libraries, 606
  Conformationally constrained, 605
  Controlling molecular properties, 20
  Cost-Effective, 301, 327
  Database Searches, 508
  Delphi, 521
  Descriptors, 337, 430
    Topological, 430, 613
    Evaluation, 430
    Geometry-dependent, 430
  Desirable Reagent Properties, 324
  Directed library, 575
  Dissimilarity Based Methods, 379, 381
  Dissimilarity metric, 449
  Distance-Based Selection, 314
  Diverse library, 575
  Diversity, 301

[Combinatorial Library Design, Diversity]
    3D Descriptors, 439
    3D Indices, 439
    Absolute Diversity, 355
    Atom Pair Vectors, 338
    Average Pairwise Similarity, 351
    Binning, 343
    BCUT, 341, 477, 490, 491
    Cell-based, 473, 477
    Cerius2, 478
    Chemical diversity space, 532
    Connectivity Indices, 339
    Cosine Coefficient, 346
    Distance Measures, 345
    Diverse Solutions, 477
    Diversity metrics, 452
    Diversity Space, 475
    Descriptors, 337, 474
    Finger Prints, 338
    Hamming Distance, 346
    Models (PCA, RP), 617
    Nearest-Neighbor Profiles, 352
    Neighborhood Behavior, 347
    Nonlinear Mapping, 343
    Pairwise Identity, 350
    Pharmacophore triplets, 341
    Physical Properties, 341
    Principal Component Analysis,
        PCA, 342, 491
    Property space, 615
    Relative Diversity, 349
    Scope, 337
    Substitution Patterns, 465
    Substructural Keys, 338
    Topological torsions, 339
    Union Bit Sets, 349
  Diversity Selection, 307
  Druglike, 301, 319, 327, 515
  Flexible docking, 520
  FPT (Ras farnesyl transferase), 38
  Fragment Positioning, 506
  GA-PLS QSAR, 366, 479
  GRID, 506
  Genetic Function Approximation,
    479

[Combinatorial Library Design]
  HARPICK Application, 410
  HARPICK Program, 406
  High Throughput Technologies, 565
  HOOK, 508, 521
  Iterative library design, 572, 578
  Large Combinatorial Libraries, 515
  Library Comparison, 325
  Library Diversity, 19
  Library Profiling Functions, 414
  Library Optimization, 36
  Library Size, 475
  LUDI, 508, 520
  Knowledge-Based Approaches, 267
  Maximum Dissimilarity Algorithm,
    382
  Molecular fingerprint, 440
  Molecular property improvement, 36
  Molecular recognition, 571
  Monte Carlo Optimization, 305
  Multiple Copy Simultaneous Search
    (MCSS), 200, 503, 514, 517
  Objective Function, 306
  Optimization Algorithms, 386
  Organic compound libraries, 515
  Partition-Based Selection, 310, 380
  Peptidomimetic, 605
  Pharmacophore-Based Approaches,
    399, 404
  Pharmacophore Descriptor Validation,
    420
  Pharmacophore Fingerprint, 414
  Pharmacophore Hypothesis, 516
  Pharmacophore Profiling Alterations,
    407
  Pharmacophores in, 402
  Privileged Substructures, 415
  Profile-Based Biasing, 322
  Product Based, 474
  Property Space, 475
  QSAR Model Development, 367, 369
  Rational core design, 609
  Rational Database Mining, 369
  Reagent Selection, 417
  Reagent Based, 474

[Combinatorial Library Design]
  Reagent vs Product Diversity, 308
  Relative Diversity/Similarity, 415
  Rule-Based Biasing, 320
  Scaffolds, 579, 598, 609
  Scoring Function, 408, 520
  Search Space, 304
  SELECT, 388
  Selection of Reagents, 284
  Small Cyclic Analogs of Protein
      Loops (SCAPL), 605, 620
  Sphere-Exclusion Algorithms, 384
  Stochastic Optimization, 407
  Structure-based, 503, 563, 567
  Theory and Methodology, 303
  Ugi Library, 327
Conformational
  Sampling, 429
  Search, 575
  Space, 437
  3Dmodels, 437
Continuum Electrostatics Method, 197
Continuum Solvation, 206
CORINA, 586
CXCR2 receptor, 481
Cyclodextrin Glycosyltransferase
  Maltose Complex, 179

Database Mining, 375
  Estrogen Receptor, 375
Deadend Elimination Algorithm, 611
Delphi, 521
DHFR, 173
  Methotrexate (MTX) Complex, 173
DHYDM (Distance hyperspace distance
    measurement), 56
Dihydrofolate Reductase (see DHFR)
Dissimilarity-Based Methods, 379
Distance matrix
  Inter-atomic, 441
Diversity, 301
DOCK, 249, 518, 578, 594
Docking studies, 585
Drugability, 279
  Quantitative Score, 280

Druglikeness, 291
  Classification Algorithm, 293
  Descriptors, 293
  Knowledge Sources, 292
  Statistical Assessment, 293
Druglike Molecules, 107
  AlogP, 269
  Computational Filter, 295
  DURGA, 269
  Functional Group Profiling, 269
  HlogP, 270
  Physicochemical Property
      Profiling, 269
Discovery Libraries, 25
  Efficient Construction, 27
  Branching Strategies, 28
DMPK (Distribution, metabolism,
    pharmacokinetics), 233, 251

FlexX, 249
FLO, 520
Fragment Docking, 205
Fragment Desolvation, 210
Functional Genomics, 504, 531

Genetic Algorithm, 531, 533
  Chromosome, 532
  Crossover, 533
  Darwinian evolution, 532
  Fitness, 532
  Generation, 541
  Genes, 532
  Genesis, 534
  Genomes, 536
  Mutation, 533, 537
  Selection, 533
  Stochastic method, 544
Ghost Database Mechanism, 457
Gold, 249
GRID, 506
Graphical Interfaces, 432

Hepatitis Delta Antigen, 514
HLOGP, 270
HOOK, 508, 521

HTS (High throughput screening or
synthesis), 3, 532
Parallel synthesis, 11
Pooled synthesis, 14
Solid phase organic synthesis
(SPOS), 4
Solution phase synthesis, 8
Hydrogen bond acceptor, 440
Hydrogen bond donor, 440
Hydrophobic Interactions, 246
Hydrophobicity, 238, 440

Information Content, 462
Inhibitors, 451
Interaction Sites, 512
Farnesyl protein transferase
inhibitors, 451
Intermolecular Interactions, 246
Inverse QSAR, 110, 363
Activity Prediction, 365
Molecular Descriptors, 364
QSAR Models, 365

Knowledge-Based Approaches, 267

Lead Discovery, 565
Lead Generation, 531
GA directed, GA driven, 531, 552
Peptoid-based, 543
Lead Optimization, 454, 554, 565
Library Construction Strategy, 11
Parallel synthesis, 11
Pooled synthesis, 14
Library Diversity, 19
Ligand Design, 546
Ligand-Protein Binding, 159
Computer Simulations, 167
Thermodynamics, 159
Dynamics, 161
Energetic Aspects, 164
Monte Carlo Simulations, 167
Similarity Clustering, 172
Weighted Histogram Analysis, 169
Ligand-Protein Complexes, 157

Ligand-Receptor Interface, 247
LUDI, 508, 520, 568, 586

MAP Kinase, 211
Matrix Metalloproteinases (MMP),
589, 590
MCSS (Multiple Copy Simultaneous
Search), 200, 503, 514, 517
Pharmacophore Site Points, 517
Merck Molecular Force Field
(MMFF), 611
Methotrexate, 173
Molecular Diversity, 109, 532, 569
Molecular Docking, 157, 161
Validation Experiment, 185
Molecular Fingerprint, 440
Conformational, 440
Molecular Recognition, 158, 571
Molecular Similarity, 57
Monte Carlo Optimization, 305
MTX (*see* Methotrexate)
Multiple Copy Simultaneous Search (*see*
MCSS)

Octanol/Water Partition Coefficient, 98

P38, 212, 217
Pharmacophore, 51
3D QSAR, 65
Bipolar, 441
Combinatorial library design, 63
Critical Aspects, 67
Descriptor Analysis, 417
Descriptor Validation, 420
Features, 440
Fingerprint, 414
General classification, 52
Geometry, 54
Experimental validation, 60
Geometry of Natural Substrate, 60
Hypothesis, 60, 516
Identification, 53
MCSS-generated, 505
Model, 62
3D QSAR, 65
Use in medicinal chemistry, 62

[Pharmacophore]
  Molecular topologies, 455
  Protein binding site exploration, 60
  Pharmacophoric substituents, 456
  Traditional approach to identify, 52
  Types, 441
Pharmacophoric Groups, 53
  Search for, 62
Pharmacophoric Hypothesis, 60
  Consensus pharmacophore hypothesis, 450
  Ligand-based hypothesis, 450
Plasmepsin II, 483, 577
Pka, 100
Polio virus capsid protein, 507
Principal components (PCs), 446, 546, 613
Protein Flexibility, 248

QSAR (Quantitative Structure-Activity Relationships), 73, 482
  Applications, 98
    LogP, 98
    Octanol-water partition coefficients, 98
    Physicochemical property calculation, 98
    Pka, 100
    QSPR, 98
    Biological activity, 101
    Antiulcer compounds, 101
    Calcium Channel Activator, 102
    Carcinogenicity, 103
    Mutagenicity, 103
    Phenol Toxicity, 104
    Pharmacokinetics, 105
  Artificial Neural Network (ANN), 83, 111
  Bilinear Model, 81
  Binary QSAR, 85, 111
  Classical, 235
  Database Mining, 375
  Databases, 113
    C-QSAR, 115
  Free-Wilson Method, 76
  Fujita-Ban Method, 76
  GA-PLS, 366

[QSAR (Quantitative Structure-Activity Relationships)]
  Hansch-Fujita Method, 77
  HQSAR, 84
  KNN (k Nearest-Neighbors), 371
    Applications, 373
    Estrogen Receptor Ligands, 373
  Linear Model, 78
  Methodologies, 75
  MCASE (Multi-CASE), 82
  Parabolic Model, 79
  Parameters, 85
    Electronics, 85
    Hydrogen-Bonding, 94
    Hydrophobic, 88
    Steric, 91
    Topological, 95
  Software, 113
    ACD Physicochemical Laboratory, 115
    ADAPT, 115
    Cerius2, 115
    ChemPlus, 115
    ClogP, 115
    CODESSA, 115
    Galaxy, 115
    Hint, 115
    Molconn-Z, 115
    Kow Win, 115
    M-CASE, KLOGP, 115
    pKalc, PrologP, 115
    PCModels, 115
    POLLY, 116
    Prochemists, 116
    QlogP, 116
    QuaSAR-Binary, 116
    ScilogP, 116
    SIMCA, 116
    HQSAR, 116
    Tsar, 116
    XlogP, 116
  Statistical Techniques, 97
    Multiple Linear Regression, 97
    Partial Least Squares (PLS), 97
QSPR, 98

Quantitative Structure-Activity
  Relationships (*see* QSAR)
Quantitative Structure-Property
  Relationships (see QSPR)
Quo Vadis, Scoring Function, 233

Receptor Model, 548
Receptor Desolvation, 208

SCAPL (Small Cyclic Analogs of
  Protein Loops), 605, 620, 621
Similarity, 453, 460, 569, 614
  Similarity Search, 453
  Paradigm, 460
  Structural similarity, 460
  Property similarity, 460
SEED, 200, 203, 214
SLIDE, 249
STERIMOL, 93
Stromelysin, 540
Structure-Activity
  Models, 617
Structure-Based Ligand Design, 198
  Binding Energy, 202
  CAVEAT, 200
  CCLD, 200
  Docked Fragments, 198
    Acetate, 223
    Apolar Vectors, 218

[Structure-Based Ligand Design]
  Benzene, 218
  Charged Groups, 223
  Cyclohexane, 220
  Diethylether, 221
  Methylammonium, 224
  Polar Groups, 221
  Pyridine, 222
  Pyrrole, 222
  MCSS, 200, 510, 511
  SEED, 200

Targeted Libraries, 31
Thrombin, 539, 581
Thymidylate Synthase, 593
Trypsin, 542
Tumor necrosis factor, 606

VALIDATE, 239
Virtual Library, 614
  Property profiles, 616
  Screening, 505, 509, 517, 543, 546
  Sidechain rotamers, 611
Virtual Screening, 449, 451, 459, 509,
    517, 568
  Small Molecule Databases, 568
  Structure-based, 517
  Algorithm, 459
VolSurf, 252